Mastering Linux Administration

A comprehensive guide to installing, configuring, and maintaining Linux systems in the modern data center

Alexandru Calcatinge

Julian Balog

Packt>

BIRMINGHAM—MUMBAI

Mastering Linux Administration

Copyright © 2021 Packt Publishing

Group Product Manager: Wilson D'souza

Associate Publishing Product Manager: Preet Ahuja

Senior Editor: Rahul D'souza

Content Development Editor: Nihar Kapadia

Technical Editor: Nithik Cheruvakodan

Copy Editor: Safis Editing

Project Coordinator: Ajesh Devavaram

Proofreader: Safis Editing

Indexer: Tejal Soni

Production Designer: Nilesh Mohite

First published: June 2021

Production reference: 1190521

Published by Packt Publishing Ltd.

Livery Place

35 Livery Street

Birmingham

B3 2PB, UK.

ISBN 978-1-78995-427-2

www.packt.com

Contributors

About the authors

Alexandru Calcatinge is an open-minded architect with a background in computer science and mathematics, and is constantly eager to learn new things. He is a senior university lecturer with a PhD in urban planning and regional development. With 13 years of experience in architectural programming and development, Alex teaches students about smart cities, big data, and open source technologies. He has authored four books and numerous scientific articles on urban and regional planning, with an emphasis on open source technologies in urban and rural development.

Alex started using Linux in 2009 and never looked back. He has also been a certified Linux trainer since 2017 and teaches Ubuntu Server administration to students in Bucharest, Romania. Alex loves the DevOps philosophy and the possibilities that cloud technologies bring to the future. He is a certified programming analyst, computer network administrator, trainer, and designer.

For my wife, Rodica. You always stand by my side and offer me support, compassion, and guidance. For my sister and my parents. In memory of Bogdan Paul, who left too early to read the book.

Julian Balog is a senior software engineer with more than 15 years of experience in the industry. His primary work currently focuses on application delivery controllers, containerized workflows, networking, and security. With a never-ending passion for Linux and open source technologies, Julian is always in pursuit of learning new things while solving problems and making things work through simple, efficient, and practical engineering.

He lives with his wife, two children, and an Aussie-doodle in the greater Seattle area, Washington.

For Adelina, Anouk, and Indio. You graciously endured my absence while I was working on the book. Remembering Eduard, who sparked my love for Linux.

The authors would like to thank the wonderful editorial and production team at Packt for their professional leadership, dedication, and guidance throughout the writing of this book. We are indebted to Neil D'mello, Nihar Kapadia, Preet Ahuja, Hemangi Lotlikar, Rahul D'souza, Sulagna Mohanti, for many helpful suggestions and the comprehensive revision of the drafts. We are also grateful to Marcus Patman for his thorough reviews and critical comments. We could not hope for a better team and support.

About the reviewer

Marcus Patman is a senior systems engineer with over 20 years' experience as an IT professional. He has worked in many environments, ranging from small startups to large enterprises with thousands of production servers. During this time, he has worked with multiple platforms, including Linux, Windows, AWS, and Azure. He currently manages an enterprise Linux environment for a non-profit, providing automation and management of their on-premises and cloud systems.

Thanks to all my colleagues past and present for allowing me to do what I do best.

Love to my friends Mary and Jay for encouraging me pursue a career in technology. I couldn't have done it without you.

Table of Contents

3

Linux Software Management

4

Managing Users and Groups

5

Working with Processes, Daemons, and Signals

Section 2: Advanced Linux Server Administration

6

Working with Disks and Filesystems

7

Networking with Linux

Section 3: Cloud Administration

11

Working with Containers and Virtual Machines

12

Cloud Computing Essentials

13
Deploying to the Cloud with AWS and Azure

14
Deploying Applications with Kubernetes

15

Automating Workflows with Ansible

Preface

In recent years, Linux has become increasingly prevalent across a wide variety of computing platforms, including desktop computers, enterprise servers, smartphones, IoT devices, and on-premises and cloud infrastructures. Consequently, the complexity of related administrative tasks, configuration management, and DevOps workloads has also grown considerably. Now, perhaps more than ever, Linux administration skills are highly relevant.

But there's an evolutionary twist to the making of a modern-day Linux administrator. Tedious manual administrative operations are gradually replaced by orchestrated and automated workflows. One-off magic commands and scripts are replaced by declarative manifests invoked on demand while scaling up or down system configurations based on computing needs. Yesterday's Linux administrator progressively morphs into a DevOps persona.

There are countless books on Linux administration, and some of them will still be relevant for years to come. But *Mastering Linux Administration* is written with the admin-turned-DevOps in mind. We'll start with the basic concepts and commands addressing the most common areas of everyday Linux administration tasks. You'll learn how to install Linux on a desktop PC and virtual machine. Next, we'll introduce you to the Linux filesystem, package managers, users and groups, processes, and daemons. After a quick primer on networking and application security, we'll make the leap from on-premises to the cloud, exploring containerized workloads using Docker and Kubernetes. Together, we'll walk in the clouds by deploying Linux in AWS and Azure. You'll work with hands-on application deployments using EKS and AKS, the Kubernetes fabrics of AWS and Azure, respectively. Finally, we conclude our journey with Ansible and configuration management automation, taking us full circle to a DevOps-centric view of Linux.

We can only hope that by the end of the book, you'll be on the path to becoming a proficient Linux administrator with a versatile DevOps mindset.

Who this book is for

The intended audience for this book includes users with beginner- to intermediate-level Linux administration skills who are not shy of rolling up their sleeves and getting dirty with the Linux command-line terminal, working with scripts and CLI tools, on-premises and in the cloud. For most of the book, a regular desktop computer or laptop will suffice. Some of the chapters, such as those exploring Kubernetes and Ansible, may require a relatively powerful machine to set up the related lab environments.

The public cloud sections require AWS and Azure accounts if you want to follow along with the practical examples. Both cloud providers provide free subscription tiers, and we highly encourage you to sign up for their services.

What this book covers

Chapter 1, Installing Linux, provides a practical guide to installing Linux on a personal computer or virtual machine. The chapter includes hands-on workshops for working with Ubuntu and RHEL/CentOS Linux distributions.

Chapter 2, The Linux Filesystem, explores the Linux shell and filesystem, with the related commands for manipulating files and directories.

Chapter 3, Linux Software Management, introduces some of the most common Linux package managers, including **DEB**, **RPM**, **APT**, **YUM**, **Snap**, and **Flatpak**.

Chapter 4, Managing Users and Groups, looks at working with users and groups and managing the related system permissions.

Chapter 5, Working with Processes, Daemons, and Signals, takes a deep dive into Linux processes and daemons and the related inter-process communication mechanisms.

Chapter 6, Working with Disks and Filesystems, introduces some of the most common Linux filesystem types, such as **Ext4**, **XFS**, and **btrfs**. The chapter also covers disks, partitions, and logical volume management in Linux.

Chapter 7, Networking with Linux, is a concise primer on Linux networking internals, including the OSI and TCP/IP models, networking protocols, and services. The chapter also briefly touches upon network security.

Chapter 8, Configuring Linux Servers, explores some of the most common Linux networking servers and services, such as **DNS**, **DHCP**, **NFS**, **Samba**, **FTP**, and web servers.

Chapter 9, Securing Linux, looks at the Linux application security frameworks, including **SELinux** and **AppArmor**. The chapter also covers different firewalls and firewall managers, such as **Netfilter**, **iptables**, **nftables**, **firewalld**, and **ufw**.

Chapter 10, *Disaster Recovery, Diagnostics, and Troubleshooting*, provides a high-level overview of Linux disaster recovery and troubleshooting practices, including backup-restore and troubleshooting common system issues.

Chapter 11, *Working with Containers and Virtual Machines*, explores virtual and containerized Linux environments, focusing on different hypervisors and Docker Engine.

Chapter 12, *Cloud Computing Essentials*, is a brief overview of cloud technologies, describing the SaaS, PaaS, and IaaS solutions and service providers.

Chapter 13, *Deploying to the Cloud with AWS and Azure*, looks at Linux deployments in the cloud, using AWS EC2 instances and Azure VMs.

Chapter 14, *Deploying Applications with Kubernetes*, provides a practical guide to using Kubernetes on-prem and in the cloud with EKS and AKS.

Chapter 15, *Automating Workflows with Ansible*, explores automated configuration management workloads using Ansible.

To get the most out of this book

We used Ubuntu 20.04 LTS and CentOS 8 throughout the book, and we advise you do the same. We also expect most of the command-line examples and code to work with newer versions of Ubuntu following the publication of this book. With CentOS 8 reaching the end of support in December 2021, you can still use Fedora or CentOS Stream as highly similar distributions. You may also try the RHEL distribution available for free with a Red Hat developer account.

If you are using the digital version of this book, we advise you to type the code yourself or access the code via the GitHub repository (link available in the next section). Doing so will help you avoid any potential errors related to the copying and pasting of code.

Download the example code files

You can download the example code files for this book from GitHub at `https://github.com/PacktPublishing/Mastering-Linux-Administration`. In case there's an update to the code, it will be updated on the existing GitHub repository.

We also have other code bundles from our rich catalog of books and videos available at `https://github.com/PacktPublishing/`. Check them out!

Download the color images

We also provide a PDF file with color images of the screenshots/diagrams used in this
book. You can download it here: `http://www.packtpub.com/sites/default/`
`files/downloads/9781789954272_ColorImages.pdf`.

Conventions used

There are a number of text conventions used throughout this book.

`Code in text`: Indicates code words in text, database table names, folder names,
filenames, file extensions, pathnames, dummy URLs, user input, and Twitter handles. Here
is an example: "One of the most well-known disk backup commands is the `dd` command."

A block of code is set as follows:

```
(parted) print
Error: /dev/sda: unrecognised disk label
Model: ATA ST1000LM048-2E71 (scsi)
Disk /dev/sda: 1000GB
Sector size (logical/physical): 512B/4096B
Partition Table: unknown
```

When we wish to draw your attention to a particular part of a code block, the relevant
lines or items are set in bold:

```
exit
sudo unmount /mnt
```

Any command-line input or output is written as follows:

```
man vmstat
```

Bold: Indicates a new term, an important word, or words that you see onscreen. For
example, words in menus or dialog boxes appear in the text like this. Here is an example:
"Select the **Try Ubuntu** option from the window on the screen."

> **Tips or important notes**
> Appear like this.

Get in touch

Feedback from our readers is always welcome.

General feedback: If you have questions about any aspect of this book, mention the book title in the subject of your message and email us at customercare@packtpub.com.

Errata: Although we have taken every care to ensure the accuracy of our content, mistakes do happen. If you have found a mistake in this book, we would be grateful if you would report this to us. Please visit www.packtpub.com/support/errata, selecting your book, clicking on the Errata Submission Form link, and entering the details.

Piracy: If you come across any illegal copies of our works in any form on the Internet, we would be grateful if you would provide us with the location address or website name. Please contact us at copyright@packt.com with a link to the material.

If you are interested in becoming an author: If there is a topic that you have expertise in and you are interested in either writing or contributing to a book, please visit authors.packtpub.com.

Reviews

Please leave a review. Once you have read and used this book, why not leave a review on the site that you purchased it from? Potential readers can then see and use your unbiased opinion to make purchase decisions, we at Packt can understand what you think about our products, and our authors can see your feedback on their book. Thank you!

For more information about Packt, please visit packt.com.

Section 1: Linux Basic Administration

In this first section, you will master the Linux command line and basic administrative tasks such as managing users, packages, files, services, processes, signals, and disks.

This part of the book comprises the following chapters:

- *Chapter 1, Installing Linux*
- *Chapter 2, The Linux Filesystem*
- *Chapter 3, Linux Software Management*
- *Chapter 4, Managing Users and Groups*
- *Chapter 5, Working with Processes, Daemons, and Signals*

1
Installing Linux

The recent years have marked a significant rise in the adoption of Linux as the operating system of choice for both server and desktop computing platforms. From enterprise-grade servers and large-scale cloud infrastructures to individual workstations and small-factor home appliances, Linux has become an ever-present platform for a wide range of applications.

The prevalence of Linux, perhaps now more than ever, brings into the spotlight much-needed administration skills for a growing community of system administrators and developers. In this book, we take a practical approach to Linux administration essentials, with the modern-day system administrator, DevOps, or developer in mind.

In this first chapter, we'll guide you through the Linux installation process, either on physical hardware (bare metal) or using a **Virtual Machine** (**VM**). We'll take you further with a few case studies on choosing a Linux distribution based on functional requirements. Along the way, we introduce you to the Linux graphical user interface with some hands-on examples of configuring GNOME and KDE. Finally, we build a Linux workstation suitable for our daily computing needs.

Here are the topics we cover in this chapter:

- The Linux operating system
- Linux distributions
- Choosing the right Linux distribution
- Installing Linux - the basics
- The **Windows Subsystem for Linux** (**WSL**)
- Installing Linux graphical user interfaces
- Setting up and using the Linux workstation

Technical requirements

We will use the following platforms and technologies in this chapter:

- Linux distributions: Ubuntu Server, Ubuntu Desktop, CentOS
- Linux package managers: DEB, RPM
- VM hypervisors: Oracle VM VirtualBox, VMware Workstation
- VM host platforms: Windows, macOS X
- The Bash **Command-Line Interface** (**CLI**)
- GitHub (`https://github.com/`)

The Linux operating system

Linux is a relatively modern operating system created in 1991 by Linus Torvalds, a Finnish computer science student at the time, from Helsinki. Originally released as a free and open source platform prohibiting commercial redistribution, Linux eventually adopted the GNU **General Public Licensing** (**GPL**) model in 1992. This move played a significant role in its wide adoption by the developer community and commercial enterprises alike.

It is important to note that the Free Software Foundation community distinctly refers to Linux operating systems (or distributions) as **GNU/Linux** to emphasize the importance of GNU or free software.

Initially made for Intel x86 processor-based computer architectures, Linux has since been ported to a wide variety of platforms, becoming one of the most popular operating systems currently in use.

The genesis of Linux might be considered the open source alternative of its mighty predecessor, Unix. This system was a commercial-grade operating system developed at AT&T Bell Labs research center by Ken Thompson and Dennis Ritchie in 1969.

Linux distributions

A Linux operating system is typically referred to as a **distribution**. A Linux distribution, or **distro**, is the installation bundle (usually an ISO image) of an operating system that has a collection of tools, libraries, and additional software packages installed on top of the Linux kernel.

A kernel is the core interface between a computer's hardware and its processes, controlling the communication between the two and managing the underlying resources as efficiently as possible.

The software collection bundled with the Linux kernel usually consists of a bootloader, shell, package management system, graphical user interface, and various software utilities and applications.

The following diagram is a simplified illustration of a generic Linux distribution architecture:

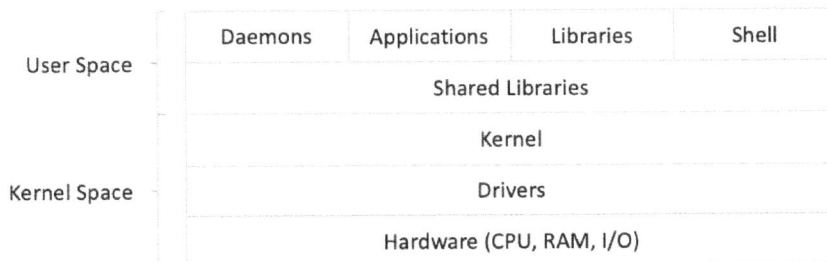

	Daemons	Applications	Libraries	Shell
User Space	Shared Libraries			
	Kernel			
Kernel Space	Drivers			
	Hardware (CPU, RAM, I/O)			

Figure 1.1 – Simplified view of a generic Linux architecture

There are hundreds of Linux distributions currently available. Among the most popular are **Debian, Fedora, openSUSE, Arch Linux**, and **Slackware**, with many other Linux distributions either based upon or derived from them. Some of these distros are divided into commercial and community-supported platforms.

One of the key differences between Linux distributions is the package management system they use and the related Linux package format. We'll get into more detail on this topic in later chapters. For now, the focus is on choosing the right Linux distribution based on our needs.

Choosing the right Linux distribution

There are many aspects involved in choosing a Linux distribution, based on various functional requirements. A comprehensive analysis would be far beyond the scope of this chapter. However, considering a few essential points may help with making the right decision:

- **Platform**: The choice between a server, a desktop, or an embedded platform is probably one of the top decisions in selecting a Linux distribution. Linux server platforms and embedded systems are usually configured with the core operating system services and essential components required for specific applications (such as networking, HTTP, FTP, SSH, and email), mainly for performance and optimization considerations. On the other hand, Linux desktop workstations are loaded (or pre-loaded) with a relatively large number of software packages, including a graphical user interface for a more user-friendly experience. Some Linux distributions come with server and desktop flavors (such as **Ubuntu**, **Fedora**, and **openSUSE**), but most distros have a minimal operating system, with further configuration needed (such as **CentOS**, and **Debian**). Usually, such distributions would be good candidates for Linux server platforms. There are also Linux distributions specifically designed for desktop use, such as **elementary OS**, **Pop!_OS**, or **Deepin**. For embedded systems, we have highly optimized Linux distros, such as **Raspbian** and **OpenWRT**, to accommodate small-form factor devices with limited hardware resources.

- **Infrastructure**: Today, we see a vast array of application and server platform deployments, ranging from hardware and local (on-premises) data centers to hypervisors, containers, and cloud infrastructures. Weighing a Linux distribution against any of these types of deployments should take into consideration the resources and costs involved. For example, a multi-CPU, large-memory, and generally high-footprint Linux instance may cost more to run in the cloud or a **Virtual Private Server** (**VPS**) hosting infrastructure. Lightweight Linux distributions take fewer resources and are easier to scale in environments with containerized workloads and services (for instance, with Kubernetes and Docker). Most Linux distributions now have their cloud images available for all major public cloud providers (for instance, Amazon AWS, Microsoft Azure, and Google Compute Engine). Docker container images for various Linux distributions are available for download on Docker Hub (`https://hub.docker.com`). Some Docker images are larger (heavier) than others. For example, the **Ubuntu Server** Docker image outweighs the **Alpine Linux** Docker image considerably, and this may tip the balance when choosing one distribution over the other. Also, to address the relatively new shift to containerized workflows and services, some Linux distributions offer a streamlined or more optimized version of their operating system to support the underlying application infrastructure. For example, Fedora features the **Fedora CoreOS** (for containerized workflows) and **Fedora IoT** (for Internet of Things ecosystems). CentOS has the **Atomic** project, as a lean CentOS for running Docker containers.

- **Performance**: Arguably, all Linux distributions can be tweaked to high-performance benchmarks in terms of CPU, GPU, memory, and storage. Performance should be regarded very closely with the platform and the application of choice. An email backend won't perform very well on a Raspberry Pi, while a media streaming server would do just fine (with some external storage attached). The configuration effort for tuning the performance should also be taken into consideration. **CentOS**, **Debian**, and **Ubuntu** all come with server and desktop versions reasonably optimized for their use. The server versions can be easily customized for a particular application or service by only limiting the software packages to those that are essential for the application. To further boost performance, some would go to the extent of recompiling a lightweight Linux distro (for instance, **Gentoo**) to benefit from compiler-level optimizations in the kernel for specific subsystems (for instance, the networking stack or user permissions). As with any other criteria, choosing a Linux distribution based on some application or platform performance is a balancing act, and most of the time, common Linux distros will perform exceptionally well.

- **Security**: When considering security, we have to keep in mind that a system is as secure as its weakest link. An insecure application or system component would put the entire system at risk. Therefore, the security of a Linux distribution should be scrutinized in close relation to the related application and platform environment. We can talk about *desktop security* for a Linux distro serving as a desktop workstation, for example, with the user browsing the internet, downloading media, installing various software packages, and running different applications. The safe handling of all these operations (against malware, viruses, and intrusions) would make for a good indicator of how secure a system can be. There are Linux distros that are highly specialized in application security and isolation, well suited for desktop use: **Qubes OS**, **Kali Linux**, **Whonix**, **Tails**, and **Parrot Security OS**. Some of these distributions are developed for penetration testing and security research.

  On the other hand, we may consider the *server security* aspect of Linux server distributions. In this case, regular operating system updates with the latest repositories, packages, and components would go a long way to securing the system. Removing unused network-facing services and configuring stricter firewall rules are further steps of reducing the possible attack surface. Most Linux distributions are well equipped with the required tools and services to accommodate the preceding. Opting for a distro with *frequent* and *stable* upgrades or release cycles is generally the first prerequisite for a secure platform (for instance, **Centos**, **RHEL**, **Ubuntu LTS**, or **SUSE Enterprise Linux**).

- **Reliability**: Linux distributions with aggressive release cycles and a relatively large amount of new code added in each release are usually less stable. For such distros, it's essential to choose a *stable* version. **Fedora**, for example, has rapid releases, being one of the fastest progressing Linux platforms. Yet, we should not heed myths claiming that Fedora or other similar fast-evolving Linux distros are less reliable. Don't forget, one of the most reliable Linux distributions out there, **Red Hat Enterprise Linux** (**RHEL**), is derived from Fedora.

There's no magic formula for choosing a Linux distribution. In most cases, the choice of platform (server or desktop) combined with a couple of the data points mentioned previously and some personal preferences would decide a Linux distribution. With production-grade environments, most of the previously enumerated criteria become critical, and the available options for our Linux platform of choice would be reduced to a few industry-proven solutions. In the following section, we enumerate some of the most popular Linux distributions.

Common Linux distributions

This section summarizes the most popular and common Linux distributions at the time of this writing, with emphasis on their package manager. Most of these distros are free and open source platforms. Their commercial-grade variations, if any, are noted.

CentOS and RHEL

CentOS and its derivatives use **RPM** as their package manager. CentOS is based on the open source Fedora project. It is suited to both servers and workstations. RHEL is a commercial-grade version of CentOS, designed to be a stable platform with long-term support.

Debian

The package manager for Debian and most of its derivatives is **Debian Package** (**DPKG**). Debian is releasing at a much slower pace than other Linux distributions, such as Linux Mint or Ubuntu, for example, but it's relatively more stable.

Ubuntu

Ubuntu uses **Advanced Package Tool** (**APT**) and DKPG as package managers. Ubuntu is one of the most popular Linux distributions, releasing every 6 months, with more stable **Long Term Support** (**LTS**) releases every other year.

Linux Mint

Linux Mint uses APT as its package manager. Built on top of Ubuntu, Linux Mint is mostly suitable for desktop use, with a lower memory usage than Ubuntu (with the Cinnamon desktop environment, compared to Ubuntu's GNOME). There's also a version of Linux Mint built directly on top of Debian, called **Linux Mint Debian Edition** (**LMDE**).

openSUSE

openSUSE uses **RPM**, **Yet another Setup Tool** (**YaST**), and **Zypper** as package managers. openSUSE is a bleeding-edge Linux distribution, suited to both desktop and server environments. SUSE Linux Enterprise Server is the commercial-grade platform. openSUSE was regarded as one of the most user-friendly desktop Linux distributions before the days of Ubuntu.

> **Important note**
> In this book, our focus is mainly on two Linux distributions, widely used in both community and commercial deployments, **Ubuntu** and **CentOS**.

The following section presents some hands-on use cases, where, depending on specific functional requirements, we choose the right Linux distribution.

Linux distributions – a practical guide

The following use cases are inspired by real-world problems, taken mostly from the authors' own experience in the software engineering field. Each of these scenarios presents the challenge of choosing the right Linux distribution for the job.

Case study – development workstation

This case study is based on the following scenario made from the perspective of a software developer:

> *I'm a backend/frontend developer, writing mostly in Java, Node.js, Python, and Golang, and using mostly IntelliJ and VS Code as my primary IDE. My development environment makes heavy use of Docker containers (building and deploying) and I occasionally use VMs (with VirtualBox) to deploy and test my code locally. I need a robust and versatile development platform.*

Functional requirements

The requirements suggest a relatively powerful day-to-day development platform, either as a PC/desktop or a laptop computer. The developer relies on local resources to deploy and test the code (for instance, Docker containers and VMs), perhaps frequently in an offline (airplane mode) environment if on the go.

System requirements

The system would be primarily using the Linux desktop environment and window manager, with frequent context switching between the **Integrated Development Environment** (**IDE**) and terminal windows. The required software packages for the IDE, Docker, hypervisor (VirtualBox), and tools should be readily available from open source or commercial vendors, ideally always being up to date and requiring minimal installation and customization effort.

Linux distribution

The choice here would be the **Ubuntu Desktop Long Term Support** (**LTS**) platform. Ubuntu LTS is relatively stable, runs on virtually any hardware platform, and it's mostly up to date with hardware drivers. Software packages for the required applications and tools are generally available and stable, with frequent updates. Ubuntu LTS is an enterprise-grade, cost-effective, and secure operating system suitable for organizations and home users alike.

Case study – secure web server

This case study is based on the following scenario made from the perspective of a DevOps engineer:

> *I'm looking for a robust platform running a secure, relatively lightweight, and enterprise-grade web server. This web server handles HTTP/SSL requests, offloading SSL before routing requests to other backend web servers, websites, and API endpoints. No load-balancing features are needed.*

Functional requirements

When it comes to open source, secure, and enterprise-grade web servers, the top choices are usually NGINX, Apache HTTP Server, Node.js, Apache Tomcat, and Lighttpd. Without going into the details of choosing one web server over another, let's just assume we pick Apache HTTP Server. It has state-of-the-art SSL/TLS support, excellent performance, and is relatively easy to configure.

We deploy this web server in VPS environments, in local (*on-premises*) data centers, or the public cloud. The deployment form factor is either a VM or a Docker container. We are looking for a relatively low-footprint, enterprise-grade Linux platform.

Linux distribution

Our choice is **CentOS**. Most of the time, CentOS and Apache HTTP Server are a perfect match. CentOS is relatively lightweight, coming only with barebone server components and an operating system networking stack. It is widely available as a VPS deployment template in private and public cloud vendors. There is also CentOS Atomic Host, a Linux distribution designed to run Docker containers. Our Apache HTTP Server can run as a Docker container on top of CentOS Atomic, as we may horizontally scale to multiple web server instances.

Use case – personal blog

This case study is based on the following scenario made from the perspective of a software engineer and blogger:

> *I want to create a software engineering blog. I'll be using the Ghost blogging platform, running on top of Node.js, with MySQL as the backend database. I'm looking for a Virtual Private Server (VPS) solution hosted by one of the major cloud providers. I'll be installing, maintaining, and managing the related platform myself. Which Linux distribution should I use?*

Functional requirements

We are looking for a self-managed publicly hosted VPS solution. The related hosting cost is a sensitive matter. Also, the maintenance of the required software packages should be relatively easy. We foresee frequent updates, including the Linux platform itself.

Linux distribution

Our pick is **Ubuntu Server LTS**. As previously highlighted, Ubuntu is a robust, secure, and enterprise-class Linux distribution. The platform maintenance and administration efforts are not demanding. The required software packages – Node.js, Ghost, and MySQL – are easily available and are well maintained. Ubuntu Server has a relatively small footprint. We can run our required software stack for blogging well within the Ubuntu system requirements so that the hosting costs would be reasonable.

Use case – media server

This case study is based on the following scenario made from the perspective of a home theater aficionado:

> *I have a moderately large collection of movies (personal DVD/Blu-ray backups), videos, photos, and other media, stored on Network Attached Storage (NAS). The NAS has its own media server incorporated, but the streaming performance is rather poor. I'm using Plex as a media player system, with Plex Media Server as the backend. What Linux platform should I use?*

Functional requirements

The critical system requirements of a media server are speed (for a high-quality and smooth streaming experience), security, and stability. The related software packages and streaming codecs are subject to frequent updates, so platform maintenance tasks and upgrades are quite frequent. The platform is hosted locally, on a PC desktop system, with plenty of memory and computing power in general. The media is being streamed from the NAS, over the in-house **Local Area Network** (**LAN**), where the content is available via a **Network File System** (**NFS**) share.

Linux distribution

Both **Debian** and **Ubuntu** would be excellent choices for a good media server platform. Debian's *stable* release is regarded as rock solid and very reliable by the Linux community, although it's somewhat outdated. Both feature advanced networking and security, but what may come as a decisive factor in choosing between the two is that Plex Media Server has an ARM-compatible package for Debian. The media server package for Ubuntu is only available for Intel/AMD platforms. If we owned a small-factor, ARM processor-based appliance, Debian would be our choice. Otherwise, **Ubuntu LTS** would serve our purpose just as well.

Installing Linux – the basics

This section serves as a quick guide for the basic installation of an arbitrary Linux distribution. For hands-on examples and specific guidelines, we use Ubuntu and CentOS. We also take a brief look at different environments hosting a Linux installation. There is an emerging trend of hybrid cloud infrastructures, with a mix of on-premises data center and public cloud deployments, where a Linux host can either be a bare-metal system, a hypervisor, a VM, or a Docker container.

In most of these cases, the same principles apply when performing a Linux installation. For Docker containerized Linux deployments, we reserve a separate chapter.

How to install Linux

Here are the essential steps usually required for a Linux installation.

Step 1 – download

We start by downloading our Linux distribution of choice. Most distributions are typically available in ISO format on the distribution's website. For example, we can download Ubuntu Desktop at `https://ubuntu.com/download/desktop`, or CentOS at `https://www.centos.org/download/`.

Using the ISO image, we can create the bootable media required for the Linux installation. We can also use the ISO image to install Linux in a VM (see the *Linux in a VM* section).

Step 2 – create a bootable media

If we install Linux on a PC desktop or workstation (*bare-metal*) system, the bootable Linux media is generally a CD/DVD or a USB device. With a DVD writeable optical drive at hand, we can simply burn a DVD with our Linux distribution ISO. But, as modern-day computers, especially laptops, rarely come equipped with a CD or a DVD unit of any kind, the more common choice for a bootable media is a USB drive.

There's also a third possibility of using a so-called PXE boot server. **PXE** (pronounced *pixie*) stands for **Preboot eXecution Environment**, which is a client-server environment where a PXE-enabled client (PC/BIOS) loads and boots a software package over a local or wide area network from a PXE-enabled server. PXE eliminates the need for physical boot devices (CD/DVD, USB) and reduces the installation overhead, especially for a large number of clients and operating systems. Probing the depths of PXE internals is beyond the scope of this chapter. A good starting point to learn more about PXE is `https://en.wikipedia.org/wiki/Preboot_Execution_Environment`.

A relatively straightforward way to produce a bootable USB drive with a Linux distribution of our choice is to use the open source tool **UNetbootin** (`https://unetbootin.github.io`). UNetbootin is a cross-platform utility, running on Windows, Linux, and macOS:

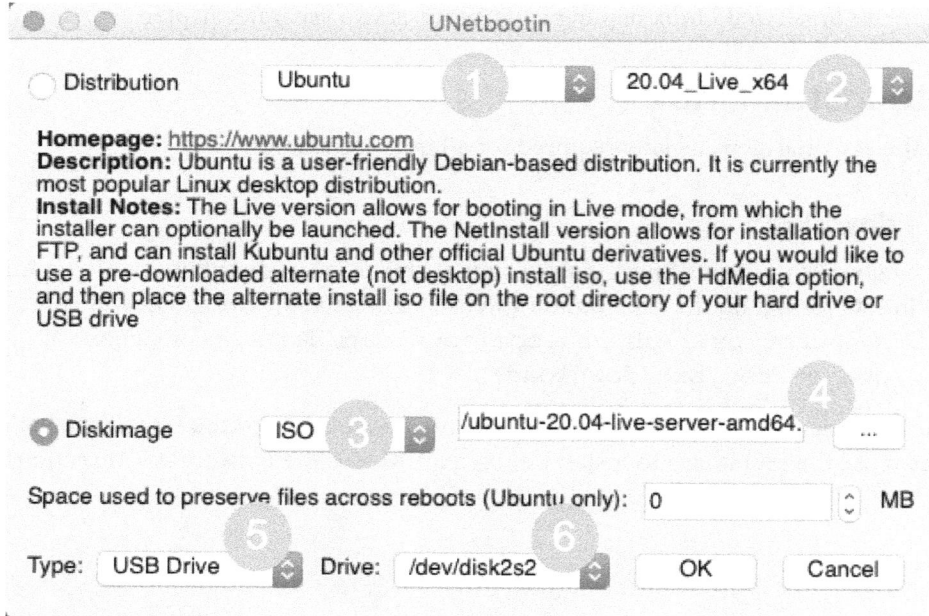

Figure 1.2 – Creating a bootable USB drive with UNetbootin

Here are the steps for creating a bootable USB drive with Ubuntu Desktop using UNetbootin. We assume the Ubuntu Desktop ISO image has been downloaded and UNetbootin is installed (in our case on macOS):

1. Choose our Linux distribution (Ubuntu).

2. Specify the version of our Linux distribution (20.04).

3. Select the disk image type that matches our download (ISO).

4. Browse to the location of our downloaded ISO image (ubuntu-20.04-live-server-amd64.iso).

5. Specify the media format of our bootable drive (USB).

6. Choose the filesystem mount of our USB drive (/dev/disk2s2)

Now, let's look at how we can take the bootable media for a spin.

Step 3 – try it out in live mode

This step is optional.

Most Linux distributions have their ISO image available for download as *live* media. Once we have the bootable media created with our Linux distribution of choice, we can run a live environment of our Linux platform without actually installing it. In other words, we can evaluate and test the Linux distribution before deciding whether we want to install it. The live Linux operating system is loaded in the system memory (RAM) of our PC, without using any disk storage. We should make sure the PC has enough RAM to accommodate the minimum required memory of our Linux distribution.

We can run Linux in live mode in either of the following ways:

*   Booting a PC/Mac workstation from our bootable media

*   Launching a VM created with our Linux distribution ISO

When booting the PC from a bootable media, we need to make sure the boot order in the BIOS is set to read our drive with the highest priority. On a Mac, we need to press the *Option* key immediately after the reboot start up chime and select our USB drive to boot from.

Upon reboot, the first splash screen of our Linux distribution should provide the option of running in live mode, as seen in the following illustration for Ubuntu Desktop (**Try Ubuntu**):

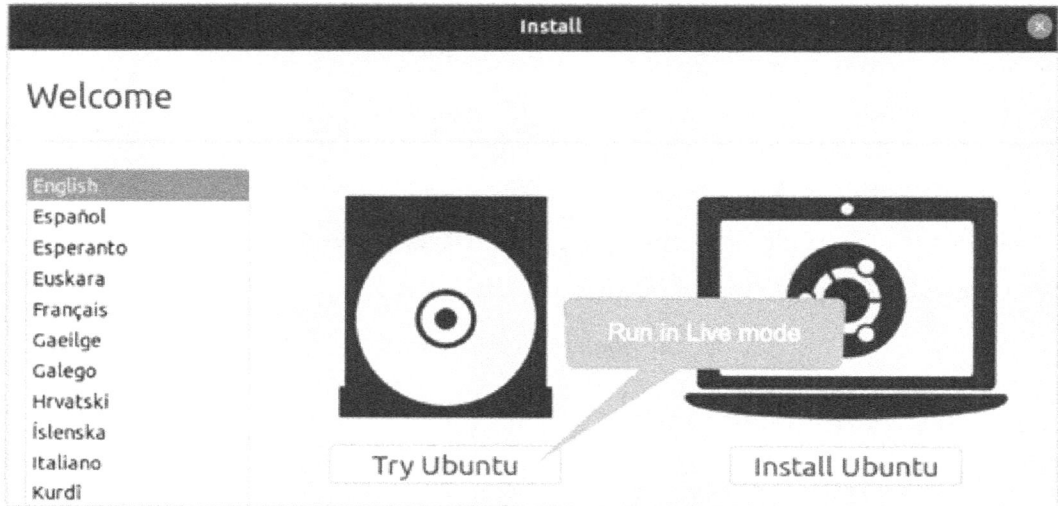

Figure 1.3 – Choosing live mode for Ubuntu

Next, let's take a look at the installation procedure of our Linux distro, using the bootable media.

Step 4 – perform the installation

We start the installation of our Linux distribution by booting the PC from the bootable media previously created. To ensure the system can boot from our drive (DVD or USB), we are sometimes required to change the boot order in the BIOS, especially if we boot from a USB drive. Most of the time, entering a system BIOS on a PC or laptop computer is done by pressing a *Function* key (or even the *Delete* key) immediately after powering on or restarting the machine. This key is usually mentioned at the bottom of the initial bootup screen.

In the following sections, we showcase the installation process of Ubuntu and CentOS using their ISO images. We choose the Desktop and Server versions for Ubuntu and highlight the main differences. CentOS comes in a single flavor, in essence, a server platform with an optional graphical user interface.

Linux in a VM

In each of the Linux installation sections, we also provide a brief guide on how to prepare a VM environment for the related Linux platform.

A VM is an isolated software abstraction of a physical machine. VMs are deployed on top of a **hypervisor**. A hypervisor provides the runtime provisioning and resource management of VMs. For the simple illustration of Linux VM installations, in this section, we limit ourselves to a couple of general-purpose hypervisors:

- **Oracle VM VirtualBox** (`https://www.virtualbox.org`)
- **VMware Workstation** (`https://www.vmware.com/products/workstation-pro.html`)

Both these hypervisors are cross-platform virtualization applications, and they run on Intel or AMD processor architectures on Windows, Linux, and macOS.

The difference between installing Linux on a VM compared to a physical machine is minor. The notable distinction is related to the VM sizing and configuration steps, making sure that the minimum system requirements of the Linux distribution are met.

Installing Ubuntu

In this section, we briefly illustrate the installation of Ubuntu Server LTS. If we plan to install Ubuntu in a VM, there are some preliminary steps required for provisioning the VM environment. Otherwise, we proceed directly to the *Installation* section.

VM provisioning

In the following steps, we will create a VM based on Ubuntu Server—using VMware Workstation on macOS:

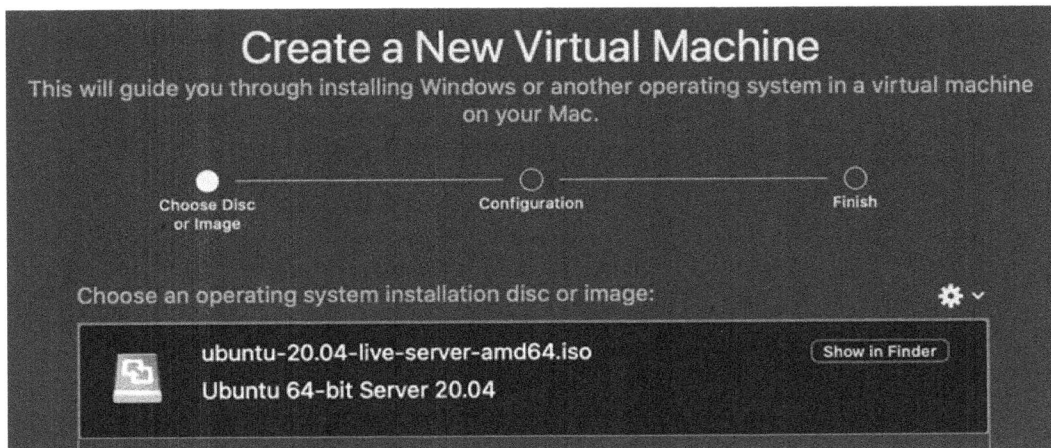

Figure 1.4 – Creating a new VM based on the Ubuntu ISO image

Let's look at the steps:

1. We start by creating a new VM based on the Ubuntu Server ISO. For illustration purposes, we use VMware Workstation. *Figure 1.4* shows the initial screen of creating a new VM for an Ubuntu Server instance.

2. Following the VM deployment wizard, we get to the final step summarizing the VM provisioning information (*Figure 1.5*):

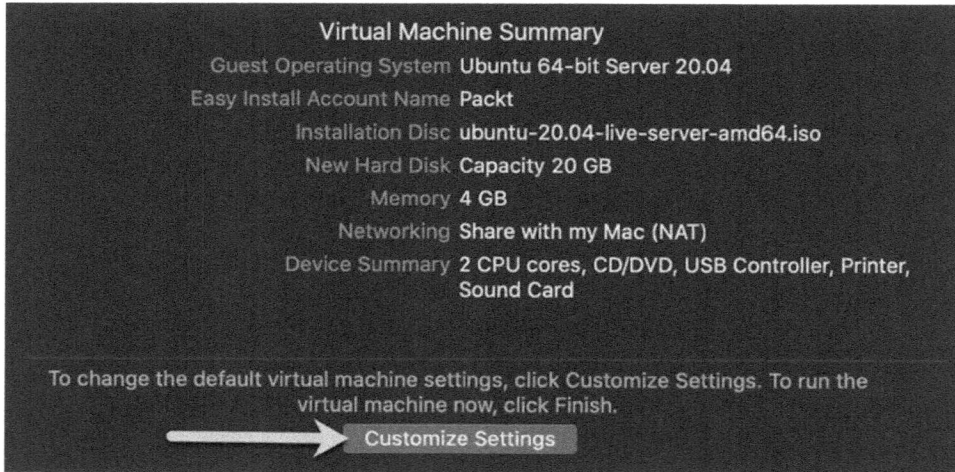

Figure 1.5 – Customizing the VM settings

3. Sometimes the default VM sizing may have to be changed to accommodate the minimum system requirements of our Linux distribution. In our case, Ubuntu Server requires a minimum hard disk capacity of 25 GB. We can further customize the VM settings and increase the disk capacity, for example, to 30 GB (*Figure 1.6*):

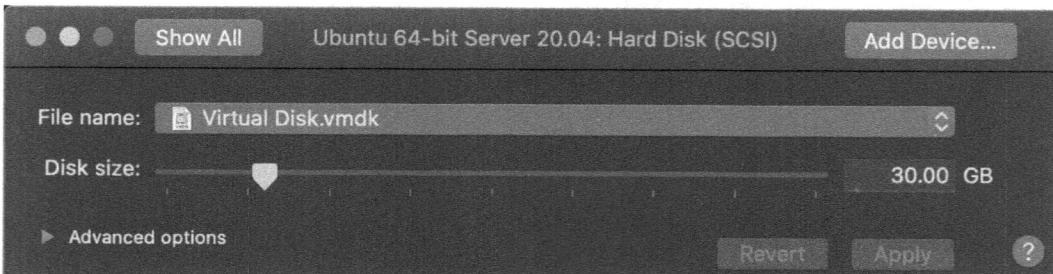

Figure 1.6 – Customizing the VM disk size

The remaining part of the Linux VM installation is identical to standard physical machine installation, shown in the following sections.

Installation

Here's the normal installation process for Ubuntu Server LTS, following the initial boot into setup mode:

1. Make sure we choose **Install Ubuntu** in the initial setup screen (we assume that *live* mode has already been visited, according to *Figure 1.3*):

2. The initial welcome screen prompts for the language of our choice (**English**), followed by the keyboard layout (**English (US)**):

3. Next, we need to set up the server profile, which requires a display name (Packt), a server name (neptune), a username (packt), and the password (*Figure 1.7*):

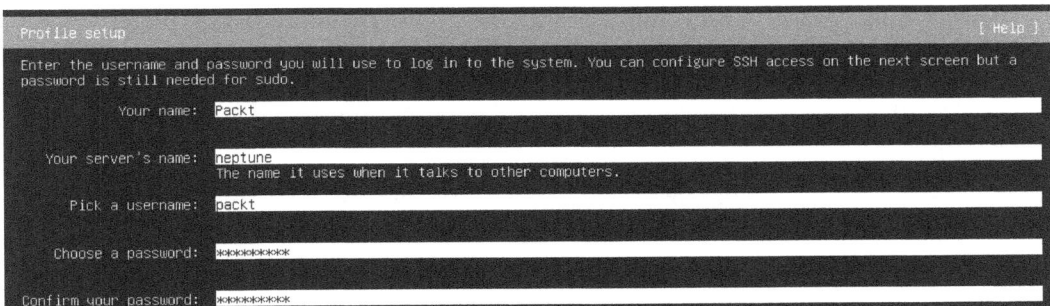

Figure 1.7 – Setting up the server profile

4. The next screen is asking for the **OpenSSH** server package installation (*Figure 1.8*). OpenSSH enables secure remote access to our server. We choose to check (using **[X]**) the option for **Install OpenSSH server**. Optionally, we can import our existing SSH keys for passwordless authentication:

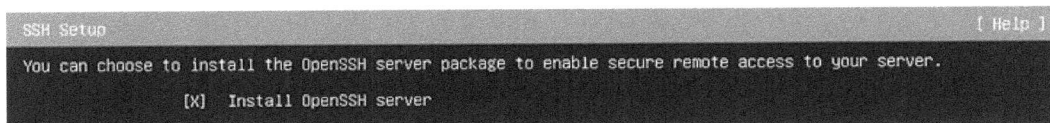

Figure 1.8 – Enabling the OpenSSH server

5. An additional screen presents us with popular software packages (**Snaps**) that we may want to install (*Figure 1.9*). Among them, there are a few we'll be covering in later chapters (`microk8s`, `docker`, and `aws-cli`):

```
Featured Server Snaps                                                                    [ Help ]

These are popular snaps in server environments. Select or deselect with SPACE, press ENTER to see more details of the package,
publisher and versions available.

[ ] microk8s              canonical√       Lightweight Kubernetes for workstations and appliances                          ▶
[ ] nextcloud             nextcloud√       Nextcloud Server - A safe home for all your data                                ▶
[ ] wekan                 xet7             Open-Source kanban                                                              ▶
[ ] kata-containers       katacontainers√  Lightweight virtual machines that seamlessly plug into the containers ecosyste ▶
[ ] docker                canonical√       Docker container runtime                                                        ▶
[ ] canonical-livepatch   canonical√       Canonical Livepatch Client                                                      ▶
[ ] rocketchat-server     rocketchat√      Group chat server for 100s, installed in seconds.                               ▶
[ ] mosquitto             mosquitto√       Eclipse Mosquitto MQTT broker                                                   ▶
[ ] etcd                  canonical√       Resilient key-value store by CoreOS                                             ▶
[ ] powershell            microsoft-powershell√ PowerShell for every system!                                              ▶
[ ] stress-ng             cking-kernel-tools A tool to load, stress test and benchmark a computer system                  ▶
[ ] sabnzbd               safihre           SABnzbd                                                                        ▶
[ ] wormhole              snapcrafters      get things from one computer to another, safely                               ▶
[ ] aws-cli               aws√              Universal Command Line Interface for Amazon Web Services                       ▶
```

Figure 1.9 – Enabling additional software packages

6. If everything goes well, a few minutes later, we get the **Installation complete!** screen, prompting us to reboot.

After the system reboot, the login screen appears. We have completed the Ubuntu Server installation.

Next, let's take a look at a similar installation procedure, this time with the CentOS Linux distribution.

Installing CentOS

In this section, we briefly illustrate the installation of CentOS. If we plan to install CentOS in a VM, there are some preliminary steps required for provisioning the VM environment. Otherwise, we proceed directly with the *Installation* section.

VM provisioning

In the following steps, we show the setup of a CentOS VM using Oracle VM VirtualBox on macOS. The choice of VirtualBox over VMware Workstation (used in the previous section with Ubuntu Server) is simply for showcasing the use of an alternative hypervisor:

1. The VirtualBox setup wizard guides us through the following configuration steps of our VM (we specify our choices as shown):

    a) Hostname and operating system (*jupiter*, *CentOS Red Hat, 64-bit*)

b) Memory size (*4 GB*)

c) Hard disk size (*30 GB*)

d) Hard disk file type (*VDI VirtualBox Disk Image*)

e) Storage on the physical hard disk (*dynamically allocated*)

f) File location and size (*path to .vdi file, 30 GB*)

2.   After a few steps, we end up with a VirtualBox configuration window of our VM, similar to the following (*Figure 1.10*):

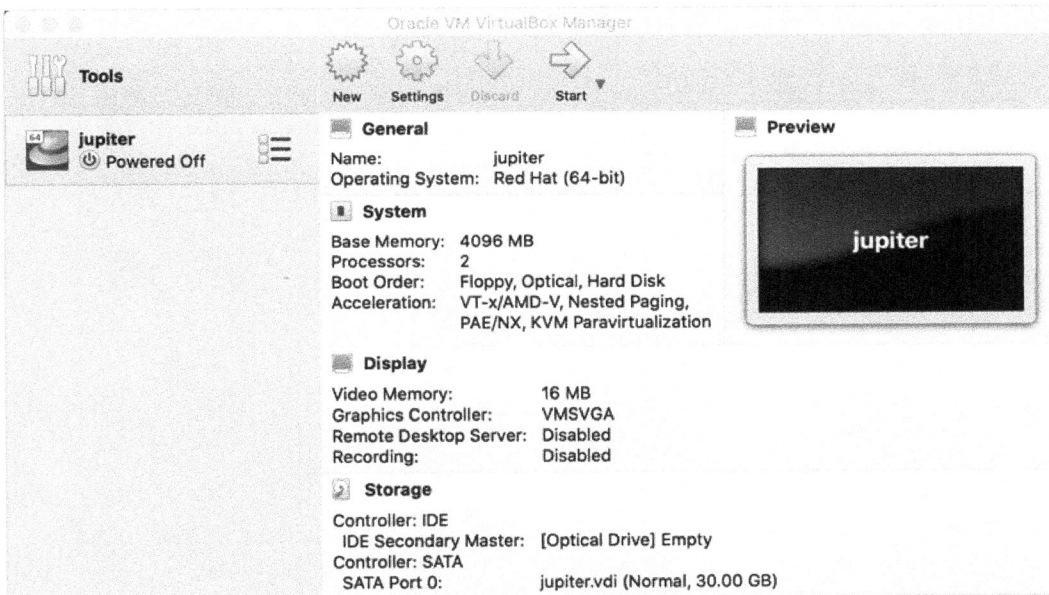

Figure 1.10 – VirtualBox VM configuration

3. We can do further customization by choosing **Settings** in the VirtualBox Manager window of our VM. Next, we should point the IDE controller of our VM to the CentOS ISO image we want to install. We choose **Settings | Storage | Optical Drive | IDE Secondary Master** and click the disk icon. Here, we browse to the location of our CentOS image file. We want the VM to use the IDE Secondary Master ISO file, to boot from and install our operating system (*Figure 1.11*):

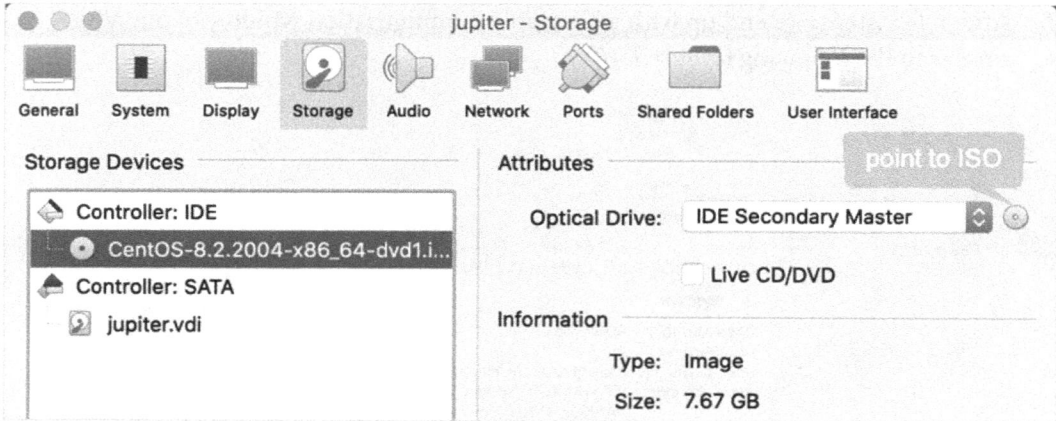

Figure 1.11 – Virtual Box VM storage settings

At this stage, starting the VM initiates the CentOS installation.

Installation

Here's the normal installation process for CentOS, following the initial boot into setup mode:

1. First, we get the welcome screen with the choice of either installing or testing (in live mode) the CentOS Linux platform (*Figure 1.12*):

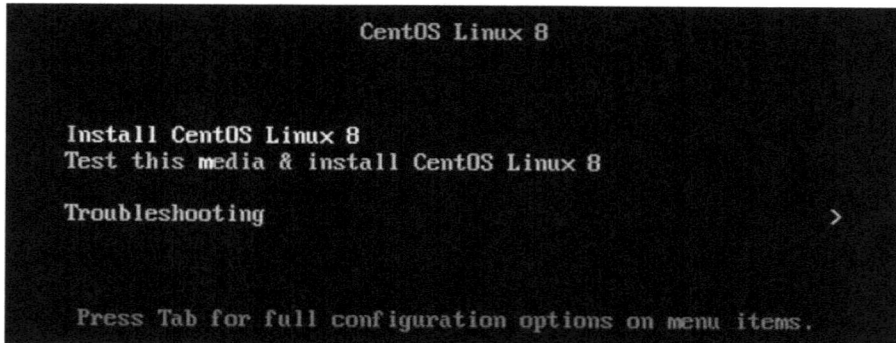

Figure 1.12 – The CentOS welcome screen

2. In the next few steps of the installation process, we get to choose our options for the following:

a) Language support and localization (*English, US*)

b) Software selection (*Server with GUI*)

c) Device selection and storage configuration (*Local media*)

The following screenshot summarizes all this. The settings mostly reflect default values (*Figure 1.13*):

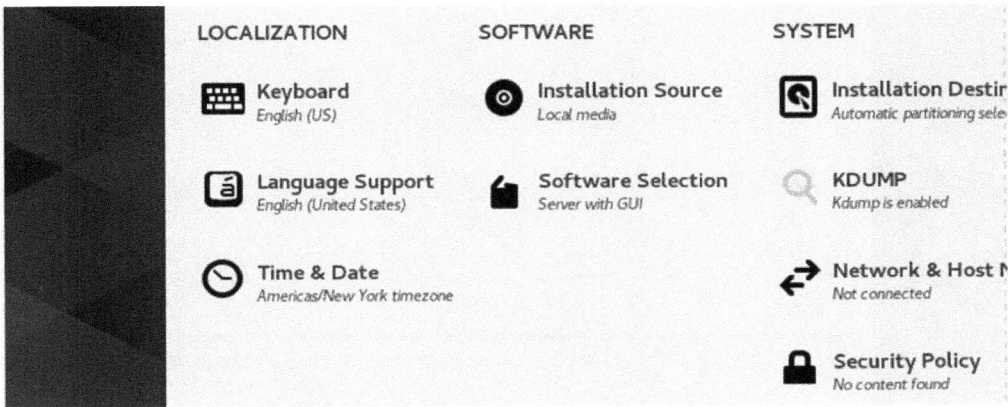

Figure 1.13 – The CentOS installation summary

3. The final step of the CentOS installation is about configuring the local user accounts on our Linux platform (*Figure 1.14*):

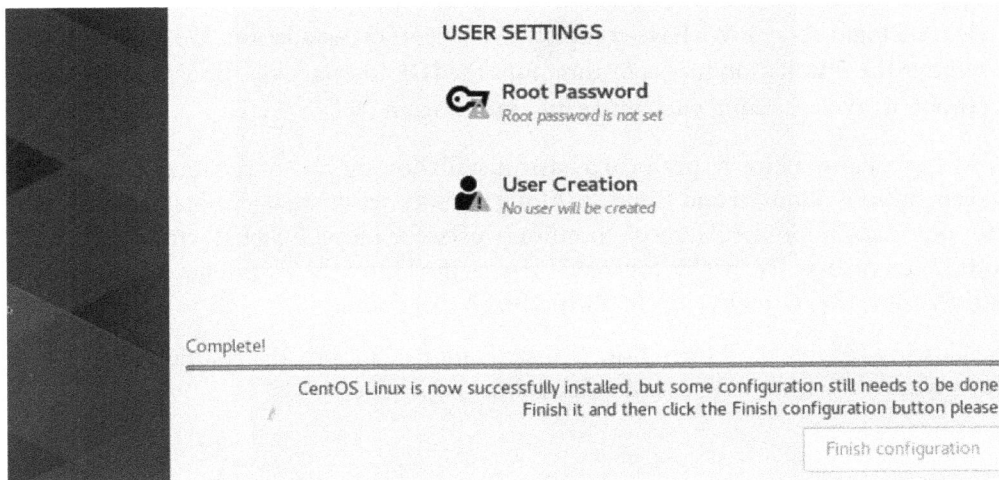

Figure 1.14 – The CentOS user settings

4. The user configuration screen prompts for a *root* password and a new user account. In our case, we create a user account (*packt*) with administrator privileges. This account has full administrative privileges over the system, yet it doesn't have root access. In later chapters, we'll show how to give **superuser** (**sudo**) privileges to our administrator account (*Figure 1.15*):

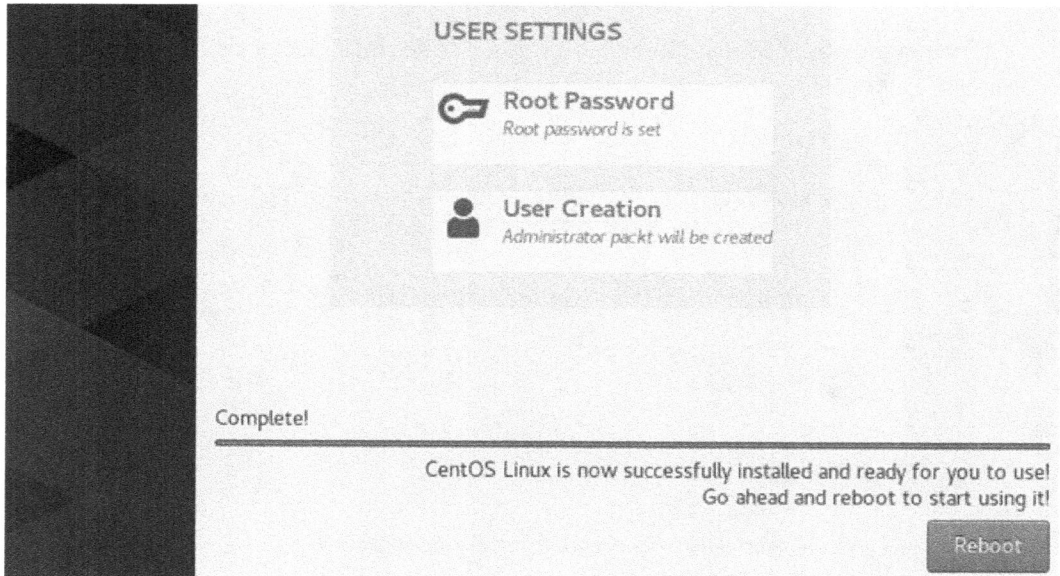

Figure 1.15 – Rebooting CentOS after installation

5. Upon finishing the user configuration, we reboot the system, and this takes us to the CentOS login screen. We have completed the CentOS installation. We may have to remove the installation media or unmount the IDE interface on the VM before the reboot, to avoid circling back into setup mode again.

So far, we have learned how to perform a basic installation for two of the most common Linux distributions, Ubuntu and CentOS. Along the way, we created a bootable USB flash drive for our installation media, most commonly used for Linux PC platform installations. For both Linux distros, we briefly covered VM-specific Linux environments using the VMware Workstation and Oracle VM VirtualBox hypervisors.

In the following section, we'll learn how to install and run a Linux distribution on a Windows platform without the use of a standalone hypervisor.

The Windows Subsystem for Linux (WSL)

Software developers and system administrators often face a tough decision in choosing the appropriate hardware and operating system platform for the specific requirements of their work or environment. In the past, Windows professionals were frequently reminded that some standard development tools, frameworks, or server components were available at large on Linux or macOS platforms, while lacking native support on Windows. **Windows Subsystem for Linux** (**WSL**) attempts to close this gap.

WSL is a Windows 10 platform feature that provides a native GNU/Linux runtime along with the Windows desktop environment. WSL enables the seamless deployment and integration of select Linux distributions on top of the Windows kernel, without the need for a dedicated hypervisor. With WSL enabled, you can easily install and run Linux as a native Windows application.

Important note

Without WSL, we could only deploy and run a Linux distribution on a Windows platform by using a standalone hypervisor, such as Hyper-V, Oracle VM VirtualBox, or VMware Workstation. WSL eliminates the need for a dedicated hypervisor. At the time of writing, WSL is a Windows kernel extension with a hypervisor embedded.

In this section, we provide the steps required to enable WSL and run an Ubuntu distribution on Windows. The following commands are executed on a Windows machine (or VM) in a PowerShell CLI with administrator privileges:

1. First, we need to enable the *Windows Subsystem for Linux* optional feature in Windows:

   ```
   dism.exe /online /enable-feature /featurename:Microsoft-
   Windows-Subsystem-Linux /all /norestart
   ```

   We should get the result shown in *Figure 1.16*:

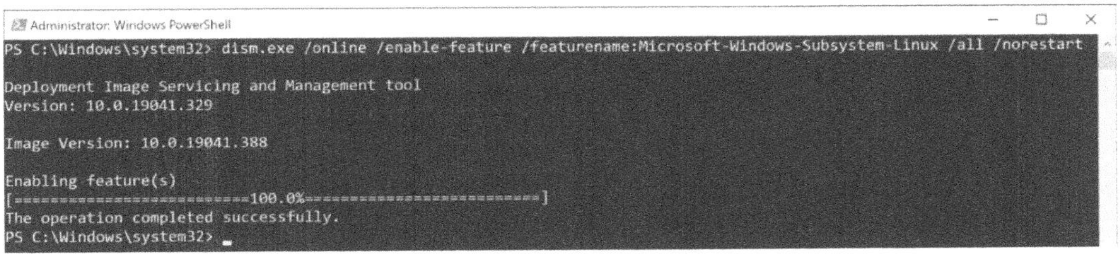

Figure 1.16 – Enabling the WSL optional feature

2. Next, we want to make sure WSL 2 is supported on our Windows platform. We need Windows 10, version 2004, build 19041 or higher.

    WSL 2 uses hypervisor technology. We need to enable the *Virtual Machine Platform* optional feature for Windows:

    ```
    dism.exe /online /enable-feature /
    featurename:VirtualMachinePlatform /all /norestart
    ```

    The result should be similar to *Figure 1.17*:

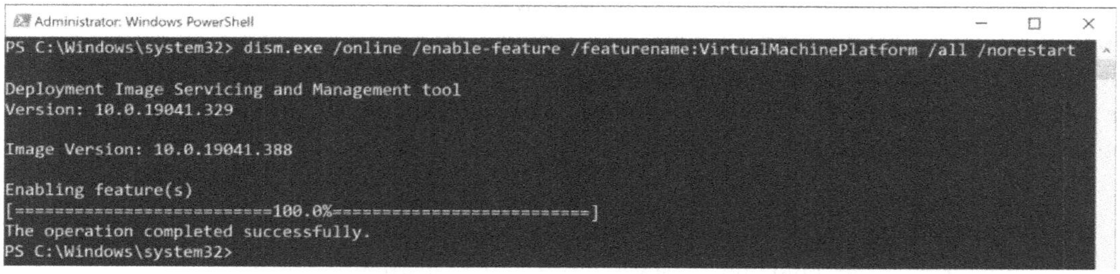

Figure 1.17 – Enabling the VM platform optional feature

3. At this point, we need to restart our Windows machine to complete the WSL installation and upgrade to WSL 2.

    > **Important note**
    > The Windows restart may feature an *Update and Restart* option. We need to make sure that the system gets updated upon restart.

4. After the restart, we need to set WSL 2 as our default version:

    ```
    wsl --set-default-version 2
    ```

    This command may yield the message shown in *Figure 1.18*:

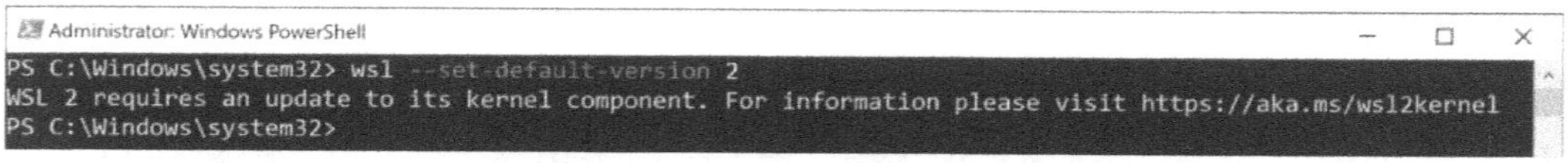

Figure 1.18 – Further update needed for WSL 2

5. Just follow the instructions in the link to circumvent the problem. After proceeding with these steps, we have to rerun the preceding command. If successful, the command output is similar to *Figure 1.19*:

Figure 1.19 – Setting the default version to WSL 2

6. With WSL 2 active, we are ready to install our Linux distribution of choice. Open the Microsoft Store and simply search for `Linux` (*Figure 1.20*):

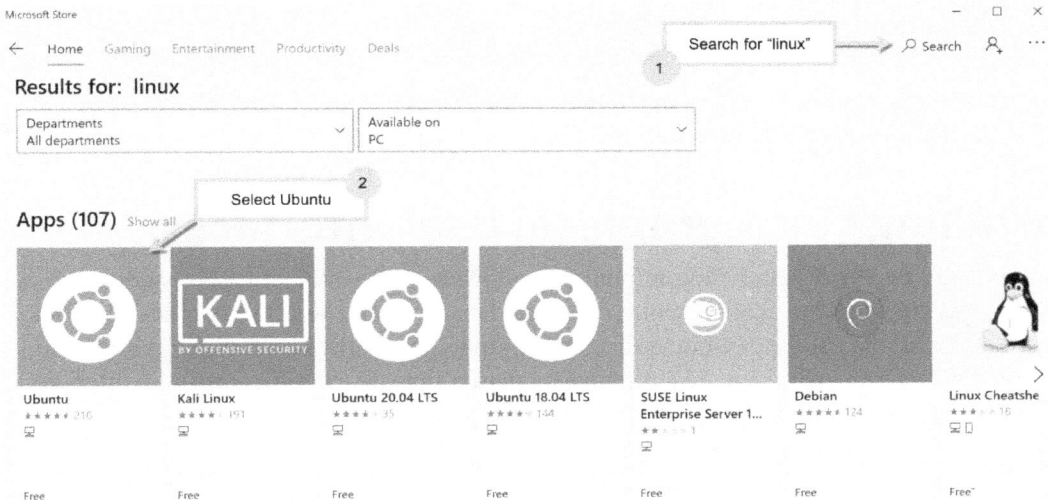

Figure 1.20 – Searching for Linux distro apps in the Microsoft Store

7. We choose to install Ubuntu (*Figure 1.21*) and then follow the installation procedure driven by the Windows Store app. Ubuntu is installed just like any regular Windows application. When finished, we run Ubuntu by launching the application from the Windows Start menu:

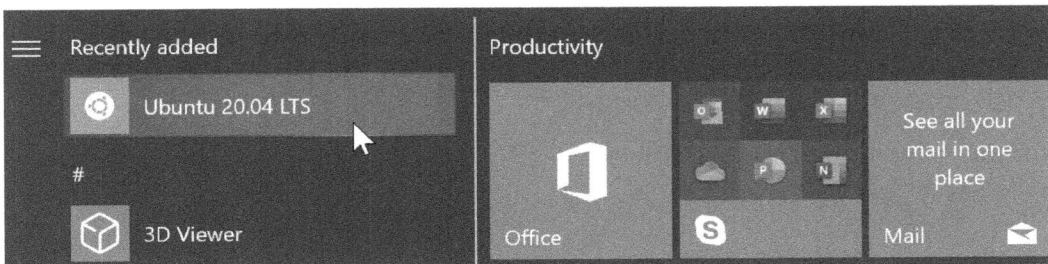

Figure 1.21 – Installing the Ubuntu app

After installation, Ubuntu runs as a traditional Windows desktop command-line application.

WSL enables a swift adoption of Linux for a growing number of Windows professionals. As shown in this section, WSL is relatively easy to configure, and with WSL, there's no need for a dedicated hypervisor to run a Linux instance.

> **Important note**
> WSL currently doesn't support a GUI-driven Linux runtime. With WSL, we are exclusively interacting via the CLI.

In the following section, we take a quick look at the Linux graphical user interface and two of the most prominent Linux desktop environments: GNOME and KDE.

Installing Linux graphical user interfaces

The Linux GUI is the desktop environment that allows users to interact with system-level components via windows, icons, menus, or other visual elements. For Linux users who look beyond the CLI, the choice of a Linux distribution may start with the desktop environment. Choosing a desktop environment is ultimately a matter of taste. And for some, the ultimate GUI represents the very extension of their eyes and hands at work.

Among the multitude of Linux desktop environments, there are two that stand out. Let's take a brief look at them.

GNOME

With its current GNOME 3 (or GNOME Shell) iteration, this GUI platform is one of the most common Linux desktop environments. Nowadays, almost every major Linux distribution comes with GNOME as the default GUI. The Linux open source community also created GNOME Extensions, which overcomes some of the infamous shortcomings of GNOME and extends the desktop functionality to suit a variety of needs. When it's not the default desktop environment (such as with Linux Mint's Cinnamon desktop), GNOME can easily be installed and adapted.

At the time of writing, the latest distributions of Ubuntu (20.04 LTS) and CentOS (8) have GNOME as their default GUI.

Installing the GNOME desktop

Let's look at a real-world scenario requiring the installation of the latest GNOME desktop.

Ubuntu Server administrator:

> *I installed the latest version of Ubuntu Server LTS (20.04), and it looks like the GUI desktop is missing. How do I install the GNOME desktop on my Ubuntu Server?*

Let's look at how to install it on Ubuntu:

1. We start by making sure the current package lists and installed software packages are up to date. The following command updates the local package repository metadata for all configured sources:

    ```
    sudo apt-get update -y
    ```

    Now, we can upgrade with this:

    ```
    sudo apt-get upgrade -y
    ```

2. To browse all the available ubuntu-desktop packages, run this:

    ```
    apt-cache search ubuntu-desktop
    ```

3. We choose to install the ubuntu-desktop package (the first option in the preceding list):

    ```
    sudo apt-get install ubuntu-desktop -y
    ```

4. The command will take a few minutes to complete. When done, we need to check the status of the **GNOME Display Manager** (**GDM**) service and make sure it shows an active (running) status:

    ```
    systemctl status gdm
    ```

    The expected response should be similar to *Figure 1.22*:

```
packt@neptune:~$ systemctl status gdm
• gdm.service - GNOME Display Manager
    Loaded: loaded (/lib/systemd/system/gdm.service; static; vendor preset: enabled)
    Active: active (running) since Thu 2020-08-27 18:32:48 UTC; 53min ago
  Main PID: 1182 (gdm3)
     Tasks: 3 (limit: 4581)
    Memory: 6.7M
    CGroup: /system.slice/gdm.service
            └─1182 /usr/sbin/gdm3
```

Figure 1.22 – Checking the status of GDM

The GNOME 3 desktop is now installed and active on Ubuntu Server (*Figure 1.23*):

Figure 1.23 – The Ubuntu GNOME desktop login screen

Next, let's take a look at the KDE desktop and a similar case study of enabling it on a Linux server platform.

KDE

Linux administrators usually look for a desktop environment that is relatively easy to use, lightweight, and efficient. KDE combines all of these attributes into a reliable and speedy desktop interface. Users familiar with Windows (up to version 7) would feel very much at home with KDE.

KDE has become a very robust desktop environment over the last few iterations, and versions of KDE have been released for almost every major Linux distribution.

If there's one ideal desktop for Linux administrators, KDE comes very close.

Installing the KDE desktop

In this section, we take a fresh CentOS 8 installation with the default GNOME desktop enabled and replace it with KDE.

CentOS 8 administrator:

> *I installed the latest version of CentOS 8, choosing the "Server with GUI"*
> *option during setup. It looks like the CentOS 8 GUI runs the GNOME*
> *desktop. How do I install the KDE desktop on my CentOS 8 server?*

When we click the cogwheel in the CentOS 8 login screen, we get a list of the currently installed display servers. The default is *Standard*, a GNOME implementation of the *Wayland* compositor protocol for Linux desktop management. Wayland also has a KDE implementation in its latest *KDE Plasma* iteration at the time of writing (*Figure 1.24*):

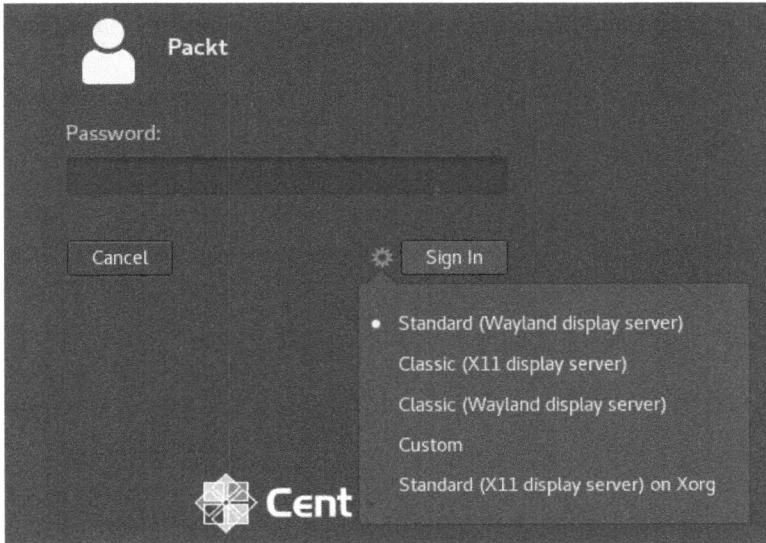

Figure 1.24 – The default CentOS 8 GNOME login screen

Let's add the KDE Plasma desktop to our CentOS 8 server:

1. We log in to the CentOS 8 GUI and open the terminal, as suggested in the following illustration (*Figure 1.25*). Alternatively, we can simply SSH into the CentOS 8 server and run the commands in a similar CLI:

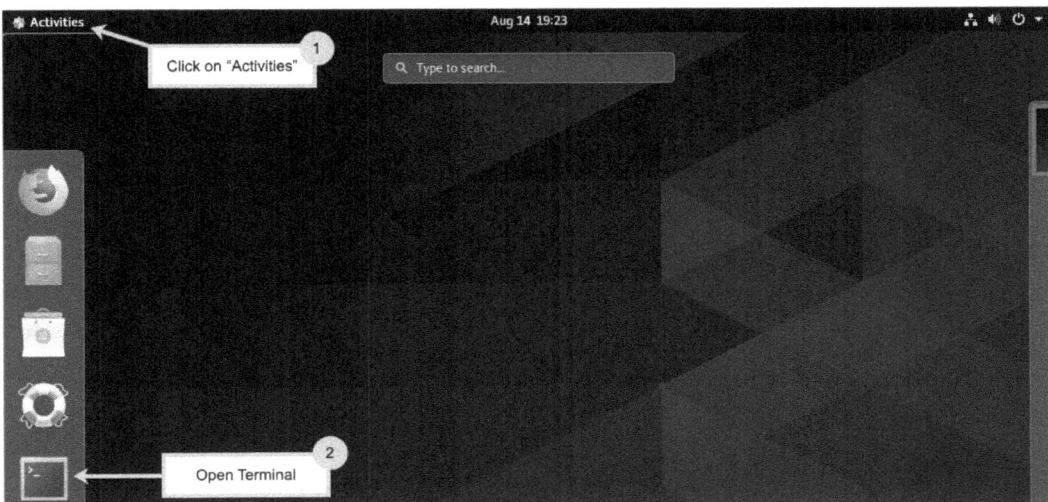

Figure 1.25 – Opening the terminal in CentOS

2. The following commands have to be executed either as *root* or by a user account with *sudoer* privileges. Let's add our administrator user account (packt) to the sudoers group. We need to switch to *root* to run the required command:

```
su
```

3. We get prompted for the *root* password (specified during the CentOS installation). See the *Installing CentOS* section for the related information. To add our packt user to the sudoers group, run this:

```
usermod -aG wheel packt
```

4. When the command finishes, switch to our newly minted sudoer account:

```
su - packt
```

5. Verify that our packt account does indeed have *sudoer* privileges:

```
sudo whoami
```

The command should prompt for the packt user password and then yield root.

6. Next, proceed with the installation of the EPEL configuration files, enabling the related repositories:

```
sudo rpm -Uvh https://dl.fedoraproject.org/pub/epel/epel-
release-latest-8.noarch.rpm
```

7. Follow that by enabling the CentOS PowerTools repository (needed by KDE). We use dnf (*Dandified YUM*), a package manager CLI for RPM-based Linux distributions:

```
sudo dnf -y config-manager --enable PowerTools
```

8. Finally, we install the KDE Plasma desktop package group. The next command downloads a relatively large number of files (approximately 400 MB), and it may take a while:

```
sudo dnf -y group install kde-desktop
```

When the `kde-desktop` installation completes successfully, the output ends with a `Complete!` message prompt.

9. Now that the KDE Plasma desktop has been installed, we can either reload the GNOME Desktop Manager (`gdm`) to account for the changes or simply reboot:

```
sudo systemctl reload gdm
```

10. If we have a CentOS 8 VM environment, a reboot is recommended:

```
reboot
```

We are now able to log in to CentOS 8 using the KDE Plasma desktop. Clicking the cogwheel in the login window, we can see the **Plasma** desktop available (*Figure 1.26*):

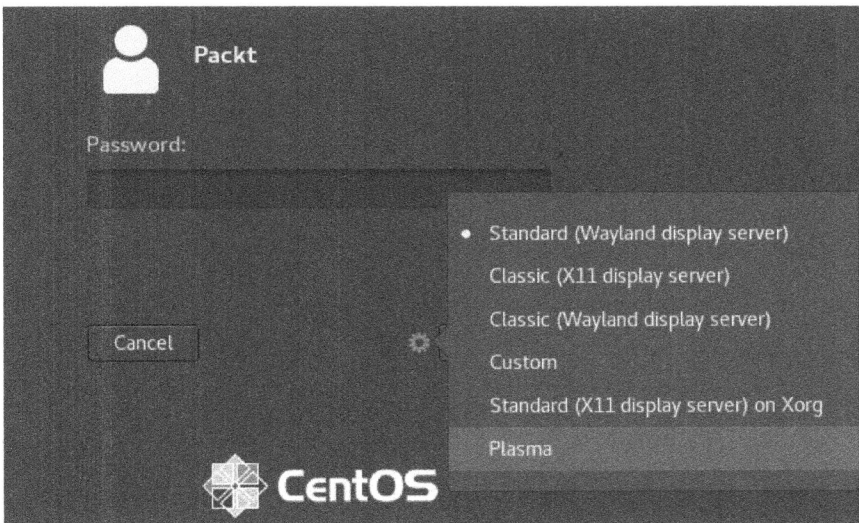

Figure 1.26 – Choosing the KDE Plasma desktop

11. Once we're logged in, we can customize the KDE Plasma desktop to our preference (*Figure 1.27*):

Figure 1.27 – Customizing the KDE Plasma desktop

In this section, we briefly looked at the Linux GUI and showcased the installation of the GNOME and KDE desktop environments on Ubuntu Server and CentOS, respectively. Next, we provide a quick guide on setting up a Linux workstation with some insight into various software packages and applications that could benefit our everyday work.

Setting up and using the Linux workstation

In this section, we will learn how to set up and use a Linux platform as our primary workstation for day-to-day work. We choose the Ubuntu Desktop LTS distribution, but any other modern-day Linux distribution would fit the bill. We aim to showcase some basic user operations and workflows for building and using a general-purpose Linux desktop.

As we did in the previous sections about installing a Linux distribution, we start with a brief installation guide of Ubuntu Desktop LTS.

Installing Ubuntu Desktop

If we plan to install Ubuntu Desktop on a VM, there are some preliminary steps required for provisioning the VM environment. Otherwise, we proceed directly with the *Installation* section.

VM provisioning

The VM provisioning procedure is very similar to the procedure described in the *Installing Ubuntu* section, with Ubuntu Server LTS.

We have to pay attention to the minimum system requirements of our Linux distribution and size the VM accordingly. In our case, Ubuntu Desktop requires a dual-core CPU, at least 4 GB RAM, and a minimum hard disk capacity of 25 GB. Since we plan to install a handful of additional software packages, we set the hard disk capacity to 60 GB (at least).

Installation

Ubuntu Desktop installation is relatively straightforward and requires very few user actions:

1. As with the most modern Linux distributions, the very first choice we get is between trying out Ubuntu in live mode or simply proceeding with the installation. The related screen is similar to *Figure 1.3*.

2. Beyond the welcome screen, there are a few more steps with the following configuration settings:

a) Keyboard layout

b) Installation type (normal versus minimal)

c) Localization and time zone

d) User configuration (account credentials) (*Figure 1.28*):

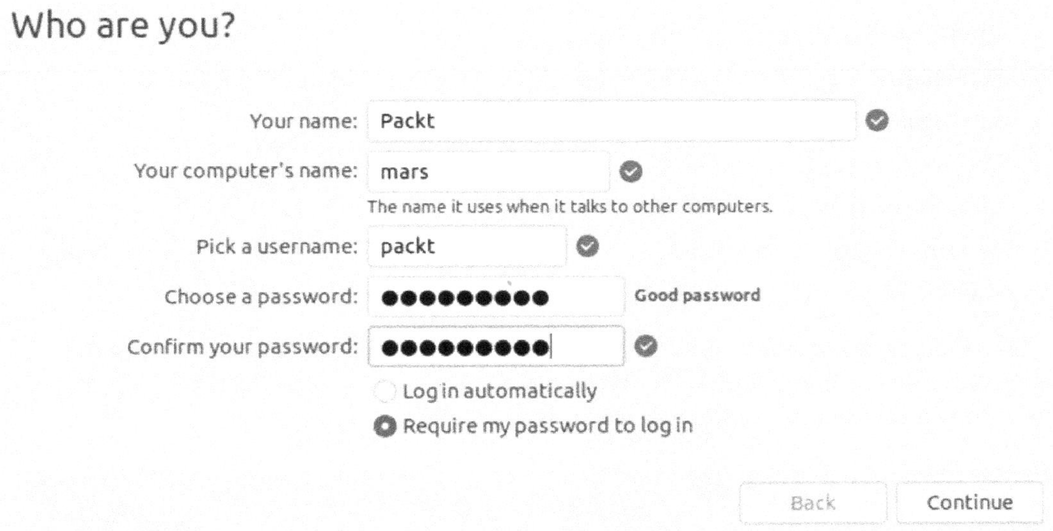

Who are you?

Your name: Packt ✓

Your computer's name: mars ✓

The name it uses when it talks to other computers.

Pick a username: packt ✓

Choose a password: ●●●●●●●●   Good password

Confirm your password: ●●●●●●●● ✓

◯ Log in automatically
◉ Require my password to log in

Back   Continue

Figure 1.28 – Setting up the hostname and user credentials

3. Past this point, the installation files are unpackaged and copied, while we are briefly introduced to the most attractive features of the Ubuntu platform.

After the installation completes, the system reboot takes us to the Ubuntu Desktop login screen. At this point, we have completed the Ubuntu Desktop installation.

Default software packages

The default Ubuntu Desktop installation gets us a handful of software packages and productivity tools, enough for our general-purpose, day-to-day work. Here are just a few examples:

- **Calculator**: Arithmetic, scientific, and financial calculations.
- **Calendar**: Access and manage your calendars.
- **Disks**: Manage drives and media.

- **Files**: Access and organize files.
- **Firefox**: Web browser.
- **Image Viewer**: Browse and rotate images.
- **LibreOffice**: Productivity suite – **Calc** (spreadsheet), **Draw** (graphics), **Impress** (presentation), and **Writer** (word processor).
- **Logs**: System logs.
- **Remmina**: Remote desktop connection.
- **Rhythmbox**: Organize and play music.
- **Screenshot**: Capture and save images of the screen or individual windows.
- **Settings**: GNOME desktop configuration utility.
- **Shotwell**: Organize photos.
- **Thunderbird Mail**: Send and receive mail.
- **To Do**: Manage personal tasks.
- **Videos**: Play movies.

To view all currently installed applications, we click on the *Ubuntu Software* icon in the taskbar and then on the **Installed** tab. We could choose to remove (uninstall) any of these applications if we wanted to save disk space (*Figure 1.29*):

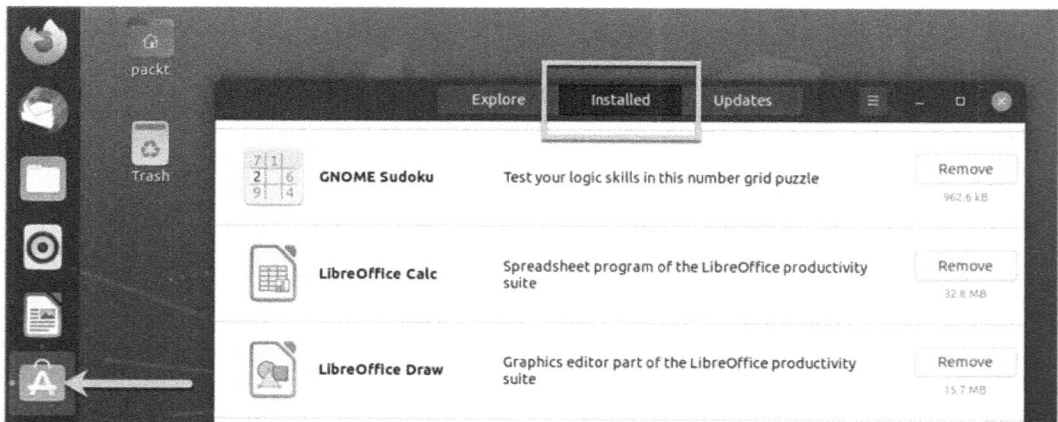

Figure 1.29 – Adding or removing Ubuntu applications

To browse or open any of these applications, click the grid icon (*Show applications*) at the bottom of the taskbar and then search for our application by beginning to type its name (*Figure 1.30*):

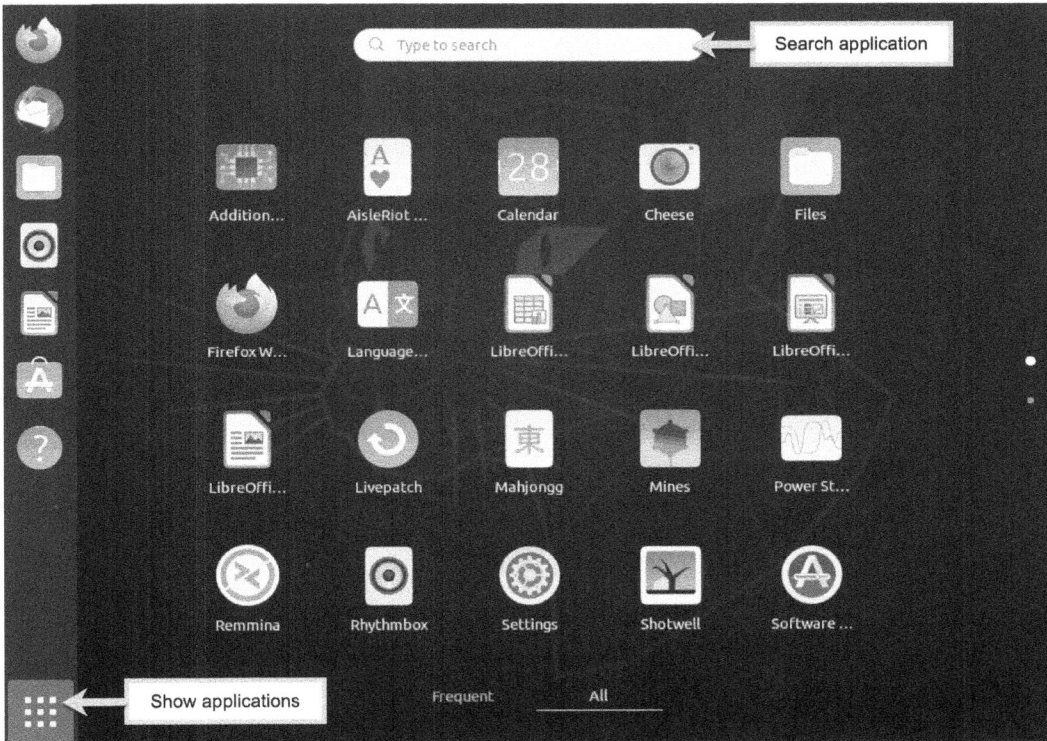

Figure 1.30 – Searching for installed applications

Next, let's take a look at how to install additional applications on our Ubuntu Linux workstation.

Additional software packages

A power user may need additional software, tools, or utilities beyond the ones provisioned with the default Ubuntu Desktop installation. To add new applications, we click on the *Ubuntu Software* icon in the taskbar, followed by **Explore**, and then on the *search* icon. In the example illustrated in the following figure, we look for **Visual Studio Code**, a powerful code editor.

In general, when we decide on a particular application we want to install, we select the application, and this opens up a window for installing the related Ubuntu software package. Once we have completed the installation, the application shows up in the **Installed** section, where we can later uninstall it, if we choose to do so (*Figure 1.31*):

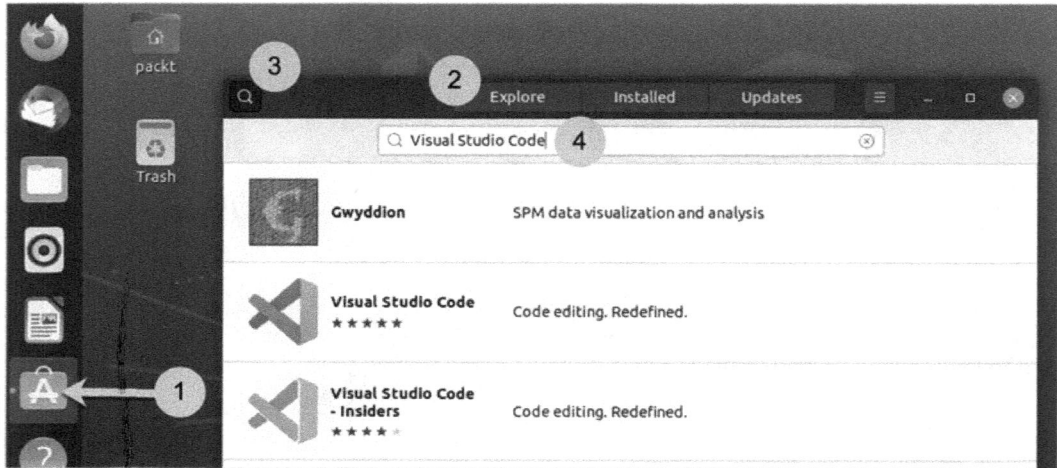

Figure 1.31 – Installing new applications

Different power users may look for a particular category of applications, tools, or utilities. Here are just a few of the most common types of productivity tools and applications used by the Linux user community. Some of them are readily available for download via the *Ubuntu Software* package management interface. Others can be downloaded from their related vendor websites.

Here are some commonly used applications for email and collaboration available for Ubuntu and other major Linux distributions:

- **Hiri**: Cross-platform email client, a real alternative to Microsoft Outlook.
- **Thunderbird Mail**: Send and receive emails (installed by default).
- **Slack for Linux**: Channel-based instant messaging platform.
- **Teams for Linux**: Instant messaging; unofficial Microsoft Teams client for Linux.

Image editing tools abound for Linux. Here are just a few:

- **GIMP**: Image manipulation program
- **digiKam**: Photo editing software
- **KRITA**: Free image editing and artistry

Most software development IDEs and tools are cross-platform these days. Ubuntu and many other Linux distros are not an exception in offering a vast array of such applications:

- **Visual Studio Code**: Code editing
- **Sublime Text**: Code editing, markup
- **Atom**: Highly customizable code editor
- **Eclipse**: Extensible tool platform and Java IDE
- **IDEA Ultimate**: Capable and ergonomic Java IDE for enterprise, web, and mobile development
- **Terminator**: Multiple terminals in one window

Virtualization and containerized workflows are in high demand now more than ever. All major hypervisors and container orchestration platforms are available for Linux as well:

- **VirtualBox**: VM manager by Oracle.
- **VMware Workstation**: VM manager by VMware.
- **Docker Engine**: Open source containerization platform.
- **MicroK8s**: Lightweight Kubernetes for workstations and appliances.
- **Minikube**: Run Kubernetes locally.

Small-factor Linux PCs and appliances are often the ideal platforms for media servers. These are the most common media server applications generally available for almost any Linux distribution:

- **Plex Media Server**: Organize media libraries and stream to any device.
- **Kodi**: Manage and view media.

In our highly collaborative world, screen capture and recording tools always come in handy. Here are just a few:

- **Screenshot**: Capture and save images of the screen or individual windows (installed by default).
- **Kazam**: Record a video or take a screenshot.
- **OBS Studio**: Live streaming and screen recording.

Some of these applications may require a CLI for installation. Other applications are available either as a software package installer in the *Ubuntu Software* management user interface or via the APT package manager CLI (`apt-get`).

As we have previously shown, installing an application via the *Ubuntu Software* management interface is straightforward. Next, we take a look at Ubuntu's APT package manager platform and CLI.

Managing software packages with APT

APT is the default package manager in Ubuntu, and it's a platform interface for installing and removing software packages. As a core system library of Ubuntu, APT is a collection of tools and CLI programs helping with package and repository management: `apt`, `apt-cache`, and `apt-get`.

> **Important note**
> APT and DPKG: In essence, **APT** is a shim (frontend) to **DPKG** – the package manager of Debian Linux distributions. The `apt` CLI invokes the `dpkg` CLI under the hood.

Let's look at a few examples of using the `apt-get` and `dpkg` CLI tools for installing and managing software packages.

Case study – installing Java (OpenJDK)

This scenario is from the perspective of a Java developer:

> *I have a powerful developer machine running Ubuntu Desktop. I work on a project that requires Java development. How do I install OpenJDK on my machine?*

We assume an Ubuntu Desktop platform and OpenJDK version 14 for our case study.

Let's look at how to install OpenJDK (Java Development Kit):

1. Before installing OpenJDK, let's make sure the current package lists and installed software packages are up to date. Start by updating the local package repository metadata for all configured sources:

    ```
    sudo apt-get update -y
    ```

    Next, proceed with the upgrade:

    ```
    sudo apt-get upgrade -y
    ```

    The preceding commands require *superuser* (*sudo*) privileges. We have to enter the password for our privileged (administrator) account. The `-y` option runs the `apt-get` command without requiring further user interaction.

2. Now, we are ready to install OpenJDK 14:

```
sudo apt-get install -y openjdk-14-jre-headless
```

3. Let's make sure the OpenJDK environment is ready. Get the OpenJDK version:

```
java --version
```

This command should yield the following response for our freshly installed OpenJDK (*Figure 1.32*):

```
packt@neptune:~$ curl -s http://localhost | grep Welcome
<title>Welcome to nginx!</title>
<h1>Welcome to nginx!</h1>
packt@neptune:~$ []
```

Figure 1.32 – Testing the OpenJDK installation

4. We can verify the status of the OpenJDK installation with this:

```
dpkg --list openjdk-14-jre-headless
```

The preceding command yields a response similar to *Figure 1.33*:

```
packt@neptune:~$ dpkg --list nginx
Desired=Unknown/Install/Remove/Purge/Hold
| Status=Not/Inst/Conf-files/Unpacked/halF-conf/Half-inst/trig-aWait/Trig-pend
|/ Err?=(none)/Reinst-required (Status,Err: uppercase=bad)
||/ Name           Version            Architecture Description
+++-==============-==================-============-=====================================
ii  nginx          1.18.0-0ubuntu1 all             small, powerful, scalable web/proxy server
packt@neptune:~$ []
```

Figure 1.33 – Querying the OpenJDK package using dpkg

Alternatively, we can use the apt-cache CLI to verify the status of the OpenJDK package:

```
apt-cache policy openjdk-14-jre-headless
```

The output is similar to *Figure 1.34*:

```
packt@neptune:~$ apt-cache policy nginx
nginx:
  Installed: 1.18.0-0ubuntu1
  Candidate: 1.18.0-0ubuntu1
  Version table:
 *** 1.18.0-0ubuntu1 500
        500 http://us.archive.ubuntu.com/ubuntu focal-updates/main amd64 Packages
        100 /var/lib/dpkg/status
     1.17.10-0ubuntu1 500
        500 http://us.archive.ubuntu.com/ubuntu focal/main amd64 Packages
packt@neptune:~$ []
```

Figure 1.34 – Querying the OpenJDK package using apt-cache

5. To remove OpenJDK, run the following:

```
sudo apt-get remove openjdk-14-jre-headless
```

Or, you can use dpkg:

```
sudo dpkg --remove openjdk-14-jre-headless
```

To find out more about the dpkg CLI options, run dpkg --help. To get more information on the apt CLI tools, run man apt.

Case study – installing Plex Media Server

This scenario is from the perspective of a home theater enthusiast:

I'm using Plex as a media player system, with Plex Media Server as the backend. I want to migrate my Plex Media Server to Linux. How do I install Plex Media Server on Ubuntu?

To install **Plex Media Server (PMS)** on Ubuntu, follow these steps:

1. Make sure the current package lists and installed software packages are up to date. Start by updating the local package repository for all configured sources:

```
sudo apt-get update -y
```

Next, run the upgrade:

```
sudo apt-get upgrade -y
```

According to the vendor's website, there are specific prerequisites regarding enabling the related repository, which we don't detail here.

2. The vendor also recommends the use of the dpkg CLI for installing PMS; use the following command:

```
sudo dpkg -i plexmediaserver_1.19.4.2935-79e214ead_amd64.
deb
```

3. We can verify the status of our plexmediaserver installation with the following:

```
dpkg --list plexmediaserver
```

The preceding command results in a response similar to *Figure 1.35*:

```
julian@plex:~$ dpkg --list plexmediaserver
Desired=Unknown/Install/Remove/Purge/Hold
| Status=Not/Inst/Conf-files/Unpacked/halF-conf/Half-inst/trig-aWait/Trig-pend
|/ Err?=(none)/Reinst-required (Status,Err: uppercase=bad)
||/ Name            Version        Architecture  Description
+++-===============-==============-=============-========================================
ii  plexmediaserver 1.19.4.2935-7 amd64         Plex organizes all of your personal med
julian@plex:~$ []
```

Figure 1.35 – Querying the PMS package using dpkg

4. To remove the `plexmediaserver` package, run this:

```
sudo dpkg --remove plexmediaserver
```

To find out more about the `dpkg` CLI options, run `dpkg --help`.

Summary

In this chapter, we learned about Linux distributions, with a practical emphasis on choosing the right platform for our needs and performing the related installation procedures. Along the way, we showcased some hands-on, real-world scenarios that we thought are relevant for the topics covered, and to better capture the *why* or *how* of what we learned.

Throughout the chapter, the main emphasis has been on the Ubuntu and CentOS Linux distributions. In the spirit of the practical approach, we covered both physical and VM environments running Linux. We also took a short route through the Windows realm, where we touched upon WSL, a modern-day abstraction of Linux as a native Windows application.

We took the exercise of building a Linux workstation with some hands-on examples of how to customize the applications and tools we need for our everyday work.

With the skills learned in this chapter, we hope you'll have a better understanding of how to choose different flavors of Linux distros based on your needs. You've learned how to install and configure Linux on a variety of platforms: server, desktop, VM, and WSL. We also started exploring using the Linux command-line terminal for some of the tasks described in our case studies. You will use some of these skills throughout the rest of the book, but most importantly, you'll now be comfortable quickly deploying the Linux distribution of your choice and testing with it.

In today's increasingly rapid and agile development environments, **Continuous Integration and Continuous Delivery (CI/CD)** infrastructures make heavy use of Linux distros. Future chapters will introduce you to containerized workflows, and the knowledge gained in this chapter will help you with the design and deployment efforts of Linux containers.

Starting with the next chapter, we'll take a closer look at the various Linux subsystems, components, services, and applications. *Chapter 2, The Linux Filesystem*, will familiarize you with the Linux filesystem internals and related tools.

Questions

Here are a few questions and thought experiments that you may ponder, some based on the skills you have learned in this chapter, and others revealed in later parts of the book:

1. *My Linux workstation is running low on disk space. How do I uninstall or remove applications that I don't use anymore?*

   Hint: Software packages installed with GUI package management tools can be uninstalled from the same GUI (for instance, see *Figure 1.36*). Others can be removed using the package manager CLI tools (for instance, apt, dpkg, and rpm).

2. *If I had a relatively large number of Linux VM instances or distros deployed and running at the same time, how would I make it easier to manage them?*

   Hint: Use Vagrant, a tool for building and managing VM environments.

3. *How do I enable multiple GUI desktops (GNOME, KDE) for different users on the same Linux system?*

   Hint: In the *Installing Linux graphical user interfaces* section, we showcased the addition of KDE side by side with GNOME. Each user can log in with their GUI desktop of choice (see *Figure 1.31*).

4. *I'm looking for a robust email client and team collaboration solution for my Linux workstation.*

   Hint: The *Additional software packages* section may provide some guidance.

5. *Can I run multiple Linux instances in WSL?*

   Hint: Yes. Use wsl --import.

6. *How do I upgrade my Linux machine to a new version of the distro?*

   Hint: Create a bootable media of your new Linux distro and run the installer. Choose to upgrade when prompted. You may also use the command line. Check out the related CLI tools for your Linux distro (for instance, do-release-upgrade for Ubuntu).

2

The Linux Filesystem

Understanding the Linux filesystem, file management fundamentals, and the basics of the Linux shell and command-line interface is essential for a modern-day Linux professional.

In this chapter, you will learn how to use the Linux shell and some of the most common commands in Linux. You will learn about the structure of a basic Linux command and how the Linux filesystem is organized. We'll explore various commands for working with files and directories. Along the way, we'll introduce you to the most common command-line text editors. We hope that by the end of this chapter, you'll be comfortable using the Linux command-line terminal and be ready for future, more advanced explorations.

We're going to cover the following main topics:

- Introducing the Linux shell
- The Linux filesystem
- Working with files and directories
- Using text editors to create and edit files

Technical requirements

This chapter requires a working installation of a standard Linux distribution, on either server, desktop, PC, or **Virtual Machine** (**VM**). Our examples and case studies use the Ubuntu and CentOS platforms, but the commands and examples explored are equally suitable for any other Linux distribution.

Introducing the Linux shell

Linux has its roots in the Unix operating system, and one of its main strengths is the command-line interface. In the old days, this was called *the shell*. In Unix, the shell is invoked with the sh command. The shell is a program that has two streams: an *input stream* and an *output stream*. The input is a command given by the user, and the output is the result of that command, or an interpretation of it. In other words, the shell is the primary interface between the user and the machine.

The main shell in major Linux distributions is called **Bash,** which is an acronym for **Bourne Again Shell**, named after Steve Bourne, the original creator of the shell in Unix. Ubuntu, Fedora, CentOS, RHEL, Debian, and openSUSE all use Bash as their default shell. Alongside Bash, there are other shells available in Linux, such as **ksh**, **tcsh**, and **zsh**. In this chapter, we will cover the Bash shell, as it is the most widely used shell in modern Linux distributions.

One shell can be assigned to each user. Users on the same system can use different shells. One way to check the default shell is by accessing the /etc/passwd file. More details about this file and user accounts will be discussed in *Chapter 4, Managing Users and Groups*. For now, it is important to know where to look for the default shell. In this file, the last characters from each line represent the user's default shell. The /etc/passwd file has the users listed one on each line, with details about their PID, GID, username, home directory, and basic shell. To see the default shell for your current user, execute the following command using your user name (in our case packt):

```
cat /etc/passwd | grep packt
```

The output should be similar to what is shown *Figure 2.1*, only that you will see your username listed, not ours:

```
packt@neptune:~$ cat /etc/passwd | grep packt
packt:x:1000:1000:Packt:/home/packt:/bin/bash
```

Figure 2.1 – Showing the user's default shell

The /etc/passwd file has many lines for all the users, but we only extracted the line for our user.

An easier way to see the current shell is by running the following command:

```
packt@neptune:~$ echo $0
-bash
```

Figure 2.2 – A simpler way to show the shell

This shows what is running your command, which is the shell. $0 is a bash special parameter and refers to the current running process. If you have other shells installed, you can easily assign another shell to your user, depending on your preference. Nevertheless, if you know Bash, you will be comfortable with all of them.

> **Important note**
> *The Linux shell is case sensitive.* This means that everything you type inside the command line should respect this. For example, the cat command used earlier used lowercase. If you type Cat or CAT, the shell will not recognize it as being a command. The same rule applies to file paths. You will notice that default directories in your home directory use uppercase for the first letter, as in ~/Documents, ~/Downloads, and so on. Those names are different to ~/documents or ~/downloads.

Bash shell features

The shell not only runs commands. It has many more features that make a system administrator's life more comfortable while at the command line.

Wildcards and metacharacters

In Linux, wildcards are used to match filenames. There are three main types of wildcards:

- The asterisk (*): This is used to match any string of none or more characters.
- The question mark (?): This is used to match a single character.
- The brackets ([ ]): This is used to match any of the characters inside the brackets.

Metacharacters are special characters that are used in Linux and any Unix-based system. Those metacharacters are as follows:

- >: Output redirection.
- >>: Output redirection – append.
- <: Input redirection.
- <<: Input redirection.
- *: File substitution wildcard (explained previously).
- ?: File substitution wildcard (explained previously).
- [ ]: File substitution wildcard (explained previously).
- |: Pipe for using multiple commands.

- `;`: Command execution sequence.
- `( )`: Group of commands in the execution sequence.
- `||`: Conditional execution (OR).
- `&&`: Conditional execution (AND).
- `&`: Run a command in the background.
- `#`: Use a command directly in the shell.
- `$`: Variable value expansion.
- `\`: The escape character.
- `` `cmd` ``: Command substitution.
- `$(cmd)`: Command substitution.

Following are some examples using the metacharacters from the preceding list. In the first example, we use the output of one command inside another command, by long listing the output of the `which ls` command inside the `ls -l` command:

```
packt@neptune:~$ ls -l `which ls`
-rwxr-xr-x 1 root root 142144 Sep  5 2019 /usr/bin/ls
packt@neptune:~$ ls -l $(which ls)
-rwxr-xr-x 1 root root 142144 Sep  5 2019 /usr/bin/ls
```

Figure 2.3 – Example of command execution and substitution

You can use the pipe to combine two commands, to pipe the output of the first command as the input for the second command. In the following example, we will use the `less` command piped with the `ls -l` command when showing the long listing of a large directory such as `/etc`:

```
ls -l /etc | less
```

Let's execute some commands in a sequence. After that, we will use metacharacters to group commands and redirect the output to a file. All this is shown in the following screenshot:

```
packt@neptune:~$ who; pwd;
packt    :0        2020-08-28 12:00 (:0)
packt    pts/0     2020-08-28 12:00 (172.16.232.1)
/home/packt
packt@neptune:~$ (who; pwd) > users
packt@neptune:~$ cat users
packt    :0        2020-08-28 12:00 (:0)
packt    pts/0     2020-08-28 12:00 (172.16.232.1)
/home/packt
```

Figure 2.4 – Example of command sequence execution

As you can see in the preceding output, the two commands executed first can easily be grouped using brackets.

Brace expansion

Curly brackets can also be used to expand the arguments of a command. Braces are not just limited to filenames, like a wildcard. They work with any type of string. Inside the braces, you can use a single string, a sequence, or several strings separated by commas.

Here are some examples of using this type of expansion. First, we will use brace expansion to delete two files from one directory. Second, we will show you how to create multiple new files using brace expansion. Let's say that we have two files named `report` and `new-report` and we want to delete them both at once. We will use the following command:

```
rm ~/xpackt/{report,new-report}
```

To create multiple files (five of them, for example) that share parts of their name, as in `file1, file2, ... file n`, we will use the following command:

```
packt@neptune:~/xpackt$ touch file{1..5}
packt@neptune:~/xpackt$ ls
file1  file2  file3  file4  file5
```

Figure 2.5 – Creating multiple files using brace expansion

Brace expansion is a powerful tool that adds flexibility and power to any system administrator's workflow.

Bash shell variables

The Bash shell has some built-in variables and offers the possibility to define your own variables as well. Here is a shortlist of some of the standard built-in variables:

- HOME: The user's home directory (for example, /home/packt)
- LOGNAME: The user's login name (for example, packt)
- PWD: The shell's current working directory
- OLDPWD: The shell's previous working directory
- PATH: The shell's search path (list of directories separated by colons)
- SHELL: The path to the shell
- USER: The user's login name
- TERM: The type of the terminal

To call a variable while in the shell, all you have to do is to place a dollar sign, $, in front of the variable's name. Try the following commands on your machine:

```
echo $SHELL; echo $USER; echo $TERM; echo $PATH; echo $HOME;
echo $PWD
```

You can also assign your own shell variables, as in the following example:

```
packt@neptune:~$ MYVAR=sysadmin
packt@neptune:~$ echo $MYVAR
sysadmin
```

Figure 2.6 – Assigning a new variable

To see all the shell variables, use the printenv command. If the list is too long, you can redirect it to a file, as in the following example. Now your variables list is inside the variables file, and you can see it by concatenating or by editing inside a text editor such as vim:

```
printenv > ~/variables
```

The shell's variables are only available inside the shell. If you want some variables to be known to other programs that are run by the shell, you must export them by using the export command. Once a variable is exported from the shell, it is known as an *environment variable*.

The shell's search path

The PATH variable is an essential one in Linux. It helps the shell know where all the programs are located. When you enter a command into your Bash shell, it first has to search for that command through the Linux filesystem. Some directories are already listed inside the PATH variable, but you can also add new ones. Your addition can be temporary or permanent, depending on how you do it. To make a directory's path available temporarily, simply add it to the PATH variable. In the following example, we will add the /home/alex directory to PATH:

```
packt@neptune:~$ echo $PATH
/usr/local/sbin:/usr/local/bin:/usr/sbin:/usr/bin:/sbin:/bin:/usr/games:/usr/loc
al/games:/snap/bin
packt@neptune:~$ PATH=$PATH:/home/packt
packt@neptune:~$ echo $PATH
/usr/local/sbin:/usr/local/bin:/usr/sbin:/usr/bin:/sbin:/bin:/usr/games:/usr/loc
al/games:/snap/bin:/home/packt
```

Figure 2.7 – Adding a new location to PATH

To make any changes permanent, you must modify the PATH variable inside a file called `~/.bash_profile` or `~/.bashrc`.

The shell's aliases

The Linux shell supports aliases. They are a very convenient way to create shorter commands as aliases for longer ones. For example, in Ubuntu, there is a predefined alias called `ll` that is a shorthand for `ls  -alF`. You can define your own aliases too. You can make them temporary or permanent, similar to variables. In the following example, we changed the alias for the `ll` command, as follows:

```
packt@neptune:~$ alias ll
alias ll='ls -alF'
packt@neptune:~$ alias ll='ls -l'
packt@neptune:~$ alias ll
alias ll='ls -l'
```

Figure 2.8 – Changing the alias of a command

This modification is only temporary, and it will revert to the default one after reboot or shell restart. If you want to make it permanent, you should edit the `~/.bashrc` file.

The shell connection

Connections to the shell are made by using two different types: `tty` and `pts`. The `tty` connection is considered a native one, with ports that are direct connections to your computer. The link between the user and the computer is mainly found to be through a keyboard, which is considered to be a native terminal device. The name `tty` stands for *teletypewriter*, which was a type of terminal used at the beginning of computing.

The `pts` connection is generated by SSH or Telnet types of links. Its name stands for *pseudo terminal slave* and it is an emulated connection made by a program, in most cases `ssh` or **xterm**. It is the slave of the *pseudo-terminal device*, which is `pty`.

Virtual consoles/terminals

The terminal was thought of as a device that manages the input strings (which are commands) between a process and other I/O devices such as a keyboard and a screen. There are also pseudo terminals, which are emulated terminals that behave the same way as a classical terminal. The difference is that it does not interact with devices directly, as it is all emulated by the Linux kernel, which transmits the I/O to a program called the shell.

Virtual consoles are accessible and run in the background, even though there is no open terminal. To access those virtual consoles, you can use the commands *Ctrl + Alt + F1 / Ctrl + Alt + F2, (…) Ctrl + Alt + F5,* and *Ctrl + Alt + F6*. These will open `tty1`, `tty2`, (…) `tty 5`, and `tty6`, respectively, on your computer.

We will explain this using an Ubuntu 20.04.1 LTS Server VM installation, but it is identical in CentOS 8 too. After starting the VM and being prompted to log in with your username and password, the first line on the screen will be something similar to the following output:

```
Ubuntu 20.04.1 LTS ubuntu-server tty1
```

If you press any of the preceding key combinations, you will see your terminal change from `tty1` to any of the other `tty` instances. For example, if you press *Ctrl + Alt + F6*, you will see this:

```
Ubuntu 20.04.1 LTS ubuntu-server tty6
```

As we were using the server edition of Ubuntu, we did not have the GUI installed. But if you were to use a desktop edition, you will be able to use *Ctrl + Alt + F7* to enter X `graphical` mode.

If you are not able to use the preceding keyboard combinations, there is a dedicated command for changing virtual terminals. The command is called `chvt`. Even though we have not discussed shell commands yet, we will show you an example of how to use it and other related commands. This action can only be performed by an administrator account or by using **sudo** (more details about this in *Chapter 4, Managing Users and Groups*).

Briefly, `sudo` stands for *superuser do* and allows any user to run programs with administrative privileges or with the privileges of another user.

In the following example, we will use CentOS 8 in a graphical environment. First, we will see which virtual terminal we are currently using to change it to another one without using the *Ctrl + Alt + Fn* keys.

The `who` command will show you information about the users currently logged in to the computer. In our case, it will show that the user `packt` is currently using virtual terminal two (`tty2`):

```
who
packt tty2 2020-08-05 16:17 (tty2)
```

Now, by using the chvt command, we will show you how to change to the sixth virtual terminal. After running sudo chvt 6, you will be prompted to provide your password and immediately be switched to virtual terminal number six without a GUI (remember that we are doing this exercise from inside CentOS 8 with a GUI). Running who once more will show you all logged-in users and the virtual terminals they use.

The command-line prompt

The command-line or shell prompt is the place where you type in the commands. Usually, the command prompt will show the username, hostname, present working directory, and a symbol that indicates the type of user running the shell.

Here is an example from Ubuntu 20.04.1 LTS:

```
packt@neptune:~$ _
```

Here is an example for CentOS 8:

```
[packt@localhost ~]$ _
```

Here is a short explanation of the output:

- packt is the name of the user currently logged in.
- neptune and localhost are the hostnames.
- ~ represents the home directory (it is called a tilde).
- $ shows that the user is a regular user (when you are logged in as an administrator, the sign changes into a hashtag, #).

Let's look at the shell command types next.

Shell command types

The shell works with commands, and there are two types that it uses: internal ones and external ones. Internal commands are ones that are built inside the shell. External ones are installed separately. If you want to check the type of the command you are using, there is the type command. For example, you can check what type of command cd (change directory) is:

```
packt@neptune:~$ type cd
cd is a shell builtin
```

Figure 2.9 – Command type output

The output shows that the cd command is an internal one, built inside the shell. If you are curious, you could find out the types of other commands that we will show you in the following sections.

Command structure

We have already used some commands, but we did not explain the structure of a Linux command. We will do that now for you to be able to understand how to use commands. In a nutshell, Unix and Linux commands have the following form:

- The command's name
- The command's options
- The command's arguments

Inside the shell, you will have a general structure such as the following:

```
command [-option(s)] [argument(s)]
```

A short example would be the use of the ls command (ls comes from *list*). This command is one of the most-used commands in Linux. It lists files and directories and can be used with both options and arguments, or as ls.

In the preceding example, we used ls in its simplest form. It lists the contents of your present working directory (pwd). In our case, it is the home directory, indicated by the ~ tilde character in the shell's prompt. Your output should be similar if you are inside your home directory.

The ls command with the -l option (lowercase L) uses a long listing format, giving you extra information about files and directories from your present working directory (pwd):

```
ls -l ; ls -l xpact/
```

In the preceding example, we used ls -l xpackt/ to show the contents of the ~/xpackt directory. Shown here is a way to use the command with both options and attributes, without changing our present working directory. If we would like to see the contents of the to ~/Downloads directory, you could use the command with the path to that directory as an argument.

Help from the manual

Any Linux system administrator's best friend is the manual. Each command in Linux has a manual page that gives the user detailed information about its use, options, and attributes. If you know the command you want to learn more about, simply use the `man` command to explore. For the `ls` command, for example, you use `man ls`.

The manual organizes its command information into different sections, with each section being named by convention to be the same on all distributions. Briefly, those sections are `name`, `synopsis`, `configuration`, `description`, `options`, `exit status`, `return value`, `errors`, `environment`, `files`, `versions`, `conforming to`, `notes`, `bugs`, `example`, `authors`, `copyright`, and `see also`.

When you use the manual, keep in mind that it is not a step-by-step how-to guide. It is technical documentation that might be confusing at first. Our advice is to use the man pages as much as you can. Before you search for anything on the internet, try to read the manual first. This will be a good exercise, and you will become proficient with Linux commands in no time.

Similar to the manual pages, almost all commands in Linux have a `-help` option. You can use this for quick reference.

For more information about the help and man pages, you can check each command's help or manual page. Try the following commands:

```
$ man man
$ help help
```

We'll now learn about the Linux filesystem in the following section.

The Linux filesystem

The Linux filesystem consists of a logical collection of files that are stored on a partition or a disk. Your hard drive can have one or many partitions. Those partitions usually contain only one filesystem, and it can extend to an entire disk. One filesystem could be the `/` (`root`) filesystem and another the `/home` filesystem. Or, there can be just one that contains all filesystems.

Generally, using one filesystem per partition is considered to be good practice by allowing for logical maintenance and management. As everything in Linux is a file, physical devices such as hard drives, DVD drives, USB devices, and floppy drives are treated as files too.

Directory structure

Linux uses a hierarchical filesystem structure. It is similar to an upside-down tree, with the root (` / `) at the base of the filesystem. From that point, all the branches (directories) spread throughout the filesystem.

The **Filesystem Hierarchy Standard** (**FHS**) defines the structure of Unix-like filesystems. However, Linux filesystems also contain some directories that aren't yet defined by the standard.

Exploring the Linux filesystem from the command line

Feel free to explore the filesystem yourself by using the `tree` command. In CentOS 8, it is already installed, but if you use Ubuntu, you will have to install it by using the following command:

```
$ sudo apt install tree
```

Do not be afraid to explore the filesystem, because no harm will be done by just looking around. Use only the `ls` command to list the contents of directories.

We will show you an example here, as we will explore some of the directories of the filesystem. Our test machine is using Ubuntu 20.04.1 LTS. We will use the `tree` command by invoking the `-L` option, which tells the command how many levels down to go, and the last attribute states which directory to start with. Thus, the command will go down one level, starting from the `root` directory (represented by the forward slash):

```
$ tree -L 1 /
```

Start exploring the directories from the structure by using the `ls` command, as shown here. Remember that some of the directories you are about to open will contain a large number of files and/or other directories, which will clutter your terminal window.

The following are the directories that exist on almost all versions of Linux. Here's a quick overview of the Linux root filesystem:

- `/`: Root directory: the root for all other directories.
- `/bin`: Essential command binaries: the place where binary programs are stored.
- `/boot`: Static files of the boot loader: the place where the kernel, bootloader, and `initramfs` are stored.
- `/dev`: Device files: nodes to the device equipment, a kernel device list.

- `/etc`: Host-specific system configuration: essential config files for the system, boot time loading scripts, `crontab`, `fstab` device storage tables, and the `passwd` user accounts file.

- `/home`: the user's home directory: the place where the user's files are stored.

- `/lib`: Essential shared libraries and kernel modules: shared libraries are similar to **Dynamic Link Library (DLL)** files in Windows DLLs.

- `/media`: Mount point for removable media: for external devices and USB external media.

- `/mnt`: Mount point for mounting a filesystem temporarily: used for legacy systems.

- `/opt`: Add-on application software packages: the place where *optional* software is installed.

- `/proc`: Virtual filesystem managed by the kernel: a special directory structure that contains files essential for the system.

- `/sbin`: Essential system binaries: vital programs for the system's operation.

- `/srv`: Data for services provided by this system.

- `/tmp`: Temporary files.

- `/usr`: Secondary hierarchy: the largest directory in Linux that contains support files for regular system users; `/usr/bin` – system executable files; `/usr/lib` – shared libraries from `/usr/bin`; `/usr/local` – source compiled programs not included in the distribution; `/usr/sbin` – specific system administration programs; `/usr/share` – data shared by the programs in `/usr/bin` like config files, icons, wallpapers or sound files; `/usr/share/doc` – documentation for the system-wide files.

- `/var`: Variable data: only data that is modifiable by the user is stored here, such as databases, printing spool files, user mail, and others; `/var/log` – contains log files that register system activity.

Next, we're going to see how to work with these files and directories.

Working with files and directories

Remember that everything in Linux is a file. A directory is a file too. As such, it is essential to know how to work with them. Working with files in Linux implies the use of several commands for basic file and directory operations, file viewing, file creation, file location, file properties, and linking. Some of the commands, which will not be covered here but their use is closely related to files, will be covered in the following section.

Understanding file paths

Each file in the FHS has a *path*. The path is the file's location represented in an easily readable representation. In Linux, all the files are stored in the root directory by using the FHS as a standard to organize them. Relations between files and directories inside this system are expressed through the forward-slash character (/). Throughout computing history, this was used as a symbol that described addresses. Paths are, in fact, addresses for files.

There are two types of paths in Linux, *relative* ones and *absolute* ones. An absolute path always starts with the root directory and follows the branches of the system up to the desired file. A relative path always refers to the current working directory and represents the relative path to it. Thus, a relative path is always a path that is relative to your present working directory.

The absolute path is useful to know about when you work with files. After some practice, you will come to learn the paths to the most-used files. For example, one file that you will need to learn the path for is the passwd file. It resides in the /etc directory. Thus, when you will refer to it, you will use its absolute path, /etc/passwd. Using a relative path to that file would imply that you are either inside its parent directory or somewhere close in the FHS.

Working with relative paths involves knowing two special characters used to work with the FHS. One special character is the dot (.), and it refers to the current directory. The other is two consecutive dots (. .) and refers to the parent directory of the current directory. When working with relative paths, make sure that you always check what directory you are in. Use the pwd command to show your present working directory.

A good example of working with relative paths is when you are already inside the parent directory and need to refer to it. If you need to see the accounts list of your system, which is stored inside the passwd file, you can refer to it by using a relative path. For this exercise, we are inside our home directory:

```
packt@neptune:~$ pwd
/home/packt
packt@neptune:~$ cat passwd
cat: passwd: No such file or directory
packt@neptune:~$ cat ../../etc/pa
pam.conf    pam.d/      papersize   passwd      passwd-
packt@neptune:~$ cat ../../etc/passwd_
```

Figure 2.10 – File paths related to the current working directory

First, we checked our present working directory with the pwd command, and the output was our home directory's path, /home/packt. Secondly, we tried to show the contents of the passwd file using the cat (concatenate) command right from the home directory, but the output was an error message saying that there is no such file or directory inside our home directory. We used the relative path, which is always relative to our present working directory, hence the error. Thirdly, we used the double-consecutive dots special characters to refer to the file with its relative path.

> **Tip**
> Always use the *Tab* key on your keyboard for autocompletion and to check whether the path you typed is correct or not. In the preceding example, we typed ../../etc and pressed *Tab*, which autocompleted with a forward slash. Then, we typed the first two letters of the file we were looking for and pressed *Tab* again. This showed us a list of files inside the /etc directory that started with pa. Seeing that passwd was in there, we knew that the path was right, so we typed two more s characters and pressed *Tab* again. This completed the command for us and we pressed *Enter/Return* to execute the command.

The path in the final command is relative to our home directory and it translates as follows: *concatenate the file with the name* passwd *that is located in the* /etc *directory in the parent directory (first two dots) of the parent directory (second two dots) of our current directory (home)*. Therefore, the /etc/passwd absolute path is translated into a relative path to our home directory like this: ../../etc/passwd.

Basic file operations

Daily, as a system administrator, you will manipulate files. This includes creating, copying, moving, listing, deleting, linking, and so on. The basic commands for these operations have already been discussed throughout this chapter, but now it is time to get into more details about their use, options, and attributes. Other more advanced commands will be detailed in the following sections.

Creating files

There are situations when you will need to create new files. You have one option to create a new empty file with the touch command. When you use it, it will create a new file with you as the file owner and with a size of zero, because it is an empty file. In the following example, we created a new file called new-report inside the ~/xpackt/ directory.

The `touch` command is also used to change the modification time of a file without changing the file itself. Notice the difference between the initial time when we first created the `new-report` file and the new time after using the `touch` command on it:

```
packt@neptune:~/xpackt$ touch new-report
packt@neptune:~/xpackt$ ls -l new-report
-rw-rw-r-- 1 packt packt 0 Aug 28 16:15 new-report
packt@neptune:~/xpackt$ touch new-report
packt@neptune:~/xpackt$ ls -l new-report
-rw-rw-r-- 1 packt packt 0 Aug 28 16:16 new-report
```

Figure 2.11 – Using the touch command to create and alter files

You can also change the access time by using the `-a` option of the `touch` command. By default, the long listing of the `ls` command shows only the modification/creation time. If you want to see the access time, there is the `atime` parameter you can use with the `-time` option:

```
packt@neptune:~/xpackt$ touch -a new-report
packt@neptune:~/xpackt$ ls -l --time=atime new-report
-rw-rw-r-- 1 packt packt 0 Aug 28 16:18 new-report
```

Figure 2.12 – Using touch to alter the access time

The modification, creation, and access time stamps are very useful, especially when using commands such as `find`. They give you a more *granular* search pattern.

The echo command

Files can also be created by using redirection and the `echo` command. `echo` is a command that prints the string given as a parameter to the standard output (the screen). For example, if you use the command as shown in the following screenshot, it will print the text to the screen:

```
packt@neptune:~$ echo text
text
packt@neptune:~$ echo "text"
text
packt@neptune:~$ echo 'text'
text
```

Figure 2.13 – Using the echo command to print to the standard output (screen)

In the preceding screenshot, you can see three different ways to print text to the screen. All have the same output. The output of the echo command can be written directly to a file by using the output redirection:

```
packt@neptune:~/xpackt$ echo this is a presentation file > presentation
packt@neptune:~/xpackt$ cat presentation
this is a presentation file
packt@neptune:~/xpackt$ echo this is a new line >> presentation
packt@neptune:~/xpackt$ cat presentation
this is a presentation file
this is a new line
```

Figure 2.14 – Using echo with output redirection

In the preceding example, we redirected the text from the echo command to the presentation file. It did not exist at the beginning, so it was automatically created by the command. The first echo command added a line to the file by using the > operator. The second echo command appended a new line of text to the end of the file, by using the >> operator.

Listing files

We have already used some examples with the ls command before, so you are somewhat familiar with it. We covered the -l option as an example of the command's structure. Thus, we will not cover it any further here. We will explore new options for this essential and useful command:

- ls -lh: The -l option lists the files in an extended format, while the -h option shows them in a human-readable form, with the size in kilobytes or megabytes rather than bytes.

- ls -la: The -a option shows all the files, including hidden ones. Combined with the -l option, the output will be a list of all the files and their details.

- ls -ltr: The -t option sorts files by their modification time, showing the newest first; the -r option reverses the order of the sort.

- ls - lS: The -S option sorts the files by their size, with the largest file first.

- ls -R: The -R option shows the contents of the current or specified directory in recursive mode.

Let's look at long listing in the next section.

Long listing

The long listing output of the ls command should be explained just a little bit now, even if more details will be given in *Chapter 4, Managing Users and Groups*. Here are some extracts from long listing different file types in our home directory:

```
ls -la
total 80
drwxr-xr-x 16 packt  packt  4096 sep 13 01:15 .
drwxr-xr-x  3 root   root   4096 sep  8 11:58 ..
-rw-------  1 packt  packt   945 sep 18 23:38 .bash_history
-rw-r--r--  1 packt  packt   220 sep  8 11:58 .bash_logout
-rw-r--r--  1 packt  packt  3771 sep  8 11:58 .bashrc
[...]
```

In the output, the first row after the command shows the number of blocks inside the directory listed. After that, each line represents one file or subdirectory, with the following detailed information:

- The first character is the type of the file: d for directory, : for file, l for a link, c for character device, and b for the block device.

- The following nine characters represent the permissions (detailed in *Chapter 4, Managing Users and Groups*).

- The hard link for that file.

- The owner's PID and GID (details in *Chapter 4, Managing Users and Groups*).

- The size of the file (the number depends on whether it is in human-readable format of not).

- The last modification time of the file.

- The name of the file or directory.

The next section is about copying and moving files.

Copying and moving files

To copy files in Linux, the cp command is used. The mv command moves files around the filesystem. This command is also used to rename files.

To copy a file, you can use the `cp` command in the simplest way:

```
cp oldfile newfile
```

Here `oldfile` is the name of the file to be copied, and `newfile` is the name of the destination file. You can also copy multiple files inside a directory that already exists. If the destination directory does not exist, the shell will signal you that the target is not a directory.

Now let's look at some variations.

cp -a

The `-a` option copies an entire directory hierarchy in recursive mode by preserving all the attributes and links. In the following example, we copied the entire `xpackt` directory to a newly created directory called `backup` by using the `-a` option:

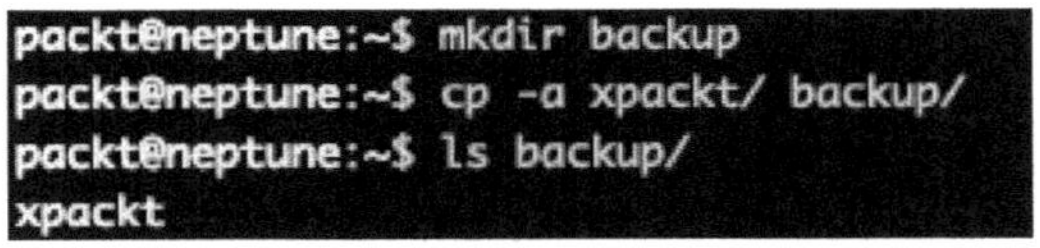

```
packt@neptune:~$ mkdir backup
packt@neptune:~$ cp -a xpackt/ backup/
packt@neptune:~$ ls backup/
xpackt
```

Figure 2.15 – Using the copy command with the -a option

The next option is `cp  -r`.

cp -r

This option is similar to `-a`, but it does not preserve attributes, only symbolic links.

cp -p

The `-p` option retains the file's permissions and timestamps. Otherwise, just by using `cp` in its simplest form, copies of the files will be owned by your user with a timestamp of the time you did the copy operation.

Moving files around is done with the `mv` command. It is either used to move files and directories from one destination to another or to rename a file.

cp -R

The -R option allows you to copy a directory recursively. In the following example, we will use the ls command to show you the contents of the ~/xpackt/ directory, and then the cp -R command to copy the contents of the /files directory to the /new-files one. The /new-files directory did not exist. The cp -R command created it:

```
packt@neptune:~/xpackt$ ls
files  new-report  presentation
packt@neptune:~/xpackt$ ls files/
file1  file2  file3  file4  file5
packt@neptune:~/xpackt$ cp -R files/ new-files
packt@neptune:~/xpackt$ ls
files  new-files  new-report  presentation
packt@neptune:~/xpackt$ ls new-files/
file1  file2  file3  file4  file5
```

Figure 2.16 – Using the cp -R command

There are many other options that you could learn about just by visiting the manual pages. Feel free to explore them and use them in your daily tasks.

Working with links

Links are a compelling option in Linux. They can be used as a means of protection for the original files, or just as a tool to keep multiple copies of a file without having separate hard copies. Consider it as a tool to create alternative names for the same file.

The command is ln, and it can be used to create two types of links:

- Symbolic links
- Hard links

Those two links are different types of files that point to the original file. In the case of a symbolic link, it is a physical file that points to the original file, and a hard link is a virtual file that points to the original file.

A **symbolic link** is used for an original file that already exists; they are linked and have the same content. Also, it can span different filesystems and physical media, meaning that it can link to original files that are on other drives or partitions with different types of filesystems. The command used is as follows:

```
ln -s [original_filename] [link_filename]
```

Here is an example in which we listed the contents of the ~/xpackt directory and then created a symbolic link to the new-report file using the ln -s command and then listed the contents again:

```
packt@neptune:~/xpackt$ ls
files  new-files  new-report  presentation
packt@neptune:~/xpackt$ ln -s new-report new-report-link
packt@neptune:~/xpackt$ ls -l
total 8
drwxrwxr-x 2 packt packt 4096 Aug 28 16:42 files
drwxrwxr-x 2 packt packt 4096 Aug 28 16:44 new-files
-rw-rw-r-- 1 packt packt    0 Aug 28 16:16 new-report
lrwxrwxrwx 1 packt packt   10 Aug 28 17:09 new-report-link -> new-report
-rw-rw-r-- 1 packt packt    0 Aug 28 16:36 presentation
```

Figure 2.17 – Using symbolic links

You can see that the link created is named new-report-link and is visually represented with an arrow, ->, that shows the original file that it points to. You can also distinguish the difference in size between the two files, the link and the original one. The permissions are different too. This is a way to know that they are two different physical files. To double-check that they are different physical files, you can use the ls -i command to show the **inode** for every file. In the following example, you can see that new-report and new-report-link have different inodes:

```
packt@neptune:~/xpackt$ ls -li
total 8
412853 drwxrwxr-x 2 packt packt 4096 Aug 28 16:42 files
412855 drwxrwxr-x 2 packt packt 4096 Aug 28 16:44 new-files
393533 -rw-rw-r-- 1 packt packt    0 Aug 28 16:16 new-report
393872 lrwxrwxrwx 1 packt packt   10 Aug 28 17:09 new-report-link -> new-report
412821 -rw-rw-r-- 1 packt packt    0 Aug 28 16:36 presentation
```

Figure 2.18 – Comparing the inodes for the symbolic link and original file

If you want to know where the link points to and you do not want to use ls -l, there is the readlink command. It is available in both Ubuntu and CentOS. The output of the command is simply the name of the file the symbolic link points to. It only works in the case of symbolic links:

```
packt@neptune:~/xpackt$ readlink new-report-link
new-report
```

Figure 2.19 – The readlink command output

In the preceding example, you can see that the output shows that the new-report-link file is a symbolic link to the file named new-report.

A **hard link** is a different virtual file that points to the original file. They are physically the same. The command is simply `ln` without any options:

```
ln [original-file] [linked-file]
```

In the following example, we created a hard link for the `new-report` file, and we named it `new-report-hl`. In the output, you will see that they have the same size, the same inode, and after altering the original file using `echo` and output redirection, the changes were available to both files. The two files have a different representation than symbolic links. They appear as two different files in your listing, with no visual aids to show which file is pointed to:

```
packt@neptune:~/xpackt$ ls
files   new-files   new-report   new-report-link   presentation
packt@neptune:~/xpackt$ ln new-report new-report-hl
packt@neptune:~/xpackt$ echo this is a new report line > new-report
packt@neptune:~/xpackt$ ls -l
total 16
drwxrwxr-x 2 packt packt 4096 Aug 28 16:42 files
drwxrwxr-x 2 packt packt 4096 Aug 28 16:44 new-files
-rw-rw-r-- 2 packt packt   26 Aug 28 17:22 new-report
-rw-rw-r-- 2 packt packt   26 Aug 28 17:22 new-report-hl
lrwxrwxrwx 1 packt packt   10 Aug 28 17:09 new-report-link -> new-report
-rw-rw-r-- 1 packt packt    0 Aug 28 16:36 presentation
packt@neptune:~/xpackt$ ls -li
total 16
412853 drwxrwxr-x 2 packt packt 4096 Aug 28 16:42 files
412855 drwxrwxr-x 2 packt packt 4096 Aug 28 16:44 new-files
393533 -rw-rw-r-- 2 packt packt   26 Aug 28 17:22 new-report
393533 -rw-rw-r-- 2 packt packt   26 Aug 28 17:22 new-report-hl
393872 lrwxrwxrwx 1 packt packt   10 Aug 28 17:09 new-report-link -> new-report
412821 -rw-rw-r-- 1 packt packt    0 Aug 28 16:36 presentation
```

Figure 2.20 – Working with hard links

Essentially, a hard link is linked to the inode of the original file. You can see it as a new name for a file, similar but not identical to renaming it.

Deleting files

In Linux, you have the remove (`rm`) command for deleting files. In its simplest form, the `rm` command is used without an option. For more control over how you delete items, you could use the `-i`, `-f`, and `-r` options.

rm -i

This option enables interactive mode by asking you for acceptance before deleting:

```
packt@neptune:~/xpackt$ ls
files  new-files  new-report  new-report-hl  new-report-link  presentation
packt@neptune:~/xpackt$ rm -i new-report-hl
rm: remove regular file 'new-report-hl'? y
packt@neptune:~/xpackt$ ls
files  new-files  new-report  new-report-link  presentation
```

Figure 2.21 – Removing a file interactively

In the preceding example, we deleted the hard link created in the previous section by using the -i option. When asked to interact, you have two options. You can approve the action by typing y (yes), or n (no) to cancel the action.

rm -f

The -f option deletes the file by force, without any interaction from the user:

```
packt@neptune:~/xpackt$ rm -f new-report-link
packt@neptune:~/xpackt$ ls
files  new-files  new-report  presentation
```

Figure 2.22 – Force remove a file

We deleted the symbol link created earlier by using the rm -f command. It did not ask for our approval and deleted the file directly.

rm -r

This option deletes the files in recursive mode, and it is used to delete multiple files and directories. For example, we will try to delete the new-files directory inside our xpackt directory. When using the rm command in its simplest way, the output will show an error saying that it cannot delete a directory. But when used with the -r option, the directory is deleted right away:

```
packt@neptune:~/xpackt$ ls
files  new-files  new-report  presentation
packt@neptune:~/xpackt$ rm new-files/
rm: cannot remove 'new-files/': Is a directory
packt@neptune:~/xpackt$ rm -r new-files/
packt@neptune:~/xpackt$ ls
files  new-report  presentation
```

Figure 2.23 – Remove a directory recursively

> **Important note**
>
> We advise *extra caution* when using the `remove` command. The most destructive mode is to use `rm  -rf`. This will delete anything, files and directories, without warning. Pay attention, as there is no going back from this. Once used, the damage will be done.

Most of the time, removing files is like a one-way street, with no turning back. This makes the process of deleting files a very important one, and a backup before a deletion could save you a lot of unnecessary stress.

Creating directories

In Linux, you can create a new directory with the `mkdir` command:

```
packt@neptune:~/xpackt$ ls
files  new-report  presentation
packt@neptune:~/xpackt$ mkdir new-directory
packt@neptune:~/xpackt$ ls
files  new-directory  new-report  presentation
```

Figure 2.24 – Creating a new directory

If you want to create more directories and sub-directories at once, you will need to use the -p option (p from the parent):

```
 mkdir -p reports/month/day
```

Then use the `ls  -R` command to see the directory structure:

```
 ls -R reports/
```

Directories are files too in Linux, only that they have special attributes. They are essential to organizing your filesystem. For more options with this useful tool, feel free to visit the manual pages.

Deleting directories

The Linux command for removing directories is called `rmdir`. It is designed to default by deleting only empty directories. Let's see what happens if we try to delete a directory that is not empty:

```
packt@neptune:~/xpackt$ rmdir reports/
rmdir: failed to remove 'reports/': Directory not empty
```

Figure 2.25 – Using the rmdir command

This is a precautionary measure from the shell, as deleting a directory that is not empty could have disastrous consequences, as we've seen when using the `rm` command. The `rmdir` command does not have an `-i` option like `rm`. The only way to delete the directory using the `rmdir` command is to delete files inside it first manually.

Commands for file viewing

As everything in Linux is a file, being able to view and work with file contents is an essential asset for any system administrator. In this sub-chapter, we will learn commands for file viewing, as almost all files contain text that, at some point, should be readable.

The cat command

This command was shortly used in some of our previous examples. It is short from **conCATenate** and is used to print the contents of the file to the screen. Here is yet another example in which we concatenate the `/etc/papersize` file:

```
packt@neptune:~/xpackt$ cat /etc/papersize
letter
```

Figure 2.26 – Example of using the cat command

The `cat` command has several options available, which we will not cover here, as most of the time its purest form will be the most used. For more details, see the manual pages.

The less command

There are times when a file has so much text that it will cover many screens, and it will be difficult to view on your terminal using just `cat`. This is where the `less` command is handy. It shows one screen at the time. How much a screen means, it all depends on the size of your terminal window. Let's take, for example, the `/etc/passwd` file. It could have multiple lines that you would not be able to fit in just one screen. You could use the following command:

```
$ less /etc/passwd
```

When you press *Enter*, the contents of the file will be shown on your screen. To navigate through it, you could use the following keys:

- *Spacebar*: Move forward one screen.
- *Enter*: Move forward one line.
- *b*: Move backward one screen.
- */*: Enter search mode; this searches forward in your file.

- *?*: Search mode; this searches backward in your file.

- *v*: Edit your file with the default editor.

- *g*: Jump to the beginning of the file.

- *G*: Jump to the end of the file.

- *q*: Exit the output.

The `less` command has a multitude of options that could be used. We advise you to consult the manual pages for this command.

The head command

This command is handy when you only want to print to the screen the beginning (the head) of a text file. By default, it will print only the first 10 lines of the file. You can use the same `/etc/passwd` file for the head exercise and execute the following command. Watch what happens. It prints the first 10 lines and then exits the command, taking you back to the shell prompt:

```
head /etc/passwd
```

One useful option of this command is to print more or less than 10 lines of the file. For this, you can use the `-n` argument or simply just – with the number of lines you want to print:

```
packt@neptune:~$ head -2 /etc/passwd
root:x:0:0:root:/root:/bin/bash
daemon:x:1:1:daemon:/usr/sbin:/usr/sbin/nologin
packt@neptune:~$ head -n 2 /etc/passwd
root:x:0:0:root:/root:/bin/bash
daemon:x:1:1:daemon:/usr/sbin:/usr/sbin/nologin
```

Figure 2.27 – Using the head command

Many other options can prove useful for your work as a system administrator, but we will not cover them here. Feel free to explore them yourself.

The tail command

The `tail` command is similar to the `head` command, only it prints the last 10 lines of a file, by default. The tail is actively used for actively watching log files that are constantly changing. It can print the last lines of the file as other applications are writing it. The options are similar to the ones for the `head` command:

```
tail -f /var/log/syslog
```

Using the -f option will make the command watch the /var/log/syslog file as it is being written. It will show you the contents of the file to the screen in an effective manner. To exit that screen, you will need to press *Ctrl + C* to go back to the shell prompt.

Commands for file properties

There could be times when just viewing the contents of a file is not enough, and you need extra information about that file. There are other handy commands that you could use, and we describe them as follows.

The stat command

This command gives you more information than the ls command does. The following example shows a comparison between the ls and stat outputs for the same file:

```
packt@neptune:~/xpackt$ ls
files  new-directory  new-report  presentation  report  reports
packt@neptune:~/xpackt$ ls -l new-report
-rw-rw-r-- 1 packt packt 26 Aug 28 17:22 new-report
packt@neptune:~/xpackt$ stat new-report
  File: new-report
  Size: 26          Blocks: 8          IO Block: 4096    regular file
Device: fd00h/64768d   Inode: 393533      Links: 1
Access: (0664/-rw-rw-r--)  Uid: ( 1000/   packt)   Gid: ( 1000/   packt)
Access: 2020-08-28 16:36:18.087118022 +0000
Modify: 2020-08-28 17:22:27.386285828 +0000
Change: 2020-08-28 17:37:49.899332845 +0000
 Birth: -
```

Figure 2.28 – Using the stat command

This gives you more information about the name, size, number of blocks, type of file, inode, number of links, permissions, UID and GID, and atime, mtime, and ctime. For more information about it, please refer to the manual page.

The file command

This command simply reports on the type of file. Here is an example of a text file and a command file:

```
packt@neptune:~/xpackt$ file new-report
new-report: ASCII text
packt@neptune:~/xpackt$ file /usr/bin/who
/usr/bin/who: ELF 64-bit LSB shared object, x86-64, version 1 (SYSV), dynamicall
y linked, interpreter /lib64/ld-linux-x86-64.so.2, BuildID[sha1]=a91ee1d8ec92841
5ca785c63401fba542199b89a, for GNU/Linux 3.2.0, stripped
```

Figure 2.29 – Using the file command

Linux does not rely on file extensions and types as some other operating systems do. In this respect, the `file` command determines the file type more by its contents than anything else.

> **Important note**
>
> There are some other vital commands such as `umask`, `chown`, `chmod`, and `chgrp`, which are used to change or set the default creation mode, owner, mode (access permissions), and group, respectively. They will be briefly introduced here as they involve setting the file's properties, but for a more detailed description, please refer to *Chapter 4, Managing Users and Groups*.

File ownership and permissions

In Linux, file security is set by ownership and permissions. File ownership is determined by the file's owner and the owner's group. Judging by the owner, a file's ownership has three types assigned to it: *user*, *group*, and *other*. The user is, most of the time, the owner of a file. Whoever created the file is its owner. The owner can be changed using the `chown` command. When setting up group ownership, you determine the permissions for everyone in that group. This is set up using the `chgrp` command. When it comes to other users, the reference is to everyone else on that system, those who did not create the file and are not not the owner, and who do not belong to the owner's group. *Other* is also known as the *world*.

Besides setting user ownership, the system must know how to determine user behavior, and it does that through the use of permissions. We will do a quick review of a file's properties by using the `ls -l` command:

```
packt@neptune:~/xpackt$ ls -l
total 20
drwxrwxr-x 2 packt packt 4096 Aug 28 16:42 files
drwxrwxr-x 2 packt packt 4096 Aug 28 17:56 new-directory
-rw-rw-r-- 1 packt packt   26 Aug 28 17:22 new-report
-rw-rw-r-- 1 packt packt    0 Aug 28 16:36 presentation
-rw-rw-r-- 1 packt packt   79 Aug 28 18:14 report
drwxrwxr-x 3 packt packt 4096 Aug 28 18:01 reports
```

Figure 2.30 – Long listing output

In the preceding examples, you see two different types of permissions for the files inside the ~/xpackt directory. Each line has 12 characters reserved for special attributes and permissions. Out of those 12, only 10 are used in the preceding examples. 9 of them represent the permissions, and the first one the file type. There are three easy-to-remember abbreviations for permissions:

- r is for read permission.
- w is for write permission.
- x is for execute permission.
- - is for no permission.

The nine characters are divided into three regions, each consisting of three characters. The first three characters are reserved for the user permissions, the following three characters are reserved for group permissions, and the last three characters represent other, or world, permissions.

File types also have their codes, as follows:

- d: A directory
- -: A file
- l: A symbolic link
- p: A named pipe; a special file that facilitates the communication between programs
- s: A socket, similar to the pipe but with bi-directional and network communications
- b: A block device; a file that corresponds to a hardware device
- c: A character device; similar to a block device

The permission string is a 10-bit string. The first bit is reserved for the file type. The next nine bits determine the permissions by dividing them into 3-bit packets. Each packet is expressed by an octal number (because an octal number has 3 bytes). Thus, permissions are represented using a power of two:

- read is $2 \wedge 2$ (two to the power of two), which equals 4.
- write is $2 \wedge 1$ (two to the power of one), which equals 2.
- execute is $2 \wedge 0$ (two to the power of zero), which equals 1.

In this respect, file permissions should be represented according to the following diagram:

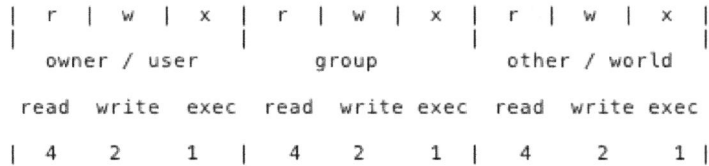

```
| r | w | x | r | w | x | r | w | x |
|   |   |   |   |   |   |   |   |   |
   owner / user          group        other / world

  read  write  exec  read  write exec  read  write exec

|  4     2      1  |  4    2     1  |  4     2     1  |
```

Figure 2.31 – File permissions explained

In the preceding diagram, you have the permissions shown as a string of nine characters, just like you would see them in the ls -la output. The row is divided into three different sections, one for owner/user, one for group and one for other/world. Those are shown on the first two rows. The other two rows show you the types of permissions (read, write, and execute) and the octal numbers down below.

This is useful as it relates the octal representations to the character representations of permissions. Thus, if you were to translate a permission shown as rwx r-x into octal, based on the preceding diagram, you could easily say it is 755. This is because for the first group, the owner, you have all of them active (rwx), which translates into 4+2+1=7. For the second group, you have only two permissions active, r and x, which translates into 4+1=5. Finally, for the last group, you have also two permissions active, similar to the second group (r and x), which translates to 4+1=5. Now you know that the permission in octal is 755.

As an exercise, you should try to translate into octal the following permissions:

- rwx rwx
- rwx r-x
- rwx r-x - - -
- rwx - - - - - -
- rw- rw- rw-
- rw- rw- r - -
- rw- rw- - - -
- rw- r- - r- -
- rw- r- - - - -
- rw- - - - - - -
- r - - - - - - - -

More details about the chown, chgrp, and chmod commands will be given in *Chapter 4, Managing Users and Groups.*

Commands for file compression and archiving

In Linux, the standard tool for archiving is called tar, from tape archive. It was initially used in Unix to write files to external tape devices for archiving. Nowadays, in Linux, it is also used to write to a file in a compressed format. Other popular formats, except tar archives, are gzip and bzip for compressed archives, together with the popular zip from Windows.

The tar command

This command is used with options and does not offer compression by default. To use compression, we would need to use specific options. Here are some of the most useful arguments available for tar:

- tar -c: Creates an archive
- tar -r: Appends files to an already existing archive
- tar -u: Appends only changed files to an existing archive
- tar -A: Appends an archive to the end of another archive
- tar -t: Lists the contents of the archive
- tar -x: Extracts the archive contents
- tar -z: Uses gzip compression for the archive
- tar -j: Uses bzip2 compression for the archive
- tar -v: Uses verbose mode by printing extra information on the screen
- tar -p: Restores original permission and ownership for the extracted files
- tar -f: Specifies the name of the output file

There is a chance that in your daily tasks, you will have to use these arguments in combination with each other.

For example, to create an archive of the files directory, we used the -cvf arguments combined, as here:

```
tar -cvf files-archive.tar files/
```

The archive created is not compressed. To use compression, we would need to add the -z or -j arguments. Next, we will use the -z option for the gzip compression algorithm. Use the following command and then compare the size of the two archive files using the ls -l command. As a general rule, it is advised to use an extension for the archive files:

```
tar -czvf gzipped-archived.tar.gz files
```

There are other useful archiving tools in Linux, but tar is still the most commonly used one. Feel free to explore the others.

Commands for locating files

Locating files in Linux is an essential task for any system administrator. As Linux systems contain vast numbers of files, finding files might be an intimidating task. Nevertheless, you have handy tools at your disposal, and knowing how to use them will be one of your greatest assets. Among those commands, we will show you locate, which, whereis, and find.

The locate command

The locate command is not installed by default on Ubuntu. To install it, use the following command:

```
sudo apt install mlocate
```

This command creates an index of all the file locations on your system. Thus, when you execute the command, it searches for your file inside the database. It uses the updatedb command as its partner.

Before starting to use the locate command, you should execute updatedb to update the location database. After you do that, you can start locating files. In the following example, we will locate any file that has new-report in its name:

```
packt@neptune:~/xpackt$ locate new-report
/home/packt/backup/xpackt/new-report
/home/packt/xpackt/new-report
```

Figure 2.32 – Using the locate command

If we were to search for a file with a more generic name, such as presentation, the output would be too long and irrelevant. Here is an example where we used output redirection to a file and the wc (word count) command to show only the number of lines, words, and bytes of the file to the standard output:

```
packt@neptune:~/xpackt$ locate presentation > ~/xpackt/locate-search && wc ~/xpa
ckt/locate-search
   348    348 27271 /home/packt/xpackt/locate-search
```

Figure 2.33 – Using the locate command with output redirection and the wc command

In the preceding output, the file has 348 lines. The exact number is used for the words inside the file, as there are no spaces between the paths, so every line is detected as a single word. The file has 27,271 bytes. For more options, please refer to the manual pages.

The which command

This command locates an executable file (program or command) in the shell's search path. For example, to locate the ls command, type the following:

```
which ls
```

The output will show the location of the ls command. Now try it with the cd command:

```
which cd
```

You will see that there is no output. This is due to the fact that the cd command is built inside the shell and has no other location for the command to show.

The whereis command

This command finds only executable files, documentation files, and source code files. Therefore, it might not find what you want, so use it with caution. Lets search for two commands, cd and ls:

```
whereis cd; whereis ls
```

The output for the cd command shows nothing relevant, as it is a built-in shell command. As for the ls command, the output will show the location of the command itself and of the manual pages.

The find command

This command is one of the most powerful commands in Linux. It can search for files in directories and subdirectories based on certain criteria. It has more than 50 options. Its main drawback is the syntax, as it is somehow different from other Linux commands. The best way to learn how the find command works is by example. This is why we will show you a large number of examples using this command, hoping that you will become proficient in using it. To see its powerful options, please refer to the manual pages.

> **Important note**
>
> As we will use all the following examples to search for files in the root directory,
> you will receive many permission denied errors. You could overcome those by
> using sudo or the root user, but we advise doing that only if you feel confident
> in your actions. You might come across one specific error, even when running
> as root: `find: '/run/user/1000/gvfs' : Permission
> denied`. Do not be afraid, you are not doing anything wrong. The error
> points out that the mount point for GNOME's virtual filesystem GVFS in the
> FUSE cannot be accessed. Feel free to Google it if you need more information
> on this matter.

Find, inside the root directory, all the files that have the `e100` string in the name and print
them to the standard output:

```
find / -name e100 -print
/usr/lib/firmware/e100
```

Find, inside the root directory, all the files that have the `file` string in their name and are
of type `file`, and print the results to the standard output:

```
find / -name file -type f -print
/usr/share/bash-completion/completions/file
/usr/bin/file
/usr/lib/apt/methods/file
/snap/core18/1705/usr/share/bash-completion/completions/file
/snap/core18/1885/usr/share/bash-completion/completions/file
```

Find all the files that have the `print` string in their name, by looking only inside the
`/opt`, `/usr`, and `/var` directories:

```
find /opt /usr /var -name print -type f -print
```

Find all the files in the root directory that have the `.conf` extension:

```
find / type f -name "*.conf"
```

Find all the files in the root directory that have the `file` string in their name and
no extension:

```
find / -type f -name "file*.*"
```

Find, in the root directory, all the files with the following extensions: `.c`, `.sh`, and `.py`, and add the list to a file named `findfile`:

```
find / -type f \( -name "*.c" -o -name "*.sh" -o -name "*.py"
\) > findfile
```

Find, in the root directory, all the files with the `.c` extension, sort them, and add them to a file:

```
find / -type f -name "*.c" -print | sort > findfile2
```

Find all the files in root directory, with the permission set to `0664`:

```
find / -type f -perm 0664
```

Find all the files in root directory that are read-only (have read-only permission) for their owner:

```
find / -type f -perm /u=r
```

Find all the files in the root directory that are executable:

```
find / -type f -perm /a=x
```

Find all the files inside the root directory that were modified 2 days ago:

```
find / -type f -mtime 2
```

Find all the files in the root directory that have been accessed in the last 2 days:

```
find / -type f -atime 2
```

Find all the files that have been modified in the last 2 to 5 days:

```
find / -type f -mtime +2 -mtime -5
```

Find all the files that have been modified in the last 10 minutes:

```
find / -type f -mmin -10
```

Find all the files that have been created in the last 10 minutes:

```
find / -type f -cmin -10
```

Find all the files that have been accessed in the last 10 minutes:

```
find / -type f -amin -10
```

Find all the files that are 5 MB in size:

```
find / -type f -size 5M
```

Find all the files that have a size between 5 and 10 MB:

```
find / -type f -size +5M -size -10M
```

Find all the empty files and empty directories:

```
find / -type f -empty
find / -type d -empty
```

Find all the largest files in the /etc directory and print to the standard output the first five. Please take into account that this command could be very resource heavy. Do not try to do this for your entire root directory, as you might run out of system memory:

```
find /etc -type f -exec ls -l {} \; | sort -n -r | head -5
```

Find the smallest first five files in the /etc directory:

```
find /etc -type f -exec ls -s {} \; | sort -n | head -5
```

Feel free to experiment with as many types of find options as you want. The command is very permissive and powerful. Use it with caution.

Commands for text manipulation

Text manipulation is probably the best asset of Linux. It gives you a plethora of tools to work with text at the command line. Some of the more important and widely used ones are grep, tee, and the more powerful ones such as sed and awk.

The grep command

This is one of the most powerful commands in Linux. It is also an extremely useful one. It has the power to search for strings inside text files. It has many powerful options too:

- grep -v: Show the lines that are not according to the search criteria.
- grep -l: Show only the filenames that match the criteria.

- grep -L: Show only the lines that do *not* comply with the criteria.
- grep -c: A counter that shows the number of lines matching the criteria.
- grep -n: Show the line number where the string was found.
- grep -i: Searches are case insensitive.
- grep -R: Search recursively inside directory structure.
- grep -E: Use Extended Regular Expressions.
- grep -F: Use a strict list of strings instead of regular expressions.

Here are some examples of how to use the grep command.

Find out the last time the sudo command was used:

```
sudo grep sudo /var/log/auth.log
```

Search for the packt string inside text files from the /etc directory:

```
grep -R packt /etc
```

Show the exact line where the match was found:

```
grep -Rn packt /etc
```

If you don't want to see the filename of each file where the match was found, use the -h option. Then, grep will only show you the lines where the match was found:

```
grep -Rh packt /etc
```

To show only the name of the file where the match was found, use -l:

```
grep -Rl packt /etc
```

grep will most likely be used in combination with shell pipes. Here are some examples.

If you want to see only the directories from your current working directory, you could pipe the ls command output to grep. In the following example, we listed only the lines that start with the letter d, which represent directories:

```
ls -la | grep '^d'
```

If you want to display the model of your CPU, you could use the following command:

```
cat /proc/cpuinfo | grep -i 'Model'
```

You will find `grep` to be one of your closest friends as a Linux system administrator, so don't be afraid to dig deeper into its options and hidden gems.

The tee command

This command is very similar to the `cat` command. Basically, it does the same thing, by copying the standard input to standard output with no alteration, but it also copies that into one or more files.

In the following example, we used the `wc` command to count the number of lines inside the /home/packt/ variables file. We piped the output to the `tee` command using the `-a` option (append if the file already exists), which wrote it to a new file called no-variables and printed it to the standard output at the same time:

```
wc -l variables | tee -a no-variables
```

`tee` is more of an underdog of file-manipulating commands. While it is quite powerful, its use can easily be overlooked. Nevertheless, we encourage you to use its powers as often as you can.

The sed and awk commands

`sed` is more than a simple command. It is a data stream editor that edits files based on a strict set of rules supplied beforehand. Based on the rules, the command reads the file line by line and the data inside the file is then manipulated. `sed` is a non-interactive stream editor that makes changes based on a script, and in this respect, it is well suited for editing more files at once or for doing mundane repetitive tasks. The `sed` command structure is as follows:

```
sed 's/regex/replacement/flag'
```

One of the common use cases of `sed` is for text substitution. There are many other use cases that we will not discuss here, but if you feel the need to learn more about the `sed` tool, there are plenty of great materials online and in print.

Here are some examples of the most common use cases of `sed`.

Replace one name with another inside a text file. For this example, we will use a new file called poem in our ~/xpackt/ directory. Inside it, we generated a random poem. The task is to replace the name Jane with Elane from within the file. The letter g, as a flag of the command, specifies that the operation should be global, as in applying to the entire text document. Here is the result:

```
packt@neptune:~/xpackt$ cat poem
Jane, Jane
Happy Birthday, Jane!
packt@neptune:~/xpackt$ sed 's/Jane/Elane/g' poem
Elane, Elane
Happy Birthday, Elane!
```

Figure 2.34 – Using the sed command to replace a string in a text file

If you check the original file using the cat command, you will see that sed only delivered the changed name result to the standard output and did not make any changes to the original file. To make the changes to the file permanent, you will have to use the -i attribute.

In the following example, we will add three new spaces at the beginning of each line and redirect the output to a new file. We will use the same poem file as before. The beginning of a file is represented by the ^ character:

```
packt@neptune:~/xpackt$ cat poem
Jane, Jane
Happy Birthday, Jane!
packt@neptune:~/xpackt$ sed 's/^/ /g' poem > poem-spaces
packt@neptune:~/xpackt$ cat poem-spaces
 Jane, Jane
 Happy Birthday, Jane!
```

Figure 2.35 – Using sed to add spaces

We will use sed to show only line number two from the poem file and to show all the lines except for line number two:

```
packt@neptune:~/xpackt$ cat poem-spaces
 Jane, Jane
 Happy Birthday, Jane!
packt@neptune:~/xpackt$ sed -n 2p poem
Happy Birthday, Jane!
packt@neptune:~/xpackt$ sed 2d poem
Jane, Jane
```

Figure 2.36 – Using sed to show specific lines in a file

Show only lines between four and six from a file, in our case, the `/etc/passwd` file:

```
packt@neptune:~$ sed -n 4,6p /etc/passwd
sys:x:3:3:sys:/dev:/usr/sbin/nologin
sync:x:4:65534:sync:/bin:/bin/sync
games:x:5:60:games:/usr/games:/usr/sbin/nologin
```

Figure 2.37 – Using sed to show a specific number of lines in a text file

Here is a more practical exercise. We will show the contents of `/etc/apt/sources.list` from Ubuntu without the commented lines. To do this, use the following command:

```
sed '/^#/g' /etc/apt/sources.list
```

As an exercise, inspect the output yourself.

`awk` is much more than a simple command; it is a pattern-matching language. It is a full-fledged programming language that was the base for Perl. It is used for data extraction from text files, with a syntax similar to C. It sees a file as being composed of fields and records. The general structure of the `awk` command is as follows:

```
awk '/search pattern 1/ {actions} /search pattern 2/ {actions}'
file
```

The true power of `awk` cannot be shown in this book, so we will show no more than a few simple examples of its use that could prove practical for a future systems administrator.

As a first example, we will generate a list with the names of all the packages installed by Ubuntu. We only want to print the name of each package, not all the other details. For this, we will use the following command:

```
packt@neptune:~/xpackt$ sudo dpkg -l | awk '{print $2}' > package-list
packt@neptune:~/xpackt$ tail package-list
yaru-theme-gtk
yaru-theme-icon
yaru-theme-sound
yelp
yelp-xsl
zenity
zenity-common
zerofree
zip
zlib1g:amd64
```

Figure 2.38 – Using awk to generate a list of package names

Generally, to see the installed packages in Ubuntu, we would run the `dpkg -l` command. In the preceding example, we piped the output of that command to the `awk` command, which printed the second column (field) from the `dpkg -l` output (`'{print $2}'`). Then, we redirected everything to a new file called `package-list` and used the `tail` command to see the last 10 lines of the newly created file.

Both `sed` and `awk` are very powerful tools and we have merely scratched the surface of what they can do. Please feel free to dig deeper into these two awesome tools.

Using text editors to create and edit files

Linux has several command-line text editors that you can use. There are **nano**, **emacs**, and **vim**, among others. Those are the most used ones. There are also **pico**, **joe**, and **ed** as text editors that are less frequently used than the aforementioned ones. We will cover vim, as there is a very good chance that you will find it on any Linux system that you work with. Nevertheless, the current trend is to replace vim with nano as the default text editor. Ubuntu, for example, does not have vim installed by default, but CentOS does. Fedora is currently looking to make nano the default text editor. Therefore, you might want to learn nano, but for legacy purposes, vim is a very useful tool to know.

Using vim to edit text files

Vim is the improved version of vi, the default text editor from Unix. It is a very powerful editing tool. This power comes with many options that can be used to ease your work, and this can be overwhelming. In this sub-chapter, we will introduce you to the basic commands of the text editor, just enough to help you be comfortable using it.

Vim is a mode-based editor, as its operation is organized around different modes. In a nutshell, those modes are as follows:

- `command` mode is the default mode, waiting for a command.
- `insert` mode is the text insert mode.
- `replace` mode is the text replace mode.
- `search` mode is the special mode for searching a document.

Let's see how we can switch between these modes.

Switching between modes

When you first open vim, you will be introduced to an empty editor that only shows information about the version used and a few help commands. You are in command mode. This means that vim is waiting for a command to operate.

To activate insert mode, press *i* on your keyboard. You will be able to start inserting text at the current position of your cursor. You can also press *a* (for append) to start editing to the right of your cursor's position. Both *i* and *a* will activate insert mode. To exit the current mode, press the *Esc* key. It will get you back to command mode.

If you open a file that already has text in it, while in command mode, you can navigate the file using your arrow keys. As vim inherited the vi workflow, you can also use *h* (to move left), *j* (to move down), *k* (to move up), and *l* (to move right). Those are legacy keys from a time when terminal keyboards did not have separate arrow keys.

While still in command mode (the default mode), you can activate replace mode by pressing *r* on your keyboard. You can replace the character that is right at the position of your cursor.

search mode is activated, while in command mode, by pressing the / key. Once in your mode, you can start typing a search string and then press *Enter*.

There is also a last line mode, or ex command mode. This mode is activated by pressing :. This is an extended mode where commands like w for saving the file, q for quitting, or wq for saving and quitting at the same time.

Basic vim commands

Here is a list of common vim commands. They can be used while in command mode:

- yy: Copy a block of text (yank).
- p: Paste the copied block.
- u: Undo the last operation.
- x: Delete the next character to the right (relative to the cursor's position)
- X: Delete the preceding character (relative to the cursor's position).
- dd: Delete the entire line on which the cursor is positioned.
- h: Move the cursor left.
- l: Move the cursor right.
- k: Move the cursor up.

- `j`: Move the cursor down.
- `w`: Move the cursor right to the beginning of the next word.
- `b`: Move the cursor left to the beginning of the previous word.
- `^`: Move the cursor to the beginning of the line.
- `$`: Move the cursor to the end of the line.
- `gg`: Move the cursor to the beginning of the document.
- `G`: Move the cursor to the end of the document.
- `: n`: Move the cursor to line number *n*.
- `i`: Insert before the cursor's position.
- `I`: Insert at the beginning of the line.
- `a`: Append at the right of the cursor.
- `A`: Append at the end of the line.
- `o`: Insert at the beginning of next line.
- `/`: Activate `search` mode and look for strings.
- `?`: Search backward from the cursor's position.
- `n`: Show the next position of the searched string.
- `N`: Show the previous position of the searched string.
- `:wq`: Save changes and exit vim (write and quit).
- `:q!`: Enforced exit (quit) without saving.
- `:w!`: Enforced save (write) without exiting.
- `ZZ`: Save and quit the file.
- `:w file`: Save as a new file (`file` will be the new filename).

Vim can be quite intimidating for newcomers to Linux. There is no shame if you prefer other editors, as there are plenty to choose from. Now, we will show you a glimpse of nano.

The nano text editor

Vim is a powerful text editor and knowing how to use it is an important thing for any system administrator. Nevertheless, there are other text editors that are equally powerful and even easier to use.

This is the case with nano, which is installed by default in Ubuntu and CentOS and can be used right out of the box on both. The default editor is not set up in the `.bashrc` file by using the `$EDITOR` variable. However, in Ubuntu, you can check the default editor on your system by using the following command:

```
packt@neptune:~/xpackt$ sudo update-alternatives --config editor
[sudo] password for packt:
There are 4 choices for the alternative editor (providing /usr/bin/editor).

  Selection    Path                 Priority   Status
------------------------------------------------------------
* 0            /bin/nano             40        auto mode
  1            /bin/ed              -100       manual mode
  2            /bin/nano             40        manual mode
  3            /usr/bin/vim.basic    30        manual mode
  4            /usr/bin/vim.tiny     15        manual mode

Press <enter> to keep the current choice[*], or type selection number:
```

Figure 2.39 – Checking the default text editor on Ubuntu

You can invoke the nano editor by using the `nano` command on both Ubuntu and CentOS. When you type the command, the nano editor will open, with a very straightforward interface.

In this section, we learned about the different editors and their commands.

Summary

In this chapter, you have learned how to work with the most commonly used commands in Linux. You now know how to manage (create, delete, copy, and move) files, how the filesystem is organized, how to work with directories, and how to view file contents. You now understand the shell and basic permissions. The skills you have learned will help you manage files in any Linux distribution and edit text files. You have learned how to work with vim, one of the most widely used command-line text editors in Linux. Those skills will help you to learn how to use other text editors such as nano and emacs. You will use these skills in almost every chapter of this book, as well as in your everyday job as a systems administrator.

In the next chapter, you will learn how to manage packages, including how to install, remove, and query packages in both Debian- and Red Hat-based distributions. This skill is important for any administrator and must be part of any basic training.

Questions

In our second chapter, we covered the Linux filesystem and the basic commands that will serve as the foundation for the entire book. Here are some questions for you to test your knowledge and for further practice:

1. What is the command that creates a compressed archive with all the files inside the /etc directory that use the .conf extension?

   Hint: Use the tar command just as shown in this chapter.

2. What is the command that lists the first five files inside /etc and sorts them by dimension in descending order?

   Hint: Use find combined with sort and head.

3. What command creates a hierarchical directory structure?

   Hint: Use mkdir just as shown in this chapter.

4. What is the command that searches for files with three different extensions inside root?

   Hint: Use the find command.

5. Find out which commands inside Linux have the **Set owner User ID (SUID)** set up.

   Hint: Use the find command with the -perm parameter.

6. What is the command that lists all the installed packages on a system?

   Hint: Combine dpkg with awk and output redirection.

7. Which command is used to create a file with 1,000 lines of randomly generated words (one word per line)?

   Hint: Use the shuf command (not shown in this chapter).

8. Perform the same exercise as before, but this time generate a file with 1,000 randomly generated numbers.

   Hint: Use a for loop.

9. How do you find out when sudo was last used and which commands were executed by it?

   Hint: Use the grep command.

Further reading

For more information about what was covered in this chapter, please refer to the following:

- *Fundamentals of Linux*, Oliver Pelz, Packt Publishing
- *Mastering Ubuntu Server – Second Edition*, Jay LaCroix, Packt Publishing

3
Linux Software Management

Software management is an important aspect of Linux system administration. Knowing how to work with software packages is an asset that you will master after finishing this chapter.

In this chapter, you will learn how to use specific software management commands as well as learn how software packages work depending on your distribution of choice. You will learn about the latest snap and flatpak package types and how to use them on modern Ubuntu and CentOS distributions. By the end of the chapter, you will also have a glimpse of how to build your own package.

In this chapter, we're going to cover the following main topics:

- Linux software package types
- Managing software packages
- Building a package from source

Technical requirements

No special technical requirements are needed, just a working installation of Linux on your system. Ubuntu and CentOS are equally suitable for this chapter's exercises, as we will cover both types of package managers.

Linux software package types

As you've already learned by now, a Linux distribution comes packed with a Kernel and applications on top of it. Although plenty of applications are already installed by default, there will certainly be occasions when you will need to install some new ones or to remove ones that you don't need.

In Linux, applications come bundled into **repositories**. A repository is a centrally managed location that consists of software packages maintained by developers. Those packages could contain individual applications or operating system-related files. Each Linux distribution comes with several official repositories, but on top of those, you can add some new ones. The way to add them is specific to each distribution, and we will get into more details soon.

Linux has several types of packages, but as we are only covering Ubuntu and CentOS, we will mostly refer to the ones those two distributions use. Ubuntu uses deb packages, as it is based on Debian, and CentOS uses rpm packages, as it is based on Red Hat Enterprise Linux. Besides those, there are two new package types that have been recently introduced, the snap packages developed by Ubuntu, and the flatpak packages, developed by a large community of developers and organizations, including GNOME, Red Hat, and Endless.

The DEB and RPM package types

DEB and RPM are the oldest types of packages and are used by Ubuntu and CentOS, respectively. They are still widely used, even though the two new types mentioned earlier are starting to gain ground on Linux on desktops.

Both package types are compliant with the **Linux Standard Base (LSB)** specifications. The last iteration of LSB is version 5.0, released in 2015. You can find more information about it at the following address: https://refspecs.linuxfoundation.org/lsb.shtml#PACKAGEFMT.

The DEB package anatomy

DEB was introduced with the Debian distribution back in 1993 and has been in use ever since on every Debian and Ubuntu derivative. A deb package is a binary package. This means that it contains the files of the program itself, as well as its dependencies and meta-information files, all contained inside an archive.

To check the contents of a binary deb package, you can use the `ar` command. It is not installed by default in Ubuntu 20.04.1 LTS, so you will have to install it yourself using the following command:

```
$ sudo apt install binutils
```

Once `ar` is installed, you can check the contents of any deb package. For this exercise, we downloaded the Slack deb package and checked its contents. To download the package, perform the following steps:

1. Use the `wget` command as follows, and the file will be downloaded inside your current working directory:

    ```
    $ wget https://downloads.slack-edge.com/linux_releases/
    slack-desktop-4.8.0-amd64.deb
    ```

2. After that, use the `ar t slack-desktop-4.8.0-amd64.deb` command to view the contents of the binary package. The `t` option will display a table of contents for the archive:

    ```
    packt@neptune:~/Downloads$ ar t slack-desktop-4.8.0-amd64.deb
    debian-binary
    control.tar.gz
    data.tar.xz
    _gpgorigin
    ```

    Figure 3.1 – Using the ar command to view the contents of a deb file

    As you can see here, the output listed four files, from which two are archives. You can also investigate the package with the `ar` command.

3. Use the `ar x slack-desktop-4.8.0-amd64.deb` command to extract the contents of the package to your present working directory:

    ```
    $ ar x slack-desktop-4.8.0-amd64.deb
    ```

4. Use the `ls` command to list the contents of your directory. The four files are now extracted and ready to inspect. The `debian-binary` file is a text file that contains the version of the package file format, which in our case is 2.0. You can concatenate the file to verify on your package, with the help of the following command:

```
$ cat debian-binary
```

5. The `control.tar.gz` archive contains meta-information packages and scripts to be run during the installation or before and after, depending on the case. The `data.tar.xz` archive contains the executable files and libraries of the program that are going to be extracted during the installation. You can check the contents with the following command:

```
$ tar tJf data.tar.xz | head
```

6. The last file is a `gpg` signature file.

Meta-information for each package is a collection of files that are essential for the programs to run. They contain information about certain package pre-requisites, all their dependencies, conflicts, and suggestions. Feel free to explore everything that a package is made of, using just the packaging-related commands.

The RPM package anatomy

The RPM packages were developed by Red Hat and are used in Fedora, CentOS, RHEL, and SUSE. The name is an acronym from Red Hat Package Manager. RPM binary packages are similar to the DEB binary packages. They are packaged as an archive, too.

We will test the rpm package of Slack, just as we did with the deb package in the previous section. Download the `rpm` package with the following command:

```
# wget https://downloads.slack-edge.com/linux_releases/slack-
4.8.0-0.1.fc21.x86_64.rpm
```

If you want to use the same `ar` command, you will see that in the case of `rpms`, the archiving tool will not recognize the file format. Nevertheless, there are other more powerful tools to use. We will use the `rpm` command, the designated low-level package manager for `rpms`. We will use the `-q` (query), `-p` (package name), and `-l` (list) options:

```
# rpm -qpl slack-4.8.0-0.1.fc21.x86_64.rpm
```

The output, contrary to the `deb` package, will be a list of all the files related to the application, with their installation locations for your system.

To see the meta-information for the package, run the `rpm` command with the `-q`, `-p`, and `-i` (install) options. For example, use it on the slack package:

```
rpm -qpi slack-4.8.0-0.1.fc21.x86_64.rpm
```

The output will contain information about the application name, version, release, architecture, installation date, group, size, license, signature, source RPM, build date and host, URL, relocation, and summary.

To see which other dependencies the package will require at installation, you can run the same `rpm` command with the `-q`, `-p`, and `–requires` options:

```
$ rpm -qp - -requires slack-4.8.8-0.1.fc21.x86_64.rpm
```

DEB and RPM packages are not the only ones. As we stated earlier, there are new packages available for cross-platform Linux distributions. Those packages are flatpaks and snaps, and we will detail them in the following section.

The snap and flatpak package types

Snap and flatpak are relatively new package types, and they are considered to be the future of apps on Linux. They both build and run applications in isolated containers, for more security and portability. Both have been created to overcome the need for desktop applications' ease of installation and portability.

Even though major Linux distributions have large application repositories, distributing software for so many types of Linux distributions, each with its own kind of package types, can become a serious issue for **Independent Software Vendors** (**ISVs**) or community maintainers. This is where both snaps and flatpaks come to the rescue, aiming to reduce the weight of distributing software.

Let's consider that we are independent software developers, aiming to develop our product on Linux. Once a new version of our software is available, we need to create at least two types of packages to be directly downloaded from our website – a `.deb` package for Debian/Ubuntu, and an `.rpm` package for Fedora/RHEL/SUSE.

But, if we want to overcome this and make our app available cross-distribution for most of the existing Linux distributions, we can distribute it as a flatpak or snap. The flatpak would be available through `flathub`, the centralized flatpak repository, and the snap would be available through the `snapcraft` store, the centralized snap repository. Either one is equally suitable for our aim to distribute the app for all major Linux distributions with minimal resource consumption and centralized effort.

The moral of this situation is that the effort to distribute software for Linux is higher than in the case of the same app packaged for Windows or macOS. Hopefully, in the near future, there will be only one universal package for distributing software for Linux, and this will all be for the better for both users and developers alike.

The snap package anatomy

The snap file is actually a `SquashFS` file. This means that it has its own filesystem encapsulated in an immutable container. It has a very restrictive environment, with specific rules for isolation and confinement. Every snap file has a meta-information directory that stores files that control its behavior.

Snaps, as opposed to flatpaks, are used not only for desktop applications, but also for a wider range of server and embedded apps. This is because snap has its origins in Ubuntu Snappy for IoT and phones, the distribution that emerged as the beacon of convergence effort from Canonical, Ubuntu's developer.

The flatpak package anatomy

Flatpak is based on a technology called OSTree. The technology was started by developers from GNOME and Red Hat, and it is now heavily used in Fedora Silverblue in the form of `rpm-ostree`. It is a new upgrade system for Linux that is meant to work alongside existing package management systems. It was inspired by Git, as it operates in a similar manner. Consider it as a version control system at the OS level. It uses a content-addressed object store, allows you to share branches, and offers transactional upgrades and rollback and snapshot options for the operating system.

Currently, the project has changed its name to `libostree`, for smooth focus on projects that already use the technology. Among many projects that use it, we will bring just two into the discussion: flatpak and `rpm-ostree`. The `rpm-ostree` project is considered to be a next generation hybrid package system for distributions such as Fedora and CentOS/RHEL. They are based on the Atomic project developed by Fedora and Red Hat teams, which brings immutable infrastructure for servers and desktops alike. The openSUSE developers had a similar technology developed called Snapper, which was an OS snapshot tool for its `btrfs` filesystems.

Flatpak uses `libostree` just like `rpm-ostree`, but is solely used for desktop application containers, with no bootloader management. Flatpak uses sandboxing based on a project named `Bubblewrap`, which allows unprivileged users to access user namespaces and use container features.

Both snaps and flatpaks have full support for graphical installations, but also have commands for easier installations and setup from the shell. In the following sections, we will focus solely on command operations for both package types.

Managing software packages

Each distribution has its own package managers. There are two types of package managers for each distribution, one for low-level and one for high-level package management. For an RPM-based distribution such as CentOS or Fedora, the low-level tool is the rpm command, and the high-level tools are the yum and dnf commands. For openSUSE, another major RPM-based distribution, the low-level tool is the same rpm command, and for high-level tools, there is the zypper command. For DEB-based distributions, the low-level command is dpkg and the high-level command is apt (or the now deprecated apt-get).

What is the difference between low-level and high-level package managers in Linux? The low-level package managers are responsible for the backend of any package manipulation, and are capable of unpacking packages, running scripts, and installing apps. The high-end managers are responsible for dependency resolution, installing and downloading packages (and groups of packages), and metadata searching.

Managing DEB packages

Usually, for any distribution, package management is handled by the administrator or by a user with root privileges (sudo). Package management implies any type of package manipulation, such as installation, search, download, and removal. For all these types of operations, there are specific Linux commands, and we will show you how to use them in the following sections.

The repositories

The Ubuntu official repositories consist of about 60,000 packages. Those take the form of binary .deb packages or as snap packages. The configuration of the system repositories is stored in one file, the /etc/apt/sources.list file. Ubuntu has four main repositories, and you will see them detailed inside the sources.list file. Those repositories are as follows:

- Main – Contains free and open source software supported by Canonical
- Universe – Contains free and open source software supported by the community
- Restricted – Contains proprietary software
- Multiverse – Contains software restricted by copyright

All the repositories are enabled by default in the `sources.list` file. If you would like to disable some of them, feel free to edit the file.

The APT-related commands

Until 4 years ago, packages in any Debian-based distribution were implemented using the `apt-get` command. Since then, a new and improved command called `apt` (from Advanced Packaging Tool) is used as the high-level package manager. The new command is more streamlined and better structured then `apt-get`, thus offering a more integrated experience.

Before doing any kind of work with the `apt` command, you should update the list of all available packages. This is done with the following command:

```
$ sudo apt update
```

The output of the command will show you if any updates are available. The number of packages that require updates will be shown, together with a command that you could run if you want more details about them.

Before going any further, we will encourage you to use the `apt - -help` command, as this will show you the most commonly used `apt`-related commands.

Let's go into more detail on some of the most commonly used commands in the next sub-sections.

Installing and removing packages

Basic system administration tasks will include installing and removing packages. In this sub-chapter, we will show you how to install and remove using the apt commands.

To install a new package, you will use the `apt install` command. We have already used this one before in this book. Remember that we had to install the `ar` command as an alternative to inspect `.deb` packages. Then we used the following command:

```
$ sudo apt install binutils
```

This command installed several packages on the system, and among them the one that we need to fulfill our action. The `apt` command automatically installed any requisite dependencies, too.

To remove a package, you can use the apt remove and apt purge commands. The first one removes the installed packages and all its dependencies installed by the apt install command. The latter will perform the uninstallment, just like apt remove, but it also deletes any configuration files created by the applications.

In the following example, we will remove the binutils applications installed previously:

```
$ sudo apt remove binutils
```

The output will show you a list of packages that are no longer needed and will be removed from the system and will ask for your confirmation to continue. This is a very good safety measure, giving you the opportunity to review the files that are going to be deleted. If you feel confident about the operation, you can add a -y option parameter at the end of the command, which tells the shell that the answer to any question provided by the command will automatically be *Yes*.

Here is the apt purge command:

```
$ sudo apt purge binutils
```

The output is similar to the apt remove command, showing you which packages will be removed, how much space will be freed on the disk, and your confirmation to continue the operation.

Therefore, as stated earlier, only by using the apt remove command, some configuration files are left behind, in case the operation was an accident, and the user wants to revert to the previous configuration. The files that are not deleted by the remove command are small user configuration files that can easily be restored. If the operation was not an accident and you still want to get rid of all the files, you can still use the apt purge command to do that, by using the same name as those of the already removed packages.

Upgrading the system

Every now and then, you will need to perform a system upgrade to ensure that you have all the latest security updates and patches installed. The command to do this is as follows:

```
$ sudo apt upgrade
```

It should always be preceded by this command:

```
$ sudo apt update
```

This is so as to make sure that all the package lists and repository information is updated. The update command will sometimes show you which packages are no longer required with a message similar to the following:

```
The following packages were automatically installed and are no
longer required:
  libfprint-2-tod1 libllvm9
Use 'sudo apt autoremove' to remove them.
```

You can use the sudo apt autoremove command to remove the unneeded packages, after you perform the upgrade. The autoremove command's output will show you which packages will be removed and how much space will be freed on the disk and will ask for your approval to continue the operation.

Let's say that during our work with Ubuntu, a new distribution is released, and we would like to use that, as it has newer packages of the software we use. Using the command line, we can make a full distribution upgrade. The command for this action is as follows:

```
$ sudo apt dist-upgrade
```

Similar to this, we can also use the following command:

```
$ sudo apt full-upgrade
```

Upgrading to a newer distribution version should be a straightforward process, but this is not always a guarantee. It all depends on your custom configurations. No matter the case, we advise you to do a full system backup before upgrading to a new version.

Managing package information

Working with packages sometimes implies the use of information gathering tools. Simply installing and removing packages is not enough. You will need to search for certain packages, to show details about them, to create lists based on specific criteria, and so on.

To search for a specific package, use the apt search command. It will list for you all the packages that have the searched string in their name, as well as others that use the string in various ways. For example, let's search for the package named nmap:

```
$ sudo apt search nmap
```

The output will show a considerably long list of packages that use the nmap string in various ways. You will still have to scroll up and down the list to find the package you wanted. For better results, you can pipe the output to the grep command, but you will notice a warning, like the one in the following screenshot:

Managing software packages   101

```
packt@neptune:~$ sudo apt search nmap | grep nmap

WARNING: apt does not have a stable CLI interface. Use with caution in scripts.

libnmap-parser-perl/focal,focal 1.37-1 all
  module to parse nmap scan results with perl
  Library for doing location lookup based on free openwlanmap.org data
  Library for doing location lookup based on free openwlanmap.org data
nmap/focal 7.80+dfsg1-2build1 amd64
nmap-common/focal,focal 7.80+dfsg1-2build1 all
  Architecture independent files for nmap
nmapsi4/focal 0.5~alpha1-3build1 amd64
  graphical interface to nmap, the network scanner
python-libnmap-doc/focal,focal 0.7.0-2 all
python3-libnmap/focal,focal 0.7.0-2 all
python3-nmap/focal,focal 0.6.1-1.1 all
  post-processor for TopHat unmapped reads
```

Figure 3.2 – Output of the apt search command

Following the warning, the output shows a short list of packages that contain the string nmap, and among them, is the actual package we are looking for, highlighted in *Figure 3.5*.

To overcome that warning, you can use a legacy command called apt-cache search. By running it, you will have a list of packages as output, but not as detailed as the output of the apt search command:

```
packt@neptune:~$ sudo apt-cache search nmap | grep nmap
libnmap-parser-perl - module to parse nmap scan results with perl
libwlocate-dev - Library for doing location lookup based on free openwlanmap.org
 data
libwlocate0 - Library for doing location lookup based on free openwlanmap.org da
ta
nmap - The Network Mapper
nmap-common - Architecture independent files for nmap
nmapsi4 - graphical interface to nmap, the network scanner
python-libnmap-doc - Python NMAP Library (common documentation)
python3-libnmap - Python 3 NMAP library
python3-nmap - Python3 interface to the Nmap port scanner
tophat-recondition - post-processor for TopHat unmapped reads
```

Figure 3.3 – The output of the apt-cache command

Now that we know that the nmap package exists in the Ubuntu repositories, we can investigate it further by showing more details using the apt show command:

```
$ apt show nmap
```

The output will show a detailed description including the package name, version, priority, origin and section, maintainer, size, dependencies, suggested extra packages, download size, APT sources, and description.

Apt also has a useful `list` command, which can list packages based on certain criteria. For example, if we use the `apt list` command alone, it will list all the packages available. But if we use different options, the output will be personalized.

To show the installed packages, we will use the `-- installed` option:

```
$ sudo apt list --installed
```

To list all the packages, use the following command:

```
$ sudo apt list
```

For comparative reasons, we will redirect each output to a different file, and then compare the two files. This is an easier task to do in order to see that there are differences between the two outputs, as the lists are reasonably large. We will now run the specific commands, as follows:

```
packt@neptune:~/Documents$ sudo apt list > list
[sudo] password for packt:

WARNING: apt does not have a stable CLI interface. Use with caution in scripts.

packt@neptune:~/Documents$ sudo apt list --installed > list-installed

WARNING: apt does not have a stable CLI interface. Use with caution in scripts.

packt@neptune:~/Documents$ ls -la
total 3148
drwxr-xr-x  2 packt packt    4096 Sep  1 05:43 .
drwxr-xr-x 15 packt packt    4096 Sep  1 04:10 ..
-rw-rw-r--  1 packt packt 3098245 Sep  1 05:42 list
-rw-rw-r--  1 packt packt  113167 Sep  1 05:43 list-installed
```

Figure 3.4 – Comparison of installed packages

There are other ways in which to compare the two outputs, and we would like to let you discover them by yourself, as an exercise for this sub-chapter. Feel free to use any other apt-related commands you would like, and practice with them enough to get familiar with their use. The APT is a powerful tool, and it is essential for any system administrator to know how to use it in order to sustain a usable and well-maintained Linux system. Usability is closely related to the apps used and their system-wide optimization.

Managing RPM packages

RPM packages are the equivalent packages for Linux distributions such as Fedora, CentOS, RHEL, and openSUSE/SLES. They have dedicated high-level tools, including dnf, yum, and zypper. The low-level tool is the rpm command.

In CentOS 8, the default package manager is yum (from Yellow Dog Updater, Modified) and it is actually based on dnf (Dandified YUM), the default package manager in Fedora. If you use both Fedora and CentOS, for ease of use, you can use only one of those, as they are basically the same command. For consistency, we will use the YUM name for all the examples in this chapter.

Yum is the default high-level manager. It manages installation, removal, update and package queries, and resolves dependencies. Yum can manage both packages installed from repositories or from local .rpm packages.

The repositories

Repositories are all managed from the /etc/yum.repos.d/ directory, with configuration available inside the /etc/yum.conf file. If you do a listing with the ls -l command for the repos directory, the output will show you a list of all the repository files:

```
ls -l /etc/yum.repos.d/
```

All those files listed here contain vital information pertaining to the repository, such as the name, the mirror list, the gpg key location, and enabled status. All the ones listed are official repositories.

The YUM-related commands

Yum has many commands and options, but the most commonly used ones are related to package installation, removal, search, information query, system update, and repository listing.

Installing and removing packages

To install a package from a repository in CentOS 8, simply run the command yum install. In the following example, we will install the GIMP application from the command line:

```
$ sudo yum install gimp
```

If you already have a package downloaded and would like to install it, you can use the `yum localinstall` command. To install the previously downloaded Slack `.rpm` package, use the following command (logged in as root user):

```
yum localinstall slack-4.8.0-0.1.fc21.x86_64.rpm
```

Once running the command, you will see that it is automatically resolving the dependencies needed, and shows the source (repository) for each of them (in our case, it is the AppStream repository). A notable difference is the repository for the local package, which appears to be `@commandline`.

This is a very powerful command, which makes the use of the `rpm` command itself almost redundant in some cases. The main difference between the `yum install` and `yum localinstall` commands is that the latter is capable of solving dependencies for locally downloaded packages. The first command looks for packages inside the active repositories, while the second one looks for packages to install in the current working directory.

To remove a package from the system, use the `yum remove` command. We will remove the newly installed Slack package with the following command (logged in as root):

```
yum remove slack.x86_64
```

After running the command, you will be asked if you want to delete the packages and its dependencies or not. In this respect, please consider the next important note.

> **Important note**
>
> The default action for pressing the *Enter* or *Return* key while inside a command dialog in CentOS is *N* (for No, or Negative), while in Ubuntu, the default action is set to *Y* (for Yes). This is a precautionary safety measure, which requires extra attention and intervention.

The output, very similar to the output of the installation command, will show you which packages and dependencies will be removed if you proceed with the command.

As you can see, all the dependencies installed with the package using the `yum localinstall` command will be removed using the `yum remove` command. If asked to proceed, type y and continue with the operation.

Upgrading the system

To upgrade a CentOS 8 system, we will use the `yum upgrade` command. There is also a `yum update` command, which has the same effect by updating the installed packages:

```
$ sudo yum upgrade
```

You can use the `-y` option to automatically respond to the command's questions.

There is also an `upgrade-minimal` command, which installs only the newest security updates for packages.

Managing package information

Managing files with `yum` is very similar to managing files with `apt`. There are plenty of commands to use, and we will detail some of them, the ones we consider to be the most commonly used. To find out more about those commands and their use, run `yum - -help`.

To see an overview of the yum command history and which package was managed, use the following command:

```
$ sudo yum history
```

This will give you an output that shows every yum command ran, how many packages were altered, and the time and date when the actions were executed, as in the following example:

```
[root@localhost Downloads]# yum history
ID      | Command line              | Date and time       | Action(s)  | Altered
--------------------------------------------------------------------------------
     4 | remove slack.x86_64        | 2020-09-01 06:03 | Removed    |    6
     3 | localinstall slack-4.8.0   | 2020-09-01 06:00 | Install    |    6
     2 | update -y                  | 2020-09-01 04:36 | I, U       |  115 EE
     1 |                            | 2020-09-01 03:45 | Install    | 1376 EE
```

Figure 3.5 – Using the yum history command

To show details about a certain package, we have the `yum info` command. We will query the `nmap` package, similar to what we did in Ubuntu:

```
yum info nmap
```

The output will show you the name, version, release, source, repository, and description, very similar to what we saw with the `.deb` packages.

To list all the installed packages, or all the packages for that matter, we use the `yum list` command:

```
# yum list
# yum list installed
```

If we redirect the output of each command to specific files, and then compare the two files, we will see the differences between them, similar to what we did in Ubuntu. The output shows the name of the packages, followed by the version and release number, and the repository from which it was installed. Here is a short excerpt:

```
xorg-x11-xkb-utils.x86_64              7.7-27.el8              @AppStream
xz.x86_64                              5.2.4-3.el8             @anaconda
xz-libs.x86_64                         5.2.4-3.el8             @anaconda
yajl.x86_64                            2.1.0-10.el8            @AppStream
yelp.x86_64                            2:3.28.1-3.el8          @AppStream
yelp-libs.x86_64                       2:3.28.1-3.el8          @AppStream
yelp-tools.noarch                      3.28.0-3.el8            @AppStream
yelp-xsl.noarch                        3.28.0-2.el8            @AppStream
yum.noarch                             4.2.17-7.el8_2          @BaseOS
zenity.x86_64                          3.28.1-1.el8            @AppStream
zip.x86_64                             3.0-23.el8              @anaconda
zlib.x86_64                            1.2.11-13.el8           @anaconda
```

Figure 3.6 – Excerpt of the yum list installed command

As we have covered the most commonly used commands for both deb and rpm files, let's see how to manage flatpaks and snaps on your Linux machine.

Using snap and flatpak packages

Snaps and flatpaks are relatively new package types that are used in various Linux distributions. In this section, we will show you how to manage these types of packages. For snaps we will use Ubuntu as our test distribution, and for flatpaks we will use CentOS, even though, with a little bit of work, both package types can work on either distribution.

Managing snap packages on Ubuntu

Snap is installed by default in Ubuntu 20.04.1 LTS. Therefore, you don't have to do anything to install it. Simply start searching for the package you want and install it on your system. We will use the Slack application to show you how to work with snaps.

Searching for snaps

Slack is available in the snapcraft store, so you can install it. To make sure, you can search for it using the snap find command, as in the following example:

```
snap find "slack"
```

In the command's output, you will see many more packages that contain the string `slack` or are related to the Slack application, but we are not interested in those. Only the one that shows the Slack application is of interest to us.

> **Important note**
>
> In any Linux distribution, two apps originating from different packages and installed with different package managers can coexist. For example, Slack can be installed using the deb file provided from the website, as well as the one installed from the snap store.

The output says that the package is available, so we can proceed and install it on our system.

Installing a snap package

To install the snap package for Slack, we will use the `snap install` command. In our case, which might coincide with yours, the first attempt to install the Slack package ended with a warning. It says the package we are about to install will execute outside the sandbox of a regular snap, and we should continue only if we understand the risks:

```
packt@neptune:~$ sudo snap install slack
[sudo] password for packt:
error: This revision of snap "slack" was published using classic confinement
       and thus may perform arbitrary system changes outside of the security
       sandbox that snaps are usually confined to, which may put your system
       at risk.

       If you understand and want to proceed repeat the command including
       --classic.
```

Figure 3.7 – Output error while trying to install the Slack snap package

We understand the risks and decide to proceed, but you may not do so if you consider that the risk could be too high. The following command installed the snap package:

```
packt@neptune:~$ sudo snap install slack --classic
slack 4.8.0 from Slack✓ installed
```

Figure 3.8 – A successful attempt to install Slack

Next, let's see how we can find out more about the snap package.

Snap package information

If you want to find out more about the package, you can use the `snap info` command:

```
$ snap info slack
```

The output will show you relevant information about the name of the package, summary, publisher, description, and ID. The last information displayed will be about the available channels, which are the following in the case of our Slack package:

```
tracking:      latest/stable
refresh-date: today at 06:25 PDT
channels:
    latest/stable:      4.8.0 2020-07-31 (27) 124MB classic
    latest/candidate: ↑
    latest/beta:        ↑
    latest/edge:        3.3.1 2018-08-31  (8) 148MB classic
installed:              4.8.0             (27) 124MB classic
```

Figure 3.9 – Snap channels shown for the Slack app

Each channel has information about a specific version and it is important to know which one to choose. By default, the stable channel will be chosen by the `install` command, but if you would like a different version, you could use the `--channel` option during installation. In the preceding example, we used the default option.

Showing installed snap packages

If you want to see a list of the installed snaps on your system, use the `snap list` command. Even though we installed only Slack on the system, in the output, you will see that there are many more apps installed. Some, such as `core` and `snapd`, are installed by default from the distribution's installation, and are required by the system:

```
packt@neptune:~$ snap list
Name                Version            Rev   Tracking         Publisher   Notes
core18              20200707           1880  latest/stable    canonical✓  base
gnome-3-34-1804     0+git.3009fc7      36    latest/stable/…  canonical✓  -
gtk-common-themes   0.1-36-gc75f853    1506  latest/stable/…  canonical✓  -
slack               4.8.0              27    latest/stable    slack✓      classic
snap-store          3.36.0-80-g208fd61 467   latest/stable/…  canonical✓  -
snapd               2.45.2             8542  latest/stable    canonical✓  snapd
```

Figure 3.10 – Output of the snap list command

Now we'll learn how to update a snap package.

Updating a snap package

Snaps are automatically updated. Therefore, you won't have to do anything yourself. The least you can do is check whether an update is available and speed up its installation using the `snap refresh` command, as follows:

```
$ sudo snap refresh slack
```

Following an update, if you want to go back to a previously used version of the app, you can use the `snap revert` command, as in the following example:

```
$ sudo snap revert slack
```

In the next section, we'll see how to enable and disable snap packages.

Enabling or disabling snap packages

If we decide to not use an application for a temporary period of time, we can disable that app using the `snap disable` command. If we decide to reuse the app, we can enable it again, using the `snap enable` command:

```
packt@neptune:~$ sudo snap disable slack
slack disabled
packt@neptune:~$ sudo snap enable slack
slack enabled
```

Figure 3.11 – Enabling and disabling a snap app

If disabling is not what you are looking for, you can completely remove the snap.

Removing a snap package

When removing a snap application, the associated configuration files, users, and data are also removed. The command to use is `snap remove`, as in the following example, where we will remove the Slack package installed previously:

```
sudo snap remove slack
```

An application's internal user, configuration, and system data are saved and retained for 31 days. The files are called snapshots, they are archived and saved under `/var/lib/snapd/snapshots`, and contain the following types of files: a `.json` file containing a description of the snapshot, a `.tgz` file containing system data, and specific `.tgz` files with each system's user details. A short listing of the aforementioned directory will show the automatically created snapshot for Slack:

```
sudo ls /var/lib/snapd/snapshots/
```

If you don't want the snapshots to be created, you can use the `--purge` option for the `snap remove` command. For applications that use a large amount of data, those snapshots could have a significant size and impact the available disk space. To see the snapshots saved on your system, use the `snap saved` command:

```
packt@neptune:~$ sudo snap saved
Set  Snap   Age    Version  Rev  Size   Notes
1    slack  7m13s  4.8.0    27   123B   auto
```

Figure 3.12 – Showing the saved snapshots

The output shows the list, in our case, just the one app has been removed, with the first column indicating the ID of the snapshot (`set`). If you would like to delete a snapshot, you can do so by using the `snap forget` command. In our case, to delete the Slack application's snapshot, we can use the following command:

```
packt@neptune:~$ sudo snap saved
Set  Snap   Age    Version  Rev  Size   Notes
1    slack  23.2s  4.8.0    27   124B   auto
packt@neptune:~$ sudo snap forget 1
Snapshot #1 forgotten.
packt@neptune:~$ sudo snap saved
No snapshots found.
```

Figure 3.13 – Using the snap forget command to delete a snapshot

To verify that the snapshot was removed, we used the `snap saved` command again, as shown in *Figure 3.23*.

Snaps are really versatile packages and easy to use. This package type is the choice of Ubuntu developers, but they are not commonly used on other distributions. If you would like to install snaps on distributions other than Ubuntu, use the instructions from `https://snapcraft.io/docs/installing-snapd` and test its full capabilities.

Now we will go and test the other new kid on the block, the *flatpaks*. Our test distribution will be CentOS 8, but keep in mind that flatpaks are supported by default, out of the box, after installation, on Ubuntu-based distributions such as Linux Mint and elementary OS, Debian-based distributions such as PureOS and Endless OS, and Fedora.

Managing flatpak packages on CentOS

The same as snaps, flatpaks are isolated applications that run inside sandboxes. Each flatpak contains the needed runtimes and libraries for the application. Flatpaks offer full support for graphical user interface management tools, together with a full set of commands that can be used from the **Command Line Interface** (**CLI**). The main command is called flatpak, which has several other built-in commands to use for package management. To see all of these, use the following command:

```
$ flatpak - -help
```

In the following, we will detail some of the commonly used commands for flatpak package management. But before that, we will say a few lines about how flatpak apps are named and how they will appear on the command line. Each app has an identifier in a form similar to com.company.App. Each part of this is meant to easily identify an app and its developer. The final part identifies the application's name, as the preceding one identifies the entity that developed the app. This is an easy way for developers to publish and deliver multiple apps.

Adding repositories

Repositories must be set up in order to install applications. Flatpaks call repositories remotes, so this will be the term by which we will refer to them. The main repository for flatpaks is called flathub.

On our CentOS 8 machine, flatpak is already installed, but we will need to add the flathub repository. We will add it with the flatpak remote-add command, as in the following example:

```
$ sudo flatpak remote-add --if-not-exists flathub https://
dl.flathub.org/repo/flathub.flatpakrepo
```

We used the - -if-not-exists argument, which stops the command if the repository already exists, without showing an error. Once the repository is added, we can start installing packages from it.

Installing a flatpak application

To install a package, we need to know its name. We can either go to the website
https://flathub.org/home and search for apps there, or we could use the
flatpak search command to do that. We will use the same application, Slack, as for
the snaps. We will search for Slack as follows:

```
[packt@jupiter ~]$ flatpak search slack
Name          Description                         Application ID          Version  Branch  Remotes
Slack         Chat with your team                 com.slack.Slack         4.8.0    stable  flathub
Zulip         Zulip Desktop Client for Linux      org.zulip.Zulip         5.4.2    stable  flathub
Franz         Messenger for the desktop           com.meetfranz.Franz     5.5.0    stable  flathub
Rocket.Chat   Open Source Team Communication      chat.rocket.RocketChat  2.17.11  stable  flathub
```

Figure 3.14 – Using the flatpak search command to search for an application

The output of the command shows that Slack is available on flathub. Therefore, we
can proceed and install it. The first line shows the Slack we are looking for. To install it,
we will use the flatpak install command. In the preceding command, the name
of the repository (remote) was given as flathub, followed by the full name of the
application. Installation will ask for your approval to install the required runtime for the
application, and then the process continues. Use the following command to install Slack
from flathub and follow the onscreen messages:

```
sudo flatpak install flathub com.slack.Slack
```

On recent versions of flatpak (since version 1.2), installation can be performed with a
much simpler command. In this case, you only need the name of the app, as follows:

```
$ sudo flatpak install slack
```

The result is the same as using the first install command shown earlier.

Running, updating, removing, and listing flatpak applications

After installing an application, you can run it using the command line with the
following command:

```
$ flatpak run com.slack.Slack
```

If you want to update all the applications and runtimes, use this command:

```
$ sudo flatpak update
```

To remove a flatpak package, simply run the flatpak uninstall command:

```
$ sudo flatpak uninstall com.slack.Slack
```

To list all the installed flatpak applications and runtimes, use the `flatpak list` command:

```
[packt@jupiter ~]$ flatpak list
```

To see only the installed applications, we will use the `--app` argument:

```
[packt@jupiter ~]$ flatpak list --app
```

The commands shown above are the most commonly used for flatpak package management. Needless to say, there are many other commands that we will not cover here, but you are free to look them up and test them on your system.

Application streams in CentOS 8

Starting with RHEL 8 and CentOS 8, two principal repositories are available: the `BaseOS` and `AppStream` repositories. This was the solution found for an old problem faced by the RHEL and CentOS distributions for a long time: different life cycles for the operating system (`Kernel` and `glibc` libraries, and so on) and for the installed applications (tools, languages, and suchlike). RHEL and CentOS distributions are supported for 10 years, but this rarely happens for any other supported applications. Some of the installed applications change versions with a faster cadence every few months or so.

Before application streams were introduced, CentOS and RHEL used the software collections concept. It was similar to AppStreams in that it allowed users to install and use multiple versions of different applications on the same system. Application streams consist of components that can be either modules or rpm packages. They are delivered by default through the `AppStream` depository without any specific action required by the user. As described by the official RHEL documentation, modules are collections of packages representing a logical unit. A logical unit is represented by either an application, a language pack, database, or tool.

AppStreams are a great addition to the default repositories in CentOS. They solve a longstanding issue with application life cycles and, in our opinion, probably represent the best future-proof solution for system stability and well-tested application delivery.

Building a package from source

Building from source is an activity that is less and less needed on a day-by-day basis. Most of the distributions are already providing the most necessary packages for any task. The days of building applications from source are long gone. Repositories provide tens of thousands of packages that you can easily install.

The two scenarios where you would have to build from source would be: (1) if you need a legacy app that is no longer maintained and delivered with your current distribution; and (2) when you need to use an application developed in-house. If those scenarios apply, you will probably need to build from source.

Compiling a source package would require you to have proper development tools installed on your system. In the following example, we will use a simple script written in bash that shows the IP and network interface to the standard output. This would be our application developed in-house. Please keep in mind that the following is only a gentle introduction to building packages.

In this sub-chapter, we will package the app as an RPM for our CentOS 8 distribution, leaving the DEB packaging as an exercise for you to look for and try. As Ubuntu and Debian have larger repositories, the need to build from source for those distributions is minimal. You can also build a snap package for Ubuntu, as this is the trend nowadays. Find out more about this in the *Further reading* section.

The source code file

The in-house application is a bash script that has some simple lines of code to show us the IP address of the working system. Finding the IP address, extracting it, and using it is a common task for any system administrator. First of all, you need to know that there are several ways in which to find, extract, and use the IP; this is just one of them. The default command to show the IP address in modern Linux distributions is called `ip`.

The `ip` command is a powerful one and we do not intend to discuss it in detail here. We will only focus on the tools needed to show the IP address with our small bash script built in-house. Generally, to find the IP address of our local machine, we would use the following command:

```
ip addr show
```

The output of the `ip addr show` command is quite large and makes it difficult to find the IP address quickly. To automate it, you will need to separate the IP, which could be difficult. Another command that we could use is `ip route`.

As this is not the place to talk about networking, we will only give you a brief glimpse in order to understand the script we will use as our in-house built app. By default, the `ip route` command shows the IP routing table for your system, as shown in the following screenshot:

```
[packt@jupiter Documents]$ ip route
default via 192.168.0.1 dev enp0s25 proto dhcp metric 100
192.168.0.0/24 dev enp0s25 proto kernel scope link src 192.168.0.113 metric 100
192.168.122.0/24 dev virbr0 proto kernel scope link src 192.168.122.1 linkdown
```

Figure 3.15 – Showing the IP routing table using the ip route command

To get route information, we will use the `ip route get` command. For the sake of simplicity, we will use it to display the route taken to reach Google's default DNS, `8.8.8.8`. Remember that you can replace Google's DNS from the command with any IP from your network if you have different subnets in use. Furthermore, we will use the `awk` command to extract only the relevant IP address. We will use the awk's `-F` option to determine the field separator, and the NR variable to show the total number of records being processed.

The `ip route get` command's output is as follows:

```
[packt@jupiter Documents]$ ip route get 8.8.8.8
8.8.8.8 via 192.168.0.1 dev enp0s25 src 192.168.0.113 uid 1000
    cache
```

Figure 3.16 – The ip route get output

By analyzing the output, we see that the IP is shown after the `src` string. First is shown the DNS address, the default gateway IP, the interface name after the `dev` string, and the system's IP address, followed by the UID.

Now we will create the script inside a locally saved file. We will go to the `/Documents` directory and create a new file called `ip-script`, using the following commands:

```
[packt@jupiter ~]$ cd Documents/
[packt@jupiter Documents]$ vim ip-script _
```

Figure 3.17 – Creating the ip-script file

As we said earlier, we will use `awk` to extract the IP and interface name. The code of the bash script will look similar to the one in the following screenshot:

```
#!/bin/bash
printf "your IP is: \n"
ip route get 8.8.8.8 | awk -F"src " 'NR==1{split($2,ip," ");print ip[1]}'
printf "your interface name is: \n"
ip route get 8.8.8.8 | awk -F"dev " 'NR==1{split($2,inter," ");print inter[1]}'
```

Figure 3.18 – Scripting code used in our app built in-house

A short explanation of the preceding code is surely needed, as this is your first encounter with a bash script inside this book. Each bash script begins with a first line that defines the environment that would run the script. That first line always starts with the # character followed by ! (the shebang) and the path of the environment. As this will be a bash script, the path to the /bin/bash #!/bin/bash. This will ensure that the command-line interpreter knows which language is used to read the file.

The second line invokes the printf command, which prints directly to the standard output (the screen). The message that will be displayed on the screen is written between quotations marks: "your IP is: \n". The \n represents the newline character. It will bring the command line's cursor to a new line after the message is displayed.

The third line is the one that really displays the IP from the ip route get command. This redirects the ip route get command's output to the awk command, which is the one that extracts only the IP address from it. As we have already shown you the power of awk in *Chapter 2*, *The Linux Filesystem*, we'd expect that you already have a certain knowledge of it.

In our in-house developed app, awk is used with the following options:

- The -F option is used to determine the field separator that awk will first look for in the command's output. As we noticed that the IP is shown after the src string in the ip route get command's output, we will use src as the field separator, followed by a blank space: -F"src "

- Then we will use the NR built-in variable, which shows the total number of records being processed. As we only have one line, we assign NR==1 followed by the extraction command

- To extract the IP, we use the split function of awk. Its syntax is as follows: split(SOURCE, DESTINATION, DELIMITER). In our case, the source is $2, which represents the second column. As you may already know, awk is considering files as composed of lines and columns. A column is defined as the characters surrounded by spaces. Thus, starting from the field separator, the second column is the string between spaces representing the IP address. The destination is a variable we called ip, and the delimiter is a space:

```
'NR==1{split($2,ip," ");print ip[1]}'
```

- In the end, we print the destination using the print ip[1] command. As the NR variable is limited to one record, as there is only one line, the command will refer to it. Therefore, the entire line of code is as follows:

```
ip route get 8.8.8.8 | awk -F"src " 'NR==1{split($2,ip," ");print ip[1]}'
```

By running the script locally, you will have the following output, which is exactly what we are looking for. To run the script, we will first need to make it executable and then run it:

```
[packt@jupiter Documents]$ chmod +x ip-script
[packt@jupiter Documents]$ ./ip-script
your IP is:
192.168.0.113
your interface name is:
enp0s25
```

Figure 3.19 – Running the script after making it executable

Now that the source code is working according to our needs, we can start packaging it and make it ready for distribution. In the following, we will add an open source license and package it as an RPM.

Preparing the source code

As a common rule, software should always be distributed with a software license. This license is written inside a LICENSE file that accompanies the software. For our in-house app, we will use the GPLv3 license. We will create a LICENSE file with the following text inside:

```
This program is free software: you can redistribute it and/or modify
it under the terms of the GNU General Public License as published by
the Free Software Foundation, either version 3 of the License, or
(at your option) any later version.

This program is distributed WITHOUT ANY WARRANTY;
without even the implied warranty of MERCHANTABILITY
or FITNESS FOR A PARTICULAR PURPOSE.  See the
GNU General Public License for more details.

You should have received a copy of the GNU General Public License
along with this program.  If not, see <http://www.gnu.org/licenses/>.
```

Figure 3.20 – The GPLv3 license for our app

In the following, we need to have all the files in a single directory and archive them:

1. Move all the files to a single directory and archive the app for distribution:

```
[packt@jupiter Documents]$ ls ip-app-0.1
LICENSE  ip-app
```

Figure 3.21 – Moving all files into one directory

2. Now that all the files are in the same directory, you can archive them in a single tarball:

```
[packt@jupiter Documents]$ tar -czvf ip-app-0.1.tar.gz ip-app-0.1
ip-app-0.1/
ip-app-0.1/LICENSE
ip-app-0.1/ip-app
```

Figure 3.22 – Archiving the app in a single tarball

Next, we will need to prepare and set up our CentOS environment for RPM building. We will need to make sure that all the additional tools are installed.

Setting up the environment

First of all, we need to make sure that `rpm build` is installed on the system. To check whether it is installed, we can use the `rpmbuild - -showrc` command. The output will be a large amount of data with details of the build environment. If you get an error saying *command not found*, you will have to install it using the following command:

```
$ sudo yum install rpm-build
```

Before building the RPM, take the following into consideration.

> **Important note**
> Building RPMs should always be done using a regular, unprivileged user. It should never be done using the root user!

Next, we will create the RPM building directories.

Creating RPM building directories

After installing `rpmbuild`, you need to create the files and directory structure inside your home directory. Then you will have to create the `.rpmmacros` file to override any default location settings. To do this automatically, install the `rpmdevtools` package and then run the `rpmdev-setuptree` command:

```
[packt@jupiter ~]$ sudo yum install rpmdevtools
[sudo] password for packt:
Last metadata expiration check: 2:31:42 ago on Sun 13 Sep 2020 11:22:01 AM EEST.
Package rpmdevtools-8.10-7.el8.noarch is already installed.
Dependencies resolved.
Nothing to do.
Complete!
[packt@jupiter ~]$ rpmdev-setuptree
[packt@jupiter ~]$ ls
Desktop  Documents  Downloads  Music  Pictures  Public  Templates  Videos  rpmbuild
```

Figure 3.23 – Installing rpmdevtools and running rpmdev-setuptree

If you run the `ls  -la` command, you will see that the `.rpmmacros` file was automatically created inside your home directory. Also, the structure of the `rpmbuild` directory has already been created:

```
[packt@jupiter ~]$ ls rpmbuild/
BUILD  RPMS  SOURCES  SPECS  SRPMS
[packt@jupiter ~]$ ls -la .rpmmacros
-rw-rw-r--. 1 packt packt 228 Sep 13 13:53 .rpmmacros
```

Figure 3.24 – Structure of the rpmbuild directory

The `rpmdevtools` package we installed earlier provides some useful utilities for RPM packaging. Those utilities can be listed with the following command:

```
rpm -ql rpmdevtools | grep bin
```

As indicated, the RPM packaging workspace consists of five directories, each having a specific purpose:

- `BUILD` – `%buildroot` directories are created when a package is built.
- `RPMS` – Contains subdirectories for different architectures that have binary RPMs.
- `SOURCES` – Contains the compressed source code archives.
- `SPECS` – The location where the `SPEC` files are stored.
- `SRPMS` – When creating an SRPM instead of a binary RPM, the files are stored here.

In the following, we will detail how to edit a `SPEC` file for the application.

Defining a SPEC file

A `SPEC` file is a sort of a recipe used by the `rpmbuild` tool to create the RPM. It contains instructions in several sections defined as *Preamble* and *Body*. To create a new `SPEC` file, use the `rpmdev-newspec` command:

```
[packt@jupiter SPECS]$ rpmdev-newspec ip-app
ip-app.spec created; type minimal, rpm version >= 4.14.
```

Figure 3.25 – The rpmdev-newspec command to create a new SPEC file

If we examine the file with the `cat` command, we will see that it contains several directives. Now you can edit the `SPEC` file according to your app. Here are the details that we use for our app. The details are available inside the books GitHub repository.

The next thing to do is to copy the previously created tarball inside your ~/rpmbuild/ SOURCES/ directory using the following command:

```
$ cp ~/Documents/ip-app-0.1.tar.gz ~/rpmbuild/SOURCES/
```

In the preceding example, we have used the absolute paths for the file, and you should change these depending on your file's destination.

The next step is to build the application for use on any RPM-based distribution, such as Fedora, CentOS, or RHEL.

Building the source and binary RPM from the SPEC file

Now that the SPEC file has been created, we can build the RPM file using the rpmbuild command with the -bs and -bb options. The -bs option stands for *build source*, and the -bb option stands for *build binary*.

To build a source RPM, use the following command:

```
[packt@jupiter SPECS]$ rpmbuild -bs ip-app.spec
Wrote: /home/packt/rpmbuild/SRPMS/ip-app-0.1-1.el8.src.rpm
```

Figure 3.26 – Building a source RPM

To build a binary RPM, there are two different scenarios. One is to rebuild it from the SRPM (Source RPM), and the second is to build it from a SPEC file. To rebuild from a source RPM, use the following command:

```
$ rpmbuild -bs ip-app.spec
```

To build a package from the SPEC file, use the following command:

```
$ rpmbuild -bb ip-app.spec
```

Now the RPM is built. The location is /home/packt/rpmbuild/RPMS/noarch:

```
[packt@jupiter rpmbuild]$ ls
BUILD BUILDROOT RPMS SOURCES SPECS SRPMS
[packt@jupiter rpmbuild]$ cd RPMS/
[packt@jupiter RPMS]$ ls
noarch
[packt@jupiter RPMS]$ cd noarch/
[packt@jupiter noarch]$ ls
ip-app-0.1-1.el8.noarch.rpm
```

Figure 3.27 – Location of the newly built RPM

Building an RPM is a complicated task. What we showed you is a very simple step-by-step approach to creating an RPM file for a bash shell script that shows the IP and interface name. If you would like to build specific C or Python applications that have specific dependencies, the task will be a little more complicated.

Summary

In this chapter, you learned how to work with packages in Ubuntu and CentOS, but the skills learned will help you to manage packages in any Linux distribution. You have learned how to work with both `.deb` and `.rpm` packages, and also the newer ones such as flatpaks and snaps. You will use those skills in every chapter of the book, as well as in your day job as a systems administrator.

Furthermore, we tackled with the process of building an RPM package by walking you through the process of creating an `rpm` file from a simple bash script.

In the next chapter, we will show you how to manage user accounts and permissions, when you will be introduced to general concepts and specific tools.

Questions

Now that you have a clear idea of how to manage software packages, here are some exercises that will contribute further to your learning.

1. Make a list of all the packages installed on your system.
2. Find the sources of an open source program that you love and build it from source.
3. Add support for flatpaks on your Ubuntu system.
4. Add support for snaps on your CentOS system.
5. Test other distributions and use their package managers. We recommend that you try openSUSE and, if you feel confident, Arch Linux.

No hints this time, as we want you to discover all the ups and downs of package management on Linux.

Further reading

For more information about what was covered in this chapter, please refer to the following links:

- Red Hat 8 documentation: `https://access.redhat.com/documentation/en-us/red_hat_enterprise_linux/8/html/installing_managing_and_removing_user-space_components/index`

- Snapcraft.io official documentation: `https://snapcraft.io/docs`

- Flatpak documentation: `https://docs.flatpak.org/en/latest/`

4
Managing Users and Groups

Linux is a multiuser, multitasking operating system, which means multiple users can access the operating system at the same time while sharing platform resources, with the kernel performing tasks for each user concurrently and independently. Linux provides the required isolation and security mechanisms to avoid multiple users accessing or deleting each other's files.

When multiple users are accessing the system, permissions come into play. We'll learn how **permissions** work in Linux, with their essential read, write, and execution tenets. We'll introduce you to the concept of a *superuser* (`root`) account, with complete access to the operating system resources.

Along the way, we'll take a hands-on approach to the topics learned, further deepening the assimilation of key concepts through practical examples. This chapter covers the following topics:

- Managing users
- Managing groups
- Managing permissions

We hope that by the end of the chapter, you will be comfortable with the command-line utilities for creating, modifying, and deleting users and groups, while proficiently handling file and directory permissions.

Let's take a quick look at the technical requirements necessary for the study of this chapter.

Technical requirements

We need a working Linux distribution installed on either a **virtual machine** (**VM**) or a desktop platform. In case you don't have one already, *Chapter 1*, *Installing Linux*, takes you through the related process. In this chapter, we'll be using Ubuntu or CentOS, but most of the commands and examples used would pertain to any other Linux platform.

Managing users

In this context, a user is anyone using a computer or a system resource. In its simplest form, a Linux *user* or *user account* is identified by a name and a **unique identifier**, known as a **UID**.

From a purely technical point of view, in Linux we have the following types of users:

- **Normal** (or regular) users—General-purpose, everyday user accounts, mostly suited for personal use and for common application and file management tasks, with limited access to system-wide resources. A regular user account usually has a *login* shell and a *home* directory.

- **System** users—These are similar to regular user accounts, except they may lack a login shell or a home directory. System accounts are usually assigned to background application services, mostly for security reasons and to limit the attack surface associated with the related resources—for example, a web server daemon handling public requests should run as a system account, ideally without login or root privileges. Consequently, possible vulnerabilities exposed through the web server would remain strictly isolated to the limited action realm of the associated system account.

- **Superusers**—These are privileged user accounts, with full access to system resources, including the permission to create, modify, and delete user accounts. The root user is an example of a superuser.

In Linux, only the root user or users with sudo privileges (**sudoers**) can create, modify, or delete user accounts.

Understanding sudo

The root user is the default superuser account in Linux, and it has the ability to do anything on a system. Ideally, acting as root on a system should generally be avoided due to safety and security reasons. With sudo, Linux provides a mechanism of *promoting* a regular user account to superuser privileges, using an additional layer of security. This way, a sudo user is generally used instead of root.

sudo is a command-line utility that allows a permitted user to execute commands with the security privileges of a superuser or another user (depending on the local system's security policy). sudo originally stood for *superuser do* due to its initial implementation of acting exclusively as the superuser, but has since been expanded to support not only the superuser but also other (restricted) user impersonations. Thus, it is also referred to as *substitute user do*. Yet, more often than not, it is perceived as *superuser do* due to its frequent use in Linux administrative tasks.

Most of the command-line tools for managing users in Linux require sudo privileges, unless the related tasks are carried out by the root user. If we want to avoid using the root context, we can't genuinely proceed with the rest of this chapter—and create a user in particular—before we have a user account with superuser privileges. So, let's take this chicken-egg scenario out of the way first.

Most Linux distributions create an additional user account with superuser privileges, besides root, during installation. The reason, as noted before, is to provide an extra layer of security and safety for elevated operations. The simplest way to check if a user account has sudo privileges is to run the following command in a terminal, while logged in with the related user account:

```
sudo -v
```

According to the sudo manual (man sudo), the -v option causes sudo to update the user's cached credentials and authenticate the user if the cached credentials expired.

If the user (for example, julian) doesn't have superuser privileges on the local machine (for example, neptune), the preceding command yields the following (or a similar) error:

```
Sorry, user julian may not run sudo on neptune.
```

In recent Linux distributions, the execution of a sudo command usually grants elevated permissions for a limited time. Ubuntu, for example, has a 15-minute sudo elevation span, after which time a sudo user would need to authenticate again. Subsequent invocations of sudo may not prompt for a password if done within the sudo cache credential timeout.

If we don't have a default superuser account, we can always use the root context to create new users (see the next chapter) and elevate them to **sudoer** privileges. We'll learn more about this in the *Creating a superuser* section, later in this chapter.

Now, let's have a look at how to create, modify, and delete users.

Creating, modifying, and deleting users

In this section, we explore a few command-line tools and some common tasks for managing users. The example commands and procedures are shown for Ubuntu and CentOS, but the same principles apply for any other Linux distribution. Some user management **command-line interface** (**CLI**) tools may differ or may not be available on specific Linux platforms (for example, useradd is not available on Alpine Linux and adduser should be used instead). Please check the documentation of the Linux distribution of your choice for the equivalent commands.

Creating users

To create users, we can use either the useradd or the adduser command, although on some Linux distributions (for example, Debian or Ubuntu), the recommended way is to use the adduser command in favor of the low-level useradd utility. We'll cover both in this section.

adduser is a Perl script using useradd—basically a shim of the useradd command—with a user-friendly guided configuration. Both command-line tools are installed by default in Ubuntu and CentOS. Let's take a brief look at each of these commands.

Creating users with useradd

The syntax for the useradd command is shown here:

```
useradd [OPTIONS] USER
```

In its simplest invocation, the following command creates a user account (julian):

```
sudo useradd julian
```

The user information is stored in a /etc/passwd file. Here's the related user data for julian:

```
cat /etc/passwd | grep julian
```

In our case, this is the output:

```
packt@neptune:~$ cat /etc/passwd | grep julian
julian:x:1001:1001::/home/julian:/bin/sh
```

Figure 4.1 – The user record created with useradd

Let's analyze the related user record. Each entry is delimited by a colon (:) and is listed here:

- julian: The username
- x: The encrypted password (the password hash is stored in /etc/shadow)
- 1001: The UID
- 1001: The user **group ID (GID)**
- The (in our case, empty) **General Electric Comprehensive Operating Supervisor (GECOS)** field (for example, display name), explained next
- /home/julian: The user home folder
- /bin/sh: The default login shell for user

> **Important note**
> The GECOS field is a string of a comma-delimited attributes, reflecting general information about the user account (for example, real name; company; phone number). In Linux, the GECOS field is the fifth field in a user record. See more information at https://en.wikipedia.org/wiki/Gecos_field.

We can also use the getent command to retrieve the preceding user information, as follows:

```
getent passwd julian
```

To view the UID (uid), GID (gid), and group membership associated with a user, we can use the id command, as follows:

```
id julian
```

This command gives us the following output:

```
packt@neptune:~$ id julian
uid=1001(julian) gid=1001(julian) groups=1001(julian)
```

Figure 4.2 – The UID information

With the simple invocation of useradd, the command creates the user (julian) with some immediate default values (as enumerated), while other user-related data is empty—for example, we have no full name or password specified for the user yet. Also, while the home directory has a default value (for example, /home/julian), the actual filesystem folder would not be created unless the useradd command is invoked with the -m or --create-home option, as follows:

```
sudo useradd -m julian
```

Without a home directory, regular users would not have the ability to save their files in a private location on the system. On the other hand, some system accounts may not need a home directory since they don't have a login shell. For example, a database server (for example, PostgreSQL) may run with a non-root system account (for example, postgres) that only needs access to database resources in specific locations (for example, /var/lib/pgsql), controlled via other permission mechanisms (for example, **Security-Enhanced Linux (SELinux)**).

For our regular user, if we also wanted to specify a full name (display name), the command would change to this:

```
sudo useradd -m -c "Julian" julian
```

The -c, --comment option parameter of useradd expects a *comment*, also known as the **GECOS** field (the fifth field in our user record), with multiple comma-separated values. In our case, we specify the full name (for example, Julian). For more information, check out the useradd manual (man useradd) or useradd --help.

The user still won't have a password yet, and consequently, there would be no way for the user to log in (for example, via a **graphical user interface (GUI)** or **Secure Shell (SSH)**). To create a password for julian, we invoke the passwd command, like this:

```
sudo passwd julian
```

You should see the output, as follows:

```
packt@neptune:~$ sudo passwd julian
New password:
Retype new password:        The command above w
passwd: password updated successfully
```

Figure 4.3 – Creating or changing the user password

The `passwd` command will prompt for the new user's password. With the password set, there will be a new entry added to the `/etc/shadow` file. This file stores the secure password hashes (not the passwords!) for each user. Only superusers can access the content of this file. Here's the command to retrieve the related information for the user `julian`:

```
sudo getent shadow julian
```

You can also use the following command:

```
sudo cat /etc/shadow | grep julian
```

Once the password has been set, in normal circumstances the user can log in to the system (via SSH or GUI). If the Linux distribution has a GUI, the new user will show up on the login screen. See the following screenshot for Ubuntu:

Figure 4.4 – The new user login in Ubuntu

As noted, with the `useradd` command we have low-level granular control over how we create user accounts, but sometimes we may prefer a more user-friendly approach. Enter the `adduser` command.

Creating users with adduser

The `adduser` command is a Perl wrapper for `useradd`. The syntax for the `adduser` command is shown here:

```
adduser [OPTIONS] USER
```

sudo may prompt for the superuser password. adduser will prompt for the new user's password and other user-related information (as shown in *Figure 4.5*).

Let's create a new user account (julian) with adduser, as follows:

```
sudo adduser julian
```

The preceding command yields the following output:

```
packt@neptune:~$ sudo adduser julian
Adding user `julian' ...
Adding new group `julian' (1001) ...
Adding new user `julian' (1001) with group `julian' ...
Creating home directory `/home/julian' ...
Copying files from `/etc/skel' ...
New password:
Retype new password:
passwd: password updated successfully
Changing the user information for julian
Enter the new value, or press ENTER for the default
        Full Name []: Julian
        Room Number []:
        Work Phone []:
        Home Phone []:
        Other []:
Is the information correct? [Y/n] Y
```

Figure 4.5 – The adduser command

In CentOS, the preceding invocation of the adduser command would simply run without prompting the user for a password or any other information.

We can see the related user entry in /etc/passwd with getent, as follows:

```
getent passwd julian
```

This is the output:

```
packt@neptune:~$ getent passwd julian
julian:x:1001:1001:Julian,,,:/home/julian:/bin/bash
```

Figure 4.6 – Viewing user information with getent

In the preceding examples, we created a regular user account. Administrators or superusers can also elevate the privileges of a regular user to a superuser. Let's see how.

Creating a superuser

When a regular user is given the power to run `sudo`, they become a superuser. Let's assume we have a regular user created via any of the examples shown in the *Creating users* section.

Promoting the user to a superuser (or *sudoer*) requires a `sudo` group membership. In Linux, the `sudo` group is a reserved system group for users with elevated or `root` privileges. To make the user `julian` a sudoer, we simply need to add the user to the `sudo` group, like this:

```
sudo usermod -aG sudo julian
```

The `-aG` options of `usermod` instruct the command to append (`-a`, `--append`) the user to the specified group (`-G`, `--group`)—in our case, `sudo`.

To verify our user is now a sudoer, first make sure the related user information reflects the `sudo` membership by running the following command:

```
id julian
```

This gives us the following output:

```
packt@neptune:~$ id julian
uid=1001(julian) gid=1001(julian) groups=1001(julian),27(sudo),
1200(developers),1300(devops)
```

Figure 4.7 – Looking for the sudo membership of a user

The output shows that the `sudo` group membership (GID) in the `groups` tag is `27(sudo)`.

To verify the `sudo` access for the user `julian`, run the following command:

```
su - julian
```

The preceding command prompts for the password of the user `julian`. A successful login would usually validate the superuser context. Alternatively, the user (`julian`) can run the `sudo -v` command in their terminal session to validate the `sudo` privileges. For more information on superuser privileges, see the *Understanding sudo* section earlier in the chapter.

With multiple users created, a system administrator may want to view or list all the users in the system. In the next section, we provide a few ways to accomplish this task.

Viewing users

There are a few ways for a superuser to view all users configured in the system. As previously noted, the user information is stored in the /etc/passwd and /etc/shadow files. Besides simply viewing these files, we can parse them and extract only the usernames with the following command:

```
cat /etc/passwd | cut -d: -f1 | less
```

Alternatively, we can parse the /etc/shadow file, like this:

```
sudo cat /etc/shadow | cut -d: -f1 | less
```

In the preceding commands, we read the content from the related files (with cat). Next, we piped the result to a delimiter-based parsing (with cut, on the : delimiter) and picked the first field (-f1). Finally, we chose a paginated display of the results, using the less command.

Note the use of sudo for the shadow file, since access is limited to superusers only, due to the sensitive nature of the password hash data. Alternatively, we can use the getent command to retrieve the user information.

The following command lists all users configured in the system:

```
getent passwd
```

The preceding command reads the /etc/passwd file. Alternatively, we can retrieve the same information from /etc/shadow, as follows:

```
sudo getent shadow
```

For both commands, we can further pipe the getent output to | cut -d: -f1, to list only the usernames, like this:

```
sudo getent shadow | cut -d: -f1 | less | column
```

The output is similar to this (excerpt):

```
packt@neptune:~$ sudo getent shadow | cut -d: -f1 | less | column
root                    news                    systemd-network
daemon                  uucp                    systemd-resolve
bin                     proxy                   systemd-timesync
sys                     www-data                messagebus
```

Figure 4.8 – Viewing usernames

With new users created, administrators or superusers may want to change certain user-related information, such as password, password expiration, full name, or login shell. Next, we take a look at some of the most common ways to accomplish this task.

Modifying users

A superuser can run the `usermod` command to modify user settings, with the following syntax:

```
usermod [OPTIONS] USER
```

The examples in this section apply to a user we previously created (`julian`) with the simplest invocation of the `useradd` command. As noted in the previous section, the related user record in `/etc/passwd` has no full name for the user, and the user has no password either.

Let's change the following settings for our user (`julian`):

- Full name: To `Julian` (initially empty).
- Home folder: Move to `/local/julian` (from default `/home/julian`).
- Login shell: `/bin/bash` (from default `/bin/sh`).

The command-line utility for changing all the preceding information is shown here:

```
sudo usermod -c "Julian" -d /local/julian -m -s /bin/bash
julian
```

Here are the command options, briefly explained:

- `-c, --comment "Julian"`: The full username.
- `-d, --home local/julian`: The user's new home directory.
- `-m, --move`: Move the content of the current home directory to the new location.
- `-s, --shell /bin/sh`: The user login shell.

The related change, retrieved with the `getent` command, is shown here:

```
getent passwd julian
```

We get the following output:

```
packt@neptune:~$ getent passwd julian
julian:x:1001:1001:Julian:/local/julian:/bin/bash
```

Figure 4.9 – The user changes reflected with getent

Here are a few more examples of changing user settings with the `usermod` command-line utility.

Changing the username

The `-l`, `--login` option parameter of `usermod` specifies a new login username. The following command changes the username from `julian` to `balog` (that is, first name to last name), as illustrated here:

```
sudo usermod -l "balog" julian
```

In a production environment, the preceding command could be more involved, as we may also want to change the display name and the home directory of the user (for consistency reasons). In a previous example in the *Creating users with useradd* section, we showcased the `-d`, `--home` and `-m`, `--move` option parameters, which would accommodate such changes.

Locking or unlocking a user

A superuser or administrator may choose to temporarily or permanently lock a specific user with the `-L`, `--lock` option parameter of `usermod`, as follows:

```
sudo usermod -L julian
```

As a result of the preceding command, the login attempt for user `julian` would be denied. Should the user try to SSH into the Linux machine, they would get a **Permission denied, please try again** error message. Also, the related username will be removed from the login screen if the Linux platform has a GUI.

To unlock the user, we invoke the `-U`, `--unlock` option parameter, as follows:

```
sudo usermod -U julian
```

The preceding command restores system access for the user.

For more information on the `usermod` utility, please check out the related documentation (`man usermod`) or the command-line help (`usermod --help`).

Although the recommended way of modifying user settings is via the `usermod` command-line utility, some users may find it easier to manually edit the `/etc/passwd` file. The following section shows how.

Modifying users via /etc/passwd

A superuser can also manually edit the `/etc/passwd` file to modify user data by updating the related line. Although the editing can be done with a text editor of your choice (for example, `nano`), we recommend use of the `vipw` command-line utility for a safer approach. `vipw` enables the required locks to prevent possible data corruption—for example, in case a superuser performs a change at the same time regular users change their password.

The following command initiates the editing of the `/etc/passwd` file by also prompting for the preferred text editor (for example, `nano` or `vim`):

```
sudo vipw
```

For example, we can change the settings for user `julian` by editing the following line:

```
julian:x:1001:1001:Julian,,,:/home/julian:/bin/bash
```

The meaning of the colon (`:`)-separated fields have been previously described in the *Creating users with useradd* section. Each of these fields can be manually altered in the `/etc/passwd` file, resulting in changes equivalent to the corresponding `usermod` invocation.

For more information on the `vipw` command-line utility, you may refer to the related system manual (`man vipw`).

Another relatively common administrative task for a user account is to change a password or set up a password expiration. Although `usermod` can change a user password via the `-p` or `--password` option, it requires an encrypted hash string (and not a cleartext password). Generating an encrypted password hash would be an extra step. An easier way is to use the `passwd` utility.

A superuser (administrator) can change the password of a user (for example, `julian`) with the following command:

```
sudo passwd julian
```

Sometimes, administrators are required to remove specific users from the system. The next section shows a couple of ways of accomplishing this task.

Deleting users

The most common way to remove users from the system is to use the `userdel` command-line tool. The general syntax of the `userdel` command is shown here:

```
userdel [OPTIONS] USER
```

For example, to remove the user `julian`, a superuser would run the following command:

```
sudo userdel -f -r julian
```

Here are the command options, briefly explained:

- `-f`, `--force`: Removes all files in the user's home directory, even if not owned by the user
- `-r`, `--remove`: Removes the user's home directory and mail spool

The `userdel` command removes the related user data from the system, including the user's home directory (when invoked with the `-f` or `--force` option) and the related entries in the `/etc/passwd` and `/etc/shadow` files.

There is also an alternative way, which could be handy in some odd cleanup scenarios. The next section shows how.

Deleting users via /etc/passwd and /etc/shadow

A superuser can edit the `/etc/passwd` and `/etc/shadow` files and manually remove the corresponding lines for the user (for example, `julian`). Please note that both files have to be edited for consistency and complete removal of the related user account.

Edit the `/etc/passwd` file using the `vipw` command-line utility, as follows:

```
sudo vipw
```

Remove the following line (for user `julian`):

```
julian:x:1001:1001:Julian,,,:/home/julian:/bin/bash
```

Next, edit the `/etc/shadow` file using the `-s` or `--shadow` option with `vipw`, as follows:

```
sudo vipw -s
```

Remove the following line (for user `julian`):

```
julian:$6$xDdd7Eay/RKYjeTm$Sf.../:18519:0:99999:7:::
```

After editing the preceding files, a superuser may also need to remove the deleted user's home directory, as follows:

```
sudo rm -rf /home/julian
```

For more information on the `userdel` utility, please check out the related documentation (`man userdel`) or the command-line help (`userdel --help`).

The user management concepts and commands learned so far apply exclusively to individual users in the system. When multiple users in the system have a common access level or permission attribute, they are collectively referred to as a group. Groups can be regarded as standalone organizational units we can create, modify, or delete. We can also define and alter user memberships associated with groups. The next section focuses on group management internals.

Managing groups

Linux uses groups to organize users. Simply put, a group is a collection of users sharing a common attribute. Examples of such groups could be *employees*, *developers*, *managers*, and so on. In Linux, a group is uniquely identified by a GID. Users within the same group share the same GID.

From a user's perspective, there are two types of groups, outlined here:

- **Primary group**—The user's initial (default) login group
- **Supplementary groups**—A list of groups the user is also a member of; also known as **secondary groups**

Every Linux user is a member of a primary group. A user can belong to multiple supplementary groups or no supplementary groups at all. In other words, there is one mandatory primary group associated with each Linux user, and a user can have multiple or no supplementary group memberships.

From a practical point of view, we can look at groups as a permissive context of collaboration for a select number of users. Imagine a *developers* group having access to developer-specific resources. Each user in this group has access to these resources. Users outside the *developers* group may not have access unless they authenticate with a group password, if the group has one.

In the following section, we provide detailed examples of how to manage groups and set up group memberships for users. Most related commands require *superuser* or `sudo` privileges.

Creating, modifying, and deleting groups

While our primary focus remains on group administrative tasks, some related operations still involve user-related commands. Command-line utilities such as `groupadd`, `groupmod`, and `groupdel` are targeted strictly at creating, modifying, and deleting groups, respectively. On the other hand, the `useradd` and `usermod` commands carry group-specific options when associating users with groups. We'll also introduce you to `gpasswd`, a command-line tool specializing in group administration, combining user- and group-related operations.

With this aspect in mind, let's take a look at how to create, modify, and delete groups and how to manipulate group memberships for users.

Creating groups

To create a new group, a superuser invokes the `groupadd` command-line utility. Here's the basic syntax of the related command:

```
groupadd [OPTIONS] GROUP
```

Let's create a new group (`developers`), with default settings, as follows:

```
sudo groupadd developers
```

The group information is stored in the `/etc/group` file. Here's the related data for the `developers` group:

```
cat /etc/group | grep developers
```

The command yields the following output:

```
packt@neptune:~$ cat /etc/group | grep developers
developers:x:1002:
```

Figure 4.10 – The group with default attributes

Let's analyze the related group record. Each entry is delimited by a colon (`:`) and is listed here:

- `developers`: Group name
- `x`: Encrypted password (password hash is stored in `/etc/gshadow`)
- `1002`: GID

We can also use the `getent` command to retrieve the preceding group information, as follows:

```
getent group developers
```

A superuser may choose to create a group with a specific GID, using the `-g, --gid` option parameter with `groupadd`. For example, the following command creates the `developers` group with a GID of `1200`:

```
sudo groupadd -g 1200 developers
```

For more information on the `groupadd` command-line utility, please refer to the related documentation (`man groupadd`).

Group-related data is stored in the `/etc/group` and `/etc/gshadow` files. The `/etc/group` file contains generic group membership information, while the `/etc/gshadow` file stores the encrypted password hashes for each group.

Let's take a brief look at group passwords.

Understanding group passwords

By default, a group doesn't have a password when created with the simplest invocation of the `groupadd` command (for example, `groupadd developers`). Although `groupadd` supports an encrypted password (via the `-p, --password` option parameter), this would require an extra step to generate the secure password hash. There's a better and simpler way to create a group password: by using the `gpasswd` command-line utility.

The following command creates a password for the `developers` group:

```
sudo gpasswd developers
```

We get prompted to enter and re-enter a password.:

`gpasswd` is a command-line tool that helps with everyday group administration tasks.

The purpose of a group password is to protect access to group resources. A group password is inherently insecure when shared among group members, yet a Linux administrator may choose to keep the group password private while group members collaborate unhindered within the group's security context.

Here's a quick explanation of how it works. When a member of a specific group (for example, `developers`) logs in to that group (using the `newgrp` command), the user is not prompted for the group password. When users who don't belong to the group attempt to log in, they would be prompted for the group password.

In general, a group can have administrators, members, and a password. Members of a group who are the group's administrators may use `gpasswd` without being prompted for a password, as long as they're logged in to the group. Also, group administrators don't need superuser privileges to perform group administrative tasks for a group they are the administrator of.

We'll take a closer look at `gpasswd` in the next sections, where we further focus on group management tasks, as well as adding users to a group and removing users from a group. But for now, let's keep our attention strictly at the group level and see how we can modify a user group.

Modifying groups

The most common way to modify the definition of a group is via the `groupmod` command-line utility. Here's the basic syntax for the command:

```
groupmod [OPTIONS] GROUP
```

The most common operations when changing a group's definition are related to the GID, group name, and group password. Let's take a look at each of these changes. We assume our previously created group is named `developers`, with a GID of `1200`. To change the GID to `1002`, a superuser invokes the `groupmod` command with the `-g,` `--gid` option parameter, as follows:

```
sudo groupmod -g 1002 developers
```

To change the group name from `developers` to `devops`, we invoke the `-n,` `--new-name` option, like this:

```
sudo groupmod -n devops developers
```

We can verify the preceding changes for the `devops` group with the following command:

```
getent group devops
```

The command yields the following:

```
packt@neptune:~$ getent group devops
devops:x:1002:
```

Figure 4.11 – Verifying the group changes

To change the group password for `devops`, the simplest way is to use `gpasswd`, as follows:

```
sudo gpasswd devops
```

We are prompted to enter and re-enter a password.:

To remove the group password for `devops`, we invoke the `gpasswd` command with the `-r, --remove-password` option, as follows:

```
sudo gpasswd -r devops
```

For more information on `groupmod` and `gpasswd`, refer to the system manuals of these utilities (`man groupmod` and `man gpasswd`), or simply invoke the `-h, --help` option for each.

Next, we look at how to delete groups.

Deleting groups

To delete groups, we use the `groupdel` command-line utility. The related syntax is shown here:

```
groupdel [OPTIONS] GROUP
```

By default, Linux enforces referential integrity between a primary group and the users associated with that primary group. We cannot delete a group that has been assigned as a primary group for some users before deleting the users of that primary group. In other words, by default, Linux doesn't want to leave the users with dangling primary GIDs.

For example, assuming user `julian` has the primary group set to `devops`, attempting to delete the `devops` group results in an error, as can be seen here:

```
packt@neptune:~$ sudo groupdel devops
groupdel: cannot remove the primary group of user 'julian'
```

Figure 4.12 – Attempting to delete a primary group

A superuser may choose to *force* the deletion of a primary group, invoking `groupdel` with the `-f, --force` option, but this would be ill-advised. The command would result in users with orphaned primary GIDs and a possible security hole in the system. The maintenance and removal of such users would also become problematic.

A superuser may run the `usermod` command with the `-g, --gid` option parameter, to *change* the primary group of a user. The command should be invoked for each user. Here's an example of removing user `julian` from the `devops` primary group. First, let's get the current data for the user, as follows:

```
id julian
```

This is the output:

```
packt@neptune:~$ id julian
uid=1001(julian) gid=1002(devops) groups=1002(devops)
```

Figure 4.13 – Retrieving the current primary group for the user

The -g, --gid option parameter of the usermod command accepts both a *GID* and a group *name*. The specified group must already be present in the system, otherwise the command would fail. If we want to change the primary group (for example, to developers), we simply specify the group name in the -g, --gid option parameter, as follows:

```
sudo usermod -g developers julian
```

But let's assume we don't want to associate the user julian with a specific primary group, yet we need to specify one for usermod -g. The simplest way to address this conundrum is to create a group called julian with the GID matching the UID (in our case, 1001), as follows:

```
sudo groupadd -g 1001 julian
```

Now that we have the julian *group* defined, we can safely modify the julian *user* to have their exclusive primary group, as follows:

```
sudo usermod -g julian
```

We can verify that the primary group reflects the change with the following command:

```
id julian
```

The output now shows the **UID** of julian as the **GID**, as can be seen here:

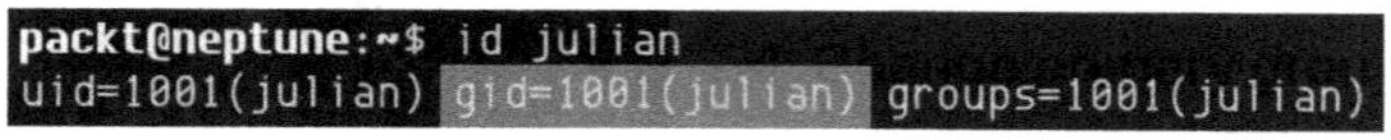

```
packt@neptune:~$ id julian
uid=1001(julian) gid=1001(julian) groups=1001(julian)
```

Figure 4.14 – Changing the primary group of the user

At this point, it's safe to delete the group (devops), as follows:

```
sudo groupdel devops
```

For more information on the groupdel command-line utility, check out the related system manual (man groupdel), or simply invoke groupdel --help.

Modifying groups via /etc/group

An administrator can also manually edit the /etc/group file to modify group data by updating the related line. Although the editing can be done with a text editor of your choice (for example, nano), we recommend the use of the vigr command-line utility for a safer approach. vigr is similar to vipr (for modifying /etc/passwd) and sets safety locks to prevent possible data corruption during concurrent changes of group data.

The following command opens the /etc/group file for editing by also prompting for the preferred text editor (for example, nano or vim):

```
sudo vigr
```

For example, we can change the settings for the developers group by editing the following line:

```
developers:x:1200:julian,alex
```

When deleting groups using the vigr command, we're also prompted to remove the corresponding entry in the group shadow file (/etc/gshadow). The related command invokes the -s or --shadow option, as illustrated here:

```
sudo vigr -s
```

For more information on the vigr utility, please refer to the related system manual (man vigr).

As with most Linux tasks, all the preceding tasks could have been accomplished in different ways. The commands chosen are the most common ones, but there might be cases when a different approach may prove more appropriate.

In the next section, we'll take a glance at how to add users to primary and secondary groups and how to remove users from these groups.

Users and groups

So far, we've only created groups that have no users associated. There is not much use for empty user groups, so let's add some users to them.

Adding users to a group

Before we start adding users to a group, let's create a few groups. In the following example, we create the groups by also specifying their GID (via the -g, --gid option parameter of the groupadd command):

```
sudo groupadd -g 1100 admin
sudo groupadd -g 1200 developers
sudo groupadd -g 1300 devops
```

Next, we create a couple of users (alex and julian) and add them to some of the groups we just created. We'll have the admin group set as the *primary group* for both users, while the developers and devops groups are defined as *secondary* (or *supplementary*) *groups*. The code can be seen here:

```
sudo useradd -g admin -G developers,devops alex
sudo useradd -g admin -G developers,devops julian
```

The -g, --gid option parameter of the useradd command specifies the (unique) primary group (admin). The -G, --groups option parameter provides the comma-separated list (without intervening spaces) of the secondary group names (developers, devops).

We can verify the group memberships for both users with the following commands:

```
id alex
id julian
```

The preceding commands yield the following output:

```
packt@neptune:~$ id alex
uid=1002(alex) gid=1100(admin) groups=1100(admin),1200(developers),1300
(devops),1400(managers)
```

Figure 4.15 – Verifying the group membership for users

As we can see, the gid attribute shows the primary group membership: gid=100(admin). The groups attribute shows the supplementary (secondary) groups: groups=1100(admin),1200(developers),1300(devops).

With users scattered across multiple groups, an administrator is sometimes confronted with the task of moving users between groups. The following section shows how to do this.

Moving and removing users across groups

Building upon the previous example, let's assume the administrator wants to move (or add) user alex to a new secondary group called managers. Please note that, according to our previous examples, user alex has admin as the primary group and developers/devops as secondary groups (see the output of the id alex command in *Figure 4.17*).

Let's create the managers group first, with GID 1400. The code can be seen here:

```
sudo groupadd -g 1400 managers
```

Next, add our existing user, alex, to the managers group. We use the usermod command with the -G, --groups option parameter, to specify the secondary groups the user is associated with.

The simplest way to *append* a secondary group to a user is by invocation of the -a, --append option of the usermod command, as illustrated here:

```
sudo usermod -a -G managers alex
```

The preceding command would preserve the existing secondary groups for user alex while adding the new managers group. Alternatively, we could run the following command:

```
sudo usermod -G developers,devops,managers alex
```

In the preceding command, we specified multiple groups (with no intervening whitespace!).

> **Important note**
> We preserved the existing secondary groups (developers/devops) and *appended* to the comma-separated list the managers additional secondary group. If we only had the managers group specified, user alex would have been *removed* from the developers and devops secondary groups.

To verify user alex is now part of the managers group, run the following command:

```
id alex
```

This is the output of the command:

```
packt@neptune:~$ id alex
uid=1002(alex) gid=1100(admin) groups=1100(admin),1200(developers),
1300(devops),1400(managers)
```

Figure 4.16 – Verifying that the user is associated with the managers group

As we can see, the `groups` attribute (highlighted) includes the related entry for the `managers` group: `1400(managers)`.

Similarly, if we wanted to *remove* user `alex` from the `developers` and `devops` secondary groups, to only be associated with the `managers` secondary group, we would run the following command:

```
sudo usermod -G managers alex
```

This is the output:

```
packt@neptune:~$ id alex
uid=1002(alex) gid=1100(admin) groups=1100(admin),1400(managers)
```

Figure 4.17 – Verifying the secondary groups for the user

The `groups` tag now shows the primary group `admin` (by default) and the `managers` secondary group.

The command to remove user `alex` from all secondary groups is shown here:

```
sudo usermod -G '' alex
```

The `usermod` command has an empty string (`''`) as the `-G`, `--groups` option parameter, to ensure no secondary groups are associated with the user. We can verify that user `alex` has no more secondary group memberships with the following command:

```
id alex
```

This is the output:

```
packt@neptune:~$ id alex
uid=1002(alex) gid=1100(admin) groups=1100(admin)
```

Figure 4.18 – Verifying the user has no secondary groups

As we can see, the `groups` tag only contains the `1100(admin)` primary GID, which by default is always shown for a user.

If an administrator chooses to remove user `alex` from a primary group or assign them to a different primary group, they must run the `usermod` command with the `-g`, `--gid` option parameter and specify the primary group name. A primary group is always mandatory for a user, and it must exist.

For example, to move user `alex` to the `managers` primary group, the administrator would run the following command:

```
sudo usermod -g managers alex
```

The related user data becomes this:

```
id alex
```

The command yields the following output:

```
packt@neptune:~$ id alex
uid=1002(alex) gid=1400(managers) groups=1400(managers),1200(developers),
1300(devops)
```

Figure 4.19 – Verifying the user has been assigned the new primary group

The gid attribute of the user record in *Figure 4.21* reflects the new primary group: gid=1400(managers).

If the administrator chooses to configure user alex without a specific primary group, they must first create an exclusive *group* (named alex, for convenience), and have the GID matching the UID of user alex (1002), as follows:

```
sudo groupadd -g 1002 alex
```

And now, we can remove user alex from the current primary group (managers) by specifying the exclusive primary group we just created (alex), like this:

```
sudo usermod -g alex
```

The related user record becomes this:

```
id alex
```

This is the output:

```
packt@neptune:~$ id alex
uid=1002(alex) gid=1002(alex) groups=1002(alex),1200(developers),
1300(devops),1400(managers)
```

Figure 4.20 – Verifying the user has been removed from primary groups

The gid attribute of the user record reflects the exclusive primary group (matching the user): gid=1002(alex). Our user doesn't belong to any other primary groups anymore.

Adding, moving, and removing users across groups may become increasingly daunting tasks for a Linux administrator. Knowing at any time which users belong to which groups is valuable information, both for reporting purposes and user automation workflows. The following section provides a few commands for viewing user and group data.

Viewing users and groups

In this section, we provide some possibly useful commands for retrieving group and group membership information. Before we get into any commands, we should keep in mind that group information is stored in the /etc/group and /etc/gshadow files. Among the two, the former has the information we're most interested in.

We can parse the /etc/group file to retrieve all groups, as follows:

```
cat /etc/group | cut -d: -f1 | column | less
```

The command yields the following output (excerpt):

```
root              irc              lxd
daemon            src              systemd-coredump
bin               gnats            packt
sys               shadow           rtkit
```

Figure 4.21 – Retrieving all group names

A similar command would use getent, like this:

```
getent group | cut -d: -f1 | column | less
```

We can retrieve the information of an individual group (for example, developers) with the following command:

```
getent group developers
```

This is the output:

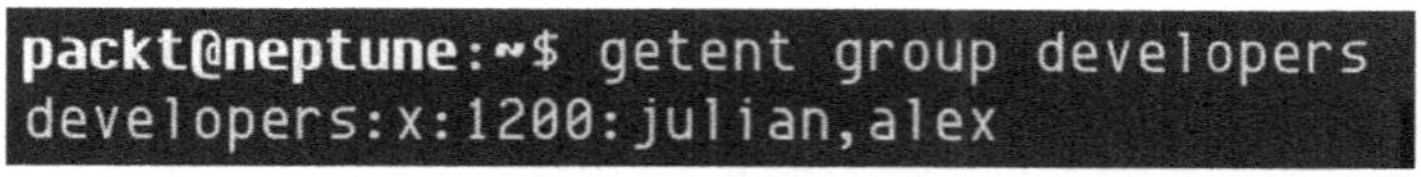

```
packt@neptune:~$ getent group developers
developers:x:1200:julian,alex
```

Figure 4.22 – Retrieving information for a single group

The output of the preceding command also reveals the members of the developers group (julian, alex).

To list all groups a specific user is a member of, we can use the groups command. For example, the following command lists all the groups that user alex is a member of:

```
groups alex
```

This is the command output:

```
packt@neptune:~$ groups alex
alex : admin developers devops managers
```

Figure 4.23 – Retrieving group membership information of the user

The output of the previous command shows the groups for user `alex`, starting with the primary group (`admin`).

A user can retrieve their own group membership using the `groups` command-line utility without specifying a group name. The following command is executed in a terminal session of user `packt`, who is also an administrator (superuser):

```
groups
```

The command yields this:

```
packt@neptune:~$ groups
packt adm cdrom sudo dip plugdev lxd
```

Figure 4.24 – The current user's groups

There are many other ways and commands to retrieve user- and group-related information. We hope that the preceding examples provide a basic idea about where and how to look for some of this information.

Next, let's look at how a user can switch or log in to specific groups.

Group login sessions

When a user logs in to the system, the group membership context is automatically set to the user's primary group. Once the user is logged in, any user-initiated task (such as creating a file or running a program) is associated with the user's primary group membership permissions. A user may also choose to access resources in other groups where they are also a member (that is, supplementary or secondary groups). To switch the group context or log in with a new group membership, a user invokes the `newgrp` command-line utility.

The basic syntax for the `newgrp` command is this:

```
newgrp GROUP
```

In the following example, we assume a user (`julian`) is a member of multiple groups—admin as the primary group, and `developers`/`devops` as secondary groups:

```
id julian
```

This is the output:

```
packt@neptune:~$ id julian
uid=1001(julian) gid=1001(julian) groups=1001(julian),1200(developers),
1300(devops)
```

Figure 4.25 – A user with multiple group memberships

Let's impersonate user `julian` for a while. When logged in as `julian`, the default login session has the following user and group context:

```
whoami
```

In our case, this is the output:

```
$ whoami
julian
```

Figure 4.26 – Getting the current user

The `whoami` command provides the current UID (see more details on the command with `man whoami` or `whoami --help`), as follows:

```
groups
```

This is the output:

```
$ groups
admin developers devops
```

Figure 4.27 – Getting the current user's groups

The `groups` command displays all groups that the current user is a member of (see more details on the command with `man groups` or `groups --help`).

The user can also view their IDs (user and GIDs) by invoking the `id` command, as follows:

```
id
```

This is the output:

```
$ id
uid=1001(julian) gid=1001(julian) groups=1001(julian),1200(developers),
1300(devops)
```

Figure 4.28 – Viewing the current user and GID information

There are various invocations of the id command that provide information on the current user and group session. The following command (with the -g, --group option) retrieves the ID of the current group session for the user:

```
id -g
1100
```

In our case, the preceding command shows 1100—the GID corresponding to the user's primary group, which is admin (see the gid attribute in *Figure 4.30*). Upon login, the default group session is always the primary group corresponding to the user. If the user were to create a file, for example, the file permission attributes would reflect the primary group's ID. We'll look at the file permissions in more detail in the *Managing permissions* section.

Now, let's switch the group session for the current user to developers, as follows:

```
newgrp developers
```

The current group session yields this:

```
id -g
1200
```

The GID corresponds to the developers secondary GID, as displayed by the groups tag in *Figure 4.30*: 1200 (developers). If the user created any files now, the related file permission attributes would have the developers GID.

If the user attempts to log in to a group they are not a member of (for example, managers), the newgrp command prompts for the managers group's password, as illustrated here:

```
newgrp managers
```

The preceding command prompts for the superuser password.

If our user had the managers group password, or if they were a superuser, the group login attempt would succeed. Otherwise, the user would be denied access to the managers group's resources.

We conclude here our topic of managing users and groups. The examples for the related administrative tasks used throughout this section are certainly all-encompassing. In many of these cases, there are multiple ways to achieve the same result, using different commands or approaches.

By now, you should be relatively proficient in managing users and groups, and comfortable using the various command-line utilities for operating the related changes. Users and groups are managed in a relational fashion, where users belong to a group, or groups are associated with users. We also learned that creating and managing users and groups requires superuser privileges. In Linux, user data is stored in the /etc/passwd and /etc/shadow files, while group information is found in /etc/group and /etc/gshadow. Besides using the dedicated command-line utilities, users and groups can also be altered by manually editing the aforementioned files.

Next, we'll turn to the security and isolation context of the multiuser group environment. In Linux, the related functionality is accomplished by a system-level access layer that controls the read, write, and execute permissions of files and directories, by specific users and groups.

The following section explores the management and administrative tasks related to these permissions.

Managing permissions

A key tenet of Linux is the ability to allow multiple users to access the system while performing independent tasks simultaneously. The smooth operation of this multiuser, multitasking environment is controlled via permissions. The Linux kernel provides a robust framework for the underlying security and isolation model. At the user level, dedicated tools and command-line utilities help Linux users and system administrators with the related permission management tasks.

For some Linux users, especially beginners, Linux permissions may appear confusing at times. This section attempts to demystify some of the key concepts about file and directory permissions in Linux. You will learn about the basic permission *rights* of accessing files and directories—the *read*, *write*, and *execute* permissions. We explore some of the essential administrative tasks for viewing and changing permissions, using system-level command-line utilities.

Most of the topics discussed in this section should be regarded closely with users and groups. The related idioms can be as simple as *a user can read or update a file*, *a group has access to these files and directories*, or *a user can execute this program*.

Let's start with the basics, introducing the file and directory permissions.

File and directory permissions

In Linux, permissions can be regarded as the *rights* or *privileges* to act upon a file or a directory. The basic rights, or *permission attributes*, are outlined here:

- **Read**—A *read* permission on a file allows users to view the content of the file. On a directory, the read permission allows users to list the content of the directory.

- **Write**—A *write* permission on a file allows users to modify the content of the file. For a directory, the write permission allows users to modify the content of the directory by adding, deleting, or renaming files.

- **Execute**—An *execute* or *executable* permission on a file allows users to run the related script, application, or service appointed by the file. For a directory, the execute permission allows users to enter the directory and make it the current working directory (using the cd command).

First, let's take a look at how to reveal the permissions for files and directories.

Viewing permissions

The most common way to view the permissions of a file or directory is by using the ls command-line utility. The basic syntax of this command is this:

```
ls [OPTIONS] FILE|DIRECTORY
```

Here is an example use of the ls command to view the permissions of the /etc/passwd file:

```
ls -l /etc/passwd
```

The command yields the following output:

```
packt@neptune:~$ ls -l /etc/passwd
-rw-r--r-- 1 root root 3056 Sep 17 07:57 /etc/passwd
```

Figure 4.29 – Viewing the permissions of /etc/passwd file

The -l option of the ls command provides a detailed output by using the *long listing format*, according to the ls documentation (man ls).

Let's analyze the output, as follows:

```
-rw-r--r-- 1 root 3056 Sep 17 07:57 /etc/passwd
```

We have nine segments, separated by single whitespace characters (delimiters). These are outlined here:

- -rw-r--r--: The file access permissions
- 1: The number of hard links
- root: The user who is the owner of the file
- root: The group that is the owner of the file
- 3056: The size of the file
- Sep: The month the file was created
- 17: The day of the month the file was created
- 07:57: The time of day the file was created
- /etc/passwd: The filename

Let's examine the file access permissions field (-rw-r--r--). File access permissions are defined as a 10-character field, grouped as follows:

- The first character (attribute) is reserved for the file type (see the *File types* section, next).
- The next nine characters represent a 9-bit field, defining the effective permissions as three sequences of three attributes (bits) each: *user owner* permissions, *group owner* permissions, and *all other users'* permissions (see the *Permission attributes* section, next).

Let's take a look at the file types.

File types

The file type attributes are listed here:

- d: A directory
- -: A regular file
- l: A symbolic link
- p: A named pipe—a special file that facilitates communication between programs
- s: A socket—similar to a pipe but with bidirectional network communications
- b: A block device—a file that corresponds to a hardware device
- c: A character device—similar to a block device

Let's have a closer look at the permission attributes.

Permission attributes

As previously noted, the access permissions are represented by a 9-bit field, a group of three sequences, each with 3 bits, defined as follows:

- **bits 1-3**: *User* owner permissions
- **bits 4-6**: *Group* owner permissions
- **bits 7-9**: *All* other users' (or *world*) permissions

Each permission attribute is a bit flag in the binary representation of the related three-bit sequence. They can be represented either as a character or as an equivalent numerical value, also known as the *octal* value, depending on the range of the bit they represent.

Here are the permission attributes with their respective octal values:

- r: *Read* permission; $2 \wedge 2 = 4$ (bit 2 set)
- w: *Write* permission: $2 \wedge 1 = 2$ (bit 1 set)
- x: *Execute* permission: $2 \wedge 0 = 1$ (bit 0 set)
- -: *No* permission: 0 (no bits set)

The resulting corroborated number is also known as the *octal value* of the file permissions (see the *File permission examples* section). Here's an illustration of the file permission attributes:

```
| r | w | x | r | w | x | r | w | x |
|           |           |           |
   owner / user        group        other / world

  read  write  exec  read  write exec  read  write exec

| 4     2      1  | 4     2      1 | 4     2      1 |
```

Figure 4.30 – The file permission attributes

Next, let's consider some examples.

File permission examples

Now, let's go back and evaluate the file access permissions for /etc/passwd: -rw-r--r--, as follows:

- -: The first character (byte) denotes the file type (a regular file, in our case).
- rw-: The next three-character sequence indicates the user owner permissions (in our case, read (r); write (w); octal value = 4 (r) + 2 (w) = 6 (rw)).

- r--: The next 3-byte sequence defines the group owner permissions (in our case, read (r); octal value = 4 (r)).

- r--: The last three characters denote the permissions for all other users in the system (in our case, read (r); octal value = 4 (r)).

According to the preceding information, the resulting octal value of the /etc/passwd file access permissions is 644. Alternatively, we can query the octal value with the stat command, as follows:

```
stat --format '%a' /etc/passwd
```

The command yields the following output:

```
packt@neptune:~$ stat --format '%a' /etc/passwd
644
```

Figure 4.31 – Getting permission attributes using the stat command

The stat command displays the file or filesystem status. The --format option parameter specifies the access rights in octal format ('%a') for the output.

Here are a few examples of access permissions, with their corresponding octal values and descriptions. The three-character sequences are intentionally delimited with whitespace for clarity. The leading file type has been omitted:

- rwx (777): Read, write, execute for all users including owner, group, and world.

- rwx r-x (755): Read, execute for all users; the file owner has write permissions.

- rwx r-x --- (750): Read, execute for owner and group; the owner has write permissions while others have no access.

- rwx --- --- (700): Read, write, execute for owner; everyone else has no permissions.

- rw- rw- rw- (666): Read, write for all users; there are no execute permissions

- rw- rw- r-- (664): Read, write for owner and group; read for others.

- rw- rw- --- (660): Read, write for owner and group; others have no permissions.

- rw- r-- r-- (644): Read, write for owner; read for group and others.

- rw- r-- --- (640): Read, write for owner; read for group; no permissions for others.

- rw- --- --- (600): Read, write for owner; no permissions for group and others.

- r-- --- --- (400): Read for owner; no permissions for others.

Read, write, and execute are the most common types of file access permissions. There are cases, particularly in user impersonation situations, when the access rights may involve some special permission attributes. Let's have a look at them.

Special permissions

In Linux, the ownership of files and directories is usually determined by the UID and GID of the user—or group—who created them. The same principle applies to applications and processes—they are owned by the users who launch them. The special permissions are meant to change this default behavior when needed.

Here are the special permission flags, with their respective octal values:

- setuid: $2 \wedge 2 = 4$ (bit 2 set)
- setgid: $2 \wedge 1 = 2$ (bit 1 set)
- sticky: $2 \wedge 0 = 1$ (bit 0 set)

When any of these special bits are set, the overall octal number of the access permissions will have an extra digit, with the leading (high-order) digit corresponding to the special permission's octal value.

Let's look at these special permission flags, with examples for each.

The setuid permission

With the setuid bit set, when an executable file is launched, it will run with the privileges of the file owner instead of the user who launched it. For example, if the executable is owned by root and launched by a *regular* user, it will run with root privileges. The setuid permission could pose a potential security risk when used inadequately, or when vulnerabilities of the underlying process could be exploited.

In the file access permission field, the setuid bit could have either of the following representations:

- s *replacing* the corresponding executable bit (x) (when the executable bit is present)
- S (capital S) for a non-executable file

The setuid permission can be set via the following chmod command (for example, for the myscript executable file):

```
chmod u+s myscript
```

The resulting file permissions are shown here (including the octal value):
-rwsrwxr-x (4775). Here is the related command-line output:

```
packt@neptune:~$ chmod u+s myscript
packt@neptune:~$ ls -l myscript
-rwsrwxr-x 1 packt packt 0 Sep 21 08:09 myscript
packt@neptune:~$ stat --format '%a' myscript
4775
```

Figure 4.32 – The setuid permission

For more information on setuid, please visit https://en.wikipedia.org/wiki/
Setuid or refer to the chmod command-line utility documentation (man chmod).

The setgid permission

While setuid controls user impersonation privileges, setgid has a similar effect for
group impersonation permissions.

An executable file with the setgid bit set runs with the privileges of the group that owns
the file, instead of the group associated with the user who started it. In other words, the
GID of the process is the same as the GID of the file.

When used on a directory, the setgid bit changes the default ownership behavior, in
a way that files created within the directory will have group ownership of the parent
directory, instead of the group associated with the user who created them. This behavior
could be adequate in file-sharing situations when files can be changed by all users
associated with the parent directory's owner group.

The setgid permission can be set via the following chmod command (for example,
for the myscript executable file):

```
chmod g+s myscript
```

The resulting file permissions are shown here (including the octal value):
-rwxrwsr-x (2775).

The command-line output is shown here:

```
packt@neptune:~$ chmod g+s myscript
packt@neptune:~$ ls -l myscript
-rwxrwsr-x 1 packt packt 0 Sep 21 07:55 myscript
packt@neptune:~$ stat --format '%a' myscript
2775
```

Figure 4.33 – The setgid permission

For more information on `setgid`, please visit `https://en.wikipedia.org/wiki/Setuid` or refer to the `chmod` command-line utility documentation (man `chmod`).

The sticky permission

The `sticky` bit has no effect on files. For a directory with `sticky` permission, only the user owner or group owner of the directory can delete or rename files within the directory. Users or groups with write access to the directory, by way of user or group ownerships, cannot delete or modify files in the directory. The `sticky` permission is useful when a directory is owned by a privileged group whose members share write access to files in that directory.

The `sticky` permission can be set via the following `chmod` command (for example, for the `mydir` directory):

```
chmod +t mydir
```

The resulting directory permissions are shown here (including the octal value): `drwxrwxr-t` (`1775`). The command-line output is shown here:

```
packt@neptune:~$ chmod +t mydir
packt@neptune:~$ ls -ld mydir
drwxrwxr-t 2 packt packt 4096 Sep 21 08:30 mydir
packt@neptune:~$ stat --format '%a' mydir
1775
```

Figure 4.34 – The sticky permission

For more information on `sticky`, please visit `https://en.wikipedia.org/wiki/Setuid` or refer to the `chmod` command-line utility documentation (man `chmod`).

So far, we have mostly focused on permission types and their representation. In the next section, we explore a few command-line tools used for altering permissions.

Changing permissions

Modifying file and directory access permissions is a common Linux administrative task. In this section, we learn about a few command-line utilities that are handy when it comes to changing permissions and ownerships of files and directories. These tools are installed with any modern-day Linux distribution, and their use is similar across most Linux platforms.

Using chmod

The `chmod` command is short for *change mode*, and it's used to set access permissions on files and directories. The `chmod` command can be used by both the current user (owner) and a superuser.

Changing permissions can be done in two different modes: relative and absolute. Let's take a look at each of them.

Using chmod in relative mode

Changing permissions in relative mode is probably the easiest of the two. It is important to remember the following:

- *For whom* we change permissions: u = user (owner), g = group, o = others
- *How* we change permissions: + = add, - = remove, = = exactly as is
- *What* permission do we change: r = read, w = write, x = execute

Let's explore a few examples of using chmod in relative mode.

In our first example, we want to add write (w) permissions for all *other* (o) users (*world*), to myfile, as follows:

```
chmod o+w myfile
```

The related command-line output is shown here:

```
packt@neptune:~$ ls -l myfile
-rw-rw-r-- 1 packt packt 16 Sep 21 09:27 myfile
packt@neptune:~$ chmod o+w myfile
packt@neptune:~$ ls -l myfile
-rw-rw-rw- 1 packt packt 16 Sep 21 09:27 myfile
```

Figure 4.35 – Setting write permissions to all other users

In the next example, we remove the read (r) and write (w) permissions for the current user owner (u) of myfile, as follows:

```
chmod u-rw myufile
```

The command-line output is shown here:

```
packt@neptune:~$ ls -l myfile
-rw-rw-rw- 1 packt packt 16 Sep 21 09:27 myfile
packt@neptune:~$ chmod u-rw myfile
packt@neptune:~$ ls -l myfile
----rw-rw- 1 packt packt 16 Sep 21 09:27 myfile
```

Figure 4.36 – Removing read-write permissions for owner

We did not use sudo in either of the preceding examples since we carried out the operations as the current owner of the file (packt).

In the following example, we assume that myfile has read, write, and execute permissions for everyone. Then, we carry out the following change:

- Remove the read (r) permission for the owner (u).

- Remove the write (w) permission for the owner (u) and group (g).

- Remove the read (r), write (w), and execute (x) permissions for everyone else (o).

This is illustrated in the following code snippet:

```
chmod u-r,ug-w,o-rwx myfile
```

The command-line output is shown here:

```
packt@neptune:~$ ls -l myfile
-rwxrwxrwx 1 packt packt 16 Sep 21 09:27 myfile
packt@neptune:~$ chmod u-r,ug-w,o-rwx myfile
packt@neptune:~$ ls -l myfile
---xr-x--- 1 packt packt 16 Sep 21 09:27 myfile
```

Figure 4.37 – A relatively complex invocation of chmod in relative mode

Next, let's look at a second way of changing permissions: using the chmod command-line utility in absolute mode, by specifying the octal number corresponding to the access permissions.

Using chmod in absolute mode

The absolute mode invocation of chmod changes all permission attributes at once, using an *octal* number. The *absolute* designation of this method is due to changing permissions without any reference to existing ones, by simply assigning the octal value corresponding to the access permissions.

Here's a quick list of the octal values corresponding to effective permissions:

- 7 rwx: Read, write, and execute

- 6 rw-: Read, write

- 5 r-w: Read, execute

- 4 r--: Read

- 3 -wx: Write, execute

- 2 -w-: Write

- 1 --x: Execute

- 0 ---: No permissions

In the following example, we change the permissions of myfile to read (r), write (w), and execute (x) for everybody:

```
chmod 777 myfile
```

The related change is illustrated by the following command-line output:

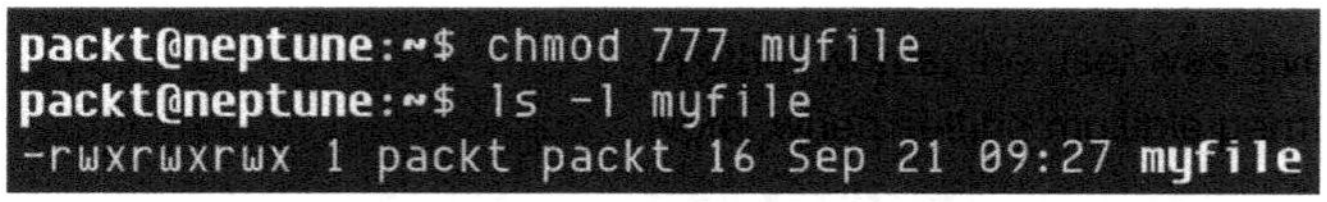

```
packt@neptune:~$ chmod 777 myfile
packt@neptune:~$ ls -l myfile
-rwxrwxrwx 1 packt packt 16 Sep 21 09:27 myfile
```

Figure 4.38 – The chmod invocation in absolute mode

For more information about the chmod command, please refer to the related documentation (man chmod).

Let's now look at our next command-line utility, specializing in file and directory ownership changes.

Using chown

The chown command is used to set the ownership of files and directories. Typically, the chmod command can only be run with *superuser* privileges (that is, by a *sudoer*). Regular users can only change the *group* ownership of their files, and only when they are a member of the target group.

The syntax of the chown command is shown here:

```
chown [OPTIONS] [OWNER][:[GROUP]] FILE
```

Usually, we invoke the chown command with both user *and* group ownerships—for example, like this:

```
sudo chown julian:developers myfile
```

The related command-line output is shown here:

```
packt@neptune:~$ sudo chown julian:developers myfile
[sudo] password for packt:
packt@neptune:~$ ls -l myfile
-rwxrwxrwx 1 julian developers 16 Sep 21 09:27 myfile
```

Figure 4.39 – A simple invocation of the chown command

One of the most common uses of chown is for *recursive mode* invocation, with the -R, --recursive option. The following example changes the ownership permissions of all files in mydir (directory), initially owned by root, to julian:

```
sudo chown -R julian:julian mydir/
```

The related changes are shown in the following command-line output:

```
packt@neptune:~$ ls -lR mydir/
mydir/:
total 12
-rwxrwxr-x 1 root root   16 Sep 21 10:49 file1
-rwxrwxr-x 1 root root   16 Sep 21 10:49 file2
drwxrwxr-x 2 root root 4096 Sep 21 10:50 subdir

mydir/subdir:
total 8
-rwxrwxr-x 1 root root 16 Sep 21 10:50 file3
-rwxrwxr-x 1 root root 16 Sep 21 10:50 file4
packt@neptune:~$
packt@neptune:~$ sudo chown -R julian:julian mydir/
packt@neptune:~$
packt@neptune:~$ ls -lR mydir/
mydir/:
total 12
-rwxrwxr-x 1 julian julian   16 Sep 21 10:49 file1
-rwxrwxr-x 1 julian julian   16 Sep 21 10:49 file2
drwxrwxr-x 2 julian julian 4096 Sep 21 10:50 subdir

mydir/subdir:
total 8
-rwxrwxr-x 1 julian julian 16 Sep 21 10:50 file3
-rwxrwxr-x 1 julian julian 16 Sep 21 10:50 file4
```

Figure 4.40 – Invoking ls and chown in recursive mode

For more information about the chown command, please refer to the related documentation (man chown).

Next, let's briefly look at a similar command-line utility that specializes exclusively in group ownership changes.

Using chgrp

The chgrp command is used to change the *group* ownership for files and directories. In Linux, files and directories typically belong to a user (owner) or a group. We can set user ownership by using the chown command-line utility, while group ownership can be set with chgrp.

The syntax for chgrp is shown here:

```
chgrp [OPTIONS] GROUP FILE
```

The following example changes the group ownership of myfile to the developers group:

```
sudo chgrp developers myfile
```

The changes are shown in the following output:

```
packt@neptune:~$ sudo chgrp developers myfile
[sudo] password for packt:
packt@neptune:~$ ls -l myfile
-rwxrwxrwx 1 packt developers 16 Sep 21 09:27 myfile
```

Figure 4.41 – Using chgrp to change group ownership

The preceding command has been invoked with superuser privileges (sudo) since the current user (packt) is not an admin for the developers group.

For more information about the chgrp utility, please refer to the tool's command-line help (chgrp --help).

Using umask

The umask command is used to view or set the default *file mode mask* in the system. The file mode represents the default permissions for any new files and directories created by a user. For example, the default file mode masks in Ubuntu are given here:

- 0002 for a regular user
- 0022 for the root user

As a general rule in Linux, the *default permissions* for new files and directories are calculated with the following formulas:

- 0666 – umask: For a new file created by a regular user
- 0777 – umask: For a new directory created by a regular user

According to the preceding formula, on Ubuntu we have the following default permissions:

- File (regular user): 0666 – 0002 = 0664
- Directory (regular user): 0777 – 0002 = 0775
- File (root): 0666 – 0022 = 0644
- Directory (root): 0777 – 0022 = 0755

In the following examples, run on Ubuntu, we create a file (`myfile`) and a directory (`mydir`), using the terminal session of a regular user (`packt`). Then, we query the `stat` command for each and verify that the default permissions match the values enumerated previously for regular users (file: `664`, directory: `775`).

Let's start with the default file permissions first, as follows:

```
touch myfile
stat --format '%a' myfile
```

The related output is shown here:

```
packt@neptune:~$ touch myfile
packt@neptune:~$ stat --format '%a' myfile
664
```

Figure 4.42 – The default file permissions for regular user (664)

Next, let's verify the default directory permissions, as follows:

```
mkdir mydir
stat --format '%a' mydir
```

The related output is shown here:

```
packt@neptune:~$ mkdir mydir
packt@neptune:~$ stat --format '%a' mydir/
775
```

Figure 4.43 – The default directory permissions for regular user (775)

Here's a list with the most typical `umask` values on Linux systems:

```
value       files                directories
0000        666 (rw-rw-rw-)      777 (rwxrwxrwx)
0002        664 (rw-rw-r--)      775 (rwxrwxr-x)
0022        644 (rw-r--r--)      755 (rwxr-xr-x)
0027        640 (rw-r-----)      750 (rwxr-x---)
0077        600 (rw-------)      700 (rwx------)
0277        400 (r--------)      500 (r-x------)
```

Figure 4.44 – Typical umask values on Linux

For more information about the `umask` utility, please refer to the tool's command-line help (`umask --help`).

File and directory permissions are critical for a secure environment. Users and processes should operate exclusively within the isolation and security constraints controlled by permissions, to avoid inadvertent or deliberate interference with the use and ownership of system resources.

Interpreting permissions can be a daunting task. This section has aimed to demystify some of the related intricacies, and we hope that you will feel more comfortable handling file and directory permissions in everyday Linux administration tasks.

Summary

In this chapter, we explored some of the essential concepts related to managing users and groups in Linux. We learned about file and directory permissions and the different access levels of a multiuser environment. For each main topic, we focused on basic administrative tasks, providing various practical examples and using typical command-line tools for everyday user access and permission management operations.

Managing users and groups, and the related filesystem permissions that come into play, is an indispensable skill of a Linux administrator. The knowledge gained in this chapter will, we hope, put you on track to becoming a proficient superuser.

In the following chapter, we continue our journey of mastering Linux internals by exploring processes, daemons, and inter-process communication mechanisms. An important aspect to keep in mind is that processes and daemons are also *owned* by users or groups. The skills learned in this chapter will help us navigate the related territory when we look at *who runs what* at any given time in the system.

Questions

Here are a few thoughts and questions that sum up the main ideas covered in this chapter:

1. Linux is a multiuser and multitasking operating system. The permissions provide a related isolation and security context to protect user resources from inadvertent access.
2. What is a superuser?
3. Think of a command-line utility for creating users. Can you think of another one?
4. What is the octal value of the -rw-rw-r— access permission?
5. What is the difference between a primary group and a secondary (supplementary) group?
6. How do you change the ownership of a user's home directory?
7. Can you remove a user from the system without deleting their home directory? How?

5
Working with Processes, Daemons, and Signals

Linux is a multitasking operating system. Multiple programs or tasks can run in parallel, each with its own identity, scheduling, memory space, permissions, and system resources. Processes encapsulate the execution context of any such program. Understanding how processes work and communicate with each other is an important skill for any seasoned Linux system administrator and developer to have.

This chapter explores the basic concepts behind Linux processes. We'll look at different types of processes, such as foreground and background processes, with special emphasis being placed on daemons as a particular type of background process. We'll closely study the anatomy of a process and various inter-process communication mechanisms in Linux – signals in particular. Along the way, we'll learn about some of the essential command-line utilities for managing processes and daemons and working with signals.

In this chapter, we will cover the following topics:

- Introducing processes
- Working with processes
- Working with daemons
- Exploring interprocess communication

As we navigate through the content, we will occasionally reference **signals** *before* their formal introduction in the second half of this chapter. In Linux, signals are almost exclusively used in association with processes, hence our approach of becoming familiar with processes first. Yet, leaving the signals out from some of the process' internals would do a disservice to understanding how processes work. We hope that the compromise taken here provides you with a better grasp of the overall picture and the inner workings of processes and daemons. Where signals are mentioned, we'll point to the related section for further reference.

Now, before we start, let's look at the essential prerequisites for our study.

Technical requirements

Practice makes perfect. Running the commands and examples in this chapter by hand would go a long way toward you learning about processes. As with any chapter in this book, we recommend that you have a working Linux distribution installed on a VM or PC desktop platform. We'll be using Ubuntu or CentOS, but most of the commands and examples would be similar on any other Linux platform.

The code files for this chapter can be found on GitHub at the following link: `https://github.com/PacktPublishing/Mastering-Linux-Administration`.

Introducing processes

A **process** represents the running instance of a program. In general, a program is a combination of instructions and data, compiled as an executable unit. When a program runs, a process is created. In other words, a process is simply a program in action. Processes execute specific tasks, and sometimes, they are also referred to as **jobs** (or **tasks**).

There are many ways to create or start a process. In Linux, every command starts a process. A command could be a user-initiated task in a Terminal session, a script, or a program (executable) that's invoked manually or automatically.

Usually, the way a process is created and interacts with the system (or user) determines its process type. Let's have a closer look at the different types of processes in Linux.

Understanding process types

At a high level, there are two major types of processes in Linux:

- **Foreground** (*interactive*)
- **Background** (*non-interactive* or *automated*)

Interactive processes assume some kind of user interaction during the lifetime of the process. Non-interactive processes are unattended, which means that they are either automatically started (for example, on system boot) or are scheduled to run at a particular time and date via job schedulers (for example, using the at and cron command-line utilities).

Our approach to exploring process types mainly pivots around the preceding classification. There are various other views or taxonomies surrounding process definitions, but they could ultimately be reduced to either foreground or background processes.

For example, batch processes and daemons are essentially background processes. Batch processes are automated in the sense that they are not user-generated but invoked by a scheduled task instead. Daemons are background processes that are usually started during system boot and run indefinitely.

There's also the concept of parent and child processes. A parent process may create other subordinate child processes.

We'll elaborate on these types (and beyond) in the following sections. Let's start with the pivotal ones – foreground and background processes.

Foreground processes

Foreground processes, also known as **interactive processes**, are started and controlled through a Terminal session. Foreground processes are usually initiated by a user via an interactive command-line interface. A foreground process may output results to the console (stdout or stderr) or accept user input. The lifetime of a foreground process is tightly coupled to the Terminal session (parent process). If the user who launched the foreground process exits the Terminal while the process is still running, the process will be abruptly terminated (via a SIGHUP signal sent by the parent process; see the *Signals* section for more details).

A simple example of a foreground process is invoking the system reference manual (man) for an arbitrary Linux command (for example, ps):

```
man ps
```

The ps command displays information about active processes. You will learn more about process management tools and command-line utilities in the *Process management* section.

Once a foreground process has been initiated, the user prompt is captured and controlled by the spawned process interface. The user can no longer interact with the initial command prompt until the interactive process relinquishes control to the Terminal session. The following screenshot shows the man ps command's invocation, switching the user control to the documentation interface:

```
PS(1)                         User Commands                         PS(1)

NAME
        ps - report a snapshot of the current processes.

SYNOPSIS
        ps [options]

DESCRIPTION
        ps displays information about a selection of the
        active processes.  If you want a repetitive update of
        the selection and the displayed information, use
        top(1) instead.

        This version of ps accepts several kinds of options:

        1    UNIX options, which may be grouped and must be
             preceded by a dash.
        2    BSD options, which may be grouped and must not be
             used with a dash.
        3    GNU long options, which are preceded by two
             dashes.

 Manual page ps(1) line 1 (press h for help or q to quit)
```

Figure 5.1 – Example of an interactive process (man ps)

Let's look at another example of a foreground process, this time invoking a long-lived task. The following command (one-liner) runs an infinite loop while displaying an arbitrary message every few seconds:

```
while true; do echo "Wait..."; sleep 5; done
```

As long as the command runs without being interrupted, the user won't have an interactive prompt in the Terminal. Using *Ctrl + C* would stop (interrupt) the execution of the related foreground process and yield a responsive command prompt:

```
packt@neptune:~$ while true; do echo "Wait..."; sleep 5; done
Wait...
Wait...
Wait...
Wait...
^C
```

Figure 5.2 – A long-lived foreground process

> **Important Note**
>
> When you press *Ctrl + C* while a foreground process is running, a SIGINT signal is sent to the running process by the current (parent) Terminal session, and the foreground process is interrupted. For more information, see the *Signals* section.

If we want to maintain an interactive command prompt in the Terminal session while running a specific command or script, we should use a background process.

Background processes

Background processes – also referred to as **non-interactive** or **automatic processes** – run independently of a Terminal session, without expecting any user interaction. A user may invoke multiple background processes within the same Terminal session without waiting on any of them to complete or exit.

Background processes are usually long-lived tasks that don't require direct user supervision. The related process may still display its output in the Terminal console, but such background tasks typically write their results to files instead.

The simplest invocation of a background process appends an ampersand (&) to the end of the related command. Building on our previous example (in the *Foreground processes* section), the following command creates a background process that runs an infinite loop, echoing an arbitrary message every few seconds:

```
while true; do echo "Wait..."; sleep 5; done &
```

Note the ampersand (&) at the end of the command. By default, a background process would still direct the output (stdout and stderr) to the console when invoked with the ampersand (&), as shown previously. However, the Terminal session remains interactive:

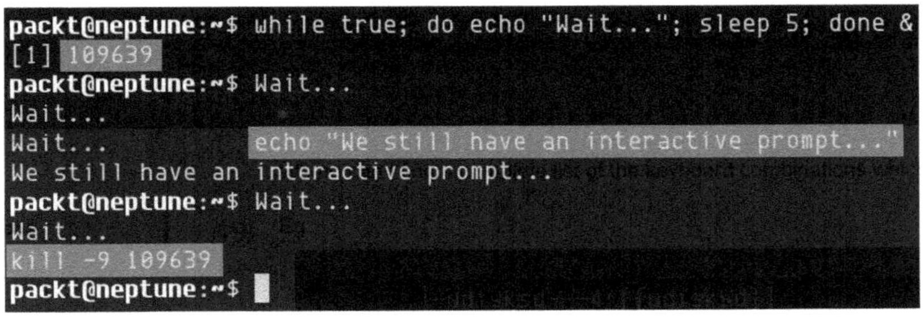

Figure 5.3 – Running a background process

As shown in the preceding screenshot, the background process is given a **process ID** (**PID**) of 109639. While the process is running, we can still control the Terminal session and run a different command, like so:

```
echo "We still have an interactive prompt..."
```

Eventually, we can force the process to terminate with the `kill` command:

```
kill -9 109639
```

The preceding command *kills* our background process (with PID 109639). The corresponding signal that's sent by the parent Terminal session to terminate this process is SIGKILL (see the *Signals* section for more information).

Both foreground and background processes are typically under the direct control of a user. In other words, these processes are created or started manually as a result of a command or script invocation. There are some exceptions to this rule, particularly when it comes to batch processes, which are launched automatically via scheduled jobs.

There's also a select category of background processes that are automatically started during system boot and terminated at shutdown without user supervision. These background processes are also known as daemons.

Introducing daemons

Daemons are a particular type of background process that are usually started upon system boot and run indefinitely or until terminated (for example, during system shutdown). A daemon doesn't have a user-controlled Terminal, even though it is associated with a system account (`root` or other) and runs with the related privileges.

Daemons usually serve client requests or communicate with other foreground or background processes. Here are some common examples of daemons, generally available on most Linux platforms:

- `systemd`: The parent of all processes (formerly known as `init`)
- `crond`: Job scheduler – runs tasks in the background
- `ftpd`: FTP server – handles client FTP requests
- `httpd`: Web server (Apache) – handles client HTTP requests
- `sshd`: Secure shell server – handles SSH client requests

Typically, system daemons in Linux are named with a d at the end, denoting a daemon process. Daemons are controlled by shell scripts usually stored in the /etc/init.d/ or /lib/systemd/ system directory, depending on the Linux platform. Ubuntu, for example, stores daemon script files in /etc/init.d/, while CentOS stores them in /lib/systemd/. The location of these daemon files depends on the platform implementation of init, a system-wide service manager for all Linux processes.

The Linux init-style startup process generally invokes these shell scripts at system boot. But the same scripts can also be invoked via service control commands, usually run by privileged system users, to manage the lifetime of specific daemons. In other words, a privileged user or system administrator can *stop* or *start* a particular daemon through the command-line interface. Such commands would immediately return the user's control to the Terminal while performing the related action in the background.

Let's have a closer look at the init process.

The init process

Throughout this chapter, we'll refer to init as the *generic* system initialization engine and service manager on Linux platforms. Over the years, Linux distributions have evolved and gone through various init system implementations, such as SysV, upstart, OpenRC, systemd, and runit. There's an ongoing debate in the Linux community about the supremacy or advantages of one over the other. For now, we will simply regard init as a system process, and we will briefly look at its relationship with other processes.

init (or systemd, and others) is essentially a system daemon, and it's among the first processes to start when Linux boots up. The related daemon process continues to run in the background until the system is shut down. init is the root (parent) process of all other processes in Linux in the overall process hierarchy tree. In other words, init is a direct or indirect ancestor of all the processes in the system.

In Linux, the pstree command displays the whole process tree, and it shows the init process at its root; in our case, systemd (on Ubuntu or CentOS):

```
Pstree
```

The output of the preceding command can be seen in the following screenshot:

```
packt@neptune:~$ pstree
systemd─┬─ModemManager───2*[{ModemManager}]
        ├─NetworkManager───2*[{NetworkManager}]
        ├─VGAuthService
        ├─accounts-daemon───2*[{accounts-daemon}]
        ├─acpid
        ├─atd
        ├─avahi-daemon───avahi-daemon
        ├─bluetoothd
        ├─colord───2*[{colord}]
        ├─cron
```

Figure 5.4 – init (systemd), the parent of all processes

The `pstree` command's output illustrates a hierarchy tree representation of the processes, where some appear as parent processes while others appear as child processes. Let's look at the parent and child process types and some of the dynamics between them.

Parent and child processes

A parent process creates other subordinate processes, also known as child processes. Child processes belong to the parent process that spawned them and usually terminate when the parent process exits (stops execution). A child process may continue to run beyond the parent process's lifetime if it's been instructed to ignore the SIGHUP signal that's invoked by the parent process upon termination (for example, via the nohup command). See the *Signals* section for more information.

In Linux, all processes except the init process (with its variations) are children of a specific process. Terminating a child process won't stop the related parent process from running. A good practice for terminating a parent process when the child is done processing is to exit from the parent process itself, after the child process completes.

There are cases when processes run unattended, based on a specific schedule. Running a process without user interaction is known as batch processing. We'll look at batch processes next.

Batch processes

A batch process is typically a script or a command that's been scheduled to run at a specific date and time, possibly in a periodical fashion. In other words, batch processing is a background process that's spawned by a job scheduler. In most common cases, batch processes are resource-intensive tasks that are usually scheduled to run during less busy hours to avoid system overload. On Linux, the most commonly used tools for job scheduling are at and cron. While cron is better suited to scheduled task management complexities, at is a more lightweight utility, better suited for one-off jobs. A detailed study of these commands is beyond the scope of this chapter. You may refer to the related system reference manuals for more information (man at and man cron).

We'll conclude our study of process types with orphan and zombie processes.

Orphan and zombie processes

When a child process is terminated, the related parent process is notified with a SIGCHILD signal. The parent can go on running other tasks or may choose to spawn another child process. However, there may be instances when the parent process is terminated before a related child process completes execution (or exits). In this case, the child process becomes an **orphan** process. In Linux, the init process – the parent of all processes – automatically becomes the new parent of the orphan process.

Zombie processes (also known as **defunct** processes) are references to processes that have completed execution (and exited) but are still lingering in the system process table (according to the ps command).

The main difference between the zombie and orphan processes is that a zombie process is dead (terminated), while an orphan process is still running.

As we differentiate between various process types and their behavior, a significant part of the related information is reflected in the composition or data structure of the process itself. In the next section, we'll take a closer look at the makeup of a process, which is mostly echoed through the ps command-line utility – an ordinary yet very useful process explorer on Linux systems.

The anatomy of a process

In this section, we will explore some of the common attributes of a Linux process through the lens of the ps and top command-line utilities. We hope that taking a practical approach based on these tools will help you gain a better understanding of process internals, at least from a Linux administrator's perspective. Let's start by taking a brief look at these commands. The ps command displays a current snapshot of the system processes. This command has the following syntax:

```
ps [OPTIONS]
```

The top command provides a live (real-time) view of all the running processes in a system. Its syntax is as follows:

```
top [OPTIONS]
```

The following command displays the processes owned by the current Terminal session:

```
ps
```

The output of the preceding command can be seen in the following screenshot:

```
packt@neptune:~$ ps
    PID TTY          TIME CMD
 171233 pts/0    00:00:00 bash
 171253 pts/0    00:00:00 ps
```

Figure 5.5 – Displaying processes owned by the current shell

Let's look at each field in the top (header) row of the output and explain their meaning in the context of our relevant process; that is, the bash Terminal session.

PID

Each process in Linux has a PID automatically assigned by the kernel when the process is created. The PID is a positive integer and is always guaranteed to be unique.

In our case, the relevant process is bash (the current shell), with a PID of 171233.

TTY

TTY is short for **teletype**, more popularly known as a controlling Terminal or device for interacting with a system. In the context of a Linux process, the TTY attribute denotes the type of Terminal the process interacts with. In our example, the bash process representing the Terminal session has pts/0 as its TTY type. **PTS** or pts stands for **pseudo terminal slave** and indicates the input type – a Terminal console – controlling the process. /0 indicates the ordinal sequence of the related Terminal session. For example, an additional SSH session would have pts/1, and so on.

TIME

The TIME field represents the cumulative CPU utilization (or time) spent by the process (in [DD-]hh:mm:ss format). Why is it zero (00:00:00) for the bash process in our example? We may have run multiple commands in our Terminal session, yet the CPU utilization could still be zero. That's because the CPU utilization measures (and accumulates) the time spent for each command, and not the parent Terminal session overall. If the commands complete within a fraction of a second, they will not amount to a significant CPU utilization being shown in the TIME field.

Let's produce some noticeable CPU utilization with the following command:

```
while true; do x=1; done
```

If we leave the command running for a few seconds and then *Ctrl + C* out of it, we will get the following results:

```
packt@neptune:~$ while true; do x=1; done
^C
packt@neptune:~$ ps
    PID TTY          TIME CMD
 171233 pts/0     00:00:06 bash
 172010 pts/0     00:00:00 ps
```

Figure 5.6 – Producing a noticeable CPU utilization (TIME)

The preceding command runs a tight while loop, where the kernel allocates the required CPU cycles for processing. We need to keep in mind that the while loop is a simple sequence of instructions that make up a command, and that it will not create a process. The ensuing (and relatively taxing) command runs in the current shell instead. Consequently, the related CPU utilization accounts for the bash process.

CMD

The CMD field stands for command and indicates the name or full path of the command (including the arguments) that created the process. For well-known system commands (for example, bash), CMD displays the command's name, including its arguments.

The process attributes we've explored thus far represent a relatively simple view of Linux processes. There are situations when we may need more information. For example, the following command provides additional details about the processes running in the current Terminal session:

```
ps -l
```

The -l option parameter invokes the so-called *long format* for the ps output:

```
packt@neptune:~$ ps -l
F S   UID     PID    PPID  C PRI  NI ADDR SZ WCHAN  TTY          TIME CMD
0 S  1000  171233  171232  0  80   0  -  2103 do_wai pts/0    00:00:06 bash
0 R  1000  174897  171233  0  80   0  -  2199 -       pts/0    00:00:00 ps
```

Figure 5.7 – A more detailed view of processes

Here are just a few of the more relevant output fields of the ps command:

- F: Process flags (for example, 0 – none, 1 – forked, 4 – superuser privileges)
- S: Process status code (for example, R – running, S – interruptible sleep, and so on)
- UID: Username or owner of the process
- PID: Process ID
- PPID: Process ID of the parent process
- PRI: Priority of the process (a higher number means lower priority)
- SZ: Virtual memory usage

There are many more such attributes and exploring them all is beyond the scope of this book. For additional information, refer to the ps system reference manual (man ps).

The ps command examples we've used so far have only displayed the processes that are owned by the current Terminal session. This approach, we thought, would add less complexity to analyzing process attributes. Many of the process output fields displayed by the ps command are also reflected in the top command, albeit some of them with slightly different notations.

Let's look at the `top` command and the meaning of the output fields that are displayed. The following command displays a real-time view of running processes:

```
top
```

The output of the preceding command can be seen in the following screenshot:

```
top - 07:34:16 up 4 days,  3:53,  1 user,  load average: 0.08, 0.02, 0.01
Tasks: 286 total,   1 running, 285 sleeping,   0 stopped,   0 zombie
%Cpu(s):  0.2 us,   0.3 sy,  0.0 ni, 99.5 id,  0.0 wa,  0.0 hi,  0.0 si,  0.0 st
MiB Mem :   3908.4 total,    779.1 free,    667.0 used,   2462.3 buff/cache
MiB Swap:   3908.0 total,   3908.0 free,      0.0 used.   2943.2 avail Mem

   PID USER      PR  NI    VIRT    RES    SHR S  %CPU  %MEM     TIME+ COMMAND
171232 packt     20   0   13952   5416   3912 S   0.3   0.1   0:02.51 sshd
181940 packt     20   0    9368   3968   3160 R   0.3   0.1   0:00.10 top
     1 root      20   0  171112  13132   8300 S   0.0   0.3   0:12.66 systemd
     2 root      20   0       0      0      0 S   0.0   0.0   0:00.19 kthreadd
     3 root       0 -20       0      0      0 I   0.0   0.0   0:00.00 rcu_gp
     4 root       0 -20       0      0      0 I   0.0   0.0   0:00.00 rcu_par_gp
```

Figure 5.8 – A real-time view of the current processes

Here are some of the output fields, briefly explained:

- USER: Username or owner of the process.
- PR: Priority of the process (a lower number means higher priority).
- NI: Nice value of the process (a sort of dynamic/adaptive priority).
- VIRT: Virtual memory size (in KB) – the total memory used by the process.
- RES: Resident memory size (in KB) – the physical (non-swapped) memory used by the process.
- SHR: Shared memory size (in KB) – a subset of the process memory shared with other processes.
- S: Process status (for example, R – running, S – interruptible sleep, I – idle, and so on).
- %CPU: CPU usage (percentage).
- %MEM: RES memory usage (percentage).
- COMMAND: Command name or command line.

Each of these fields (and many more) are explained in detail in the `top` system reference manual (`man top`).

Every day, Linux administration tasks frequently use process-related queries based on the preceding presented fields. The *Working with processes* section will explore some of the more common usages of the ps and top commands, and beyond.

An essential aspect of a process's lifetime is the **status** (or **state**) of the process at any given time and the transition between these states. Both the ps and top commands provide information about the status of the process via the S field. Let's take a closer look at these states.

Process states

During its lifetime, a process may change states according to circumstances. According to the S (status) field of the ps and top commands, a Linux process can have any of the following states:

- D: Uninterruptible sleep
- I: Idle
- R: Running
- S: Sleeping (interruptible sleep)
- T: Stopped by job control signal
- t: Stopped by debugger during trace
- Z: Zombie

At a high level, any of these states can be identified with the following process states:

- Running
- Waiting
- Stopped
- Zombie

The following sections will briefly describe each of these states. Where relevant, the corresponding S field state attributes will be provided, according to the ps and top commands.

Running state

The process is currently running (the R state) or is an idle process (the I state). In Linux, an idle process is a specific task that's assigned to every processor (CPU) in the system and is scheduled to run only when there's no other process running on the related CPU. The time that's spent in idle tasks accounts for the idle time that's reported by the top command.

Waiting state

The process is waiting for a specific event or resource. There are two types of waiting states: interruptible sleep (the S state) and uninterruptible sleep (the D state). Interruptible sleep can be disturbed by specific process signals, yielding further process execution. On the other hand, uninterruptible sleep is a state where the process is blocked in a system call (possibly waiting on some hardware conditions), and it cannot be interrupted.

Stopped state

The process has stopped executing, usually due to a specific signal – a job control signal (the T state) or a debugging signal (the t state).

Zombie state

The process is defunct or dead (the Z state) – it's terminated without being reaped by its parent. A zombie process is essentially a dead reference for an already terminated process in the system's process table. You can more information about this in the *Orphan and zombie processes* section.

To conclude our analysis of the process states, let's look at the lifetime of a Linux process. Usually, a process starts with a running state (R) and terminates once its parent has reaped it from the zombie state (Z). The following diagram provides an abbreviated view of the process states and the possible transitions between them:

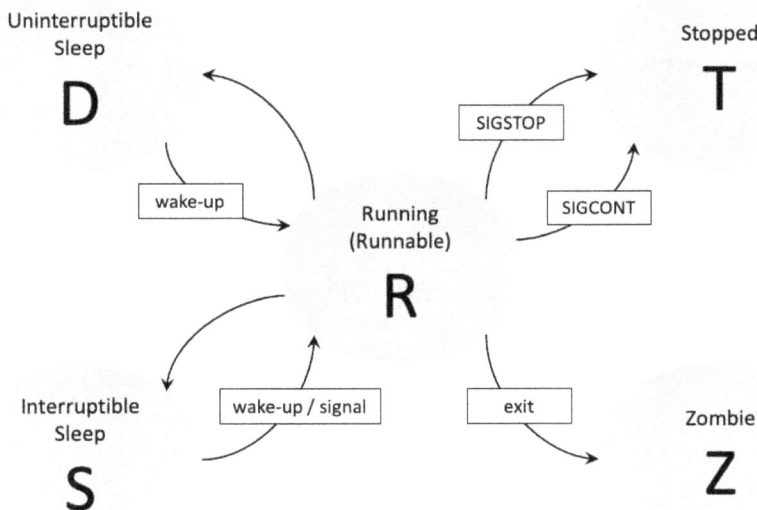

Figure 5.9 – The lifetime of a Linux process

Now that we've introduced processes and provided you with a preliminary idea of their type and structure, we're ready to interact with them. In the following sections, we will explore some standard command-line utilities for working with processes and daemons. Most of these tools operate with input and output data, which we covered in the *Anatomy of a process* section. We'll look at working with processes next.

Working with processes

This section serves as a practical guide to managing processes via resourceful command-line utilities that are used in everyday Linux administration tasks. Some of these tools have already been mentioned in previous sections (for example, ps and top), when we covered specific process internals. Here, we will summon most of the knowledge we've gathered so far and take it for a real-world spin by covering some hands-on examples.

Let's start with the ps command – the Linux process explorer.

The ps command

We described the ps command and its syntax in the *Anatomy of a process* section. The following command displays a selection of the current processes running in the system:

```
ps -e | head
```

The -e option (or -A) selects *all* the processes in the system. The head pipe invocation displays only the first few lines (10 by default):

```
packt@neptune:~$ ps -e | head
    PID TTY          TIME CMD
      1 ?        00:00:14 systemd
      2 ?        00:00:00 kthreadd
      3 ?        00:00:00 rcu_gp
      4 ?        00:00:00 rcu_par_gp
      6 ?        00:00:00 kworker/0:0H-kblockd
      9 ?        00:00:00 mm_percpu_wq
     10 ?        00:00:01 ksoftirqd/0
     11 ?        00:00:15 rcu_sched
     12 ?        00:00:02 migration/0
```

Figure 5.10 – Displaying the first few processes

The preceding information may not always be particularly useful. Perhaps we'd like to know more about each process, beyond just the PID or CMD fields in the ps command's output. (We described some of these process attributes in the *Anatomy of a process* section.)

The following command lists the processes owned by the current user in a more elaborate fashion:

```
ps -fU $(whoami)
```

The -f option specifies the full-format listing, which displays more detailed information for each process. The -U $(whoami) option parameter specifies the current user (packt) as the real user (owner) of the processes we'd like to retrieve. In other words, we want to list all the processes we own:

```
packt@neptune:~$ ps -fU $(whoami)
UID          PID    PPID  C STIME TTY          TIME CMD
packt     171072       1  0 Oct03 ?        00:00:00 /lib/systemd/systemd --user
packt     171075  171072  0 Oct03 ?        00:00:00 (sd-pam)
packt     171081  171072  0 Oct03 ?        00:00:00 /usr/bin/pulseaudio --daemon
packt     171083  171072  0 Oct03 ?        00:00:00 /usr/libexec/tracker-miner-f
packt     171092  171072  0 Oct03 ?        00:00:00 /usr/bin/dbus-daemon --sessi
packt     171107  171072  0 Oct03 ?        00:00:00 /usr/libexec/gvfsd
packt     171123  171072  0 Oct03 ?        00:00:00 /usr/libexec/gvfsd-fuse /run
packt     171124  171072  0 Oct03 ?        00:00:00 /usr/libexec/gvfs-udisks2-vo
packt     171137  171072  0 Oct03 ?        00:00:00 /usr/libexec/gvfs-gphoto2-vo
packt     171142  171072  0 Oct03 ?        00:00:00 /usr/libexec/gvfs-goa-volume
packt     171146  171072  0 Oct03 ?        00:00:00 /usr/libexec/goa-daemon
packt     171154  171072  0 Oct03 ?        00:00:00 /usr/libexec/goa-identity-se
packt     171159  171072  0 Oct03 ?        00:00:00 /usr/libexec/gvfs-mtp-volume
packt     171164  171072  0 Oct03 ?        00:00:05 /usr/libexec/gvfs-afc-volume
packt     171232  171052  0 Oct03 ?        00:00:05 sshd: packt@pts/0
packt     171233  171232  0 Oct03 pts/0    00:00:06 -bash
packt     242150  171233  0 05:05 pts/0    00:00:00 ps -fU packt
```

Figure 5.11 – Displaying the processes owned by the current user

There are situations when we may look for a specific process, either for monitoring purposes or to act upon them. Let's take a previous example, where we showcased a long-lived process and wrapped the related command into a simple script. The command is a simple while loop that runs indefinitely:

```
while true; do x=1; done
```

Using an editor of our preference (for example, nano), we can create a script file
(for example, test.sh) with the following content:

```
packt@neptune:~$ cat test.sh
#!/bin/bash
while true; do x=1; done
```

Figure 5.12 – A simple test script running indefinitely

We can make the test script executable and run it as a background process:

```
chmod +x test.sh
./test.sh &
```

Note the ampersand (&) at the end of the command, which invokes the background process:

```
packt@neptune:~$ ./test.sh &
[1] 243436
```

Figure 5.13 – Running a script as a background process

The background process running our script has a process ID (PID) of 243436. Suppose
we want to find our process by its name (test.sh). For this, we can use the ps
command with a grep pipe:

```
ps -ef | grep test.sh
```

The output of the preceding command can be seen in the following screenshot:

```
packt@neptune:~$ ps -ef | grep test.sh
packt     243436 171233 99 05:44 pts/0    00:04:39 /bin/bash ./test.sh
packt     243560 171233  0 05:49 pts/0    00:00:00 grep --color=auto test.sh
```

Figure 5.14 – Finding a process by name using the ps command

The preceding output shows that our process has a PID of 243436 and a CMD of /bin/
bash ./test.sh. The CMD field contains the full command invocation of our script,
including the command-line parameters.

We should note that the first line of the test.sh script contains #!/bin/bash, which prompts the OS to invoke bash for the script's execution. This line is also known as the **shebang** line, and it has to be the first line in a bash script. To make more sense of the CMD field, the command in our case is /bin/bash (according to the shebang invocation), and the related command-line parameter is the test.sh script. In other words, bash executes the test.sh script.

The output of the preceding ps command also includes our ps | grep command's invocation, which is somewhat irrelevant. A refined version of the same command is as follows:

```
ps -ef | grep test.sh | grep -v grep
```

The output of the preceding command can be seen in the following screenshot:

```
packt@neptune:~$ ps -ef | grep test.sh | grep -v grep
packt      243436  171233 99 05:44 pts/0      00:23:27 /bin/bash ./test.sh
```

Figure 5.15 – Finding a process by name using the ps command (refined)

The grep -v grep pipe filters out the unwanted grep invocation from the ps command's results.

If we want to find a process based on a process ID (PID), we can invoke the ps command with the -p | --pid option parameter. For example, the following command displays detailed information about our process with PID set to 243436 (running the test.sh script):

```
packt@neptune:~$ ps -fp 243436
UID          PID     PPID  C STIME TTY          TIME CMD
packt      243436  171233 99 05:44 pts/0      00:52:14 /bin/bash ./test.sh
```

Figure 5.16 – Finding a process by PID using the ps command

The -f option displays the detailed (*long-format*) process information.

There are numerous other use cases for the ps command, and exploring them all is well beyond the scope of this book. The invocations we've enumerated here should provide a basic exploratory guideline for you. For more information, please refer to the ps system reference manual (man ps).

The pstree command

pstree shows the running processes as a hierarchical, tree-like view. In some respects, pstree acts as a visualizer of the ps command. The root of the pstree command's output is either the init process or the process with the PID specified in the command. The syntax of the pstree command is as follows:

```
pstree [OPTIONS] [PID] [USER]
```

The following command displays the process tree of our current Terminal session:

```
pstree $(echo $$)
```

The output of the preceding command can be seen in the following screenshot:

```
packt@neptune:~$ pstree $(echo $$)
bash─┬─pstree
     └─test.sh
```

Figure 5.17 – The process tree of the current Terminal session

In the preceding command, echo $$ provides the PID of the current Terminal session. The PID is wrapped as the argument for the pstree command. To show the related PIDs, we can invoke the pstree command with the -p | --show-pids option:

```
pstree -p $(echo $$)
```

The output of the preceding command can be seen in the following screenshot:

```
packt@neptune:~$ pstree -p $(echo $$)
bash(246908)─┬─pstree(262214)
             └─test.sh(243436)
```

Figure 5.18 – The process tree (along with its PIDs) of the current Terminal session

The following command shows the processes owned by the current user:

```
pstree $(whoami)
```

The output of the preceding command can be seen in the following screenshot:

```
packt@neptune:~$ pstree $(whoami)
sshd───bash─┬─pstree
            └─test.sh
```

Figure 5.19 – The process tree owned by the current user

For more information about the `pstree` command, please refer to the related system reference manual (`man pstree`).

The top command

When it comes to monitoring processes in real time, the `top` utility is among the most common tool to be used by Linux administrators. The related command-line syntax is as follows:

```
top [OPTIONS]
```

The following command displays all the processes currently running in the system, along with real-time updates (on memory, CPU usage, and so on):

```
top
```

The output of the preceding command can be seen in the following screenshot:

```
top - 07:04:39 up 5 days, 21:28,  1 user,  load average: 1.02, 1.02, 1.00
Tasks: 289 total,   2 running, 287 sleeping,   0 stopped,   0 zombie
%Cpu(s): 50.0 us,   0.2 sy,  0.0 ni, 49.5 id,  0.0 wa,  0.0 hi,  0.3 si,  0.0 s
MiB Mem :   3908.4 total,    694.9 free,    674.0 used,   2539.5 buff/cache
MiB Swap:   3908.0 total,   3908.0 free,      0.0 used.   2936.2 avail Mem

    PID USER       PR  NI    VIRT    RES    SHR S  %CPU  %MEM     TIME+
 243436 packt      20   0    6892   1160   1016 R  99.3   0.0  80:02.27
    821 root       20   0  238004   8132   6704 S   0.7   0.2   9:59.74
    425 root       20   0       0      0      0 S   0.3   0.0   1:45.87
    497 root       19  -1  300472 127076 125380 S   0.3   3.2   2:02.89
    919 root       20   0  259324  18972  16136 S   0.3   0.5   0:10.34
 245819 packt      20   0    9368   4036   3268 R   0.3   0.1   0:00.04
      1 root       20   0  171112  13132   8300 S   0.0   0.3   0:15.05
```

Figure 5.20 – Displaying the running processes in real time using the top command

Pressing *Q* will exit the `top` command. By default, the `top` command sorts the output by CPU usage (shown in the `%CPU` field/column). We can see our test script process (with PID `243436`) at the very top of the output (with a CPU usage of `99.3%`).

We can also choose to sort the output of the `top` command by a different field. While `top` is running, press *Shift + F* (F) to invoke interactive mode:

```
Fields Management for window 1:Def, whose current sort field is %CPU
     Navigate with Up/Dn, Right selects for move then <Enter> or Left commits,
     'd' or <Space> toggles display, 's' sets sort.  Use 'q' or <Esc> to end!

 * PID       = Process Id        TTY      = Controlling T   vMj      = Major Faults
 * USER      = Effective Use     TPGID    = Tty Process G   vMn      = Minor Faults
 * PR        = Priority          SID      = Session Id      USED     = Res+Swap Size
 * NI        = Nice Value        nTH      = Number of Thr   nsIPC    = IPC namespace
 * VIRT      = Virtual Image     P        = Last Used Cpu   nsMNT    = MNT namespace
 * RES       = Resident Size     TIME     = CPU Time        nsNET    = NET namespace
 * SHR       = Shared Memory     SWAP     = Swapped Size    nsPID    = PID namespace
 * S         = Process Statu     CODE     = Code Size (Ki   nsUSER   = USER namespac
 * %CPU      = CPU Usage         DATA     = Data+Stack (K   nsUTS    = UTS namespace
 * %MEM      = Memory Usage      nMaj     = Major Page Fa   LXC      = LXC container
 * TIME+     = CPU Time, hun     nMin     = Minor Page Fa   RSan     = RES Anonymous
 * COMMAND   = Command Name/     nDRT     = Dirty Pages C   RSfd     = RES File-base
   PPID      = Parent Proces     WCHAN    = Sleeping in F   RSlk     = RES Locked (K
   UID       = Effective Use     Flags    = Task Flags <s   RSsh     = RES Shared (K
   RUID      = Real User Id      CGROUPS  = Control Group   CGNAME   = Control Group
   RUSER     = Real User Nam     SUPGIDS  = Supp Groups I   NU       = Last Used NUM
   SUID      = Saved User Id     SUPGRPS  = Supp Groups N
   SUSER     = Saved User Na     TGID     = Thread Group
   GID       = Group Id          OOMa     = OOMEM Adjustm
   GROUP     = Group Name        OOMs     = OOMEM Score c
   PGRP      = Process Group     ENVIRON  = Environment v
```

Figure 5.21 – The top command in interactive mode (Shift + F)

Using the arrow keys, we can select the desired field to sort by (for example, %MEM), then press *S* to set the new field, followed by *Q* to exit interactive mode. The alternative to interactive mode sorting is invoking the -o option parameter of the `top` command, which specifies the sorting field.

For example, the following command lists the top 10 processes, sorted by CPU usage:

```
top -b -o %CPU | head -n 17
```

Similarly, the following command lists the top 10 processes, sorted by CPU and memory usage:

```
top -b -o +%MEM | head -n 17
```

The -b option parameter specifies the batch mode operation (instead of the default interactive mode). The -o +%MEM option parameter indicates the additional (+) sorting field (%MEM) in tandem with the default %CPU field. The head -n 17 pipe selects the first 17 lines of the output, accounting for the seven-line header of the `top` command:

```
packt@neptune:~$ top -b -o +%MEM | head -n 17
top - 07:55:13 up 5 days, 22:19,  1 user,  load average: 1.00, 1.00, 1.00
Tasks: 287 total,   2 running, 285 sleeping,   0 stopped,   0 zombie
%Cpu(s): 51.6 us,  0.0 sy,  0.0 ni, 48.4 id,  0.0 wa,  0.0 hi,  0.0 si,  0.0 st
MiB Mem :   3908.4 total,    696.9 free,    670.3 used,   2541.2 buff/cache
MiB Swap:   3908.0 total,   3908.0 free,      0.0 used.   2939.9 avail Mem

    PID USER       PR  NI    VIRT    RES    SHR S  %CPU  %MEM     TIME+ COMMAND
   1214 gdm        20   0 3867872 167128  90460 S   0.0   4.2   2:25.77 gnome-s+
    497 root       19  -1  300472 128528 126832 S   0.0   3.2   2:03.42 systemd+
   1550 gdm        20   0  464664  57380  39512 S   0.0   1.4   0:00.57 gsd-xse+
   1547 gdm        20   0  315956  55916  38764 S   0.0   1.4   0:00.38 ibus-x11
   1386 gdm        20   0  444916  52516  33664 S   0.0   1.3   0:00.64 Xwayland
 171146 packt      20   0  539436  35696  29872 S   0.0   0.9   0:00.05 goa-dae+
   1206 gdm        20   0  539312  35636  29812 S   0.0   0.9   0:00.46 goa-dae+
    955 root       20   0 1007076  29792  15960 S   0.0   0.7   0:39.38 snapd
   1422 gdm        20   0 2522088  26716  21548 S   0.0   0.7   0:00.05 gjs
   1436 gdm        20   0  606016  24704  19468 S   0.0   0.6   0:00.44 gsd-med+
```

Figure 5.22 – The top 10 processes sorted by CPU and memory usage

The following command lists the top five processes by CPU usage, owned by the current user (`packt`):

```
top -u $(whoami) -b -o %CPU | head -n 12
```

The output of the preceding command can be seen in the following screenshot:

```
packt@neptune:~$ top -u $(whoami) -b -o %CPU | head -n 12
top - 08:06:20 up 5 days, 22:30,  1 user,  load average: 1.15, 1.05, 1.01
Tasks: 287 total,   2 running, 285 sleeping,   0 stopped,   0 zombie
%Cpu(s): 50.0 us,  3.1 sy,  0.0 ni, 46.9 id,  0.0 wa,  0.0 hi,  0.0 si,  0.0 st
MiB Mem :   3908.4 total,    695.6 free,    670.8 used,   2542.0 buff/cache
MiB Swap:   3908.0 total,   3908.0 free,      0.0 used.   2939.5 avail Mem

    PID USER       PR  NI    VIRT    RES    SHR S  %CPU  %MEM     TIME+ COMMAND
 243436 packt      20   0    6892   1160   1016 R 100.0   0.0 141:37.78 test.sh
 247657 packt      20   0    9368   3844   3276 R   6.2   0.1   0:00.01 top
 171072 packt      20   0   18780  10208   8340 S   0.0   0.3   0:00.45 systemd
 171075 packt      20   0  172364   5012     16 S   0.0   0.1   0:00.00 (sd-pam)
 171083 packt      39  19  582472  24360  16304 S   0.0   0.6   0:00.47 tracker+
```

Figure 5.23 – The top five processes owned by the current user, sorted by CPU usage

The `-u $(whoami)` option parameter specifies the current user for the `top` command.

With the `top` command, we can also monitor specific processes using the `-p` PID option parameter. For example, the following command monitors our test process (with PID `243436`):

```
top -p 243436
```

The output of the preceding command can be seen in the following screenshot:

```
top - 08:23:24 up 5 days, 22:47,  1 user,  load average: 1.02, 1.22, 1.15
Tasks:   1 total,   1 running,   0 sleeping,   0 stopped,   0 zombie
%Cpu(s): 50.0 us,   0.0 sy,   0.0 ni, 49.8 id,   0.0 wa,   0.0 hi,   0.2 si,   0.0 st
MiB Mem :   3908.4 total,    694.5 free,    671.1 used,    2542.9 buff/cache
MiB Swap:   3908.0 total,   3908.0 free,      0.0 used.   2939.2 avail Mem

   PID USER      PR  NI    VIRT    RES    SHR S  %CPU  %MEM     TIME+ COMMAND
243436 packt     20   0    6892   1160   1016 R 100.0   0.0 158:40.02 test.sh
```

Figure 5.24 – Monitoring a specific process ID (PID) with the top command

We may choose to *kill* the process by pressing *K* while using the `top` command. We'll get prompted for this by the PID of the process we want to terminate:

```
top - 08:25:27 up 5 days, 22:49,  1 user,  load average: 1.00, 1.14, 1.12
Tasks:   1 total,   1 running,   0 sleeping,   0 stopped,   0 zombie
%Cpu(s): 49.9 us,   0.0 sy,   0.0 ni, 49.8 id,   0.0 wa,   0.0 hi,   0.3 si,   0.0 st
MiB Mem :   3908.4 total,    694.2 free,    671.1 used,    2543.1 buff/cache
MiB Swap:   3908.0 total,   3908.0 free,      0.0 used.   2939.2 avail Mem
PID to signal/kill [default pid = 243436]
   PID USER      PR  NI    VIRT    RES    SHR S  %CPU  %MEM     TIME+ COMMAND
243436 packt     20   0    6892   1160   1016 R 100.0   0.0 160:43.10 test.sh
```

Figure 5.25 – Killing a process with the top command

The `top` utility can be used in many creative ways. We hope that the examples we've provided in this section have inspired you to explore further use cases based on specific needs. For more information, please refer to the system reference manual for the `top` command (`man top`).

The kill and killall commands

We use the `kill` command to terminate processes. The command's syntax is as follows:

```
kill [OPTIONS] [ -s SIGNAL | -SIGNAL ] PID [...]
```

The `kill` command sends a *signal* to a process, attempting to stop its execution. When no signal is specified, `SIGTERM` (`15`) is sent. A signal can either be specified by the signal's name without the `SIG` prefix (for example, `KILL` for `SIGKILL`) or by value (for example, `9` for `SIGKILL`).

The `kill -l` and `kill -L` commands provide a full list of signals that can be used in Linux:

```
packt@neptune:~$ kill -l
 1) SIGHUP       2) SIGINT       3) SIGQUIT      4) SIGILL       5) SIGTRAP
 6) SIGABRT      7) SIGBUS       8) SIGFPE       9) SIGKILL     10) SIGUSR1
11) SIGSEGV     12) SIGUSR2     13) SIGPIPE     14) SIGALRM     15) SIGTERM
16) SIGSTKFLT   17) SIGCHLD     18) SIGCONT     19) SIGSTOP     20) SIGTSTP
21) SIGTTIN     22) SIGTTOU     23) SIGURG      24) SIGXCPU     25) SIGXFSZ
26) SIGVTALRM   27) SIGPROF     28) SIGWINCH    29) SIGIO       30) SIGPWR
31) SIGSYS      34) SIGRTMIN    35) SIGRTMIN+1  36) SIGRTMIN+2  37) SIGRTMIN+3
38) SIGRTMIN+4  39) SIGRTMIN+5  40) SIGRTMIN+6  41) SIGRTMIN+7  42) SIGRTMIN+8
43) SIGRTMIN+9  44) SIGRTMIN+10 45) SIGRTMIN+11 46) SIGRTMIN+12 47) SIGRTMIN+13
48) SIGRTMIN+14 49) SIGRTMIN+15 50) SIGRTMAX-14 51) SIGRTMAX-13 52) SIGRTMAX-12
53) SIGRTMAX-11 54) SIGRTMAX-10 55) SIGRTMAX-9  56) SIGRTMAX-8  57) SIGRTMAX-7
58) SIGRTMAX-6  59) SIGRTMAX-5  60) SIGRTMAX-4  61) SIGRTMAX-3  62) SIGRTMAX-2
63) SIGRTMAX-1  64) SIGRTMAX
```

Figure 5.26 – The Linux signals

Each signal has a numeric value, as shown in the preceding output. For example, SIGKILL equals 9. The following command will kill our test process (with PID 243436):

```
kill -9 243436
```

The following command will also do the same as the preceding command:

```
kill -KILL 243436
```

In some scenarios, we may want to kill multiple processes in one go. The `killall` command comes to the rescue here. The syntax for the `killall` command is as follows:

```
killall [OPTIONS] [ -s SIGNAL | -SIGNAL ] NAME...
```

`killall` sends a signal to all the processes running any of the commands specified. When no signal is specified, SIGTERM (15) is sent. A signal can either be specified by the signal name without the SIG prefix (for example, TERM for SIGTERM) or by value (for example, 15 for SIGTERM).

For example, the following command terminates all the processes running the `test.sh` script:

```
killall -e -TERM test.sh
```

The output of the preceding command can be seen in the following screenshot:

```
packt@neptune:~$ killall -e -TERM test.sh
[1]    Terminated              ./test.sh
[2]-   Terminated              ./test.sh
[3]+   Terminated              ./test.sh
```

Figure 5.27 – Terminating multiple processes with killall

Killing a process will usually remove the related reference from the system process table. The terminated process won't show up anymore in the output of ps, top, or similar commands.

For more information about the kill and killall commands, please refer to the related system reference manuals (man kill and man killall).

The pgrep and pkill commands

pgrep and pkill are pattern-based lookup commands for exploring and terminating running processes. They have the following syntax:

```
pgrep [OPTIONS] PATTERN
pkill [OPTIONS] PATTERN
```

pgrep iterates through the current processes and lists the process IDs matching the selection pattern or criteria. Similarly, pkill terminates the processes matching the selection criteria.

The following command looks for our test process (test.sh) and displays the PID if the related process is found:

```
pgrep -f test.sh
```

The output of the preceding command can be seen in the following screenshot:

```
packt@neptune:~$ pgrep -f test.sh
243436
```

Figure 5.28 – Looking for a process ID based on name using pgrep

The -f | --full option enforces a full name match of the process we're looking for. We may use pgrep in tandem with the ps command to get more detailed information about the process, like so:

```
pgrep -f test.sh | xargs ps -fp
```

The output of the preceding command can be seen in the following screenshot:

```
packt@neptune:~$ pgrep -f test.sh | xargs ps -fp
UID          PID     PPID  C STIME TTY          TIME CMD
packt     243436        1 99 05:44 ?        03:22:58 /bin/bash ./test.sh
```

Figure 5.29 – Chaining pgrep and ps for more information

In the preceding one-liner, we piped the output of the pgrep command (with PID 243436) to the ps command, which has been invoked with the -f (long-format) and -p | --pid options. The -p option parameter gets the piped PID value.

To terminate our test.sh process, we simply invoke the pkill command, as follows:

```
pkill -f test.sh
```

The preceding command will *silently* kill the related process, based on the full name lookup enforced by the -f | --full option. To get some feedback from the action of the pkill command, we need to invoke the -e | --echo option, like so:

```
pkill -ef test.sh
```

The output of the preceding command can be seen in the following screenshot:

```
packt@neptune:~$ pkill -ef test.sh
test.sh killed (pid 243436)
[1]+  Terminated              ./test.sh
```

Figure 5.30 – Killing a process by name using pkill

For more information, please refer to the pgrep and pkill system reference manuals (man pgrep and man pkill).

This section covered some command-line utilities that are frequently used in everyday Linux administration tasks involving processes. Keep in mind that in Linux, most of the time, there are many ways to accomplish a specific task. We hope that the examples in this section will help you come up with creative methods and techniques for working with processes.

Next, we'll look at some common ways of interacting with daemons.

Working with daemons

As noted in the introductory sections, daemons are a special breed of background process. Consequently, the vast majority of methods and techniques for working with processes also apply to daemons. However, there are specific commands that strictly operate on daemons when it comes to managing (or controlling) the lifetime of the related processes.

As noted in the *Daemons* section, daemon processes are controlled by shell scripts, usually stored in the `/etc/init.d/` or `/lib/systemd/` system directories, depending on the Linux platform. On legacy Linux systems (for example, CentOS/RHEL 6) and Ubuntu (even in the latest distros), the daemon script files are stored in `/etc/init.d/`. On CentOS/RHEL 7 and newer platforms, they are typically stored in `/lib/systemd/`.

The location of the daemon files and the daemon command-line utilities largely depends on the `init` initialization system and service manager. In *The init process* section, we briefly mentioned a variety of `init` systems across Linux distributions. To illustrate the use of daemon control commands, we will explore a couple of `init` systems – `SysV` and `systemd` – that are extensively used across various Linux platforms.

Working with SysV daemons

On legacy Linux platforms (for example, CentOS/RHEL 6 and older), the `init` implementation typically follows the **SysV** or **System V** Linux system process initialization mechanism. `SysV` (pronounced *Sys Five* or *System Five*) is essentially a legacy service controller and service management platform, still present in all major Linux distributions, mostly for backward compatibility reasons. We won't go into the details of `SysV` due to the limited scope of this chapter.

In `SysV` environments, we typically use the `service` command to explore and control daemons. For example, the following command displays all active daemons (services):

```
service --status-all
```

To look for the status of a specific daemon (for example, `httpd`), we can use the following command:

```
service httpd status
```

The output of the preceding command can be seen in the following screenshot:

```
root:/# service httpd status
httpd (pid 6694) is running...
```

Figure 5.31 – Checking the status of a SysV daemon (httpd) in CentOS/RHEL 6

We can control a daemon (for example, `httpd`) invoking `start`, `stop`, and `restart` with the `service` command. For example, the following command will stop the `httpd` service:

```
service httpd stop
```

The output of the preceding command can be seen in the following screenshot:

```
root:/# service httpd stop
Stopping httpd:                                    [  OK  ]
```

Figure 5.32 – Stopping a SysV daemon (httpd) in CentOS/RHEL 6

For more information on the `service` command, refer to the related system reference manual (`man service`).

To *enable* or *disable* a service in CentOS/RHEL 6, we can use the `chkconfig` command. For example, the following command disables the `httpd` daemon:

```
chkconfig httpd off
```

The output of the preceding command can be seen in the following screenshot:

```
root:/# chkconfig httpd off
Note: Forwarding request to 'systemctl disable httpd.service'.
Removed symlink /etc/systemd/system/multi-user.target.wants/httpd.service.
```

Figure 5.33 – Disabling a SysV daemon (httpd) in CentOS/RHEL 6

The `chkconfig` command is also used to list the daemons that are controlled via `SysV` scripts and modify their **run level**, also known as **runlevels**. On Linux systems, runlevels are `SysV`-style `init` constructs that indicate the availability of the services at specific running stages of the system. Since runlevels became obsolete with the growing trend of `systemd init` systems, we won't cover the related topics. For more information on runlevel, please refer to the related system reference manual (`man runlevel`).

The following command lists all `SysV` services (for example, on a CentOS/RHEL 6 system):

```
chkconfig --list
```

For more information about the `chkconfig` command, please refer to the related system reference manual (`man chkconfig`).

Working with systemd daemons

The `init` system's essential requirement is to initialize and orchestrate the launch and startup dependencies of various processes when the Linux kernel is booted. These processes are also known as **userland** or **user processes**. The `init` engine also controls the services and daemons while the system is running.

Over the last few years, most Linux platforms have transitioned to `systemd` as their default `init` engine. Due to its extensive adoption, being familiar with `systemd` and its related command-line tools is of paramount importance. With that in mind, this section's primary focus is on `systemctl` – the central command-line utility for managing `systemd` daemons.

The syntax of the `systemctl` command is as follows:

```
systemctl [OPTIONS] [COMMAND] [UNITS...]
```

The actions that are invoked by the `systemctl` command are directed at units, which are system resources that are managed by `systemd`. Several unit types are defined in `systemd` (for example, service, mount, socket, and so on). Each of these units has a corresponding file. These file's types are inferred from the suffix of the related filename; for example, `httpd.service` is the service unit file of the Apache web service (daemon). For a comprehensive list of `systemd` units and detailed descriptions of them, please refer to the `systemd.unit` system reference manual (man `systemd.unit`)

The following command enables a daemon (for example, `httpd`) to start at boot:

```
systemctl enable httpd
```

The output of the preceding command can be seen in the following screenshot:

```
[packt@jupiter ~]$ systemctl enable httpd
==== AUTHENTICATING FOR org.freedesktop.systemd1.manage-unit-files =
===Ivice
Authentication is required to manage system service or unit files.
Authenticating as: Packt (packt)
Password:
==== AUTHENTICATION COMPLETE ====
==== AUTHENTICATING FOR org.freedesktop.systemd1.reload-daemon ====
Authentication is required to reload the systemd state.
Authenticating as: Packt (packt)
Password:
==== AUTHENTICATION COMPLETE ====
```

Figure 5.34 – Enabling a systemd daemon (httpd)

Typically, invoking `systemctl` commands requires superuser privileges. We should note that `systemctl` does not require the `.service` suffix when we're targeting service units. The following invocation is also acceptable:

```
systemctl enable httpd.service
```

The command to disable the `httpd` service from starting at boot is as follows:

```
systemctl disable httpd
```

To query the status of the `httpd` service, we can run the following command:

```
systemctl status httpd
```

The output of the preceding command can be seen in the following screenshot:

```
[packt@jupiter ~]$ systemctl status httpd
• httpd.service - The Apache HTTP Server
   Loaded: loaded (/usr/lib/systemd/system/httpd.service; enabled; ▶
   Active: active (running) since Mon 2020-10-05 17:44:15 PDT; 1h 2▶
     Docs: man:httpd.service(8)
 Main PID: 10747 (httpd)
   Status: "Total requests: 1; Idle/Busy workers 100/0;Requests/sec▶
    Tasks: 213 (limit: 23810)
   Memory: 28.1M
   CGroup: /system.slice/httpd.service
           ─10747 /usr/sbin/httpd -DFOREGROUND
           ─10749 /usr/sbin/httpd -DFOREGROUND
           ─10750 /usr/sbin/httpd -DFOREGROUND
           ─10753 /usr/sbin/httpd -DFOREGROUND
           ─10829 /usr/sbin/httpd -DFOREGROUND
```

Figure 5.35 – Querying the status of a systemd daemon (httpd)

Alternatively, we can check the status of the `httpd` service with the following command:

```
systemctl is-active httpd
```

The output of the preceding command can be seen in the following screenshot:

```
[packt@jupiter ~]$ systemctl is-active httpd
active
```

Figure 5.36 – Querying the active status of a systemd daemon (httpd)

The following commands stop or start the `httpd` service:

```
systemctl stop httpd
systemctl start httpd
```

For more information on `systemctl`, please refer to the related system reference manual (`man systemctl`). For more information about `systemd` internals, please refer to the corresponding reference manual (`man systemd`).

Working with processes and daemons is a constant theme of everyday Linux administration tasks. Mastering the related command-line utilities is an essential skill for any seasoned user. Yet, a running process or daemon should also be considered in relationships with other processes or daemons running either locally or on remote systems. The way processes communicate with each other could be a slight mystery to some. In the next section, we will explain how interprocess communication works.

Exploring interprocess communication

Interprocess communication (**IPC**) is a way of interacting between processes using a shared mechanism or interface. In this section, we will take a practical approach to exploring various communication mechanisms between processes. Linux processes can typically share data and synchronize their actions via the following interfaces:

- Shared storage (files)
- Shared memory
- Named and unnamed pipes
- Message queues
- Sockets
- Signals

To illustrate most of these communication mechanisms, we will build our examples using a model of **producer** and **consumer** processes. The producer and consumer share a common interface, where the producer writes some data that's read by the consumer. IPC mechanisms are usually implemented in distributed systems, built around more or less complex applications. Our examples will use simple bash scripts (`producer.sh` and `consumer.sh`), thus mimicking the producer and consumer processes. We hope that the use of such simple models will still provide a reasonable analogy for real-world applications.

Let's look at each of the IPC mechanisms we enumerated previously.

Shared storage

In its simplest form, the shared storage of an IPC mechanism can be a simple file that's been saved to disk. The producer then writes to a file while the consumer reads from the same file. In this simple use case, the obvious challenge is the integrity of the read/write operations due to possible race conditions between the underlying operations. To avoid race conditions, the file must be locked during write operations to prevent overlapping I/O with another read or write action. To keep things simple, we're not going to resolve this problem in our naive examples, but we thought it's worth calling it out.

Here are the producer (left) and consumer (right) scripts:

```
 1 #!/bin/bash                                    1 #!/bin/bash
 2                                                 2
 3 # producer.sh                                   3 # consumer.sh
 4                                                 4
 5 STORAGE_FILE="./storage"                        5 STORAGE_FILE="./storage"
 6                                                 6
 7 rm -f "${STORAGE_FILE}"                         7 while true; do
 8                                                 8     while IFS= read -r line; do
 9 while true; do                                  9         echo "${line}"
10     for (( i=1; i<=10; i++ )); do              10     done < "${STORAGE_FILE}"
11         uid="$(uuidgen)"                        11     sleep 1s
12         echo "${uid}"                          12 done
13         echo "${uid}" >> "${STORAGE_FILE}"      ~
14     done                                        ~
15     sleep 5s                                    ~
16 done                                            ~
```

Figure 5.37 – The producer (left) and consumer (right) scripts (using shared storage)

In our example, the producer writes a new set of data (10 random UUID strings) every 5 seconds to the `storage` file. The consumer reads the content of the `storage` file every second:

```
packt@neptune:~$ ./producer.sh          packt@neptune:~$ ./consumer.sh
b6bffe5e-2fe4-42a4-a95d-163ceea5ee68     b6bffe5e-2fe4-42a4-a95d-163ceea5ee68
3bac9255-78b2-4169-a322-ab2e97a67c00     3bac9255-78b2-4169-a322-ab2e97a67c00
ba2f5b56-f9eb-4b61-b43e-ea74f0276659     ba2f5b56-f9eb-4b61-b43e-ea74f0276659
405266df-0d08-4961-8683-cd98a03c856e     405266df-0d08-4961-8683-cd98a03c856e
e7bd1bbe-5fe3-41f0-b3b1-4d3300dfc217     e7bd1bbe-5fe3-41f0-b3b1-4d3300dfc217
76ea41d3-e1c7-46d1-8e10-4dc0c145baff     76ea41d3-e1c7-46d1-8e10-4dc0c145baff
676838dc-9077-4f63-9fa1-a01710cb9423     676838dc-9077-4f63-9fa1-a01710cb9423
f84b252e-f0d7-4ac9-b0c9-4679bf18caa8     f84b252e-f0d7-4ac9-b0c9-4679bf18caa8
edf47528-5e12-41e9-80af-faab36151c9a     edf47528-5e12-41e9-80af-faab36151c9a
```

Figure 5.38 – The producer (left) and consumer (right) communicating through shared storage

The producer and consumer data feeds are identical at any time. The two processes communicate via the shared `storage` file.

Shared memory

Processes in Linux typically have separate address spaces. A process can only access data in the memory of another process if the two share a common memory segment where such data would be stored. Linux provides at least a couple of **Application Programming Interfaces** (**APIs**) to programmatically define and control shared memory between processes: a legacy System V API and the more recent POSIX API. Both these APIs are written in C, though the implementation of the producer and consumer mockups is beyond the scope of this book. However, we can closely match the shared memory approach by using the /dev/shm temporary file storage system, which uses the system's RAM as its backing store (that is, the RAM disk).

With `/dev/shm` being used as shared memory, we can reuse our producer-consumer model from the *Shared storage* section example, where we simply point the storage file to `/dev/shm/storage`.

The shared memory and shared storage IPC models may not perform well with large amounts of data, especially massive data streams. The alternative would be to use IPC channels, which can be enabled through the pipe, message queue, or socket communication layers.

Unnamed pipes

Unnamed or **anonymous** pipes, also known as **regular** pipes, feed the output of a process to the input of another one. Using our producer-consumer model, the simplest way to illustrate an unnamed pipe as an IPC mechanism between the two processes would be to do the following:

```
producer.sh | consumer.sh
```

The key element of the preceding illustration is the pipe (|) symbol. The left-hand side of the pipe produces an output that's fed directly to the right-hand side of the pipe for consumption. To accommodate the anonymous pipe IPC layer, we'll make a slight modification to our producer and consumer scripts:

```
1 #!/bin/bash                       1 #!/bin/bash
2                                    2
3 # producer.sh                     3 # consumer.sh
4                                    4
5 for (( i=1; i<=10; i++ )); do     5 echo "Consumer data:"
6     uid="$(uuidgen)"              6 echo "--------------"
7     echo "${uid}"                 7
8 done                              8 if [ -t 0 ]; then
                                     9     data="$*"
                                    10 else
                                    11     data=$(cat)
                                    12 fi
                                    13
                                    14 echo "${data}"
```

Figure 5.39 – The producer (left) and consumer (right) scripts (using an unnamed pipe)

In our modified implementation, the producer prints some data to the console
(10 random UUID strings). The consumer reads and displays either the data coming
through the /dev/stdin pipe or the input arguments if the pipe is empty. Line 8 in the
consumer.sh script checks the presence of piped data in /dev/stdin (0 for fd0):

```
if [ -t 0 ]; then
```

The output of the producer-consumer communication is shown in the following screenshot:

```
packt@neptune:~$ ./producer.sh | ./consumer.sh
Consumer data:
--------------
b1da7ad8-5636-4f14-a213-a6fb2c410595
28ae3c1f-06d0-4cc4-bffa-6da07669005f
67f6b577-216a-4032-86f6-17008a934f88
fcbaa50e-8424-4180-ae2b-c63aa0b8bba5
0080c23e-8c75-4168-a3d1-ad566a02701e
38a3a60a-7454-45fc-83ef-a5c6c045ee55
b37c8be3-080f-40e9-993d-2dba1a51d388
1886b132-7769-4dcb-a480-42b30101567d
3cb86a07-63ee-473a-8698-9cb08d7eddfa
248c4bc4-7307-4b35-9363-bb7a71e015a6
```

Figure 5.40 – The producer feeding data into a consumer through an unnamed pipe

The output clearly shows the data being printed out by the consumer process. (Note the
"Consumer data:" header preceding the UUID strings.)

One of the problems with IPC anonymous pipes is that the data that's fed between the
producer and consumer is not persisted through any kind of storage layer. If the producer
or consumer processes are terminated, the pipe is gone, and the underlying data is lost.
Named pipes solve this problem.

Named pipes

Named pipes, also known as **First In, First Outs** (**FIFOs**), are similar to traditional
(unnamed) pipes but substantially different in terms of their semantics. An unnamed
pipe only persists for as long as the related process is running. However, a named pipe has
backing storage and will last as long as the system is up, regardless of the running status of
the processes attached to the related IPC channel.

Typically, a named pipe acts as a file, and it can be deleted when it's no longer being used. Let's modify our producer and consumer scripts so that we can use a named pipe as their IPC channel:

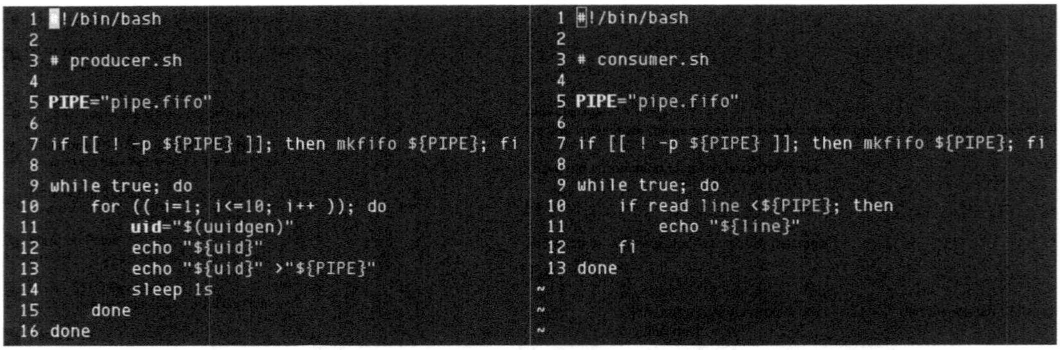

```
1 #!/bin/bash
2
3 # producer.sh
4
5 PIPE="pipe.fifo"
6
7 if [[ ! -p ${PIPE} ]]; then mkfifo ${PIPE}; fi
8
9 while true; do
10     for (( i=1; i<=10; i++ )); do
11         uid="$(uuidgen)"
12         echo "${uid}"
13         echo "${uid}" >"${PIPE}"
14         sleep 1s
15     done
16 done
```

```
1 #!/bin/bash
2
3 # consumer.sh
4
5 PIPE="pipe.fifo"
6
7 if [[ ! -p ${PIPE} ]]; then mkfifo ${PIPE}; fi
8
9 while true; do
10     if read line <${PIPE}; then
11         echo "${line}"
12     fi
13 done
```

Figure 5.41 – The producer (left) and consumer (right) scripts (using named pipe)

The named pipe is `pipe.info` (line 5 in both scripts). The pipe file is created (if it's not already present) by either the producer or consumer when they start (line 7). The related command is `mkfifo` (see `man mkfifo` for more information).

The producer writes a random UUID to the named pipe every second (line 13 in `producer.sh`), where the consumer immediately reads it (line 10-11 in `consumer.sh`):

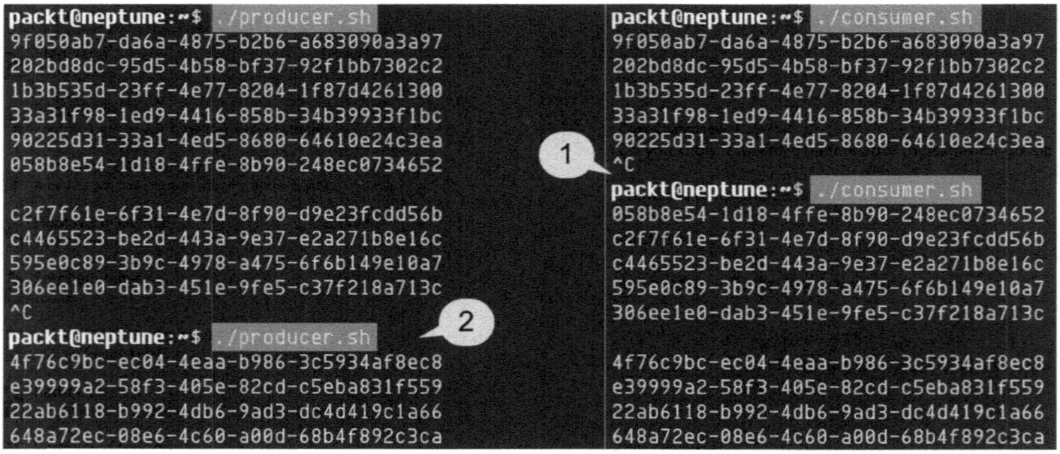

Figure 5.42 – The producer (left) and consumer (right) communicating through a named pipe

We started both scripts – the producer and the consumer – in an arbitrary order. After a while, we stopped (interrupted) the consumer (**step 1**). The producer continued to run but automatically stopped sending data to the pipe. Then, we started the consumer again. The producer immediately resumed sending data to the pipe. After a while, we stopped the producer (**step 2**). This time, the consumer became idle. After starting the producer again, both resumed normal operation, and data began flowing through the named pipe. This workflow has shown the persistence and resilience of the named pipe, regardless of the running status of the producer or consumer processes.

Named pipes are essentially queues, where data is being queued and dequeued on a first-come-first-served basis. When more than two processes communicate on the IPC named pipe channel, the FIFO approach may not fit the bill, especially when specific processes demand a higher priority for data processing. Message queues come to the rescue here.

Message queues

A message queue is an asynchronous communication mechanism that's typically used in a distributed system architecture. Messages are written and stored in a queue until they are processed and eventually deleted. A message is written (published) by a producer and is processed only once, typically by a single consumer. At a very high level, a message has a sequence, a payload, and a type. Message queues can regulate the retrieval (order) of messages (for example, based on priority or type):

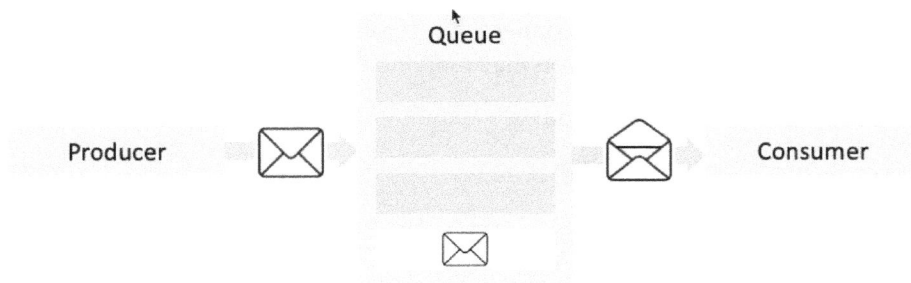

Figure 5.43 – Message queue (simplified view)

A detailed analysis of message queues or a mock implementation thereof is far from trivial, and it's beyond this chapter's scope. There are numerous open source message queue implementations available for most Linux platforms (RabbitMQ, ActiveMQ, ZeroMQ, MQTT, and so on).

IPC mechanisms based on message queues and pipes are unidirectional. One process writes the data; another one reads it. There are bidirectional implementations of named pipes, but the complexities involved would negatively impact the underlying communication layer.

For bidirectional communication, you can think of using socket-based IPC channels. The next section will show you how to do this.

Sockets

There are two types of IPC socket-based facilities:

- **IPC sockets**: Also known as Unix domain sockets
- **Network sockets**: **Transport Control Protocol (TCP)** and **User Datagram Protocol (UDP)** sockets

IPC sockets use a local file as a socket address and enable bidirectional communication between processes on the same host. On the other hand, network sockets extend the IPC data connectivity layer beyond the local machine via TCP/UDP networking. Apart from the obvious implementation differences, the IPC socket's and network socket's data communication channel behave the same.

Both sockets are configured as streams, support bidirectional communication, and emulate a client/server pattern. The socket's communication channel is active until it's closed on either end, thereby breaking the IPC connection.

Let's adapt our producer-consumer model to simulate an IPC socket (Unix domain socket) data connectivity layer. We'll use netcat to handle the underlying client/server IPC socket's connectivity. netcat is a powerful networking tool for reading and writing data using TCP, UDP, and ICP socket connections. If netcat is not installed by default on your Linux distribution of choice, you may look to install it as follows.

On Ubuntu, use the following command:

```
sudo apt-get install netcat
```

On CentOS/RHEL, use the following command:

```
sudo yum install nmap
```

For more information about netcat, please refer to the related system reference manual (man netcat):

```
 1  !/bin/bash                              1  #!/bin/bash
 2                                          2
 3  # producer.sh                           3  # consumer.sh
 4                                          4
 5  SOCKET="/var/tmp/ipc.sock"              5  SOCKET="/var/tmp/ipc.sock"
 6                                          6
 7  rm -f "${SOCKET}"                       7  nc -U "${SOCKET}"
 8                                          ~
 9  while true; do                          ~
10      uuidgen;                            ~
11      sleep 1s;                           ~
12  done \                                  ~
13  | tee /dev/tty \                        ~
14  | nc -lU "${SOCKET}"                    ~
```

Figure 5.44 – The producer (left) and consumer (right) scripts (using IPC sockets)

The producer acts as the server by initiating a `netcat` listener endpoint using an IPC socket (line 14 in `producer.sh`):

```
nc -lU "${SOCKET}"
```

The `-l` option indicates the listener (server) mode, while the `-U "${SOCKET}"` option parameter specifies the IPC socket type (Unix domain socket). The consumer connects to the `netcat` server endpoint as a client with a similar command (line 7 in `consumer.sh`). The producer and consumer both use the same (shared) IPC socket file descriptor (`/var/tmp/ipc.sock`) for communication, as defined in line 5.

The producer sends random UUID strings every second to the consumer (lines 9-12 in `producer.sh`). The related output is captured in `stdout` with the `tee` command (line 13), before being piped to `netcat` (line 14):

```
packt@neptune:~$ ./producer.sh              packt@neptune:~$ ./consumer.sh
065a3e4e-59ea-4d8e-ba66-68c4f7dbfc6a        065a3e4e-59ea-4d8e-ba66-68c4f7dbfc6a
8f608bfb-a480-4738-8505-968b64f7e294        8f608bfb-a480-4738-8505-968b64f7e294
e767025e-ba34-40ed-b42d-3af653dc37a1        e767025e-ba34-40ed-b42d-3af653dc37a1
354c5333-fcb7-43f3-8c34-2538cf9e2023        354c5333-fcb7-43f3-8c34-2538cf9e2023
HELLO!!                                     HELLO!!
a250785d-3535-4e76-b7a0-865b9c4150a6        a250785d-3535-4e76-b7a0-865b9c4150a6
94eeff88-4701-4eb9-bcde-e7a1b9574dca        94eeff88-4701-4eb9-bcde-e7a1b9574dca
5bdb16cf-5169-412c-aa8a-b6a0076fc53f        5bdb16cf-5169-412c-aa8a-b6a0076fc53f
ae1088fd-6458-4e2b-9244-69540f5e1fb2        ae1088fd-6458-4e2b-9244-69540f5e1fb2
^C                                          packt@neptune:~$
```

Figure 5.45 – The producer (left) and consumer (right) communicating through an IPC socket

The consumer gets all the messages (UUIDs) that have been generated by the producer. Also, we manually sent a HELLO!! message from the consumer to the producer to demonstrate the bidirectional communication stream between the two.

In our producer-consumer model, we used `netcat` for the IPC socket communication layer. Alternatively, we could use `socat`, a similar networking tool.

Before we wrap up our IPC toolbox, let's make a quick stop and talk about signals. We mentioned at the beginning of this section that signals are yet another IPC mechanism. Indeed, they are a somewhat limited form of IPC in the sense that through signals, processes can coordinate synchronization with each other. But signals don't carry any data payloads. They simply notify processes about events, and processes may choose to take specific actions in response to these events.

The following section will cover signals in detail.

Working with signals

In Linux, a signal is a one-way asynchronous notification mechanism that's used in response to a specific condition. A signal can act in any of the following directions:

- From the Linux kernel to an arbitrary process
- From process to process
- From a process to itself

Signals typically alert a Linux process about a specific event, such as a segmentation fault (SIGSEGV) that's raised by the kernel or execution being interrupted (SIGINT) by the user pressing *Ctrl + C*. In Linux, processes are controlled via signals. The Linux kernel defines a few dozen signals. Each signal has a corresponding non-zero positive integer value. The following command lists all the signals that have been registered in a Linux system:

```
kill -l
```

The output of the preceding command can be seen in the following screenshot:

```
packt@neptune:~$ kill -l
 1) SIGHUP       2) SIGINT       3) SIGQUIT      4) SIGILL       5) SIGTRAP
 6) SIGABRT      7) SIGBUS       8) SIGFPE       9) SIGKILL     10) SIGUSR1
11) SIGSEGV     12) SIGUSR2     13) SIGPIPE     14) SIGALRM     15) SIGTERM
16) SIGSTKFLT   17) SIGCHLD     18) SIGCONT     19) SIGSTOP     20) SIGTSTP
21) SIGTTIN     22) SIGTTOU     23) SIGURG      24) SIGXCPU     25) SIGXFSZ
26) SIGVTALRM   27) SIGPROF     28) SIGWINCH    29) SIGIO       30) SIGPWR
31) SIGSYS      34) SIGRTMIN    35) SIGRTMIN+1  36) SIGRTMIN+2  37) SIGRTMIN+3
38) SIGRTMIN+4  39) SIGRTMIN+5  40) SIGRTMIN+6  41) SIGRTMIN+7  42) SIGRTMIN+8
43) SIGRTMIN+9  44) SIGRTMIN+10 45) SIGRTMIN+11 46) SIGRTMIN+12 47) SIGRTMIN+13
48) SIGRTMIN+14 49) SIGRTMIN+15 50) SIGRTMAX-14 51) SIGRTMAX-13 52) SIGRTMAX-12
53) SIGRTMAX-11 54) SIGRTMAX-10 55) SIGRTMAX-9  56) SIGRTMAX-8  57) SIGRTMAX-7
58) SIGRTMAX-6  59) SIGRTMAX-5  60) SIGRTMAX-4  61) SIGRTMAX-3  62) SIGRTMAX-2
63) SIGRTMAX-1  64) SIGRTMAX
```

Figure 5.46 – The Linux signals

SIGHUP, for example, has a signal value of 1, and it's invoked by a Terminal session to all its child processes when it exits. SIGKILL has a signal value of 9 and is most commonly used for terminating processes. Processes can typically control how signals are handled, except for SIGKILL (9) and SIGSTOP (19), which always end or stop a process, respectively.

Processes handle signals in either of the following fashions:

- Perform the default action implied by the signal; for example, stop, terminate, core-dump a process, or do nothing.
- Perform a custom action (except for SIGKILL and SIGSTOP). In this case, the process catches the signal and handles it in a specific way.

When a program implements a custom handler for a signal, it usually defines a signal-handler function that alters the execution of the process, as follows:

- When the signal is received, the process' execution is interrupted at the current instruction.
- The process' execution immediately jumps to the signal-handler function
- The signal-handler function runs.
- When the signal-handler function exits, the process resumes execution, starting from the previously interrupted instruction.

Here's some brief terminology related to signals:

- A signal is raised by the process that generates it.
- A signal is caught by the process that handles it.
- A signal is ignored if the process has a corresponding **no operation** or **no-op** (**NOOP**) handler.
- A signal is handled if the process implements a specific action when the signal is caught.

The man signal.h system reference manual captures a detailed description of each signal. Here's an excerpt:

```
man signal.h
```

The output of the preceding command can be seen in the following screenshot:

Signal	Default Action	Description
SIGABRT	A	Process abort signal.
SIGALRM	T	Alarm clock.
SIGBUS	A	Access to an undefined portion of a memory object.
SIGCHLD	I	Child process terminated, stopped, or continued.
SIGCONT	C	Continue executing, if stopped.
SIGFPE	A	Erroneous arithmetic operation.
SIGHUP	T	Hangup.
SIGILL	A	Illegal instruction.
SIGINT	T	Terminal interrupt signal.
SIGKILL	T	Kill (cannot be caught or ignored).
SIGPIPE	T	Write on a pipe with no one to read it.
SIGQUIT	A	Terminal quit signal.
SIGSEGV	A	Invalid memory reference.
SIGSTOP	S	Stop executing (cannot be caught or ignored).
SIGTERM	T	Termination signal.
SIGTSTP	T	Terminal stop signal.
SIGTTIN	S	Background process attempting read.
SIGTTOU	S	Background process attempting write.
SIGUSR1	T	User-defined signal 1.
SIGUSR2	T	User-defined signal 2.

Figure 5.47 – Excerpt of signal definitions (from man signal.h)

The highlighted signals – SIGKILL and SIGSTOP – are the only ones that cannot be caught or ignored. The values in the Default Action column have the following significance (also captured from man signal.h):

```
The default actions are as follows:

T       Abnormal termination of the process.

A       Abnormal termination of the process with additional actions.

I       Ignore the signal.

S       Stop the process.

C       Continue the process, if it is stopped; otherwise, ignore the signal.
```

Figure 5.48 – The default actions (from man signal.h)

Let's explore a few use cases for handling signals:

- When the kernel raises a SIGKILL, SIGFPE (floating-point exception), SIGSEGV (segmentation fault), SIGTERM, or similar signals, typically, the process that receives the signal immediately terminates execution and may generate a core dump – the image of the process that's used for debugging purposes.

- When a user types *Ctrl + C* – otherwise known as an **interrupt character** (**INTR**) – while a foreground process is running, a SIGINT signal is sent to the process. The process will terminate unless the underlying program implements a special handler for SIGINT.

- Using the kill command, we can send a signal to any process based on its PID. The following command sends a SIGHUP signal to a Terminal session with PID 3741:

```
kill -HUP 3741
```

In the preceding command, we can either specify the signal value (for example, 1 for SIGHUP) or just the signal name without the SIG prefix (for example, HUP for SIGHUP).

With killall, we can signal that multiple processes are running a specific command (for example, test.sh). The following command terminates all processes running the test.sh script and outputs the result to the console (via the -e option):

```
killall -e -TERM test.sh
```

The output of this command is as follows:

```
packt@neptune:~$ killall -e -TERM test.sh
[2]+  Terminated                 ./test.sh
[1]+  Terminated                 ./test.sh
```

Figure 5.49 – Terminating multiple processes with killall

- Let's assume we have the following bash script (test.sh), which is implementing a signal handler for SIGUSR1:

```
 1 #!/bin/bash
 2
 3 trap sig_handler SIGUSR1
 4
 5 function sig_handler() {
 6     echo "Signal caught!"
 7 }
 8
 9 while true; do
10     echo "Waiting..."
11     sleep 3s
12 done
```

Figure 5.50 – Bash script implementing a signal handler for SIGUSR1

Line 3 defines a signal handler function (`sig_handler`) to catch `SIGUSR1`:

```
trap sig_handler SIGUSR1
```

We can run the script as a background process:

```
./test.sh &
```

We can let the related background process run for a while, and then send a `SIGUSR1` signal with the following command:

```
killall -e -USR1 test.sh
```

The output is as follows:

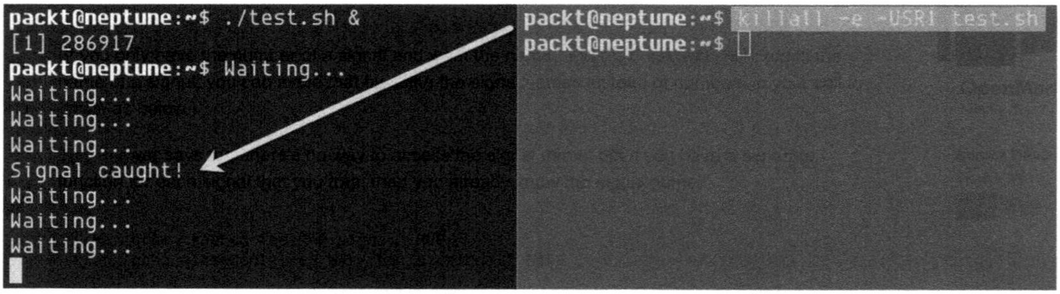

Figure 5.51 – Bash script implementing a signal handler for SIGUSR1

The bash script (`test.sh`) catches the `SIGUSR1` signal, executes the `sig_handler()` function, and then resumes execution.

Linux processes and signals are a vast domain. The information we've provided here is far from a comprehensive guide on the topic. We hope that this practical spin and hands-on approach to presenting some common use cases has inspired you to take on and possibly master more challenging issues.

Summary

A detailed study of Linux processes and daemons could be a major undertaking. Where worthy volumes on the topic have admirably succeeded, a relatively brief chapter may pale in comparison. Yet in this chapter, we tried to put on a real-world, down-to-earth, practical coat on everything we've considered to make up for our possible shortcomings in the abstract or scholarly realm.

At this point, we hope you are comfortable working with processes and daemons. The skills you've gathered so far should include a relatively good grasp on process types and internals, with a reasonable understanding of process attributes and states. Special attention has been paid to interprocess communication mechanisms, and signals in particular. For each of these topics, we did a practical dive and explored the related commands, tools, and scripts that we thought would be relevant in everyday Linux administration tasks.

The next chapter will take our journey further into working with Linux disks and filesystems. We'll explore the Linux storage, disk partitioning, and **Logical Volume Management** (**LVM**) concepts. Rest assured that everything we've learned so far will be immediately put to good use in the chapters that follow.

Questions

If you managed to skim through some parts of this chapter, you might want to recap on a few essential details about Linux processes and daemons:

1. Think of a few process types. How would they compare to each other?
2. Think of the anatomy of a process. Can you come up with a few essential process attributes (or fields in the `ps` command-line output) that you may look for when inspecting processes?
3. Can you think of a few process states and some of the dynamics or possible transitions between them?
4. If you are looking for a process that takes up most of the CPU on your system, how would you proceed?
5. Can you write a simple script and make it a long-lived background process?
6. Enumerate at least four process signals that you can think of. When or how would those signals be invoked?
7. Think of a couple of IPC mechanisms. Try to come up with some pros and cons for them.

Section 2:
Advanced
Linux Server
Administration

In this section, you will learn about advanced Linux server administration tasks by setting up different types of servers, and you will learn about and harden Linux server security.

This part of the book comprises the following chapters:

- *Chapter 6, Working with Disks and Filesystems*
- *Chapter 7, Networking with Linux*
- *Chapter 8, Configuring Linux Servers*
- *Chapter 9, Securing Linux*
- *Chapter 10, Disaster Recovery, Diagnostics, and Troubleshooting*

6
Working with Disks and Filesystems

In this chapter, you will learn to manage disks and filesystems, understand storage in Linux, learn how to use the **Logical Volume Management** (**LVM**) system, and learn how to mount and partition the hard drive. You will learn how to partition and format a disk, as well as how to create logical volumes, and you will gain a deeper understanding of filesystem types. In this chapter, we're going to cover the following main topics:

- Understanding devices in Linux
- Understanding filesystem types in Linux
- Understanding disks and partitions
- Logical Volume Management in Linux

Technical requirements

A basic knowledge of disks, partitions, and filesystems is preferred. No other special technical requirements are needed, just a working installation of Linux on your system. We will mainly use CentOS for this chapter's exercises.

The code for this chapter is available at the following link: `https://github.com/PacktPublishing/Mastering-Linux-Administration`.

Understanding devices in Linux

As already stated on several occasions in this book, everything in Linux is a file. This includes devices, too. Device files are special files in UNIX and Linux operating systems. Those special files are basically an interface to device drivers, and they are present in the filesystem as a regular file.

Linux abstraction layers

Now is as good a time as any to discuss Linux system abstraction layers and how devices fit into the overall picture. A Linux system is generally organized on three major levels: the **hardware level**, **kernel level**, and **user space level**.

The hardware level contains the hardware components of your machine, such as the memory (RAM), **Central Processing Unit** (**CPU**), and devices including disks, network interfaces, ports, and controllers. The memory is divided into two separate regions, called **kernel space** and **user space**.

The kernel is the beating heart of the Linux operating system. The kernel resides inside the memory (RAM) and manages all the hardware components. It is the interface between the software and hardware on your Linux system. The user space level is the level where user processes are executed. As presented in *Chapter 5, Working with Processes, Daemons, and Signals*, a process is a running instance of a program.

How does all this work? Well, the memory, known as RAM, consists of cells that are used to store information temporarily. Those cells are accessed by different programs that are executed and function as an intermediary between the CPU and the storage. The speeds of accessing memory are very high in order to secure a seamless process of execution. The management of user processes inside the user space is the kernel's job. The kernel makes sure that none of the processes will interfere with each other. The kernel space is usually accessed only by the kernel, but there are times when user processes need to access this space. This is done through **system calls**. Basically, a system call is the way a user process requests a kernel service through an active process inside the kernel space, for anything such as **input/output** (**I/O**) requests to internal or external devices. All those requests are transferring data to and from the CPU, through RAM, in order to get the job done.

The following diagram shows the Linux abstraction layers. Basically, there are three different layers, distributed on three main levels: the **hardware level**, which comprises all the hardware components, including CPU, RAM, controllers, hard drives, SSDs, monitors, and peripherals; the **kernel level**, which comprises millions of lines of code mostly written in C that also contain the device drivers; and the **user level**, which is the place where all the applications, services, and daemons are running, along with the GUI and the shells:

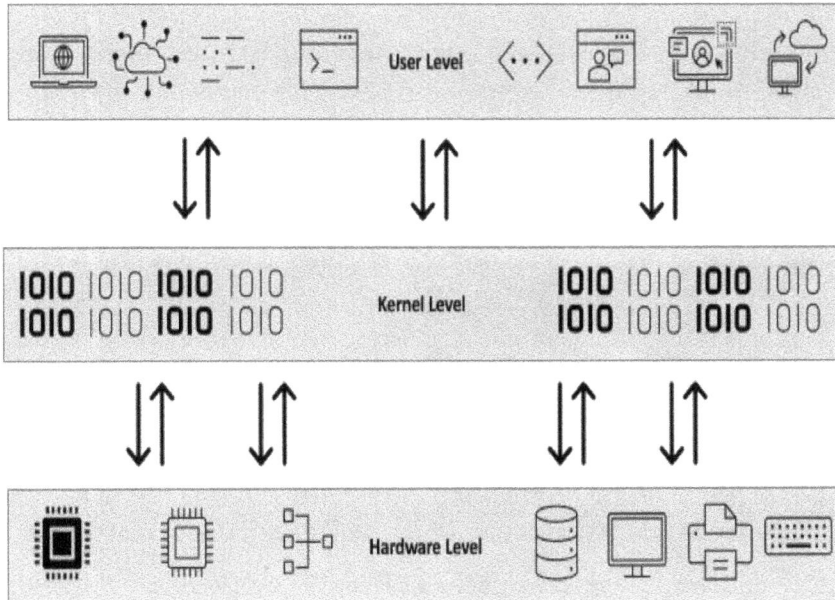

Figure 6.1 – Linux abstraction layers

Where are devices in this grand scheme of things? As shown in the preceding diagram, devices are managed by the kernel. To sum up, the kernel is in charge of managing processes, system calls, memory, and devices. When dealing with devices, the kernel is managing device drivers, which are the interface between hardware components and software. All devices are accessible only in kernel mode, for a more secure and streamlined operation.

In the following section, we will introduce you to the naming convention in Linux and how device files are managed.

Device files and naming conventions

After seeing how those abstraction layers work, you may be wondering how Linux manages devices. Well, it does that with the help of **udev** (called user space /dev), which is a device manager for the kernel. It works with **device nodes**, which are special files (also called **device files**) used as an interface to the driver.

udev runs as a daemon that listens to the user space calls that the kernel is sending, and so is aware of what kinds of devices are used and how they are used. The daemon is called udevd and its configurations are currently available under /etc/udev/udev.conf. Each Linux distribution has a default set of rules that governs udevd. Those rules are normally stored under the /etc/udev/rules.d/ directory, as shown in the following screenshot:

```
[packt@jupiter ~]$ cat /etc/udev/udev.conf
# see udev.conf(5) for details
#
# udevd is also started in the initrd.  When this file is modified you might
# also want to rebuild the initrd, so that it will include the modified configuration.

#udev_log="info"
[packt@jupiter ~]$ ls -l /etc/udev/rules.d/
total 4
-rw-r--r--. 1 root root 628 Oct  2  2019 70-persistent-ipoib.rules
```

Figure 6.2 – The udevd configuration files and rules location

Just by way of general information, the kernel sends calls for events using the **netlink** socket. The netlink socket is an interface for inter-process communication used for both user space and kernel space processes alike.

The /dev directory is the interface between user processes and devices managed by the kernel. If you were to use the ls -la /dev command, you will see a lot of files inside, each having different names. If you were to do a long listing, you would see different file types. Some of the files will start with the letters **b** and **c**, but the letters **p** and **s** may also be present, depending on your system. Files starting with those letters are device files. The ones starting with **b** are *block devices*, and those starting with the letter **c** are *character devices*. You can run the following command to see what types of device files are inside your /dev directory:

```
ls -la /dev
```

Block devices can be accessed by software only in fixed sizes. As you will see in *Figure 6.3*, the disk devices sda and sdb are represented as block devices. Block devices have a fixed size that can easily be indexed. Character devices, on the other hand, can be accessed using data streams, as they don't have a size like block devices. For example, printers are represented as character devices. In *Figure 6.3*, sg0 and sg1 are SCSI generic devices, and not assigned to any disks in our case:

```
brw-rw----. 1 root  disk       8,    0 Sep  8 12:59 sda
brw-rw----. 1 root  disk       8,    1 Sep  8 12:59 sda1
brw-rw----. 1 root  disk       8,   16 Sep  8 12:59 sdb
brw-rw----. 1 root  disk       8,   17 Sep  8 12:59 sdb1
brw-rw----. 1 root  disk       8,   18 Sep  8 12:59 sdb2
crw-rw----. 1 root  disk      21,    0 Sep  8 12:59 sg0
crw-rw----. 1 root  disk      21,    1 Sep  8 12:59 sg1
```

Figure 6.3 – Disk drives inside the /dev directory

Linux uses a device name convention that makes device management easier and consistent throughout the Linux ecosystem. udev uses a number of specific naming schemes that, by default, assign fixed names to devices. Those names are somehow standardized for device categories. For example, when naming network devices, the kernel uses information compiled from sources such as firmware, topology, and location. On a Red Hat-based system such as CentOS, there are five schemes used for naming a network interface, and we encourage you to visit these on the Red Hat customer portal official documentation website.

You could also check what udev rules are active on your system. On a CentOS 8 distribution, they are stored in the /lib/udev/rules.d/ directory:

When it comes to hard drives or external drives, the conventions are more streamlined. Here are some examples:

- **For classic IDE drivers used for ATA drives**: hda (the master device), hdb (the slave device on the first channel), hdc (the master device on the second channel), and hdd (the slave device on the second channel)

- **For NVMe drivers**: nvme0 (the first device controller – character device), nvme0n1 (first namespace – block device), and nvme0n1p1 (first namespace, first partition – block device)

- **For MMC drivers**: mmcblk (for SD cards using eMMC chips), mmcblk0 (first device), and mmcblk0p1 (first device, first partition)

- **For SCSI drivers used for modern SATA or USB**: sd (for mass storage devices), sda (for the first registered device), sdb (for the second registered device), sdc (for the third registered device), and so on, and sg (for generic SCSI layers – character device)

The devices that we are mostly interested in as regards this chapter are the mass storage devices. Those devices are usually **hard disk drives (HDD)** or **solid-state drives (SSD)** used inside your computer to store data. Those drives are most likely divided into partitions having a specific structure provided by the filesystem. We talked a little bit about filesystems earlier in this book in *Chapter 2, The Linux Filesystem*, when we referred to the Linux directory structure, but now it is time to get into more details about how disks work inside a Linux system.

Understanding filesystem types in Linux

When talking about physical media, such as hard drives or external drives, we are *not* referring to the directory structure. Here, we are talking about the structures that are created on the physical drive when formatting and/or partitioning it. Those structures, depending on their type, are known as filesystems, and they determine how the files are managed when stored on the drive.

There are several types of filesystems, some being native to the Linux ecosystem, while others are not, such as specific Windows or macOS filesystems. In this section, we will describe only the Linux-native filesystems.

The most widely used filesystems in Linux are the `Extended` filesystems, known as `Ext`, `Ext2`, `Ext3`, and `Ext4`, the `XFS` filesystem, `ZFS`, and `btrfs` (short for B-tree filesystem). Each of these have their strengths and weaknesses, but they are all able to do the job they were designed for. The `Extended` filesystems are the ones that were most widely used in Linux, and they have proven trustworthy all this time. `Ext4`, the latest iteration, is similar to `Ext3`, but better, with improved support for larger files, fragmentation, and performance. The `Ext3` filesystem uses 32-bit addressing, and `Ext4` uses 48-bit addressing, thus supporting files up to 16 TB in size. It also offers support for unlimited subdirectories, as `Ext3` only supports 32 k subdirectories. Also, support for extended timestamps was added in `Ext4`, offering two more bits for up to year 2446 AD, and online defragmentation at kernel level.

Nonetheless, `Ext4` is not a truly next-gen filesystem, more an improved, trustworthy, robust, and stable *workhorse* that failed the data protection and integrity test. Its journaling system is not suitable for detecting and repairing data corruption and degradation. That is why other filesystems, such as `XFS` and `ZFS`, started to resurface by being used in Red Hat Enterprise Linux, starting from version 7 (`XFS`) and in Ubuntu since version 16.04 (`ZFS`). The case of `btrfs` is somewhat controversial. It is considered a modern filesystem, but it is still used as a single disk filesystem, and not used in multiple disk volume managers, due to a number of performance issues compared to other filesystems. It is used in SUSE Linux Enterprise and in openSUSE, is no longer supported by Red Hat, and has been voted as the future default filesystem in Fedora, starting with version 33.

The Ext4 filesystem features

The `Ext4` filesystem was designed for Linux right from the outset. Even though it is slowly being replaced with other filesystems, this one still has powerful features. It offers block size selection, with values between 512 and 4,096 bytes. There is also a feature called inode reservation, which saves a couple of inodes when you create a directory, for improved performance when creating new files.

The layout is simple, written in little-endian order (for more details on this, visit `https://www.section.io/engineering-education/what-is-little-endian-and-big-endian/`), with block groups containing inode data for lower access times. Each file has data blocks pre-allocated for reduced fragmentation. There are also many enhancements that the `Ext4` takes advantage of. Among them, we will bring the following into the discussion: a maximum filesystem size of 1 **Exabyte** (**EB**), the ability to use multi-block allocation, splitting large files into the largest possible sizes for better performance, application of the allocate-on-flush technique for better performance, the use of the handy `fsck` command for speedy filesystem checks, the use of `checksums` for journaling and better reliability, and the use of improved timestamps.

The XFS filesystem features

Enterprise Linux is starting to change, by moving away from Ext4 to other competent filesystem types. Among those is XFS. This filesystem was first created by SGI and used in the IRIX operating system. Its most important key design element is performance, capable of dealing with large datasets. Furthermore, it is designed to handle parallel I/O tasks with a guaranteed high I/O rate. The filesystem supported is up to 16 EB with support for individual files up to 8 EB. XFS has a feature to journal quota information, together with on-line maintenance tasks such as defragmenting, enlarging, or restoring. There are also specific tools for backup and restore, including xfsdump and xfsrestore.

The btrfs filesystem features

The B-tree filesystem (btrfs) is still under development, but it addresses issues associated with existing filesystems, including the lack of snapshots, pooling, checksums, and multi-device spanning. Those are features that are required in an enterprise Linux environment. The ability to take snapshots of the filesystem and maintain its own internal framework for managing new partitions makes btrfs a viable newcomer in terms of the critical enterprise ecosystem.

There are other filesystems that we did not discuss here, including reiserFS and GlusterFS, NFS (Network File System), SMB (Samba CIFS File System), ISO9660 for CD-ROMs and Joliet extensions, and non-native Linux ones, including FAT, NTFS, exFAT, and APFS, or MacOS Extended, among others. If you want to learn about these in more detail, feel free to investigate further, and a good starting point is Wikipedia: https://en.wikipedia.org/wiki/File_system. To check the list of supported filesystems on your Linux distribution, run the following command:

```
cat /proc/filesystems
```

Linux implements a special software system that is designed to run specific functions of the filesystems. It is known as the Virtual File System, and acts as a bridge between the kernel and the filesystem types and hardware. Therefore, when an application wants to open a file, the action is delivered through the Virtual File System as an abstraction layer:

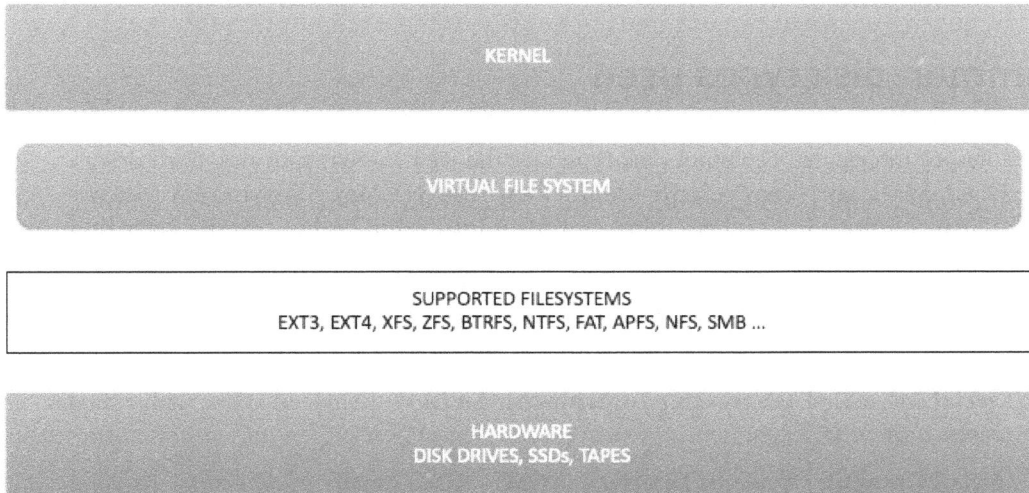

KERNEL

VIRTUAL FILE SYSTEM

SUPPORTED FILESYSTEMS
EXT3, EXT4, XFS, ZFS, BTRFS, NTFS, FAT, APFS, NFS, SMB ...

HARDWARE
DISK DRIVES, SSDs, TAPES

Figure 6.4 – The Linux Virtual File System abstraction layer

Basic filesystem functions include the provision of namespaces, metadata structures as a logical foundation for hierarchical directory structures, disk block usage, file size and access information, and high-level data for logical volumes and partitions. There is also an **application programming interface** (**API**) available for every filesystem. Thus, developers are able to access system function calls for filesystem object manipulation with specific algorithms for creating, moving, and deleting files, or for indexing, searching, and finding files. Furthermore, every modern filesystem provides a special access rights scheme, used to determine the rules governing a user's access to files.

At this point, we have already covered the principal Linux filesystems, including EXT4, btrfs, and XFS. In the next section, we will teach you the basics of disks and partition management in Linux.

Understanding disks and partitions

Understanding disks and partitions is a key asset for any system administrator. Formatting and partitioning disks is critical, starting with system installation. Knowing the type of hardware available on your system is important, and it is therefore imperative to know how to work with it.

Common disk types used

A disk is a hardware component that stores your data. It comes in various types and uses different interfaces. The main disk types are the well-known spinning hard drives (or HDDs), **SSDs**, and **Non-Volatile Memory express** (**NVMe**). SSDs and NVMes use RAM-like technologies, with better energy consumption and higher transfer rates than original spinning hard drives. The interfaces used are as follows:

- **Integrated Drive Electronics** (**IDE**) – This is an old standard used on consumer hardware with small transfer rates; now deprecated.

- **Serial Advanced Technology Attachment** (**SATA**) – Replaced IDEs; with transfer rates of up to 16 Gb/s.

- **Small Computer Systems Interface** (**SCSI**) – Used mostly in enterprise servers with RAID configurations with sophisticated hardware components.

- **Serial Attached SCSI** (**SAS**) – This is a point-to-point serial protocol interface with transfer rates similar to SATA, mostly used in enterprise environments for their reliability.

- **Universal Serial Bus** (**USB**) – Used for external hard drives and memory drives.

Each disk has a specific geometry that consists of heads, cylinders, tracks, and sectors. On a Linux system, in order to see the information regarding a disk's geometry, there is the `fdisk -l` command. On our test system, we have two disk devices installed, a SATA SSD with CentOS 8 installed on it, and a SATA HDD for data storage. We will use the following command to show disk information:

```
sudo fdisk -l
```

Disks are just a big chunk of metal if we don't format and partition them. This is why, in the next section, we will teach you what partitions are.

Partitioning disks

Commonly, disks use partitions. In order to understand partitions, knowing a disk's geometry is essential. Partitions are contiguous sets of sectors and/or cylinders, and they can be of several types: primary, extended, and logical partitions. A maximum number of 15 partitions can exist on a disk. The first four will be either primary or extended, and the next ones up to 15 are logical partitions. Furthermore, there can only be a single extended partition, but the extended partition can be divided into several logical partitions, until the maximum number is reached.

Partition types

There are two major partition types, the **Master Boot Record** (**MBR**) and the **GUID Partition Table** (**GPT**). MBR was intensively used up to around 2010. Its limitations were given by the maximum numbers of primary partitions (four) and by the maximum size of a partition (2 TB). The MBR uses hexadecimal codes for different types of partitions, such as `0x0c` for FAT, `0x07` for NTFS, `0x83` for a Linux filesystem type, and `0x82` for Swap. GPT became a part of the **Unified Extensible Firmware Interface** (**UEFI**) standard as a solution to some issues with the MBR, including partition limitations, addressing methods, using only one copy of the partition table, and so on. It supports up to 128 partitions and disk sizes of up to 75.6 **Zettabytes** (**ZB**).

The partition table

The partition table of a disk is stored inside the disk's MBR. The MBR is the first 512 bytes of a drive. Out of these, the partition table is 64 bytes and is stored after the first 446 bytes of records. At the end of the MBR, there are 2 bytes known as the end of sector marker. The first 446 bytes are reserved for code that usually belongs to a bootloader program. In the case of Linux, the bootloader is called GRUB.

When you boot up a Linux system, the boot loader is looking for the active partition. There can only be one active partition on a single disk. When the active partition is located, the boot loader loads items. The partition table has four entries, each 16 bytes in size, every one belonging to a possible primary partition on the system. Furthermore, each entry contains information regarding the beginning address of `cylinder/head/sectors`, the partition type code, the end address of `cylinder/head/sectors`, the starting sector, and the number of sectors inside one partition.

Naming partitions

The kernel interacts with the disk at a low level. It is done through device nodes that are stored inside the /dev directory. Device nodes use a simple naming convention that is helpful for knowing which disk is the one that requires your attention. Looking at the contents of the /dev directory, you can see all the available disk nodes, also referred to as disk drives, in *Figure 6.3* and *Figure 6.4* earlier in this section. A short explanation is always useful, so disks and partitions are recognized as follows:

- The first hard drive is always /dev/sda (for an SCSI or SATA device).

- The second hard drive is /dev/sdb, the third is /dev/sdc, and so on.

- The first partition of the first disk is /dev/sda1.

- The first partition of the second disk is /dev/sdb1.

- The second partition of the second disk is /dev/sdb2, and so on.

We specified that this is true in the case of SCSI and SATA, and we need to explain this in a little bit more detail. The kernel gives the letter designation, such as a, b, and c, based on the ID number of the SCSI device, and not based on the position of the hardware bus.

Partition attributes

In order to learn about your partition's attributes, you can use two programs inside Linux: blkid and lsblk. Both use the same kernel library, but the outputs are different. Here are two examples of how those utilities are used to view the key attributes of your partitions:

```
[packt@jupiter ~]$ blkid
/dev/mapper/cl-root: UUID="9cb51d4f-5979-4298-bab8-57144e5c42ac" TYPE="xfs"
/dev/sdb2: UUID="nH3IW7-DJ1D-0vnE-D4j7-ZtuZ-rQgS-jchm4V" TYPE="LVM2_member" PART
UUID="289db18a-02"
/dev/sdb1: UUID="f748abc8-285a-46f3-9410-baf57f485ddf" TYPE="ext4" PARTUUID="289
db18a-01"
/dev/mapper/cl-swap: UUID="aebd9086-3d52-4629-b50d-2414c6d55a06" TYPE="swap"
/dev/sda1: LABEL="seagate" UUID="03d5ec9d-e291-40fd-a62c-f1b9e576d305" TYPE="ext
4" PARTUUID="c4647bc5-cfa9-4056-81c3-f64157c5d734"
/dev/mapper/cl-home: UUID="0cff53d6-4c4b-463c-8172-8ad9be341fc6" TYPE="xfs"
[packt@jupiter ~]$ lsblk
NAME          MAJ:MIN RM    SIZE RO TYPE MOUNTPOINT
sda             8:0    0  931.5G  0 disk
`-sda1          8:1    0  931.5G  0 part
sdb             8:16   0  223.6G  0 disk
|-sdb1          8:17   0     1G   0 part /boot
`-sdb2          8:18   0  222.6G  0 part
  |-cl-root 253:0      0    50G   0 lvm  /
  |-cl-swap 253:1      0   7.8G   0 lvm  [SWAP]
  `-cl-home 253:2      0  164.8G  0 lvm  /home
```

Figure 6.5 – The blkid and lsblk output

In the preceding screenshot, you see that the `blkid` command shows the UUID of each partition and disk and their types. The `lsblk` command shows the device name (the node's name from `sysfs` and the `udev` database), the major and minor device number, the removable state of the device (`0` for nonremovable and `1` for a removable device), the size in human-readable format, the read-only state (again using `0` for the ones that are not read-only, and `1` for the ones that are read-only), the type of device, and the device's mount point (where available).

Partition table editors

In Linux, there are several tools to use when managing partition tables. Among the most commonly used ones are the following:

- `fdisk` – A command-line partition editor, perhaps the most widely used one
- `sfdisk` – A non-interactive partition editor, used mostly in scripting
- `parted` – The GNU (the recursive acronym for GNU is *Not Unix*) partition manipulation software
- `Gparted` – The graphical interface for `parted`

Of these, we will only detail how to use `fdisk`, as this is the most widely used command-line partition editor in Linux. It is found in both Ubuntu and CentOS, and many other distributions too. To use `fdisk`, you must be the root user. We advise you to use caution when using `fdisk` as it can damage your existing partitions and disks. `fdisk` can be used on a particular disk (`/dev/sdb` for example), by using the command:

```
sudo fdisk /dev/sdb
```

You will notice that when using `fdisk` for the first time, you are warned that changes will be done to the disk only when you decide to write them to it. You are also prompted to introduce a command, and you are shown the option m for help. We advise you to always use the help menu, even if you already know the most used commands.

When you type m, you will be shown the entire list of commands available for `fdisk`. You will see options to manage partitions, to create new boot records, to save changes, and others. As listing them would take too much of the books real estate, we encourage you to test this on your own systems:

To see the partitions that the operating system knows about, in case you are not sure about the operations you just completed, you can always visualize the contents of the /proc/ partitions file with the cat command:

```
[packt@jupiter ~]$ cat /proc/partitions
major minor #blocks  name

     8      0  976762584 sda
     8      1  976760832 sda1
     8     16  234438656 sdb
     8     17    1048576 sdb1
     8     18  233389056 sdb2
   253      0   52428800 dm-0
   253      1    8175616 dm-1
   253      2  172781568 dm-2
```

Figure 6.6 – Contents of the /proc/partitions file

Therefore, we saw how Linux allows the use of labels and UUIDs for naming disk drives and you can use fdisk to manage this. Besides fdisk, tools such as blkid and lsblk are used to view partition attributes.

Even though fdisk is the default tool in every major Linux distribution, there is another one that could be useful, called parted. Please keep in mind that parted writes all modifications directly to the disk, as opposed to fdisk. In this respect, please be careful how you use it. In the following, we will show you how to set a GPT partition table using parted. To start parted, run the parted command as sudo or as a root user:

```
sudo parted
```

To list all the partitions, run the print command inside the parted command-line interface:

```
(parted) print
Error: /dev/sda: unrecognised disk label
Model: ATA ST1000LM048-2E71 (scsi)
Disk /dev/sda: 1000GB
Sector size (logical/physical): 512B/4096B
Partition Table: unknown
```

We will now show you how to create a new GPT partition table for /dev/sda. First, we select the disk we want to work with using the select command inside the parted command-line interface:

```
(parted) select /dev/sda
Using /dev/sda
```

Then, we will set the new partition table to GPT using the mklabel command and check for the result with another print command:

```
(parted) mklabel gpt
(parted) print
Model: ATA ST1000LM048-2E71 (scsi)
Disk /dev/sda: 1000GB
Sector size (logical/physical): 512B/4096B
Partition Table: gpt
```

As you can see in the preceding output, the new partition table of the disk is set to GPT.

There are situations when you will need to back up and restore your partition tables. As partitioning could go sideways sometimes, a good backup strategy could help you. To do this, you can use the dd utility. This program is very useful and powerful, as it can clone disks or wipe data. Here is an example:

```
[packt@jupiter ~]$ pwd
/home/packt
[packt@jupiter ~]$ sudo dd if=/dev/sda
sda    sda1
[packt@jupiter ~]$ sudo dd if=/dev/sda of=mbr-backup bs=512 count=1
[sudo] password for packt:
1+0 records in
1+0 records out
512 bytes copied, 0.00182814 s, 280 kB/s
[packt@jupiter ~]$ ls
Desktop     Downloads   Pictures   Templates   mbr-backup
Documents   Music       Public     Videos      rpmbuild
```

Figure 6.7 – Backing up MBR with the dd command

The dd command has a clear syntax. By default, it uses the standard input and standard output, but you can change those by specifying new input files with the if option, and output files with the of option. We specified the input file as the device file for the disk we want to back up and gave a name for the backup output file. We also specified the block size using the bs option, and the count option to specify the number of blocks to read. To restore the boot loader, we can use the dd command as follows:

```
sudo dd if=~/mbr-backup of=/dev/sda bs=512 count=1
```

Partition table editors are important tools for managing disks in Linux. Their use is incomplete if you do not know how to format a partition. In the next section, we will show you how to format partitions.

Formatting and checking partitions

The most commonly used program for formatting a filesystem on a partition is mkfs. Formatting a partition is also known as *making* a filesystem, hence the name of the utility. It has specific tools for different filesystems, all using the same frontend utility. The following is a list of all mkfs supported filesystems:

```
[packt@jupiter ~]$ ls -lh /sbin/mkfs*
-rwxr-xr-x. 1 root root  17K Apr 24 15:39 /sbin/mkfs
-rwxr-xr-x. 1 root root  42K Apr 24 15:39 /sbin/mkfs.cramfs
-rwxr-xr-x. 4 root root 135K Apr 24 05:07 /sbin/mkfs.ext2
-rwxr-xr-x. 4 root root 135K Apr 24 05:07 /sbin/mkfs.ext3
-rwxr-xr-x. 4 root root 135K Apr 24 05:07 /sbin/mkfs.ext4
-rwxr-xr-x. 1 root root  40K May 11  2019 /sbin/mkfs.fat
-rwxr-xr-x. 1 root root  87K Apr 24 15:39 /sbin/mkfs.minix
lrwxrwxrwx. 1 root root    8 May 11  2019 /sbin/mkfs.msdos -> mkfs.fat
lrwxrwxrwx. 1 root root    8 May 11  2019 /sbin/mkfs.vfat -> mkfs.fat
-rwxr-xr-x. 1 root root 475K Apr  7  2020 /sbin/mkfs.xfs
```

Figure 6.8 – Details regarding the mkfs utility

As we stated at the outset, we are using a system with two data drives: an SSD for the OS with 240 GB, and an HDD for data, 1 TB in size. In order to format the largest disk as having the ext4 filesystem, we will use the mkfs utility. The commands to execute are shown as follows:

1. First, we will run the fdisk utility to make sure that we select the largest disk correctly. Run the following command:

    ```
    sudo fdisk -l
    ```

2. Then, check the output with extreme caution and select the correct disk name. If you have several disks on your system, pay attention to each one and choose with caution. In the output, disks will be shown with names like /dev/sda, /dev/sdb and so on.

> **Important note**
> In our case, the largest disk was also the first one to be shown by the command's output. This might be different for you, so pay attention when using the next command on a disk.

3. After carefully choosing the disk to work with, we will use mkfs to format it as an Ext4 filesystem. We will assume that our disk name is /dev/sda and we will use the following command:

```
sudo mkfs.ext4 /dev/sda
```

When using mkfs, there are several options available. To create an Ext4 type partition, you can either use the command as shown in *Figure 6.16*, or you can use the -t option followed by the filesystem type. You can also use the -v option for a more verbose output, and the -c option for bad sector scanning while creating the filesystem. You can also use the -L option if you want to add a name for the partition right from the command. The following is an example of creating an Ext4 filesystem partition with the name newpartition:

```
sudo mkfs -t ext4 -v -c -L newpartition /dev/sdb1
```

Once a partition is created, it is advised to check it for errors. Similar to mkfs, there is a tool called fsck. This is a utility that sometimes runs automatically following an abnormal shutdown or at set intervals. It has specific programs for the most commonly used filesystems, just like mkfs. The following is the output of running fsck on one of our partitions. After running, it will show whether there are any problems. In the following screenshot, the output shows that checking the partition resulted in no errors:

```
[packt@jupiter ~]$ sudo fsck -t ext4 /dev/sda
fsck from util-linux 2.32.1
e2fsck 1.45.4 (23-Sep-2019)
/dev/sda: clean, 11/61054976 files, 4114691/244190646 blocks
```

Figure 6.9 – Using fsck to check a partition

After partitions are created, they need to be mounted. Each partition will be mounted inside the existing filesystem structure. Mounting is allowed at any point in the tree structure. Each filesystem is mounted under certain directories, created inside the directory structure.

Mounting and unmounting partitions

The mounting utility in Linux is simply called mount, and the unmounting utility is called umount. To see whether a certain partition is mounted, you can simply type mount and see the output. We are looking for /dev/sda in the output, but it is not shown. This means that our second drive is not mounted.

To mount it, we need to make a new directory. For simplicity, we will show all the steps required for you to mount and use the partition:

1. Execute the fdisk command to create a new partition table:

    ```
    sudo fdisk /dev/sda
    ```

2. Select the g option from the fdisk menu to create a new, empty GPT partition table.

3. Write the changes to disk, using the w option.

4. Format the partition using the Ext4 filesystem:

    ```
    sudo mkfs.ext4 /dev/sda
    ```

5. Create a new directory to mount the partition. In our case, we created a new directory called hdd inside the /mnt directory:

    ```
    sudo mkdir /mnt/hdd
    ```

6. Mount the partition using the following command:

    ```
    sudo mount /dev/sda /mnt/hdd
    ```

7. Start using the new partition from the new location.

The mount utility has many options available. Use the help menu to see everything that it has under the hood. Now that the partition is mounted, you can start using it. If you want to unmount it, you can use the umount utility. You can use it as follows:

```
sudo umount /dev/sda
```

When unmounting a filesystem, you could receive errors if that partition is still in use. Being in use means that certain programs from that filesystem are still running in the memory, using files from that partition. Therefore, you first have to close all running applications, and if there are other processes using that filesystem, you will have to kill them, too. Sometimes, the reason a filesystem is busy is not clear at first, and to know which files are open and running, you can use the `lsof` command:

```
sudo lsof | grep /dev/sda
```

Mounting a filesystem only makes them available until the system is shut down or rebooted. If you want the changes to be persistent, you will have to edit the `/etc/fstab` file accordingly. First, open the file with your favorite text editor:

```
sudo nano /etc/fstab
```

Add a new line similar to the one that follows:

```
/dev/sda /mnt/sdb ext4 defaults 0 0
```

The `/etc/fstab` file is a configuration file for the filesystem table. It consists of a set of rules needed to control how the filesystems are used. This simplifies the need to manually mount and unmount each disk when used, by drastically reducing possible errors. The table has a six-column structure, with each column designated with a specific parameter. There is only one correct order for the parameters to work as follows:

- **Device name** – Either by using UUID or the mounted device name.
- **Mount point** – The directory where the device is, or will be, mounted.
- **Filesystem type** – The filesystem type used.
- **Options** – The options shown, with multiple ones separated by commas.
- **Backup operation** – This is the first digit from the last two digits in the file; 0 = no backup, 1 = dump utility backup.
- **Filesystem check order** – This is the last digit inside the file; 0 = no `fsck` filesystem check, with 1 for the root filesystem, and 2 for other partitions.

By updating the `/etc/fstab` file, the mounting is permanent and is not affected by any shutdown or system reboot. Usually, the `/etc/fstab` file only stores information about the internal hard drive partitions and filesystems. The external hard drives or USB drives are automatically mounted under `/media` by the kernel's **Hardware Abstraction Layer** (**HAL**).

By now, you should be comfortable with managing partitions in Linux, but there is still one type of partition we have not yet discussed: the swap partition. In the next section, we will introduce you to how swap works on Linux.

Swap

Linux uses a robust swap implementation. The virtual memory uses hard drive space when physical memory is no longer available, through swap. This additional space is made available either for the programs that do not use all the memory they are given, or when memory pressure is high. Swapping is usually done using one or more dedicated partitions, as Linux permits multiple swap areas. The recommended swap size is at least the total RAM on the system. To check the actual swap used on the system, you can concatenate the /proc/swaps file:

```
[packt@jupiter ~]$ cat /proc/swaps
Filename                    Type       Size      Used    Priority
/dev/dm-1                   partition  8175612   0       -2
```

Figure 6.10 – Checking the currently used swap

You can also check the memory usage with the free command, as follows:

```
[packt@jupiter ~]$ free
              total        used        free      shared  buff/cache   available
Mem:        7878804      625656     6410364       12552      842784     6962448
Swap:       8175612           0     8175612
```

Figure 6.11 – Memory and swap usage

If swap is not set up on your system, you can format a partition as swap and activate it. The commands to do that are as follows:

```
mkswap /dev/sda1
swapon /dev/sda1
```

The operating system is caching file contents inside the memory to prevent the use of swap as much as possible. The memory that the kernel uses is never swapped; only the memory that the user space is using gets to be swapped.

Filesystems and partitions are the bare bones of any disk management task, but there are still a number of hiccups that an administrator needs to overcome, and this can be solved by using logical volumes. This is why, in the next section, we will introduce you to **LVM**.

Logical Volume Management in Linux

Some of you may have already heard of LVM. For those who do not know what it is, we will explain it shortly in this section. Imagine a situation where your disks run out of space. You can always move it to a larger disk and then replace the smaller one, but this implies system restarts and unwanted downtimes. As a solution, you can consider LVM, which offers more flexibility and efficiency. By using LVM, you can add more physical disks to your existing volume groups, while still in use. This still offers the possibility to move data to a new hard drive, but with no downtime, everything is done while filesystems are online.

As we don't have a system with LVM, we will show you the steps necessary to create new LVM volumes by using the 1 TB drive on our system and we will make it an LVM physical volume:

1. Create the LVM physical volume with the `pvcreate` command:

```
[packt@jupiter ~]$ sudo pvcreate /dev/sda
WARNING: ext4 signature detected on /dev/sda at offset 1080. Wipe it? [y/n]: y
  Wiping ext4 signature on /dev/sda.
  Physical volume "/dev/sda" successfully created.
```

Figure 6.12 – Using pvcreate to create an LVM physical volume

2. Create a new volume group to add the new physical volume using the `vgcreate` command:

```
[packt@jupiter ~]$ sudo vgcreate newvolume /dev/sda
[sudo] password for packt:
  Volume group "newvolume" successfully created
```

Figure 6.13 – Creating a new volume group using vgcreate

3. You can see the new volume group using the `vgdisplay` command:

```
[packt@jupiter ~]$ sudo vgdisplay newvolume
  --- Volume group ---
  VG Name               newvolume
  System ID
  Format                lvm2
  Metadata Areas        1
  Metadata Sequence No  1
  VG Access             read/write
  VG Status             resizable
  MAX LV                0
  Cur LV                0
  Open LV               0
  Max PV                0
  Cur PV                1
  Act PV                1
  VG Size               931.51 GiB
  PE Size               4.00 MiB
  Total PE              238467
  Alloc PE / Size       0 / 0
  Free  PE / Size       238467 / 931.51 GiB
  VG UUID               voPPjN-gZf7-WKNT-Eg4x-S1sl-O6kL-4KrYPe
```

Figure 6.14 – See details regarding the new volume group using vgdisplay

4. Now, create a logical volume using some space from the volume group, using `lvcreate`. Use the `-n` option to add a name for the logical volume, and `-L` to set the size in a human-readable manner (we created a 5 GB logical volume named `projects`):

```
[packt@jupiter ~]$ sudo lvcreate -n projects -L 5G newvolume
[sudo] password for packt:
  Logical volume "projects" created.
```

Figure 6.15 – Creating a logical volume using lvcreate

5. Check to see whether the logical volume exists:

```
[packt@jupiter ~]$ sudo ls /dev/mapper/newvolume*
[sudo] password for packt:
/dev/mapper/newvolume-projects
```

Figure 6.16 – Checking whether the logical volume exists

6. The newly created device can only be used if it's formatted using a known filesystem and mounted afterward, in the same way as a regular partition. First, let's format the new volume:

```
[packt@jupiter ~]$ sudo mkfs -t ext4 /dev/mapper/newvolume-projects
[sudo] password for packt:
mke2fs 1.45.4 (23-Sep-2019)
Discarding device blocks: done
Creating filesystem with 1310720 4k blocks and 327680 inodes
Filesystem UUID: de5be238-5a79-417f-9488-b19bf74d44ce
Superblock backups stored on blocks:
        32768, 98304, 163840, 229376, 294912, 819200, 884736

Allocating group tables: done
Writing inode tables: done
Creating journal (16384 blocks): done
Writing superblocks and filesystem accounting information: done
```

Figure 6.17 – Formatting the new logical volume as an Ext4 filesystem

7.   Now it's time to mount the logical volume. First, create a new directory and mount the logical volume there. Then, check the size using the `df` command:

```
[packt@jupiter ~]$ sudo mkdir /mnt/projects
[packt@jupiter ~]$ sudo mount /dev/mapper/newvolume-projects /mnt/projects/
[packt@jupiter ~]$ df -h /mnt/projects/
Filesystem                      Size  Used Avail Use% Mounted on
/dev/mapper/newvolume-projects  4.9G   20M  4.6G   1% /mnt/projects
```

Figure 6.18 – Mounting the logical volume

8.   All changes implemented hitherto are not permanent. To make them permanent, you will have to edit the `/etc/fstab` file by adding the following line inside the file:

```
/dev/mapper/newvolume-projects /mnt/projects ext4
defaults 1 2
```

9.   You can now check the space available on your logical volume and grow it if you want. Use the `vgdisplay` command to see the following details:

```
sudo vgdisplay newvolume
...
VG Size              931.51 GiB
PE Size              4.00 MiB
Total PE             238467
Alloc PE / Size      1280 / 5.00 GiB
Free  PE / Size      237187 / 926.51 GiB
```

10. You can now expand the logical volume by using the `lvextend` command. We will extend the initial size by 5 GB. Here is an example:

```
[packt@jupiter ~]$ sudo lvextend -L +5G /dev/mapper/newvolume-projects
[sudo] password for packt:
  Size of logical volume newvolume/projects changed from 5.00 GiB (1280 extents)
to 10.00 GiB (2560 extents).
  Logical volume newvolume/projects successfully resized.
```

Figure 6.19 – Extending the logical volume using lvextend

11. Now, resize the filesystem to fit the new size of the logical volume using `resize2fs` and check for the size with `df`:

```
[packt@jupiter ~]$ sudo resize2fs -p /dev/mapper/newvolume-projects
resize2fs 1.45.4 (23-Sep-2019)
Filesystem at /dev/mapper/newvolume-projects is mounted on /mnt/projects; on-lin
e resizing required
old_desc_blocks = 1, new_desc_blocks = 2
The filesystem on /dev/mapper/newvolume-projects is now 2621440 (4k) blocks long
.

[packt@jupiter ~]$ df -h /mnt/projects/
Filesystem                      Size  Used Avail Use% Mounted on
/dev/mapper/newvolume-projects  9.8G   23M  9.3G   1% /mnt/projects
```

Figure 6.20 – Resizing the logical volume with resize2fs and checking for the size with df

In the following section, we will discuss a number of more advanced LVM topics, including how to take full filesystem snapshots:

LVM snapshots

What is an LVM snapshot? It is a frozen instance of an LVM logical volume. In more detail, it uses a copy-on-write technology. This technology monitors each block of the existing volume and when blocks change, due to new writings, that block's value is copied to the snapshot volume.

The snapshots are created constantly and instantly, and it persists until it is deleted. This way, you can create backups from any snapshot. As snapshots are constantly changing due to the copy-on-write technology, initial thoughts on the size of the snapshot should be given when creating one. Take into consideration, if possible, how much data is going to change during the existence of the snapshot. Once the snapshot is full, it will automatically be disabled.

To create a new snapshot, you can use the `lvcreate` command, with the `-s` option. You can also specify the size with the `-L` option and add a name for the snapshot with the `-n` option as follows:

```
[packt@jupiter mapper]$ sudo lvcreate -s -L 5G -n linux-snap01 /dev/mapper/newvo
lume-projects
  Logical volume "linux-snap01" created.
```

Figure 6.21 – Creating an LVM snapshot with the lvcreate command

In the preceding command, we set a size of 5 GB and used the name `linux-snap01`. The last part of the command contains the destination of the volume for which we created the snapshot. To list the new snapshot, use the `lvs` command:

```
[packt@jupiter dev]$ sudo lvs
  LV            VG         Attr       LSize     Pool Origin   Data%  Meta%  Move Log Cpy%Sync Convert
  home          cl         -wi-ao---- <164.78g
  root          cl         -wi-ao----   50.00g
  swap          cl         -wi-ao----  <7.80g
  linux-snap01  newvolume  swi-a-s---   5.00g        projects 0.01
  projects      newvolume  owi-aos---  10.00g
```

Figure 6.22 – Listing the available volumes and the newly created snapshot

For more information on the logical volumes, run the `lvdisplay` command. The output will show information about all the volumes, and among them, you will see the snapshot created earlier. The following is an excerpt of the output:

```
  --- Logical volume ---
  LV Path                /dev/newvolume/linux-snap01
  LV Name                linux-snap01
  VG Name                newvolume
  LV UUID                wbEIFN-t7H6-qEoW-pgeY-NS8C-UZoa-UqbZva
  LV Write Access        read/write
  LV Creation host, time jupiter, 2020-12-18 10:51:50 +0200
  LV snapshot status     active destination for projects
  LV Status              available
  # open                 0
  LV Size                10.00 GiB
  Current LE             2560
  COW-table size         5.00 GiB
  COW-table LE           1280
  Allocated to snapshot  0.01%
  Snapshot chunk size    4.00 KiB
  Segments               1
  Allocation             inherit
  Read ahead sectors     auto
  - currently set to     8192
  Block device           253:6
```

Figure 6.23 – Information about the snapshot using the lvdisplay command

When we created the snapshot, we gave it a size of 5 GB. Now, we would like to extend it to the size of the source, which was 10 GB. We will do this with the `lvextend` command:

```
[packt@jupiter mapper]$ sudo lvs
[sudo] password for packt:
  LV             VG         Attr       LSize    Pool Origin   Data% Meta% Move Log Cpy%Sync Convert
  home           cl         -wi-ao---- <164.78g
  root           cl         -wi-ao----   50.00g
  swap           cl         -wi-ao----  <7.80g
  linux-snap01   newvolume  swi-a-s---   5.00g       projects 0.01
  projects       newvolume  owi-aos---  10.00g
[packt@jupiter mapper]$ sudo lvextend -L +5G /dev/mapper/newvolume-linux--snap01
  Size of logical volume newvolume/linux-snap01 changed from 5.00 GiB (1280 extents) to 10.00 GiB (
2560 extents).
  Logical volume newvolume/linux-snap01 successfully resized.
```

Figure 6.24 – Extending the snapshot from 5 to 10 GB

The naming in the preceding output could be tricky, which is why we added another screenshot to show you why we used that name:

```
[packt@jupiter /]$ cd /dev/mapper/
[packt@jupiter mapper]$ ls
cl-home  cl-swap  newvolume-linux--snap01        newvolume-projects
cl-root  control  newvolume-linux--snap01-cow    newvolume-projects-real
[packt@jupiter mapper]$ ls -la
total 0
drwxr-xr-x.  2 root root    200 Dec 18 10:51 .
drwxr-xr-x. 21 root root   3440 Dec 18 10:51 ..
lrwxrwxrwx.  1 root root      7 Dec 18 10:20 cl-home -> ../dm-2
lrwxrwxrwx.  1 root root      7 Dec 18 10:20 cl-root -> ../dm-0
lrwxrwxrwx.  1 root root      7 Dec 18 10:20 cl-swap -> ../dm-1
crw-------.  1 root root 10, 236 Dec 18 10:20 control
lrwxrwxrwx.  1 root root      7 Dec 18 10:51 newvolume-linux--snap01 -> ../dm-6
lrwxrwxrwx.  1 root root      7 Dec 18 10:51 newvolume-linux--snap01-cow -> ../dm-5
lrwxrwxrwx.  1 root root      7 Dec 18 10:51 newvolume-projects -> ../dm-3
lrwxrwxrwx.  1 root root      7 Dec 18 10:51 newvolume-projects-real -> ../dm-4
```

Figure 6.25 – Explaining the name we used

As you can see in the preceding screenshot, the name that the snapshot volume is using is highlighted. Even though *Figure 6.30* shows that we used the name `linux-snap01` for the snapshot volume, if we do a listing of the `/dev/mapper/` directory, we will see that the name `newvolume-linux—snap01` is used instead. What is actually confusing is the second dash used for the name.

To restore a snapshot, first you would need to unmount the filesystem. To unmount, we will use the `umount` command:

```
sudo umount /mnt/projects/
```

And then we can proceed to restore the snapshot with the `lvconvert` command. This is the output:

```
[packt@jupiter ~]$ sudo lvconvert --merge /dev/mapper/newvolume-linux--snap01
  Merging of volume newvolume/linux-snap01 started.
  newvolume/projects: Merged: 100.00%
```

Figure 6.26 – Restoring the snapshot using the lvconvert command

The snapshot is merged into the source and we can check this by using the `lvs` command:

```
[packt@jupiter mapper]$ sudo lvs
[sudo] password for packt:
  LV        VG        Attr        LSize     Pool Origin Data%  Meta%  Move Log Cpy%Sync Convert
  home      cl        -wi-ao---- <164.78g
  root      cl        -wi-ao----   50.00g
  swap      cl        -wi-ao----   <7.80g
  projects  newvolume -wi-a-----   10.00g
```

Figure 6.27 – Using lvs to verify that the snapshot was merged

Following the merge, the snapshot is automatically removed. As you now know how to create snapshots to LVM volumes, we have now covered all the basics of LVM in Linux.

LVM is more complicated than normal disk partitioning. It might be intimidating to many, but it can show its strengths when needed. Nevertheless, it also comes with several drawbacks. It can add unwanted complexity in a disaster recovery scenario or in case of a hardware failure. But all this aside, it is still worth learning about.

Summary

Managing filesystems and disks is an important task for any Linux system administrator. Understanding how devices are managed in Linux, and how to format and partition disks, is essential. Furthermore, it is important to learn LVM as it offers a flexible way to manage partitions.

Mastering those skills will give you a strong foundation for any basic administration task. In the following chapter, we will introduce you to the vast domain of networking in Linux.

7
Networking with Linux

Linux networking is a vast domain. The last decades have seen countless volumes and references written about Linux network administration internals. Sometimes, the mere assimilation of essential concepts could be overwhelming for both novice and advanced users. This chapter provides a relatively concise overview of Linux networking, focusing on network communication layers, sockets and ports, networking services and protocols, **virtual private networks** (**VPNs**), and network security.

We hope that the content presented in this chapter is both a comfortable introduction to basic Linux networking principles for a novice user and a good refresher for an advanced Linux administrator.

In this chapter, we cover the following topics:

- Exploring basic networking—focusing on computer networks, networking models, protocols, network addresses, and ports. We also cover some practical aspects of configuring Linux network settings using the command-line terminal.

- Working with networking services—introducing common networking servers running on Linux, such as **Domain Host Configuration Protocol** (**DHCP**) servers, **Domain Name System** (**DNS**) servers, file-sharing servers, remote-access servers, and so on.

- Understanding network security—with a special emphasis on VPNs.

Technical requirements

Throughout this chapter, we'll be using the Linux command line to some extent. A working Linux distribution, installed on either a **virtual machine** (**VM**) or a desktop platform, is highly recommended. If you don't have one already, *Chapter 1*, *Installing Linux and Setting Up the Environment*, will guide you through the installation process. Most of the commands and examples illustrated in this chapter use Ubuntu and CentOS, but the same would apply to any other Linux platform.

Exploring basic networking

Today, it's almost inconceivable to imagine a computer not connected to some sort of network or the internet. Our ever-increasing online presence, cloud computing, mobile communications, and **Internet of Things** (**IoT**) would not be possible without the highly distributed, high-speed, and scalable networks serving the underlying data traffic, yet the basic networking principles behind the driving force of the modern-day internet are decades old. Obviously, networking and communication paradigms will continue to evolve, but some of the original primitives and concepts will still have a long-lasting effect in shaping the building blocks of future communications.

This section will introduce you to a few of these networking essentials and, hopefully, spark your curiosity for further exploration. Let's start with computer networks.

Computer networks

A computer network is a group of two or more computers (or nodes) connected via a physical medium (cable, wireless, optical) and communicating with each other using a standard set of agreed-upon communication protocols. At a very high level, a network communication infrastructure includes computers, devices, switches, routers, Ethernet or optical cables, wireless environments, and all sorts of network equipment.

Beyond the *physical* connectivity and arrangement, networks are also defined by a *logical* layout via network topologies, tiers, and the related data flow. An example of a logical networking hierarchy is the three-tiered layering of the **demilitarized zone (DMZ)**, *firewall*, and *internal* networks. The DMZ is an organization's outward-facing network, with an extra security layer against the public internet. A firewall controls the network traffic between the DMZ and the internal network.

Network devices are identified by network addresses and hostnames. Network addresses assist with locating nodes on the network using communication protocols, such as the **Internet Protocol (IP)** (see more on IP in the *TCP/IP protocols* section, later in this chapter). Hostnames are user-friendly labels associated with devices, easier to remember than network addresses.

A common classification criterion looks at the scale and expansion of computer networks. We introduce **local area networks (LANs)** and **wide area networks (WANs)** next.

LANs

A LAN represents a group of devices connected and located in a single physical location, such as a private residence, school, or office. A LAN can be of any size, ranging from a home network with only a few devices to large-scale enterprise networks with thousands of users and computers.

Regardless of the network size, a LAN's essential characteristic is that it connects devices in a single, limited area. Examples of LANs include the home network of a single-family residence or your local coffee shop's free wireless service.

For more information about LANs, you can refer to `https://www.cisco.com/c/en/us/products/switches/what-is-a-lan-local-area-network.html`.

When a computer network spans multiple regions or multiple interconnected LANs, WANs come into play.

WANs

A WAN is usually a network of networks, with multiple or distributed LANs communicating with each other. In this sense, we regard the internet as the world's largest WAN.

An example of WAN is the computer network of a multinational company's geographically distributed offices worldwide. Some WANs are built by service providers, to be leased to various businesses and institutions around the world.

WANs have several variations, depending on their type, range, and use. Typical examples of WANs include **personal area networks (PANs)**, **metropolitan area networks (MANs)**, and **cloud or internet area networks (IANs)**.

For more information about WANs, you can refer to https://www.cisco.com/c/en/us/products/switches/what-is-a-wan-wide-area-network.html.

We think that an adequate introduction to basic networking principles should always include a brief presentation of the theoretical model governing network communications in general. Let's look at this next.

The OSI model

The **Open Systems Interconnection (OSI)** model is a theoretical representation of a multilayer communication mechanism between computer systems interacting over a network. The OSI model was introduced in 1983 by the **International Organization for Standardization (ISO)** to provide a standard for different computer systems to communicate with each other.

We could regard the OSI model as a universal framework for network communications. As the following screenshot shows, the OSI model defines a stack of seven layers, directing the communication flow:

Layer	#	Unit	Description
Application Layer	7		User interaction and high-level APIs
Presentation Layer	6	data	Data translation into a usable format (such as encrypt/decrypt, encode/decode, compress/deflate)
Session Layer	5		Communication sessions (connections, sockets, ports)
Transport Layer	4	segments, datagrams	Reliable data segments using transmission control protocols (such as TCP, UDP)
Network Layer	3	packets	Data packets with addressing, routing and traffic control information (that is data path)
Data Link Layer	2	frames	Formatting the data as reliable data frames
Physical Layer	1	bits	Transmission/reception of raw bit streams over a physical medium

Figure 7.1 – The OSI model

In the layered view shown in the preceding diagram, the communication flow moves from top to bottom (on the transmitting end) or bottom to top (on the receiving end). Let's look at each of these layers and describe their functionality in shaping network communication.

The physical layer

The *physical layer* (or *Layer 1*) consists of the networking equipment or infrastructure connecting the devices and serving the communication, such as cables, wireless or optical environments, connectors, and switches. This layer handles the conversion between raw bit streams and the communication medium while regulating the corresponding bit-rate control. (The communication medium includes electrical, radio, or optical signals.)

Examples of protocols operating at the physical layer include Ethernet, **Universal Serial Bus** (**USB**), and **Digital Subscriber Line** (**DSL**).

The data link layer

The *data link layer* (or *Layer 2*) establishes a reliable data flow between two directly connected devices on a network, either as adjacent nodes in a WAN or as devices within a LAN. One of the data link layer's responsibilities is flow control, adapting to the physical layer's communication speed. On the receiving device, the data link layer corrects communication errors that originated in the physical layer. The data link layer consists of the following subsystems:

- **Media access control** (**MAC**)—This subsystem uses MAC addresses to identify and connect devices on the network. It also controls the device access permissions to transmit and receive data on the network.

- **Logical link control** (**LLC**)—This subsystem identifies and encapsulates network layer protocols and performs error checking and frame synchronization while transmitting or receiving data.

The protocol data units controlled by the data link layer are also known as *frames*. A frame is a data transmission unit acting as a container for a single network packet. Network packets are processed at the next OSI level (*network layer*). When multiple devices access the same physical layer simultaneously, frame collisions may occur. Data link layer protocols can detect and recover from such collisions and further reduce or prevent their occurrence.

An example of data link protocol is the **Point-to-Point Protocol** (**PPP**), a binary networking protocol used in high-speed broadband communication networks.

The network layer

The *network layer* (or *Layer 3*) discovers the optimal communication path (or route) between devices on a network. This layer uses a routing mechanism based on the IP addresses of the devices involved in the data exchange, to move data packets from source to destination.

On the transmitting end, the network layer disassembles the data segments that originated in the transport layer (*Layer 4*) into network packets. On the receiving end, the data frames are reassembled from the layer below (*data link layer*) into packets.

A protocol operating at the network layer is the **Internet Control Message Protocol (ICMP)**. ICMP is used by network devices to check the availability of other devices (or IP addresses) on the network. ICMP reports an error when a requested endpoint is not available.

The transport layer

The *transport layer* (or *Layer 4*) operates with data *segments* or *datagrams*. This layer is mainly responsible for transferring data from a source to a destination and guaranteeing a specific **quality of service** (**QoS**). On the transmitting end, data that originated from the layer above (*session layer*) is disassembled into segments. On the receiving end, the transport layer reassembles the data packets received from the layer below (*network layer*) into segments.

The transport layer maintains the reliability of the data transfer through flow-control and error-control functions. The flow-control function adjusts the data transfer rate between endpoints with different connection speeds, to avoid a sender overwhelming the receiver. When the data received is incorrect, the error-control function may request the retransmission of data.

Examples of transport layer protocols include the **Transmission Control Protocol** (**TCP**) and the **User Datagram Protocol** (**UDP**).

The session layer

The *session layer* (or *Layer 5*) controls the lifetime of the connection channels (or sessions) between devices communicating on a network. At this layer, sessions or network connections are usually defined by network addresses, sockets, and ports. We'll explain each of these concepts later in this chapter. The session layer is responsible for the integrity of the data transfer within a communication channel or session. For example, if a session is interrupted, the data transfer resumes from a previous checkpoint.

Some typical session layer protocols are the **Remote Procedure Call** (**RPC**) protocol used by interprocess communications, and **Network Basic Input/Output System** (**NetBIOS**), which is a file-sharing and name-resolution protocol.

The presentation layer

The *presentation layer* (or *Layer 6*) acts as a data translation tier between the *application layer* above and the *session layer* below. On the transmitting end, this layer formats the data into a system-independent representation before sending it across the network. On the receiving end, the presentation layer transforms the data into an application-friendly format. Examples of such transformations are encryption and decryption, compression and decompression, encoding and decoding, and serialization and deserialization. Usually, there is no substantial distinction between the presentation and application layers, mainly due to the relatively tight coupling of the various data formats with the applications consuming them. Standard data representation formats include **American Standard Code for Information Interchange** (ASCII), **Extensible Markup Language** (XML), **JavaScript Object Notation** (JSON), **Joint Photographic Experts Group** (JPEG), ZIP, and so on.

The application layer

The *application layer* (or *Layer 7*) is the closest to the end user in the OSI model. Layer 7 collects or provides the input or output of application data in some meaningful way. This layer does not contain or run the applications themselves. Instead, Layer 7 acts as an abstraction between applications, implementing a communication component and the underlying network. Typical examples of applications interacting with the application layer are web browsers and email clients.

A few examples of Layer 7 protocols are the DNS protocol; the **HyperText Transfer Protocol** (HTTP); the **File Transfer Protocol** (FTP); and the **Post Office Protocol** (POP), **Internet Message Access Protocol** (IMAP), **Simple Mail Transfer Protocol** (SMTP) email messaging protocols.

Before wrapping up, we should call out that the OSI model is a generic representation of networking communication layers and provides the theoretical guidelines of how network communication works. A similar—but more practical—illustration of the networking stack is the TCP/IP model, which we'll explore in the next section.

Both models are useful when it comes to network design, implementation, troubleshooting, and diagnostics. The OSI model gives network operators a good understanding of the full networking stack, from the physical medium to the application layer, each level with its **Protocol Data Units** (PDUs) and communication internals. The TCP/IP model is somewhat simplified, with a few of the OSI model layers collapsed into one, and it takes a rather protocol-centric approach to network communications.

The TCP/IP model

The *TCP/IP model* is a four-layer interpretation of the OSI networking stack, where some of the equivalent OSI layers appear consolidated, as shown in the following screenshot:

	OSI Model	TCP/IP Model	TCP/IP Protocols
7	Application Layer	Application Layer	DNS, HTTP, FTP, SMTP, SNMP, Telnet, ...
6	Presentation Layer		
5	Session Layer		
4	Transport Layer	Transport Layer	TCP, UDP, ...
3	Network Layer	Internet Layer	IP, ARP, ICMP, IGMP, ...
2	Data Link Layer	Network Interface Layer	Ethernet, Token Ring, Frame Relay, ATM, ...
1	Physical Layer		

Figure 7.2 – The OSI and TCP/IP models

Chronologically, the TCP/IP model is older than the OSI model. It was first suggested by the US **Department of Defense (DoD)** as part of an internetwork project developed by the **Defense Advanced Research Projects Agency (DARPA)**. This project eventually became the modern-day internet.

The TCP/IP model layers encapsulate similar functions to their counterpart OSI layers. Here's a brief summary of each layer in the TCP/IP model.

The network interface layer

The *network interface layer* is responsible for data delivery over a physical medium (such as wire, wireless, optical). Networking protocols operating at this layer include Ethernet, Token Ring, and Frame Relay. This layer maps to the composition of the physical and data link layers in the OSI model.

The internet layer

The *internet layer* provides *connectionless* data delivery between nodes on a network. Connectionless protocols describe a network communication pattern where a sender transmits data to a receiver without a prior arrangement between the two. This layer is responsible for disassembling data into network packets at the transmitting end and reassembling on the receiving end. The internet layer uses routing functions to identify the optimal path between the network nodes. This layer maps to the network layer in the OSI model.

The transport layer

The *transport layer* (also known as the *transmission layer* or the *host-to-host layer*) is responsible for maintaining the communication sessions between connected network nodes. The transport layer implements error-detection and correction mechanisms for reliable data delivery between endpoints. This layer maps to the transport layer in the OSI model.

The application layer

The *application layer* provides the data communication abstraction between software applications and the underlying network. This layer maps to the composition of the session, presentation, and application layers in the OSI model.

The TCP/IP model is a protocol-centric representation of the networking stack. This model served as the foundation of the internet by gradually defining and developing networking protocols required for internet communications. These protocols are collectively referred to as the *IP suite*.

The following section describes some of the most common networking protocols.

TCP/IP protocols

In this section, we describe some widely used networking protocols. The reference here should not be regarded as an all-encompassing guide. There are a vast number of TCP/IP protocols, and a comprehensive study is beyond the scope of this chapter. Nevertheless, there are a handful of protocols worth exploring that are frequently at work in everyday network communication and administration workflows.

The following sections briefly describe each TCP/IP protocol with its related **Request for Comments** (**RFC**) identifier for more information. The RFC represents the detailed technical documentation—of a protocol, in our case—usually authored by the **Internet Engineering Task Force** (**IETF**). For more information about RFC, please refer to `https://www.ietf.org/standards/rfcs/`.

IP

IP (*RFC 791*) identifies network nodes based on fixed-length addresses, also known as IP addresses. IP addresses are described in more detail later in this chapter. The IP protocol uses datagrams as the data transmission unit and provides fragmentation and reassembly capabilities of large datagrams to accommodate small-packet networks (and avoid transmission delays). The IP protocol also provides routing functions to find the optimal data path between network nodes. IP operates at the network layer (3) in the OSI model.

ARP

The **Address Resolution Protocol** (**ARP**) (*RFC 826*) is used by the IP protocol to map IP network addresses (specifically, **IP version 4 (IPv4)**) to device MAC addresses used by a data link protocol. ARP operates at the data link layer (2) in the OSI model.

NDP

The **Neighbor Discovery Protocol** (**NDP**) (*RFC 4861*) is like the ARP protocol, and it also controls **IP version 6** (**IPv6**) address mapping. NDP operates within the data link layer (2) in the OSI model.

ICMP

ICMP (*RFC 792*) is a supporting protocol for checking the availability of network devices based on IP address. When a device or node is not reachable within a given timeout, ICMP reports an error. ICMP operates at the network layer (3) in the OSI model.

TCP

TCP (*RFC 793*) is a connection-oriented, highly reliable communication protocol. TCP requires a logical connection (such as a *handshake*) between the nodes before initiating the data exchange. TCP operates at the transport layer (4) in the OSI model.

UDP

UDP (*RFC 768*) is a connectionless communication protocol. UDP has no handshake mechanism (compared to TCP). Consequently, with UDP there's no guarantee of data delivery. UDP uses datagrams as the data transmission unit, and it's suitable for network communications where error checking is not critical. UDP operates at the transport layer (4) in the OSI model.

DHCP

DHCP (*RFC 2131*) provides a framework for requesting and passing host configuration information required by devices on a TCP/IP network. DHCP enables the automatic (dynamic) allocation of reusable IP addresses and other configuration options. DHCP is considered an application layer (7) protocol in the OSI model, but the initial DHCP discovery mechanism operates at the data link layer (2).

DNS

DNS (*RFC 2929*) is a protocol acting as a network address book, where nodes in the network are identified by human-readable names instead of IP addresses. According to the IP protocol, each device on a network is identified by a unique IP address. When a network connection specifies the remote device's hostname (or domain name) before the connection is established, DNS translates the domain name (such as `dns.google.com`) to an IP address (such as `8.8.8.8`). The DNS protocol operates at the application layer (7) in the OSI model.

HTTP

HTTP (*RFC 2616*) is the vehicular language of the internet. HTTP is a stateless application-level protocol based on request and response between a client application (for example, a browser) and a server endpoint (for example, a web server). HTTP supports a wide variety of data formats, ranging from text to images and video streams. HTTP operates at the application layer (7) in the OSI model.

FTP

FTP (*RFC 959*) is a standard protocol for transferring files requested by an FTP client from an FTP server. FTP operates at the application layer (7) in the OSI model.

TELNET

The **Terminal Network protocol** (**TELNET**) (*RFC 854*) is an application-layer protocol providing a bidirectional text-oriented network communication between a client and a server machine, using a virtual terminal connection. TELNET operates at the application layer (7) in the OSI model.

SSH

Secure Shell (SSH) (*RFC 4253*) is a secure application-layer protocol, encapsulating strong encryption and cryptographic host authentication. SSH uses a virtual terminal connection between a client and a server machine. SSH operates at the application layer (7) in the OSI model.

SMTP

SMTP (*RFC 5321*) is an application-layer protocol for sending and receiving emails between an email client (for example, Outlook) and an email server (such as Exchange Server). SMTP supports strong encryption and host authentication. SMTP acts at the application layer (7) in the OSI model.

SNMP

The **Simple Network Management Protocol (SNMP)** (*RFC 1157*) is used for remote device management and monitoring. SNMP operates at the application layer (7) in the OSI model.

NTP

The **Network Time Protocol (NTP)** (*RFC 5905)* is an internet protocol used for synchronizing the system clock of multiple machines across a network. NTP operates at the application layer (7) in the OSI model.

Most of the internet protocols enumerated previously use the IP protocol to identify devices participating in a communication. Devices on a network are uniquely identified by an IP address. Let's have a closer examination of these network addresses.

IP addresses

The IP address is a fixed-length **unique identifier (UID)** of a device in a network. Devices locate and communicate with each other based on IP addresses. The concept of an IP address is very similar to a postal address of a residence, whereby a mail or a package would be sent to that destination based on its address.

Initially, IP defined the IP address as a 32-bit number known as an IPv4 address. With the growth of the internet, the total number of IP addresses in a network has been exhausted. To address this issue, a new version of the IP protocol devised a 128-bit numbering scheme for IP addresses. A 128-bit IP address is also known as an IPv6 address.

In the next sections, we'll take a closer look at the networking constructs playing an important role in IP addresses, such as IPv4 and IPv6 address formats, network classes, subnetworks, and broadcast addresses.

IPv4 addresses

An IPv4 address is a 32-bit number (4 bytes) usually expressed as four groups of 1-byte (8 bits) numbers, separated by a dot (.). Each number in these four groups is an integer between 0 and 255. Here's an example of an IPv4 address:

```
192.168.1.53
```

The following illustration shows a binary representation of an IPv4 address:

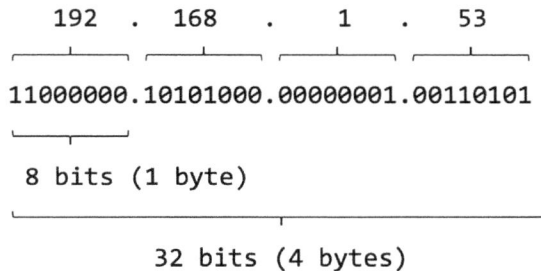

Figure 7.3 – Network classes

The IPv4 address space is limited to 4,294,967,296 (232) addresses (roughly 4 billion). Of these, approximately 18 million are reserved for special purposes (for example, private networks), and about 270 million are multicast addresses.

A multicast address is a logical identifier of a group of IP addresses. For more information on multicast addresses, please refer to *RFC 6308* (https://tools.ietf.org/html/rfc6308).

Network classes

In the early stages of the internet, the highest-order byte (first group) in the IPv4 address indicated the network number. The subsequent bytes further express the network hierarchy and subnetworks, with the lowest-order byte identifying the device itself. This scheme soon proved insufficient for network hierarchies and segregations, as it only allowed for 256 (28) networks, denoted by the leading byte of the IPv4 address. As additional networks were added, each with its own identity, the IP address specification needed a special revision to accommodate a standard model. The *Classful Network* specification introduced in 1981 addressed the problem by dividing the IPv4 address space into five classes based on the leading 4 bits of the address, as illustrated in the following screenshot:

Class	Leading bits	Start address	End address	Default Subnet Mask
Class A	0	0.0.0.0	127.255.255.255	255.0.0.0
Class B	10	128.0.0.0	192.255.255.255	255.255.0.0
Class C	110	192.0.0.0	223.255.255.255	255.255.255.0
Class D (multicast)	1110	224.0.0.0	239.255.255.255	Not defined
Class E (reserved)	1111	240.0.0.0	255.255.255.255	Not defined

Figure 7.4 – Network classes

For more information on network classes, please refer to *RFC 870* (`https://tools.ietf.org/html/rfc870`). In the preceding table, the last column specifies the default subnet mask for each of these network classes. We'll look at subnets (or subnetworks) next.

Subnetworks

Subnetworks (or *subnets*) are logical subdivisions of an IP network. Subnets were introduced with the purpose of identifying devices that belong to the same network. The IP addresses of devices in the same network have an identical most-significant group. The subnet definition yields a logical division of an IP address in two fields: the *network identifier* and a *host identifier*. The numerical representation of the subnet is called a *subnet mask* or *netmask*. The following table gives an example of a network identifier and a host identifier:

IP Address (192.168.1.53)

Network Identifier   Host Identifier

192.168.1   53

Figure 7.5 – Subnet with network and host identifiers

With our IPv4 address (192.168.1.53), we could devise a network identifier of 192.168.1 and the host identifier as 53. The resulting subnet mask is this:

```
192.168.1.0
```

We dropped the least significant group in the subnet mask, representing the host identifier (53), and replaced it with 0. The 0 in this case indicates the starting address in the subnet. In other words, any host identifier value in the range of 0 – 255 is allowed in the subnetwork. For example, the IP address of 192.168.1.92 is a valid (and accepted) IP address in the 192.168.1.0 network.

An alternative representation of subnets uses so-called **Classless Inter-Domain Routing** (**CIDR**) notation. CIDR represents an IP address as the network address (*prefix*) followed by a slash (/) and the *bit-length* of the prefix. In our case, the CIDR notation of the 192.168.1.0 subnet is this:

```
192.168.1/24
```

The first three groups in the network address make up for *3 x 8 = 24* bits, hence the /24 notation.

Usually, subnets are planned with the host identifier address as a starting point. Back to our example, suppose we wanted our host identifier addresses in the network to start with 100 and end with 125.

The binary representation of 192.168.1.100 is this:

11000000.10101000.00000001.**01100100**

The last group in the preceding sequence (highlighted) represents the host identifier (100). The closest binary value to the reserved 99 addresses that would not be permitted in our subnet is *96 = 64 + 32*. The equivalent binary value is as follows:

```
11100000
```

In other words, the three most significant bits in the host identifier are reserved. Reserved bits in the subnet representation are shown as 1. These bits would be added to the 24 already reserved bits of the network address (192.168.1), accounting in total for *27 = 24 + 3* bits. Here's the equivalent representation:

```
11111111.11111111.11111111.11100000
```

Consequently, the resulting netmask is this:

```
255.255.255.224
```

The CIDR notation of the corresponding subnet is shown here:

```
192.168.1.96/27
```

The remaining 5 bits in the host identifier's group account for 25 = 32 possible addresses in the subnet, starting with 97. This would limit the maximum host identifier value to *127 = 96 + 32 − 1*. (We subtract 1 to account for the starting number of 97 included in the total of 32). In this range of 32 addresses, the last IP address is reserved as a *broadcast address*, shown here:

```
192.168.1.127
```

A broadcast address is reserved as the highest number in a network or subnet, when applicable. Back to our example, excluding the broadcast address, the maximum host IP address in the subnet is this:

```
192.168.1.126
```

You can learn more about subnets in *RFC 1918* (https://tools.ietf.org/html/rfc1918). Since we mentioned the broadcast address, let's have a quick look at it.

Broadcast addresses

A *broadcast address* is a reserved IP address in a network or subnetwork, used to transmit a collective message (data) to all devices belonging to the network. The broadcast address is the last IP address in the network or subnet, when applicable.

For example, the broadcast address of the 192.168.1.0/24 network is 192.168.1.255. In our example in the previous section, the broadcast address of the 192.168.1.96/27 subnet is 192.168.1.127 (*127 = 96 + 32 − 1*).

For more information on broadcast addresses, please visit https://en.wikipedia.org/wiki/Broadcast_address.

IPv6 addresses

An IPv6 address is a 128-bit number (16 bytes) usually expressed as up to eight groups of 2-byte (16-bits) numbers, separated by a column (:). Each number in these eight groups is a hexadecimal number, with values between `0000` and `FFFF`. Here's an example of an IPv6 address:

```
2001:0b8d:8a52:0000:0000:8b2d:0240:7235
```

An equivalent representation of the preceding IPv6 address is shown here:

```
2001:b8d:8a52::8b2d:240:7235/64
```

In the second representation, the leading zeros are omitted, and the all-zero groups (`0000:0000`) are collapsed into an empty group (: :). The /64 notation at the end represents the *prefix length* of the IPv6 address. The IPv6 prefix length is the equivalent of the CIDR notation of IPv4 subnets. For IPv6, the prefix length is expressed as an integer value between `1` and `128`.

In our case, with the prefix length of 64 (*4 x 16*) bits, the subnet looks like this:

```
2001:b8d:8a52::
```

The subnet represents the leading four groups (`2001`, `0b8d`, `8a52`, `0000`), a total of *4 x 16 = 64* bits. In the shortened representation of the IPv6 subnet, the leading zeros are omitted and the all-zero group is collapsed to : :.

Subnetting with IPv6 is very similar to IPv4. We won't go into the details here, since the related concepts are already presented in the IPv4 section. For more information on IPv6, please refer to *RFC 2460* (`https://tools.ietf.org/html/rfc2460`).

After becoming familiar with IP addresses, it is fitting to introduce some of the related network constructs—sockets and ports—serving the software implementation of IP addresses.

Sockets and ports

A *socket* is a software data structure representing a network node for communication purposes. Although a programming concept, in Linux a network socket is ultimately a file descriptor controlled via a network **application programming interface** (**API**). A socket is used by an application process for transmitting and receiving data. An application can create and delete sockets. A socket cannot be active (sending or receiving data) beyond the lifetime of the process that created the socket.

Network sockets operate at the transport-layer level in the OSI model. There are two endpoints to a socket connection—a sender and a receiver. Both the sender and receiver have their own IP address. Consequently, a critical piece of information in the socket data structure is the *IP address* of the endpoint owning the socket.

Both endpoints create and manage their sockets via the network processes using these sockets. The sender and receiver may agree upon using multiple connections to exchange data. Some of these connections may even run in parallel. How do we differentiate between these socket connections? The IP address by itself is not sufficient, and this is where *ports* come into play.

A network *port* is a logical construct used to identify a specific process or network service running on a host. A port is an integer value in the range of 0 – 65535. Usually, ports in the range of 0 – 1024 are assigned to the most used services on a system. These ports are also called *well-known ports*. Here are a few examples of well-known ports and the related network service for each of them:

- 25—SMTP
- 21—FTP
- 22—SSH
- 53—DNS
- 67, 68—DHCP (client = 68, server = 67)
- 80—HTTP
- 443—**HTTP Secure (HTTPS)**

Port numbers beyond 1024 are for general use and are also known as *ephemeral ports*.

A port is always associated with an IP address. Ultimately, a socket is a combination of an IP address and a port. For more information on network sockets, you can refer to *RFC 147* (https://tools.ietf.org/html/rfc147). For well-known ports, see *RFC 1340* (https://tools.ietf.org/html/rfc1340).

Let's put to work the knowledge we gained so far, by looking next at how to configure the local networking stack in Linux.

Linux network configuration

This section describes the TCP/IP network configuration for Ubuntu and CentOS platforms, using their latest released versions to date. The same concepts would apply for most Linux distributions, albeit some of the network configuration utilities and files involved could be different.

We use the `ip` command-line utility to retrieve the system's current IP addresses, as follows:

```
ip addr
```

An example output is shown here:

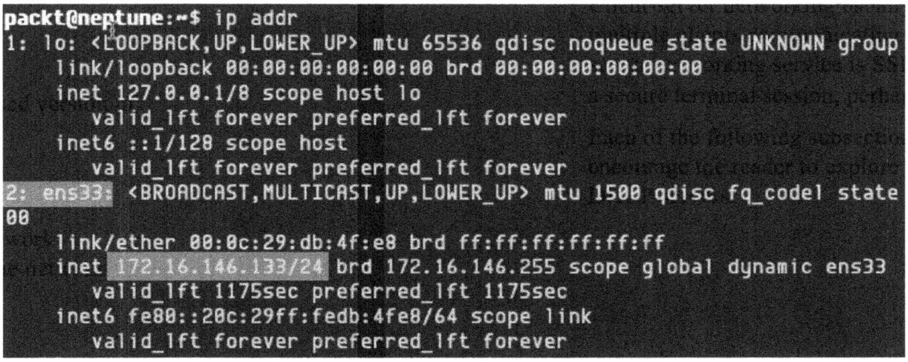

Figure 7.6 – Retrieving the current IP addresses with the ip command

We highlighted some relevant information, such as the network interface ID (`2: ens33`) and the IP address with the subnet prefix (`172.16.146.133/24`).

Let's look at Ubuntu's network configuration next. At the time of this writing, the current released version of Ubuntu is 20.04.

Ubuntu network configuration

Ubuntu 20.04 provides the `netplan` command-line utility for easy network configuration. `netplan` uses a **YAML Ain't Markup Language** (**YAML**) configuration file to generate the network interface bindings. The `netplan` configuration file(s) is in the `/etc/netplan/` directory, as shown in the following code snippet:

```
ls /etc/netplan/
```

In our case, the configuration file is `00-installer-config.yaml`, as illustrated here:

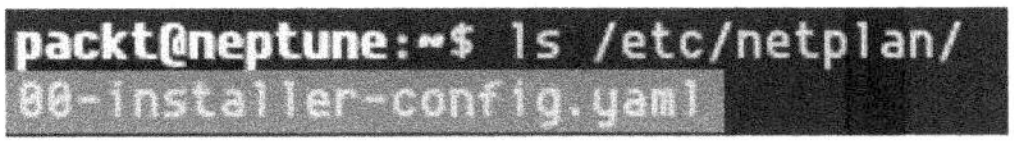

Figure 7.7 – Retrieving the netplan configuration file(s)

Changing the network configuration involves editing the `netplan` YAML configuration file. As a good practice, we should always make a backup of the current configuration file before making changes.

We'll look at dynamic IP addressing first.

Dynamic IP

To enable a dynamic (DHCP) IP address, we edit the `netplan` configuration file and set the `dhcp4` attribute to `true` for the network interface of our choice (`ens33` in our case), as follows:

```
sudo nano /etc/netplan/00-installer-config.yaml
```

Here's the related configuration excerpt, with the relevant points highlighted:

```
# This is the network config written by 'subiquity'
network:
  ethernets:
➤   ens33:
      dhcp4: true
  version: 2
```

Figure 7.8 – Enabling DHCP in the netplan configuration

After saving the configuration file, we can test the related changes with the following command:

```
sudo netplan try
```

We get the following response:

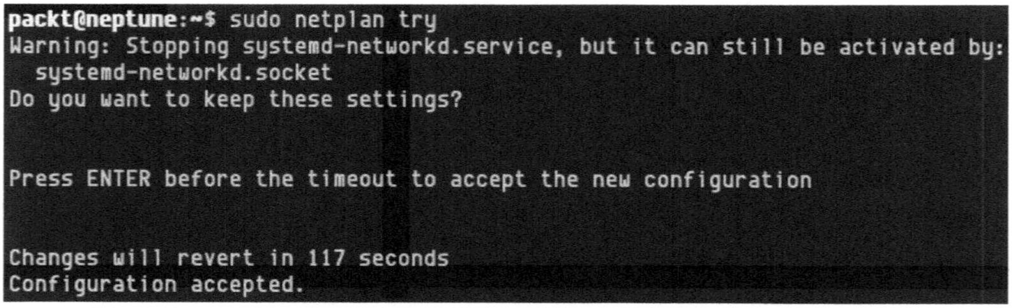

```
packt@neptune:~$ sudo netplan try
Warning: Stopping systemd-networkd.service, but it can still be activated by:
  systemd-networkd.socket
Do you want to keep these settings?


Press ENTER before the timeout to accept the new configuration


Changes will revert in 117 seconds
Configuration accepted.
```

Figure 7.9 – Testing and accepting the netplan configuration changes `netplan` validates the new configuration and prompts for accepting the changes. The following command applies the current changes to the system:

```
sudo netplan apply
```

Next, we configure a static IP address using `netplan`.

Static IP

To set the static IP address of a network interface, we start by editing the `netplan` configuration YAML file, as follows:

```
sudo nano /etc/netplan/00-installer-config.yaml
```

Here's a configuration example with a static IP address of `172.16.146.100/24`:

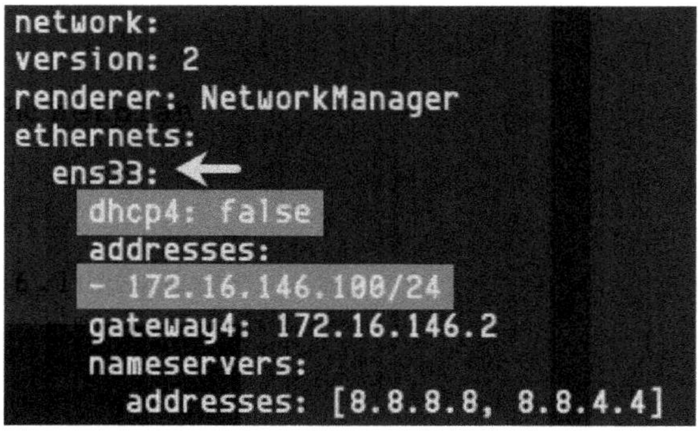

Figure 7.10 – Static IP configuration example with netplan

After saving the configuration, we can test and accept, and then apply changes, as we did in the *Dynamic IP* section, with the following commands:

```
sudo netplan try
sudo netplan apply
```

For more information on the `netplan` command-line utility, see `netplan --help` or the related system manual (`man netplan`).

We'll look at the CentOS network configuration next. At the time of this writing, the current released version of **Red Hat Enterprise Linux (RHEL)**/CentOS is CentOS 8.

CentOS network configuration

There are two ways to configure and manage network interfaces in CentOS 8, as outlined here:

- Manually editing the network interface files in /etc/sysconfig/network-scripts/
- Using the nmcli command-line utility

The network configuration files are in the /etc/sysconfig/networks-scripts/ directory, as shown in the following code snippet. They are named according to the corresponding network interface ID and prefixed with ifcfg. In our case, we retrieve the configuration file with the following command:

```
ls /etc/sysconfig/networks-scripts/
```

The output is as follows:

```
[packt@jupiter ~]$ ls /etc/sysconfig/network-scripts/
ifcfg-ens33
```

Figure 7.11 – Retrieving the network configuration files

The only network configuration file is ifcfg-ens33 and this corresponds to the ens33 network interface.

Let's look at dynamic IP addressing first.

Dynamic IP

Here's a DHCP configuration example for ifcfg-ens33:

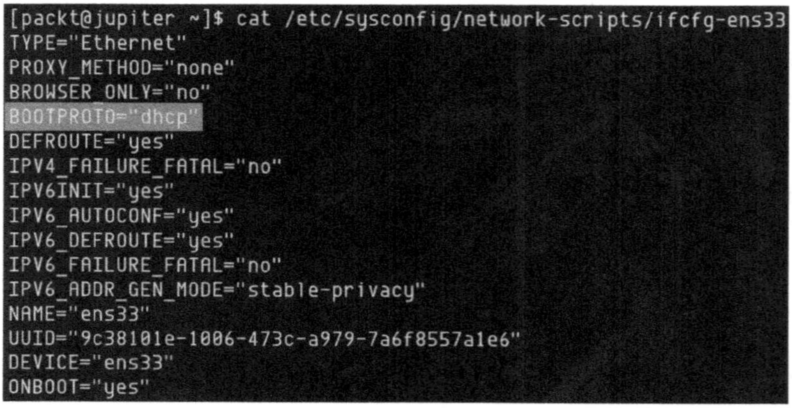

```
[packt@jupiter ~]$ cat /etc/sysconfig/network-scripts/ifcfg-ens33
TYPE="Ethernet"
PROXY_METHOD="none"
BROWSER_ONLY="no"
BOOTPROTO="dhcp"
DEFROUTE="yes"
IPV4_FAILURE_FATAL="no"
IPV6INIT="yes"
IPV6_AUTOCONF="yes"
IPV6_DEFROUTE="yes"
IPV6_FAILURE_FATAL="no"
IPV6_ADDR_GEN_MODE="stable-privacy"
NAME="ens33"
UUID="9c38101e-1006-473c-a979-7a6f8557a1e6"
DEVICE="ens33"
ONBOOT="yes"
```

Figure 7.12 – Dynamic IP configuration

The dynamic IP address is enabled with `BOOTPROTO="dhcp"`. The possible values for `BOOTPROTO` are listed as follows:

- `dhcp`—uses the DHCP protocol to set a dynamic IP address
- `bootp`—uses the **Bootstrap (BOOTP)** protocol to set a dynamic IP address
- `none`—uses a static IP address

To apply the changes, we need to restart (`down` and `up`) the related network interface (`ens33`) with the following code:

```
sudo nmcli connection down ens33
sudo nmcli connection up ens33
```

To configure a dynamic IP address using `ncmli`, we run the following command:

```
sudo nmcli connection modify ens33 IPv4.method auto
```

The `IPv4.method auto` directive enables DHCP.

Let's configure a static IP address next.

Static IP

Here's a static IP configuration example for `ifcfg-ens33`:

```
[packt@jupiter ~]$ cat /etc/sysconfig/network-scripts/ifcfg-ens33
TYPE="Ethernet"
PROXY_METHOD="none"
BROWSER_ONLY="no"
BOOTPROTO="none"
DEFROUTE="yes"
IPV4_FAILURE_FATAL="no"
IPV6INIT="yes"
IPV6_AUTOCONF="yes"
IPV6_DEFROUTE="yes"
IPV6_FAILURE_FATAL="no"
IPV6_ADDR_GEN_MODE="stable-privacy"
NAME="ens33"
UUID="9c38101e-1006-473c-a979-7a6f8557a1e6"
DEVICE="ens33"
ONBOOT="yes"
IPADDR=172.16.146.136
PREFIX=24
GATEWAY=172.16.146.2
DNS1=8.8.8.8
DNS2=8.8.4.4
```

Figure 7.13 – Static IP configuration

The relevant changes are highlighted. DHCP is disabled with BOOTPROTO="none". IPADDR and PREFIX set the static IP address (172.16.146.136/24). We also have the gateway and DNS servers specified.

The changes are saved with the following code:

```
sudo nmcli connection down ens33
sudo nmcli connection up ens33
```

To perform the equivalent static IP address changes using ncmli, we need to run multiple commands. First, we set the static IP address, as follows:

```
sudo nmcli connection modify ens33 IPv4.address
172.16.146.136/24
```

If we had no previous static IP address configured, we recommend saving the preceding change before proceeding with the next steps. The changes are saved with the following code:

```
sudo nmcli connection down ens33
sudo nmcli connection up ens33
```

Next, we set the gateway and DNS IP addresses, as follows:

```
sudo nmcli connection modify ens33 IPv4.gateway 172.16.146.2
sudo nmcli connection modify ens33 IPv4.dns 8.8.8.8
```

Finally, we disable DHCP with the following code:

```
sudo nmcli connection modify ens33 IPv4.method manual
```

After these changes, we need to restart the ens33 network interface with the following code:

```
sudo nmcli connection down ens33
sudo nmcli connection up ens33
```

Next, we'll take a look at how to change the hostname of a Linux machine.

Hostname configuration

To retrieve the current hostname on a Linux machine, we can use either the `hostname` or `hostnamectl` command, as follows:

```
hostname
```

In our case, the response is this:

```
packt@neptune:~$ hostname
neptune
```

Figure 7.14 – Retrieving the current hostname

The most convenient way to change the hostname is with the `hostnamectl` command. We can change the hostname to `jupiter` with the following code:

```
sudo hostnamectl set-hostname jupiter
```

Let's verify the hostname change with the `hostnamectl` command this time, as follows:

```
hostnamectl
```

The output of the `hostnamectl` command provides more detailed information compared to the `hostname` command, as we can see here:

```
packt@neptune:~$ hostnamectl
   Static hostname: jupiter
         Icon name: computer-vm
           Chassis: vm
        Machine ID: cde4b52a365f462ea4b64cde88ba6c0a
           Boot ID: 3375a15598074774b525334801f8fec5
    Virtualization: vmware
  Operating System: Ubuntu 20.04.1 LTS
            Kernel: Linux 5.4.0-53-generic
      Architecture: x86-64
```

Figure 7.15 – Retrieving the current hostname with the hostnamectl command

Alternatively, we can use the `hostname` command to change the hostname *temporarily*, as follows:

```
sudo hostname jupiter
```

But this change would not *survive* a reboot unless we also change the hostname in the /etc/hostname and /etc/hosts files, as follows:

```
packt@neptune:~$ cat /etc/hostname
jupiter
packt@neptune:~$ cat /etc/hosts
127.0.0.1 localhost
127.0.1.1 jupiter
```

Figure 7.16 – The /etc/hostname and /etc/hosts files

After the hostname reconfiguration, a logout followed by a login would usually reflect the changes.

Working with networking services

In this section, we enumerate some of the most common network services running on Linux. Not all the services mentioned here are installed or enabled by default in your Linux platform of choice. *Chapter 8, Configuring Linux Servers*, and *Chapter 9, Securing Linux*, go into how to install and configure some of them. Our focus in this section remains on what these networking services are, how they work, and the networking protocols they use for communication.

A network service is typically a system process implementing an application layer (OSI Layer 7) functionality for data communication purposes. Network services are usually designed as peer-to-peer or client-server architectures.

In peer-to-peer networking, multiple network nodes each run their own equally privileged instance of a network service while sharing and exchanging a common set of data. Take, for example, a network of DNS servers, all sharing and updating their domain name records.

Client-server networking usually involves one or more server nodes on a network and multiple clients communicating with any of these servers. An example of a client-server networking service is SSH. An SSH client connects to a remote SSH server via a secure terminal session, perhaps for remote administration purposes.

Each of the following subsections briefly describes a networking service, and we encourage you to explore a related topic of interest further. Let's start with DHCP servers.

DHCP servers

A DHCP server uses the DHCP protocol to enable devices on a network to request an IP address assigned dynamically. The DHCP protocol was briefly described in the *TCP/IP protocols* section earlier in this chapter.

A computer or device requesting a DHCP service sends out a broadcast message (or query) on the network to locate a DHCP server, which in turn provides the requested IP address and other information. The communication between the DHCP client (device) and the server uses the DHCP protocol.

The DHCP protocol's initial *discovery* workflow between a client and a server operates at the data link layer (2) in the OSI model. Since Layer 2 uses network frames as PDUs, the DHCP discovery packets cannot transcend the local network boundary. In other words, a DHCP client can only initiate communication with a *local* DHCP server.

After the initial *handshake* (on Layer 2), DHCP turns to UDP as its transport protocol, using datagram sockets (Layer 4). Since UDP is a connectionless protocol, a DHCP client and server exchange messages without a prior arrangement. Consequently, both endpoints (client and server) require a well-known DHCP communication port for the back-and-forth data exchange. These are the *well-known* ports 68 (for a DHCP server) and 67 (for a DHCP client).

A DHCP server maintains a collection of IP addresses and other client configuration data (such as MAC addresses and domain server addresses) for each device on the network requesting a DHCP service.

DHCP servers use a *leasing* mechanism to assign IP addresses dynamically. Leasing an IP address is subject to a *lease time*, either finite or infinite. When the lease of an IP address expires, the DHCP server may reassign it to a different client upon request. A device would hold on to its dynamic IP address by regularly requesting a lease *renewal* from the DHCP server. Failing to do so would result in the potential loss of the device's dynamic IP address. A late (or post-lease) DHCP request would possibly result in a new IP address being acquired if the previous address had already been allocated by the DHCP server.

A simple way to query the DHCP server from a Linux machine is by invoking the following command:

```
ip route
```

This is the output of the preceding command:

```
packt@neptune:~$ ip route
default via 172.16.146.2 dev ens33 proto dhcp src 172.16.146.133 metric 100
172.16.146.0/24 dev ens33 proto kernel scope link src 172.16.146.133
172.16.146.2 dev ens33 proto dhcp scope link src 172.16.146.133 metric 100
```

Figure 7.17 – Querying the IP route for DHCP information

The first line of the output provides the DHCP server (172.16.146.2).

Chapter 8, Configuring Linux Servers, will further go into the practical details of installing and configuring a DHCP server.

For more information on DHCP, please refer to *RFC 2131* (`https://tools.ietf.org/html/rfc2131`).

DNS servers

A **Domain Name Server** (**DNS**), also known as a *name server*, provides a name-resolution mechanism by converting a hostname (such as `wikipedia.org`) to an IP address (such as `208.80.154.224`). The name-resolution protocol is DNS, briefly described in the *TCP/IP protocols* section earlier in this chapter. In a DNS-managed TCP/IP network, computers and devices can also identify and communicate with each other by hostnames, not just IP addresses.

As a reasonable analogy, DNS very much resembles an address book. Hostnames are relatively easier to remember than IP addresses. Even in a local network, with only a few computers and devices connected, it would be rather difficult to identify (or memorize) any of the hosts by simply using their IP address. The internet relies on a globally distributed network of DNS servers.

There are four different types of DNS servers: **recursive servers**, **root servers**, **top-level domain (TLD) servers**, and **authoritative servers**. All these DNS server types work together to bring you the internet as you experience it in your browser.

A **recursive DNS server** is a resolver that helps you find the destination (IP) of a website you search for. When you do a lookup operation, a recursive DNS server is connected to different other DNS servers to find the IP address that you are looking for and return it to you in the form of a website. Recursive DNS lookups are faster, thanks to caching every query that they perform. In a recursive type of query, the DNS server calls itself and does the recursion while still sending the request to other DNS server to find the answer. There is also an iterative type of DNS lookup.

An **iterative DNS** lookup is done by every DNS server directly, without using caching. For example, in an iterative query, each DNS server responds with the address of another DNS server, until one of them has the matching IP address for the hostname in question and responds to the client. For more details on DNS server types, please check out the following Cloudflare learning solution: `https://www.cloudflare.com/learning/dns/what-is-dns/`.

DNS servers maintain (and possibly share) a collection of *database* files, also known as *zone* files—typically simple plain-text ASCII files, storing the name and IP address mapping. In Linux, one such DNS resolver file is `/etc/resolv.conf`.

To query the DNS server managing the local machine, we can query the `/etc/resolv.conf` file by running the following code:

```
cat /etc/resolv.conf | grep nameserver
```

The output yields the following code:

```
[packt@jupiter ~]$ cat /etc/resolv.conf | grep nameserver
nameserver 172.16.146.2
```

Figure 7.18 – Querying DNS server using /etc/resolv.conf

A simple way to query name-server data for an arbitrary host on a network is by using the `nslookup` tool. If you don't have the `nslookup` utility installed on your system, you may do so with the commands outlined next.

On Ubuntu/Debian, run the following command:

```
sudo apt-get install dnsutils
```

On CentOS, run this command:

```
sudo yum install bind-utils
```

For example, to query the name-server information for a computer named `neptune.local` in our local network, we run the following command:

```
nslookup neptune.local
```

The output is shown here:

```
[packt@jupiter ~]$ nslookup neptune.local
Server:         172.16.146.2
Address:        172.16.146.2#53

Name:   neptune.local
Address: 172.16.146.133
```

Figure 7.19 – Querying name-server information with nslookup

We can also use the `nslookup` tool interactively. For example, to query the name-server information for `wikipedia.org`, we can simply run the following command:

```
nslookup
```

Then, in the interactive prompt, we'll enter `wikipedia.org:` as illustrated here:

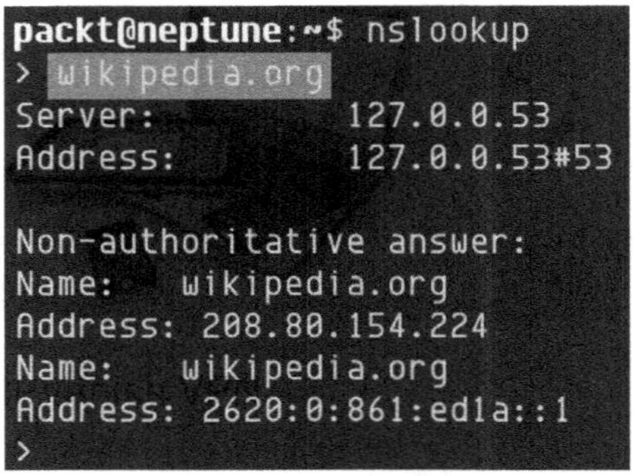

```
packt@neptune:~$ nslookup
> wikipedia.org
Server:         127.0.0.53
Address:        127.0.0.53#53

Non-authoritative answer:
Name:    wikipedia.org
Address: 208.80.154.224
Name:    wikipedia.org
Address: 2620:0:861:ed1a::1
>
```

Figure 7.20 – Using the nslookup tool interactively

To exit the interactive shell mode, press *Ctrl + C*. Here's a brief explanation of the information shown in the preceding output:

- **Server (Address)**: The loopback address (`127.0.0.53`) and port (`53`) of the DNS server running locally

- **Name**: The internet domain we're looking up (`wikipedia.org`)

- **Address**: The IPv4 (`208.80.154.224`) and IPv6 (`2620:0:861:ed1a::1`) address corresponding to the lookup domain (`wikipedia.org`)

`nslookup` is also capable of reverse DNS search when providing an IP address. The following command retrieves the name server (`dns.google`) corresponding to the IP address `8.8.8.8`:

```
nslookup 8.8.8.8
```

The command yields the following output:

```
packt@neptune:~$ nslookup 8.8.8.8
8.8.8.8.in-addr.arpa    name = dns.google.

Authoritative answers can be found from:
```

Figure 7.21 – Reverse DNS search with nslookup

For more information on the `nslookup` tool, you can refer to the `nslookup` system reference manual (`man nslookup`).

Alternatively, we can use the `dig` command-line utility. If you don't have the `dig` utility installed on your system, you can do so by installing the `dnsutils` package on Ubuntu/Debian or `bind-utils` on CentOS platforms. The related commands for installing the packages were shown previously with `nslookup`.

For example, the following command retrieves the name-server information for the computer named `jupiter.local.localdomain` in the local network:

```
dig jupiter.local.localdomain
```

This is the result (see the highlighted ANSWER SECTION):

```
packt@neptune:~$ dig jupiter.local.localdomain

; <<>> DiG 9.16.1-Ubuntu <<>> jupiter.local.localdomain
;; global options: +cmd
;; Got answer:
;; ->>HEADER<<- opcode: QUERY, status: NOERROR, id: 40239
;; flags: qr rd ra; QUERY: 1, ANSWER: 1, AUTHORITY: 0, ADDITIONAL: 1

;; OPT PSEUDOSECTION:
; EDNS: version: 0, flags:; udp: 65494
;; QUESTION SECTION:
;jupiter.local.localdomain.         IN      A

;; ANSWER SECTION:
jupiter.local.localdomain. 5    IN      A       172.16.146.136

;; Query time: 0 msec
;; SERVER: 127.0.0.53#53(127.0.0.53)
;; WHEN: Wed Nov 11 09:48:25 UTC 2020
;; MSG SIZE  rcvd: 70
```

Figure 7.22 – Querying name-server information with dig

To perform a reverse DNS lookup with `dig`, we specify the `-x` option, followed by an IP address (for example, `8.8.4.4`), as follows:

```
dig -x 8.8.4.4
```

The command yields the following output (see the highlighted ANSWER SECTION):

```
packt@neptune:~$ dig -x 8.8.4.4

; <<>> DiG 9.16.1-Ubuntu <<>> -x 8.8.4.4
;; global options: +cmd
;; Got answer:
;; ->>HEADER<<- opcode: QUERY, status: NOERROR, id: 25413
;; flags: qr rd ra; QUERY: 1, ANSWER: 1, AUTHORITY: 0, ADDITIONAL: 1

;; OPT PSEUDOSECTION:
; EDNS: version: 0, flags:; udp: 65494
;; QUESTION SECTION:
;4.4.8.8.in-addr.arpa.            IN      PTR

;; ANSWER SECTION:
4.4.8.8.in-addr.arpa.    5       IN      PTR     dns.google.

;; Query time: 12 msec
;; SERVER: 127.0.0.53#53(127.0.0.53)
;; WHEN: Wed Nov 11 10:07:10 UTC 2020
;; MSG SIZE  rcvd: 73
```

Figure 7.23 – Reverse DNS lookup with dig

For more information about the dig command-line utility, please refer to the related system manual (man dig).

The DNS protocol operates at the application layer (7) in the OSI model. The standard DNS service well-known port is 53.

Chapter 8, Configuring Linux Servers, will further go into the practical details of installing and configuring a DNS server. For more information on DNS, you can refer to RFC 1035 (https://www.ietf.org/rfc/rfc1035.txt).

The DHCP and DNS networking services are arguably the closest to the TCP/IP networking stack, while playing a crucial role when computers or devices are attached to a network. After all, without proper IP addressing and name resolution, there's no network communication.

Obviously, there's a lot more to distributed networking and related application servers than just strictly the pure network management stack performed by DNS and DHCP servers. In the following sections, we'll take a quick tour around some of the most relevant application servers running across distributed Linux systems.

Authentication servers

Standalone Linux systems typically use the default authentication mechanism, where user credentials are stored in the local filesystem (such as /etc/passwd, /etc/shadow). We explored the related user authentication internals in *Chapter 4*, *Managing Users and Groups*, earlier in this book. But as we extend the authentication boundary beyond the local machine—for example, accessing a file or email server—having the user credentials shared between the remote and localhosts would become a serious security issue.

Ideally, we should have a centralized authentication endpoint across the network, handled by a secure authentication server. User credentials should be validated using robust encryption mechanisms before users can access remote system resources.

Let's consider the secure access to a network share on an arbitrary file server. Suppose the access requires **Active Directory** (**AD**) user authentication. Creating the related mount (share) locally on a user's client machine will prompt for user credentials. The authentication request is made by the file server (on behalf of the client) to an authentication server. If the authentication succeeds, the server share becomes available to the client. The following diagram represents a simple remote authentication flow between a client and a server, using a **Lightweight Directory Access Protocol** (**LDAP**) authentication endpoint:

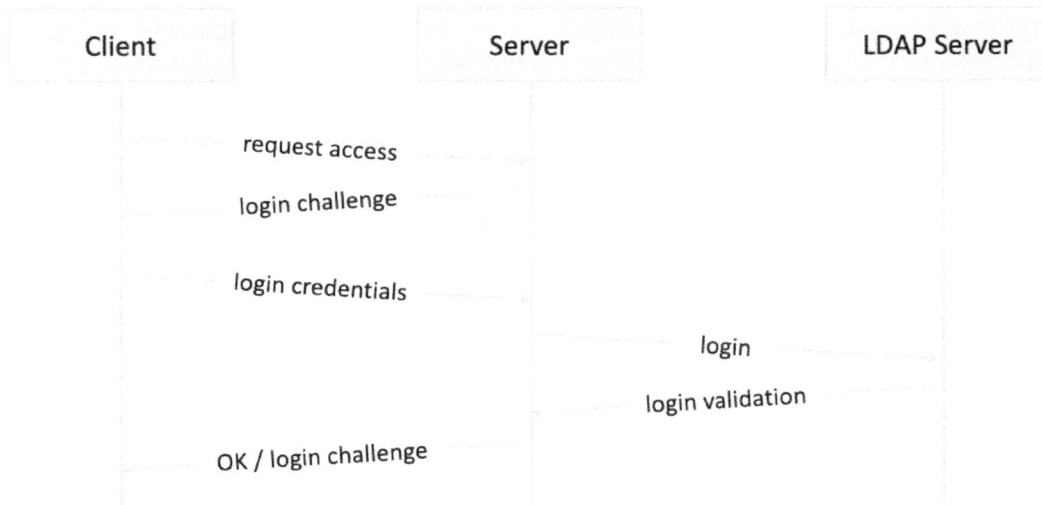

Figure 7.24 – Authentication workflow with LDAP

Examples of standard secure authentication platforms (available for Linux) include the following:

- **Kerberos** (`https://en.wikipedia.org/wiki/Kerberos_(protocol)`)
- **LDAP** (`https://en.wikipedia.org/wiki/Lightweight_Directory_Access_Protocol`)
- **Remote Authentication Dial-In User Service** (**RADIUS**) (`https://en.wikipedia.org/wiki/RADIUS`)
- **Diameter** (`https://en.wikipedia.org/wiki/Diameter_(protocol)`)
- **Terminal Access Controller Access-Control System** (**TACACS+**) (`https://datatracker.ietf.org/doc/rfc8907/`)

We'll go over the installation and configuration of a Linux LDAP authentication server (using **OpenLDAP**) in the *Configuring an LDAP server* section of *Chapter 9*, *Securing Linux*.

In this section, we illustrated the authentication workflow with an example using a file server. To remain on topic, let's look at network file-sharing services next.

File sharing

In common networking terms, file sharing represents a client machine's ability to *mount* and access a remote filesystem belonging to a server, as if it were local. Applications running on the client machine would access the shared files directly on the server. A text editor (for example) can load and modify a remote file, then save it back to the same remote location, all in a seamless and transparent operation. The underlying remoting process—the appearance of a remote filesystem acting as local—is made possible by file-sharing services and protocols.

For every file-sharing network protocol there is a corresponding client-server file-sharing platform. Although most network file servers (and clients) have cross-platform implementations, some operating system platforms are better suited for specific file-sharing protocols, as we'll see in the following subsections. Choosing between different file-server implementations and protocols is ultimately a matter of compatibility, security, and performance.

Here are some of the most common file-sharing protocols, with some brief descriptions for each.

SMB

The **Server Message Block (SMB)** protocol provides network discovery and file- and printer-sharing services. SMB also supports interprocess communication over a network. SMB is a relatively old protocol, developed by **International Business Machines Corporation (IBM)** in the 1980s. Eventually, Microsoft took over and made some considerable alterations to what became the current version through multiple revisions (SMB 1.0, 2.0, 2.1, 3.0, 3.0.2, and 3.1.1).

CIFS

The **Common Internet File System** (**CIFS**) protocol is a particular implementation of the SMB protocol. Due to the underlying protocol similarity, SMB clients would be able to communicate with CIFS servers, and vice versa. Though SMB and CIFS are idiomatically the same, their internal implementation of file locking, batch processing, and—ultimately—performance is quite different. Apart from legacy systems, CIFS is rarely used these days. SMB should always be preferred over CIFS, especially with the more recent revisions of SMB 2 or SMB 3.

Samba

As with CIFS, **Samba** is another implementation of the SMB protocol. Samba provides file- and print-sharing services for Windows clients on a variety of server platforms. In other words, Windows clients can seamlessly access directories, files, and printers on a Linux Samba server, just as if they were communicating with a Windows server.

As of version 4, Samba natively supports Microsoft AD and Windows NT domains. Essentially, a Linux Samba server can act as a domain controller on a Windows AD network. Consequently, user credentials on the Windows domain can transparently be used on the Linux server without being recreated, and then manually kept in sync with the AD users.

NFS

The **Network File System (NFS)** protocol was developed by Sun Microsystems and essentially operates on the same premise as SMB—accessing files over a network as if they were local. NFS is not compatible with CIFS or SMB, meaning that NFS clients cannot communicate directly to SMB servers, or vice versa.

Most of the time, NFS is the file-sharing protocol of choice within Linux networks. For mixed networking environments—such as Windows, Linux, and macOS interoperability—Samba and SMB are best suited for file sharing.

AFP

The **Apple Filing Protocol** (**AFP**) is a proprietary file-sharing protocol designed by Apple and exclusively operates in macOS network environments. We should note that besides AFP, macOS systems also support standard file-sharing protocols, such as SMB and NFS.

Some file-sharing protocols (such as SMB) also support print sharing and are used by print servers. Let's take a closer look at print sharing next.

Printer servers

A *printer server* (or *print server*) connects a printer to client machines (computers or mobile devices) on a network, using a printing protocol. Printing protocols are responsible for the following remote printing tasks over a network:

- Discovering printers or print servers
- Querying printer status
- Sending, receiving, queueing, or canceling print jobs
- Querying print job status

Common printing protocols include the following:

- **Line Printer Daemon** (**LPD**) protocol
- *Generic* protocols: **SMB**; **TELNET**
- *Wireless* printing protocols (such as **AirPrint** by Apple)
- *Internet* printing protocols (such as **Google Cloud Print**)

Among the generic printing protocols, SMB (also a file-sharing protocol) has been previously described in the *File sharing* section. The TELNET communication protocol is described in the *Remote access* section.

File- and printer-sharing services are mostly about *sharing* documents, digital or printed, between computers on a network. When it comes to *exchanging* documents, additional networking services come into play, such as *file transfer* and *email* services. Let's look at file transfer next.

File transfer

FTP is a standard network protocol for transferring files between computers on a network. FTP operates in a client-server environment, where an FTP client initiates a remote connection to an FTP server, and files are being transferred in either direction. FTP maintains a *control connection* and one or more *data connections* between the client and the server. The *control connection* is generally established on the FTP server's port 21, and it's used for exchanging commands between the client and the server. *Data connections* are exclusively used for data transfer and are negotiated between client and server (through the control connection). Data connections usually involve ephemeral ports for inbound traffic, and they only stay open during the actual data transfer, closing immediately after the transfer completes.

FTP negotiates data connections in one of the following two *modes*:

- **Active mode**—The FTP client sends a PORT command to the FTP server, signaling that the client *actively* provides the inbound port number for data connections.

- **Passive mode**—The FTP client sends a PASV command to the FTP server, indicating that the client *passively* awaits the server to supply the port number for inbound data connections.

FTP is a relatively "messy" protocol when it comes to firewall configurations, due to the dynamic nature of the data connections involved. The control connection port is usually well known (such as port 21 for insecure FTP) but data connections are originated on a different port (usually 20) on either side, while on the receiving end the inbound sockets are opened within a preconfigured ephemeral range (1024 - 65535).

FTP is most often implemented in a secure fashion through either of the following approaches:

- **FTP over SSL** (**FTPS**)—SSL/TLS-encrypted FTP connection. The default FTPS control connection port is 990.

- **SSH File Transfer Protocol** (**SFTP**)—**FTP** over **SSH**. The default SFTP control connection port is 22. For more information on the SSH protocol and client-server connectivity, refer to *SSH* in the *Remote access* section, later in this chapter.

Chapter 9, *Securing Linux*, looks closely at the practical implementation of a Linux FTP server.

Next, we'll look at mail servers and the underlying email exchange protocols.

Mail servers

A *mail server* (or *email server*) is responsible for email delivery over a network. A mail server can either exchange emails between clients (users) on the same network (domain)—within a company or organization—or deliver emails to other mail servers, possibly beyond the local network, such as the internet.

An email exchange usually involves the following actors:

- An *email client* application (such as Outlook or Gmail)

- One or more *mail servers* (Exchange; Gmail server)

- The *recipients* involved in the email exchange—a *sender* and one or more *receivers*

- An *email protocol* controlling the communication between the email client and the mail servers

The most used email protocols are **POP3**, **IMAP**, and **SMTP**. Let's take a closer look at each of these protocols.

POP3

POP version 3 (**POP3**) is a standard email protocol for receiving and downloading emails from a remote mail server to a local email client. With POP3, emails are available for reading offline. After download, emails are usually removed from the POP3 server, thus saving up space. Modern-day POP3 mail client-server implementations (Gmail; Outlook) also have the option of keeping email copies on the server. Persisting emails on the POP3 server becomes very important when users access emails from multiple locations (client applications).

The default POP3 ports are outlined here:

- 110—for insecure (non-encrypted) POP3 connections

- 995—for secure POP3 using SSL/TLS encryption

POP3 is a relatively old email protocol, not always suitable for modern-day email communications. When users access their emails from multiple devices, IMAP is a better choice. Let's look at the IMAP email protocol next.

IMAP

IMAP is a standard email protocol for accessing emails on a remote IMAP mail server. With IMAP, emails are always retained on the mail server, while a copy of the emails is available for IMAP clients. A user can access the emails on multiple devices, each with their IMAP client application.

The default IMAP ports are outlined here:

- `143`—for insecure (non-encrypted) IMAP connections
- `993`—for secure IMAP using SSL/TLS encryption

Both POP3 and IMAP are standard protocols for receiving emails. To send emails, SMTP comes into play. Let's take a look at the SMTP email protocol next.

SMTP

SMTP is a standard email protocol for sending emails over a network or the internet.

The default SMTP ports are outlined here:

- `25`—for insecure (non-encrypted) SMTP connections
- `465` or `587`—for secure SMTP using SSL/TLS encryption

When using or implementing any of the standard email protocols described in this section, it is always recommended to use the corresponding secure implementation with the most up-to-date TLS encryption, if possible. POP3, IMAP, and SMTP also support user authentication, an added layer of security—also recommended in commercial or enterprise-grade environments.

To get an idea about how the SMTP protocol operates, let's go through some of the initial steps for initiating a SMTP handshake with Google's Gmail SMTP server.

We start by connecting to the Gmail SMTP server, using a secure (TLS) connection via the `openssl` command, as follows:

```
openssl s_client -starttls smtp -connect smtp.gmail.com:587
```

We invoked the `openssl` command, simulating a client (`s_client`), starting a TLS SMTP connection (`-starttls smtp`), and connecting to the remote Gmail SMTP server on port `587` (`-connect smtp.gmail.com:587`).

The Gmail SMTP server responds with a relatively long TLS handshake block, ending with the following:

```
---
No client certificate CA names sent
Peer signing digest: SHA256
Peer signature type: ECDSA
Server Temp Key: X25519, 253 bits
---
SSL handshake has read 2892 bytes and written 419 bytes
Verification: OK
---
New, TLSv1.3, Cipher is TLS_AES_256_GCM_SHA384
Server public key is 256 bit
Secure Renegotiation IS NOT supported
Compression: NONE
Expansion: NONE
No ALPN negotiated
Early data was not sent
Verify return code: 0 (ok)
---
250 SMTPUTF8
```

Figure 7.25 – Initial TLS handshake with a Gmail SMTP server

Next, we initiate the SMTP communication with a `HELO` command (spelled precisely as such). Google expects the following `HELO` greeting:

```
HELO hellogoogle
```

Another handshake follows, ending with `250 smtp.gmail.com at your service`, as illustrated here:

```
     Start Time: 1605358583
     Timeout   : 7200 (sec)
     Verify return code: 0 (ok)
     Extended master secret: no
     Max Early Data: 0
---
read R BLOCK
250 smtp.gmail.com at your service
```

Figure 7.26 – Gmail SMTP server is ready for communication

Next, the Gmail SMTP server requires authentication via the `AUTH LOGIN` SMTP command. We won't go into further details, but the key point to be made here is that the SMTP protocol follows a plaintext command sequence between the client and the server. It's very important to adopt a secure (encrypted) SMTP communication channel, using TLS. The same applies to any of the other email protocols (POP3; IMAP).

So far, we've covered several network services, some of them spanning multiple networks or even the internet. Network packets carry data and destination addresses within the payload, but there are also synchronization signals between the communication endpoints, mostly to discern between sending and receiving workflows. The synchronization of network packets is based on timestamps. Reliable network communications would not be possible without a highly accurate time-synchronization between network nodes. We'll look at the network time-keepers next.

NTP servers

NTP is a standard networking protocol for clock synchronization between computers on a network. NTP attempts to synchronize the system clock on participating computers within a few milliseconds of **Coordinated Universal Time**) (**UTC**)— the world's time reference.

The NTP protocol implementation usually assumes a client-server model. The NTP server acts as a time source on the network by either broadcasting or sending updated *timestamp* datagrams to clients. An NTP server continually adjusts its system clock according to well-known accurate time servers worldwide, using specialized algorithms to mitigate network latency.

A relatively easy way to check the NTP synchronization status on our Linux platform of choice is by using the `ntpstat` utility. `ntpstat` may not be installed by default on our system. On Ubuntu, we can install it with the following command:

```
sudo apt-get install ntpstat
```

On CentOS, we install `ntpstat` with the following command:

```
sudo yum install ntpstat
```

`ntpstat` requires an NTP server running locally. To query the NTP synchronization status, we run the following command:

```
ntpstat
```

This is the output:

Figure 7.27 – Querying NTP synchronization status with ntpstat

`ntpstat` provides the IP address of the NTP server the system is synchronized with (`74.6.168.72`), the synchronization margin (`17` milliseconds), and the time-update polling interval (`1024` s). To find out more about the NTP server, we can `dig` its IP address with the following code:

```
dig -x 74.6.168.72
```

And it looks like it's one of Yahoo's time servers (`t1.time.gq1.yahoo.com`), as we can see here:

```
[packt@jupiter ~]$ dig -x 74.6.168.72

; <<>> DiG 9.11.13-RedHat-9.11.13-3.el8 <<>> -x 74.6.168.72
;; global options: +cmd
;; Got answer:
;; ->>HEADER<<- opcode: QUERY, status: NOERROR, id: 59198
;; flags: qr rd ra; QUERY: 1, ANSWER: 1, AUTHORITY: 0, ADDITIONAL: 1

;; OPT PSEUDOSECTION:
; EDNS: version: 0, flags:; MBZ: 0x0005, udp: 4096
;; QUESTION SECTION:
;72.168.6.74.in-addr.arpa.        IN        PTR

;; ANSWER SECTION:
72.168.6.74.in-addr.arpa. 5      IN        PTR        t1.time.gq1.yahoo.com.

;; Query time: 10 msec
;; SERVER: 172.16.146.2#53(172.16.146.2)
;; WHEN: Sat Nov 14 03:46:00 PST 2020
;; MSG SIZE  rcvd: 88
```

Figure 7.28 – Querying NTP synchronization status with ntpstat

The NTP client-server communication uses UDP as the transport protocol on port `123`. *Chapter 8*, *Configuring Linux Servers*, has a dedicated section for installing and configuring an NTP server. For more information on NTP, you can refer to `https://en.wikipedia.org/wiki/Network_Time_Protocol`.

Our brief journey among networking servers and protocols is coming to an end here. Everyday Linux administration tasks often require some sort of remote access to a system. There are many ways to access and manage computers remotely. Our next section describes some of the most common remote-access facilities and related network protocols.

Remote access

Most Linux networking services provide a relatively limited remote management interface, with their management **command-line interface (CLI)** utilities predominantly operating locally on the same system where the service runs. Consequently, the related administrative tasks assume local terminal access. Direct console access to the system is sometimes not possible. This is when remote-access servers come into play to enable a virtual terminal login session with the remote machine.

Let's look at some of the most common remote-access services and applications next.

SSH

SSH is perhaps the most popular secure login protocol for remote access. SSH uses strong encryption, combined with user authentication mechanisms, for secure communication between a client and a server machine. SSH servers are relatively easy to install and configure, and *Chapter 8, Configuring Linux Servers*, has a dedicated section describing the related steps.

The default network port for SSH is 22.

SSH supports the following authentication types:

- Public-key authentication
- Host-based authentication
- Password authentication
- Keyboard-interactive authentication

The following sections provide brief descriptions of these SSH authentication forms.

Public-key authentication

Public-key (or SSH-key) authentication is arguably the most common type of SSH authentication.

> **Important note**
> This section will use the terms *public-key* and *SSH-key* interchangeably, mostly to reflect the related SSH authentication nomenclature in the Linux community.

The SSH-key authentication mechanism uses a *certificate/key* pair—a *public* key (*certificate*) and a *private* key. An SSH certificate/key pair is usually created with the `ssh-keygen` tool, using standard encryption algorithms such as the **Rivest–Shamir–Adleman** algorithm (**RSA**) or the **Digital Signature Algorithm (DSA)**.

SSH public-key authentication supports either *user-based authentication* or *host-based authentication* models. The two models differ in the ownership of the certificate/key pairs involved. With client authentication, each user has its own certificate/key pair for SSH access. On the other hand, host authentication involves a single certificate/key pair per system (host).

Both SSH-key authentication models are illustrated and explained in the following sections. The basic SSH handshake and authentication workflows are the same for both models. First, the SSH client generates a secure certificate/key pair and shares its public key with the SSH server. This is a one-time operation for enabling the public-key authentication.

When a client initiates the SSH handshake, the server asks for the client's public key and verifies it against its allowed public keys. If there's a match, the SSH handshake succeeds, the server shares its public key with the client, and the SSH session is established.

Further client-server communication follows standard encryption/decryption workflows. The client encrypts the data with its private key, while the server decrypts the data with the client's public key. When responding to the client, the server encrypts the data with its own private key, and the client decrypts the data with the server's public key.

SSH public-key authentication is also known as *passwordless authentication*, and it's frequently used in automation scripts where commands are executed over multiple remote SSH connections without prompting for a password.

Let's take a closer look at the user-based and host-based public-key authentication mechanisms next.

User-based key authentication

User-based authentication is the most common SSH public-key authentication mechanism. According to this model, every user connecting to a remote SSH server has its own SSH key. Multiple user accounts on the same host (or domain) would have different SSH keys, each with its own access to the remote SSH server, as suggested in the following diagram:

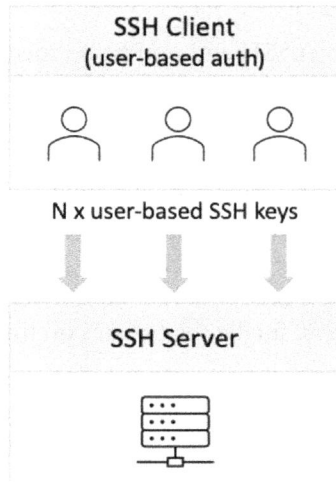

Figure 7.29 – User-based key authentication

A somewhat similar approach to user-based SSH key authentication is the host-based authentication mechanism, described next.

Host-based key authentication

Host-based authentication is another form of SSH public-key authentication and involves a single SSH key per system (host) connecting to a remote SSH server, as illustrated in the following diagram:

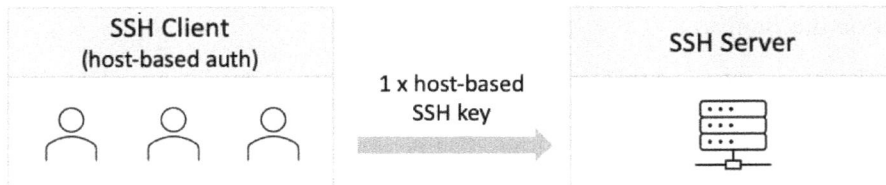

Figure 7.30 – Host-based key authentication

With host-based authentication, the underlying SSH key can only authenticate SSH sessions that originated from a single client host. Host-based authentication allows multiple users to connect from the same host to a remote SSH server. If a user attempts to use a host-based SSH key from a different machine than the one allowed by the SSH server, access would be denied.

Sometimes, a mix of the two public-key authentications is used—user- and host-based authentication—an approach that provides an increased security level to SSH access.

When security is not critical, simpler SSH authentication mechanisms could be more suitable. Password authentication is one such mechanism.

Password authentication

Password authentication requires a simple set of credentials from the SSH client, as a username and password. The SSH server validates the user credentials, either based on the local user accounts (in /etc/passwd) or select user accounts defined in the SSH server configuration (/etc/ssh/sshd_config). The SSH server configuration described in *Chapter 8*, *Configuring Linux Servers*, further elaborates on this subject.

Besides local authentication, SSH can also leverage remote authentication methods such as Kerberos, LDAP, RADIUS, and so on. In such cases, the SSH server delegates the user authentication to a remote authentication server, as described in the *Authentication servers* section earlier in this chapter.

Password authentication requires either user interaction or some automated way to provide the required credentials. Another similar authentication mechanism is keyboard-interactive authentication, described next.

Keyboard-interactive authentication

Keyboard-interactive authentication is based on a dialog of multiple challenge-response sequences between the SSH client (user) and the SSH server. The dialog is a plain-text exchange of questions and answers, where the server may prompt the user for any number of challenges. In some respect, password authentication is a single-challenge interactive authentication mechanism.

The *interactive* connotation of this authentication method could lead us into thinking that user interaction would be mandatory for the related implementation. Not really. As a matter of fact, keyboard-interactive authentication could also serve implementations of authentication mechanisms based on custom protocols, where the underlying message exchange would be modeled as an authentication protocol.

Before moving on to other remote access protocols, we should again call out the wide use of SSH due to its security, versatility, and performance. But SSH connectivity may not always be possible or adequate in specific scenarios. In such cases, *TELNET* may come to the rescue. Let's take a look at it next.

TELNET

TELNET is an application-layer protocol for bidirectional network communication using a plain-text CLI with a remote host. Historically, TELNET was among the first remote-connection protocols, but it always lacked a secure implementation. SSH eventually became the standard way to log in from one computer to another, yet TELNET has its own advantages over SSH when it comes to troubleshooting various application-layer protocols, such as web- or email-server communication.

Let's look at an example to get a sense of how TELNET works. We'll be simulating a simple HTTP request/response connecting to an Apache web server using TELNET. The general syntax of TELNET is shown here:

```
telnet HOST PORT
```

In our case, Apache runs on the `jupiter.local` host and port `80`, as shown here:

```
telnet jupiter.local 80
```

We get the following response:

```
packt@neptune:~$ telnet jupiter.local 80
Trying 172.16.146.136...
Connected to jupiter.local.localdomain.
Escape character is '^]'.
```

Figure 7.31 – Connecting with TELNET to a remote web server

Next, we initiate a web communication following the `HTTP/1.1` protocol by typing the following command:

```
GET / HTTP/1.1
```

The `HTTP/1.1` protocol requires a mandatory `Host` HTTP header, so we continue with the following code:

```
Host: localhost
```

After each of the preceding lines, we hit *Enter* (new line), and after the Host header, we hit *Enter* twice. The Apache web server responds as follows:

```
packt@neptune:~$ telnet jupiter.local 80
Trying 172.16.146.136...
Connected to jupiter.local.localdomain.
Escape character is '^]'.

GET / HTTP/1.1
Host: localhost

HTTP/1.1 403 Forbidden
Date: Sun, 15 Nov 2020 15:44:42 GMT
Server: Apache/2.4.37 (centos)
Content-Location: index.html.zh-CN
Vary: negotiate,accept-language
TCN: choice
Last-Modified: Fri, 14 Jun 2019 03:37:43 GMT
ETag: "fa6-58b405e7d6fc0;5b0f5dc6d9c8c"
Accept-Ranges: bytes
Content-Length: 4006
Content-Type: text/html; charset=UTF-8
Content-Language: zh-cn
```

Figure 7.32 – HTTP request/response with TELNET

We truncated the response for brevity. The TELNET session we just ran shows the interactive step-by-step HTTP communication between a web client—our local terminal window—and a remote Apache web server (jupiter.local).

TELNET and SSH are command-line-driven remote-access interfaces. There are cases when a direct desktop connection is needed to a remote machine through a **graphical user interface (GUI)**. We'll look at desktop sharing next.

VNC

Virtual Network Computing (VNC) is a desktop-sharing platform that allows users to access and control a remote computer's GUI. VNC is a cross-platform client-server application. A VNC server running on a Linux machine, for example, allows desktop access to multiple VNC clients running on Windows or macOS systems. The VNC network communication uses the **Remote Framebuffer (RFB)** protocol, defined by *RFC 6143*.

Setting up a VNC server is relatively simple. VNC assumes the presence of a graphical desktop system. You can refer to the *Installing Linux graphical user interfaces* section in *Chapter 1*, *Installing Linux*, to set up a GNOME or **K Desktop Environment** (**KDE**) desktop in Linux. Let's take an RHEL/CentOS 8 system with a GNOME desktop and configure VNC. We start by installing a VNC server, as follows:

```
sudo dnf install tigervnc-server tigervnc-server-module -y
```

Next, we create a VNC password for the current user, like this:

```
vncpasswd
```

The vncpasswd utility prompts for a password and asks if we want to use VNC in view-only mode. We choose full-control access. This is the output of the vncpasswd command:

```
[packt@jupiter ~]$ vncpasswd
Password:
Verify:
Would you like to enter a view-only password (y/n)? n
A view-only password is not used
```

Figure 7.33 – Setting up the VNC password

In the following step, we specify GNOME as the VNC desktop of our choice by running the following code:

```
printf 'gnome-session &\ngnome-terminal &' > ~/.vnc/xstartup
```

The ~/.vnc/xstartup file stores the current VNC configuration. Optionally, we can enable *clipboard sharing* with VNC, between the client and the server, with the following line added to the ~/.vnc/xstartup file:

```
vncconfig -iconic &
```

Here's the final content of the ~/.vnc/xstartup file:

```
gnome-session &
gnome-terminal &
vncconfig -iconic &
```

We are now ready to start the VNC server, so we run the following command:

```
vncserver -geometry 1920x1080
```

In the `vncserver` command, we specified a screen resolution for the VNC sessions (`1920x1080`). The command yields the following output:

```
[packt@jupiter ~]$ vncserver -geometry 1920x1080

New 'jupiter:1 (packt)' desktop is jupiter:1

Creating default config /home/packt/.vnc/config
Starting applications specified in /home/packt/.vnc/xstartup
Log file is /home/packt/.vnc/jupiter:1.log
```

Figure 7.34 – Starting the VNC server

In the `vncserver` command's output, we should note the VNC desktop ID (`jupiter:1`). This ID will be used as the VNC client hostname. The default network port range for the VNC server starts with `5901`. Multiple VNC clients connect to incremental ports.

Using a VNC client application—such as *VNC Viewer* (by *RealVNC*) on Mac OS X—we can now remotely access our CentOS 8 Linux machine, as shown in the following screenshot:

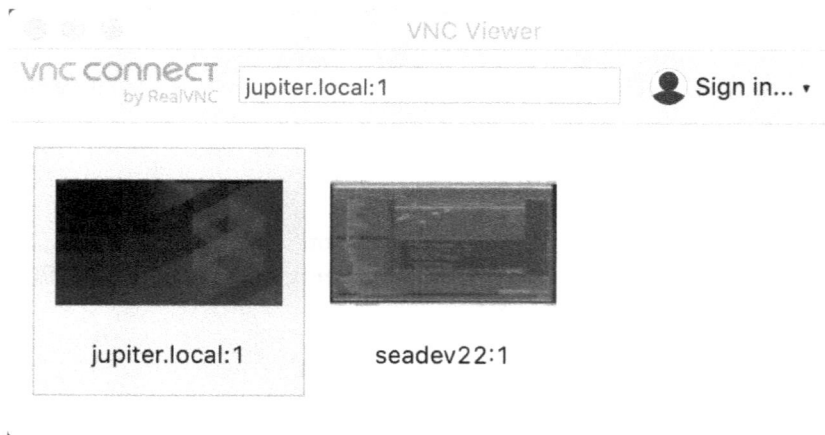

Figure 7.35 – Using a VNC client

For simplicity and space considerations, we described a relatively raw and straightforward way of running VNC. Obviously, we could be more creative about controlling the lifetime of the VNC process. The complementary source code of this chapter shows such a script.

This concludes our section about networking services and protocols. We tried to cover the most common concepts about general-purpose network servers and applications, mostly operating in a client-server or distributed fashion. With each network server, we described the related network protocols and some of the internal aspects involved. *Chapter 8, Configuring Linux Servers*, and *Chapter 9, Securing Linux*, will showcase practical implementations for some of these network servers.

In the next section, our focus turns to network-security internals.

Understanding network security

Network security represents the processes, actions, and policies to prevent, monitor, and protect unauthorized access into computer networks. Network security paradigms span a vast array of technologies, tools, and practices. Here are a few important ones:

- **Access control**—Selectively restricting access based on user authentication and authorization mechanisms. Examples of access control include users, groups, and permissions. Some of the related concepts have been covered in *Chapter 4, Managing Users and Groups*.

- **Application security**—Securing and protecting server and end-user applications (email, web, and mobile apps). Examples of application security include **Security-Enhanced Linux** (**SELinux**), strongly encrypted connections, antivirus, and anti-malware programs. We'll cover **SELinux** in *Chapter 9, Securing Linux*.

- **Endpoint security**—Securing and protecting servers and end-user devices (smartphones, laptops, and desktop PCs) on the network. Examples of endpoint security include *firewalls* and various intrusion-detection mechanisms. We'll look at *firewalls* in *Chapter 9, Securing Linux*.

- **Network segmentation**—Partitioning computer networks into smaller segments or **virtual LANs** (**VLANs**). This is not to be confused with subnetting, which is a logical division of networks through addressing.

- **VPNs**—Accessing corporate networks using a secure encrypted tunnel from public networks or the internet. We'll look at VPNs in *Chapter 9, Securing Linux*.

In everyday Linux administration, setting up a network security perimeter should always follow the paradigms enumerated previously, roughly in the order listed. Starting with access-control mechanisms and ending with VPNs, securing a network takes an *inside-out* approach, from local systems and networks, to firewalls, VLANs, and VPNs. The next section explores VPNs.

VPNs

A VPN is a networking technology that provides online security, privacy, and anonymity to users by creating a private network over a public internet connection. VPNs are generally used to accommodate the following scenarios:

- Establishing a secure encrypted connection between a device and a private or corporate network
- Enabling access to region-restricted websites (prevent geo-blocking)
- Shielding internet activity from prying eyes

VPNs essentially create a data tunnel within a network—usually the internet—to securely access network resources remotely, bypass internet censorship, mask IP addresses, and more.

Let's take a look at how VPNs work.

Working with VPNs

VPNs are based on a client-server architecture routing a client device's network communication through a VPN server. The VPN server provides an encrypted communication tunnel with the client device, acting as an intermediary of sorts between the client and the internet, or a private (or corporate) network. VPN implementations typically use OSI Layer 2 and 3 network extensions with SSL/TLS protocols.

Commercial or enterprise-grade VPN solutions usually ship a proprietary VPN client app that users install on their computers and mobile devices. The VPN client is configured to use specific VPN endpoints provisioned by the app. Notable commercial VPN products include *ExpressVPN*, *NordVPN*, *Surfshark*, *Norton Secure VPN*, and *IPVanish*.

Most operating systems have integrated client support for general-purpose VPN. In the following sections, we describe the configuration of a VPN environment using Open VPN —an open source VPN solution.

Setting up OpenVPN

In this section, we'll guide you through the process of setting up OpenVPN on Ubuntu Server 20.04 **Long-Term Support (LTS)**. We'll use a **virtual private server (VPS)** for our VPN endpoint, featuring a public network interface. The main reason for a VPS environment hosted in a public cloud (such as **Amazon Web Services (AWS)/Elastic Compute Cloud (EC2)**, DigitalOcean, Linode, and so on) is the commodity of having a directly accessible public IP address over the internet. Alternatively, we could use an arbitrary host within a private network, with the required **network address translation (NAT)** configuration settings in the firewall or router, to make it accessible from the public internet.

A quick way to set up a VPN is by using an open source utility assisting with the OpenVPN configuration, available at `https://git.io/vpn`. The following procedure will help you set up a VPN in a matter of minutes. Here are the steps we'll be following:

- Identifying the network interface for VPN connections
- Downloading and running the VPN installer script
- Connecting to the VPN using a Linux client

Let's start with the first step.

Identifying the VPN network interface

A typical host—acting as VPN server—has either of the following network configurations:

- A private static IP address behind a NAT/router/firewall with a public IP address. Examples include AWS/EC2 instances or home network computers behind a general-purpose router.
- A public static IP address routable from the internet. Examples of such include VPS instances by DigitalOcean, Linode, and others.

In our example, we use a DigitalOcean VPS instance. Let's look at our network interfaces available on the system, as follows:

```
ip addr
```

This is the output of the preceding command:

```
root@vpn:~# ip addr
1: lo: <LOOPBACK,UP,LOWER_UP> mtu 65536 qdisc noqueue state UNKNOWN group default qlen 1000
    link/loopback 00:00:00:00:00:00 brd 00:00:00:00:00:00
    inet 127.0.0.1/8 scope host lo
       valid_lft forever preferred_lft forever
    inet6 ::1/128 scope host
       valid_lft forever preferred_lft forever
2: eth0: <BROADCAST,MULTICAST,UP,LOWER_UP> mtu 1500 qdisc fq_codel state UP group default qlen 1000
    link/ether b6:5d:80:be:25:fa brd ff:ff:ff:ff:ff:ff
    inet 138.68.19.158/20 brd 138.68.31.255 scope global eth0
       valid_lft forever preferred_lft forever
    inet 10.46.0.5/16 brd 10.46.255.255 scope global eth0
       valid_lft forever preferred_lft forever
    inet6 fe80::b45d:80ff:febe:25fa/64 scope link
       valid_lft forever preferred_lft forever
```

Figure 7.36 – Identifying network interfaces

In our case, the eth0 interface has a publicly facing IP address of 138.68.19.158 with a corresponding local (internal) IP address of 10.46.0.5. These network addresses are relevant in our next step for configuring the VPN, using the VPN installer script.

Configuring VPN

We start by making sure our system is up to date by running the following commands:

```
sudo apt-get update
sudo apt-get upgrade
```

Next, we download and run the VPN installer script with the following command:

```
wget https://git.io/vpn -O vpnsetup && bash vpnsetup
```

We chose the arbitrary name of vpnsetup for the installer script. The script provides step-by-step guided assistance, as can be seen in the following screenshot:

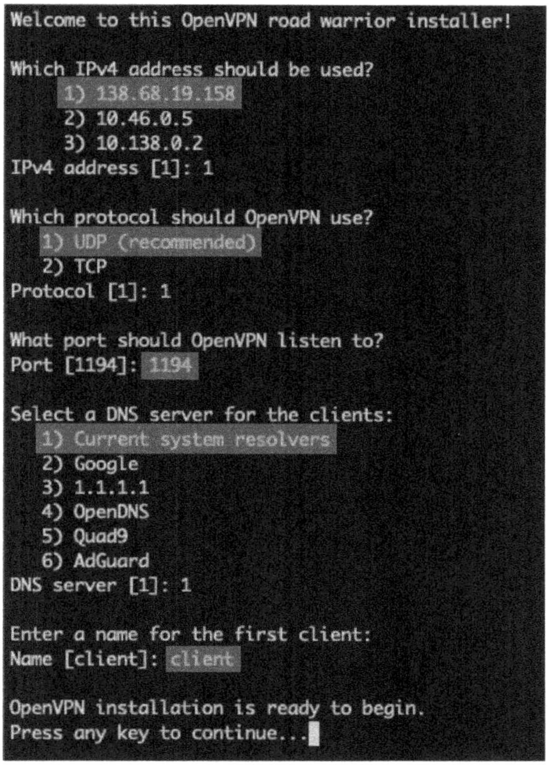

```
Welcome to this OpenVPN road warrior installer!

Which IPv4 address should be used?
    1) 138.68.19.158
    2) 10.46.0.5
    3) 10.138.0.2
IPv4 address [1]: 1

Which protocol should OpenVPN use?
    1) UDP (recommended)
    2) TCP
Protocol [1]: 1

What port should OpenVPN listen to?
Port [1194]: 1194

Select a DNS server for the clients:
    1) Current system resolvers
    2) Google
    3) 1.1.1.1
    4) OpenDNS
    5) Quad9
    6) AdGuard
DNS server [1]: 1

Enter a name for the first client:
Name [client]: client

OpenVPN installation is ready to begin.
Press any key to continue...
```

Figure 7.37 – Running the VPN installer

We highlighted the relevant choices, as follows:

- `138.68.19.158`—The IP address of the network interface we dedicate for VPN connections
- `UDP`—The recommended OpenVPN protocol
- `1194`—The VPN connection port
- `Current system resolvers`—The default DNS subsystem on the host
- `client`—The name of the VPN client instances

Upon successfully running the VPN installer, we can verify the running status of the OpenVPN server with the following command:

```
sudo systemctl status openvpn-server@server.service
```

We should get the following `active(running)` status:

```
root@vpn:~# sudo systemctl status openvpn-server@server.service
● openvpn-server@server.service - OpenVPN service for server
     Loaded: loaded (/lib/systemd/system/openvpn-server@.service; enabled; vendor preset: enabled)
     Active: active (running) since Fri 2021-01-08 04:32:18 UTC; 8min ago
       Docs: man:openvpn(8)
             https://community.openvpn.net/openvpn/wiki/Openvpn24ManPage
             https://community.openvpn.net/openvpn/wiki/HOWTO
   Main PID: 29739 (openvpn)
     Status: "Initialization Sequence Completed"
      Tasks: 1 (limit: 1137)
     Memory: 1.2M
     CGroup: /system.slice/system-openvpn\x2dserver.slice/openvpn-server@server.service
             └─29739 /usr/sbin/openvpn --status /run/openvpn-server/status-server.log --status-versi
```

Figure 7.38 – Querying the OpenVPN server status

The script also generates a default OpenVPN client profile as a file named according to the name of the VPN client chosen in the final step of the VPN installer script (that is, `client`). In our case, the file is `~/client.ovpn`, as we can see here:

```
cat ~/client.ovpn
```

Here's an excerpt from it:

```
root@vpn:~# cat ~/client.ovpn
client
dev tun
proto udp
remote 138.68.19.158 1194
resolv-retry infinite
nobind
persist-key
persist-tun
remote-cert-tls server
auth SHA512
cipher AES-256-CBC
ignore-unknown-option block-outside-dns
block-outside-dns
verb 3
<ca>
-----BEGIN CERTIFICATE-----
```

Figure 7.39 – The OpenVPN client profile (.ovpn file)

The OpenVPN client profile is shared with a specific VPN client, as we'll see next. We can generate multiple such distinct client profiles by running the VPN installer script multiple times. Since we downloaded the VPN installer script, we can also invoke it locally. We need to make the script executable, first by running the following commands:

```
chmod a+x ./vpnsetup
./vpnsetup
```

Subsequent runs of the script provide us with the following options:

```
OpenVPN is already installed.

Select an option:
    1) Add a new client
    2) Revoke an existing client
    3) Remove OpenVPN
    4) Exit
Option:
```

Figure 7.40 – A subsequent invocation of the VPN installer

Choosing option 1) will generate a new client profile. The other options are also obvious, according to their descriptions. Client profiles are shared exclusively with the OpenVPN clients using our VPN.

Next, let's take a look at how to configure VPN clients using our OpenVPN server. OpenVPN clients are supported on all major operating system platforms. In the following examples, we showcase OpenVPN clients running on Linux and Android platforms.

Configuring a Linux OpenVPN client

The instructions in this section will help you configure OpenVPN client connectivity on Ubuntu and CentOS platforms. Using your client platform of choice, start by installing the openvpn client package, as follows:

- On Ubuntu, run the following command:

```
sudo apt-get install -y openvpn
```

- On CentOS, run the following command:

```
sudo yum install -y openvpn
```

Next, we copy the OpenVPN client profile (generated in the *Configuring VPN* section) to `/etc/openvpn/client/client.conf`. We assume the `client.ovpn` profile has been copied over to the client machine (such as in `/home/packt/client.ovpn`), and run the following code:

```
sudo cp /home/packt/client.ovpn /etc/openvpn/client/client.conf
```

At this point, we can immediately test the VPN client connectivity by running the following code:

```
sudo openvpn --client --config /etc/openvpn/client/client.conf
```

Here's an excerpt from the output of a successful VPN connection:

```
Fri Jan  8 20:50:20 2021 TUN/TAP device tun0 opened
Fri Jan  8 20:50:20 2021 TUN/TAP TX queue length set to 100
Fri Jan  8 20:50:20 2021 /sbin/ip link set dev tun0 up mtu 1500
Fri Jan  8 20:50:20 2021 /sbin/ip addr add dev tun0 10.8.0.2/24 broadcast 10.8.0.255
Fri Jan  8 20:50:20 2021 /sbin/ip route add 138.68.19.158/32 via 172.16.191.1
Fri Jan  8 20:50:20 2021 /sbin/ip route add 0.0.0.0/1 via 10.8.0.1
Fri Jan  8 20:50:20 2021 /sbin/ip route add 128.0.0.0/1 via 10.8.0.1
Fri Jan  8 20:50:20 2021 WARNING: this configuration may cache passwords in memory -- use the auth-no
cache option to prevent this
Fri Jan  8 20:50:20 2021 Initialization Sequence Completed
```

Figure 7.41 – Testing the OpenVPN client connectivity

You may exit with *Ctrl + C* from the preceding process. To enable the OpenVPN client on the system, we start the related daemon, as follows:

```
sudo systemctl start openvpn-client@client
```

The `@client` invocation of the `openvpn-client` daemon denotes the related OpenVPN client configuration file (in `/etc/openvpn/client/client.conf`). If the configuration file is named differently, we should adjust the preceding command accordingly.

The status of the OpenVPN client should show as `active`, as can be seen in the following screenshot:

```
packt@neptune:~$ sudo systemctl status openvpn-client@client
● openvpn-client@client.service - OpenVPN tunnel for client
     Loaded: loaded (/lib/systemd/system/openvpn-client@.service; disabled; vendor preset: enabled)
     Active: active (running) since Fri 2021-01-08 21:34:40 UTC; 10s ago
       Docs: man:openvpn(8)
             https://community.openvpn.net/openvpn/wiki/Openvpn24ManPage
             https://community.openvpn.net/openvpn/wiki/HOWTO
   Main PID: 5978 (openvpn)
     Status: "Initialization Sequence Completed"
      Tasks: 1 (limit: 4581)
     Memory: 1.6M
     CGroup: /system.slice/system-openvpn\x2dclient.slice/openvpn-client@client.service
             └─5978 /usr/sbin/openvpn --suppress-timestamps --nobind --config client.conf
```

Figure 7.42 – Querying the OpenVPN client status

After successfully establishing the VPN client connection, your client machine's public IP address should match the VPN server's public IP (`138.68.19.158`). You can check this by running the following command:

```
dig TXT +short o-o.myaddr.l.google.com @ns1.google.com
```

This is the output of the preceding command:

```
packt@neptune:~$ dig TXT +short o-o.myaddr.l.google.com @ns1.google.com
"138.68.19.158"
```

Figure 7.43 – Retrieving the public IP address of the client machine

To stop VPN client connectivity, we run the following command:

```
sudo systemctl stop openvpn-client@client
```

We can also configure OpenVPN clients on a variety of other operating system platforms. Let's look at a mobile platform next.

Configuring an Android OpenVPN client

First, install the *OpenVPN Connect* app from the Android App store. Next, open the app and import the OpenVPN client profile. We generated the client profile previously, as described in the *Configuring VPN* section. You have a couple of options to import the profile, either by specifying the **Uniform Resource Locator** (**URL**) of the file (such as a Google Drive direct link, OneDrive, or Dropbox) or just pointing to the downloaded file, if the mobile platform supports access to downloads.

The process is illustrated in the following screenshot:

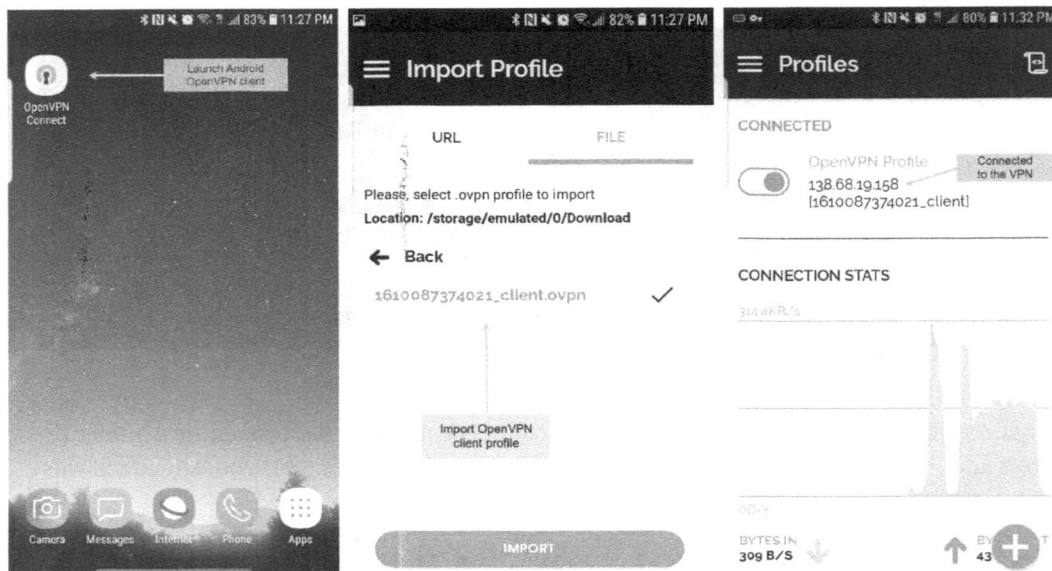

Figure 7.44 – Using the Android OpenVPN client app

After importing the OpenVPN client profile, you can connect to the VPN. The following illustrations suggest the steps described previously. Upon connecting to the VPN, our mobile device's public IP address becomes `138.68.19.158`—the VPN server's IP address.

For more information about the OpenVPN project and related product downloads, please visit `https://openvpn.net/download-open-vpn/`.

Summary

This chapter represents a relatively condensed view of basic Linux networking principles. We have learned about network communication layers and protocols, IP addressing schemes, TCP/IP configurations, well-known network application servers, and VPN. A good grasp of networking paradigms will give Linux administrators a more comprehensive view of the distributed systems and underlying communication between the application endpoints involved.

Some of the theoretical aspects covered in this chapter are taken for a practical spin in *Chapter 8, Configuring Linux Servers*, focusing on real-world implementations of network servers. *Chapter 9, Securing Linux*, will further explore network security internals and practical Linux firewalls. Everything we have learned so far will serve as a good foundation for the assimilation of these upcoming chapters.

Questions

Here's a quick quiz to outline and attest some of the essential concepts covered in this chapter:

1. How does the OSI model compare to the TCP/IP model?

2. Think of a couple of TCP/IP protocols and try to see where and how they operate in some of the network administration tasks or applications you are familiar with.

3. At what networking layer does the HTTP protocol operate? How about DNS?

4. What is the network class for IP address 192.168.0.1?

5. What is the network prefix corresponding to network mask 255.255.0.0?

6. How do you configure a static IP address using the nmcli utility?

7. How do you change the hostname of a Linux machine?

8. What is the difference between the POP3 and IMAP email protocols?

9. How does SSH host-based authentication differ from user-based SSH key authentication?

10. What is the difference between SSH and TELNET?

8
Configuring Linux Servers

In this chapter, you will learn how to configure different types of Linux servers, from **Domain Name System** (**DNS**) and **Domain Host Configuration Protocol** (**DHCP**) servers to web servers, Samba file servers, **File Transfer Protocol** (**FTP**) servers, and **Network File System** (**NFS**) servers. All these servers, in one way or another, are powering the backbone of the **World Wide Web** (**WWW**). The reason your computer is showing the exact time is because of a well-implemented **Network Time Protocol** (**NTP**) server. You can shop online and transfer files between your friends and colleagues thanks to good working DHCP, web, and file servers. Configuring different types of Linux services that power all these servers represents the knowledge base for any Linux system administrator.

In this chapter, we're going to cover the following main topics:

- Introduction to Linux services
- Setting up a DNS server
- Setting up a DHCP server
- Setting up an NTP server

- Setting up an NFS server
- Setting up a Samba file server
- Setting up an FTP server
- Setting up a web server
- Setting up a printing server

Technical requirements

Basic knowledge of networking and Linux commands is required. No special technical requirements are needed—just a working installation of Linux on your system. We will use Ubuntu 20.04.1 **Long-Term Support** (**LTS**) as the distribution of choice for this chapter's exercises and examples. Nevertheless, any other major Linux distribution—such as CentOS, openSUSE, or Fedora—is equally suitable for the tasks detailed in this chapter.

GitHub

You can read the full chapter on GitHub, in the book's complementary source code repository: https://github.com/PacktPublishing/Mastering-Linux-Administration/blob/main/08/B13196-08.pdf.

Questions

Now that you have a clear view of how to manage some of the most widely used services in Linux, here are some exercises that will further contribute to your learning:

1. Try using a VPS for all the services detailed in this chapter, not on your local network.
2. Try setting up a LEMP stack on Ubuntu.
3. Exercise with all the services described in this chapter, using the CentOS 8 distribution.

Further reading

For more information about the topics covered in the chapter, you can check the following link:

- Ubuntu 20.04 official documentation: https://ubuntu.com/server/docs

9
Securing Linux

Securing a Linux machine is usually a balancing act. The endgame is essentially protecting data from unwanted access. While there are many ways to achieve this goal, we should adopt the methods that yield maximum protection, along with the most efficient system administration. Gauging the attack and vulnerability surfaces, both internal and external, is always a good start. The rest of the work is building fences and putting on armor – not too high and not too heavy. The outer fence is a network firewall. Internally, at the system level, we build application security policies. This chapter introduces both, albeit the art of the balancing act is left to you.

In the first part of this chapter, we'll look at access control mechanisms and the related security modules – SELinux and AppArmor. In the second part, we will explore packet filtering frameworks and firewall solutions.

After completing this chapter, you will have become acquainted with the tools for designing and managing application security frameworks and firewalls – a first solid step to securing a Linux system.

Here's a brief overview of the topics that will be covered in this chapter:

- Understanding Linux security – an overview of the access control mechanisms available in the Linux kernel

- Introducing SELinux – an in-depth look at the Linux kernel security framework for managing access control policies

- Introducing AppArmor – a relatively new security module that controls application capabilities based on security profiles

- Working with firewalls – a comprehensive overview of firewall modules, including `netfilter`, `iptables`, `nftables`, `firewalld`, and `ufw`

Technical requirements

This chapter covers a relatively vast array of topics, some of which will be covered with extensive command-line operations. We recommend that you use both a CentOS and an Ubuntu platform with Terminal or SSH access. Direct console access to the systems is highly preferable due to the possibly disruptive way of altering firewall rules.

Understanding Linux security

One of the significant considerations for securing a computer system or network is the means for system administrators to control how users and processes can access various resources, such as files, devices, and interfaces, across systems. The Linux kernel provides a handful of such mechanisms, collectively referred to as **Access Control Mechanisms** (**ACMs**). We will describe them briefly next.

Discretionary Access Control

Discretionary Access Control (**DAC**) is the typical ACM related to filesystem objects, including files, directories, and devices. Such access is at the discretion of the object's owner when managing permissions. DAC controls the access to *objects* based on the identity of users and groups (*subjects*). Depending on a subject's access permissions, they could also pass permissions to other subjects – an administrator managing regular users, for example.

Access Control Lists

Access Control Lists (**ACLs**) provide control over which subjects (such as users and groups) have access to specific filesystem objects (such as files and directories).

Mandatory Access Control

Mandatory Access Control (**MAC**) provides different access control levels to subjects over the objects they own. Unlike DAC, where users have full control over the filesystem objects they own, MAC adds additional labels, or categories, to all filesystem objects. Consequently, subjects must have the appropriate access to these categories to interact with the objects labeled as such. MAC is enforced by **Security-Enhanced Linux** (**SELinux**) on RHEL/CentOS and AppArmor on Ubuntu/Debian.

Role-Based Access Control

Role-Based Access Control (**RBAC**) is an alternative to the permission-based access control of filesystem objects. Instead of permissions, a system administrator assigns *roles* that have access to a specific filesystem object. Roles could be based on some business or functional criteria and may have different access levels to objects.

In contrast to DAC or MAC, where subjects have access to objects based strictly on the permissions involved, the RBAC model represents a logical abstraction over MAC or DAC, as the subjects must be members of a specific group or role before interacting with objects.

Multi-Level Security

Multi-Level Security (**MLS**) is a specific MAC scheme where the *subjects* are processes and the *objects* are files, sockets, and other similar system resources.

Multi-Category Security

Multi-Category Security (**MCS**) is an improved version of SELinux that allows users to label files with *categories*. MCS reuses much of the MLS framework in *SELinux*.

Wrapping up our brief presentation of ACMs, we should note that we covered some of the internals of DAC and ACL in *Chapter 4*, *Managing Users and Groups*, in the *Managing permissions* section in particular. Next, we'll turn our attention to SELinux – a first-class citizen for MAC implementations.

Introducing SELinux

Security-Enhanced Linux (**SELinux**) is a security framework in the Linux kernel for managing the access control policies of system resources. It supports a combination of the MAC, RBAC, and MLS models that were described in the previous section. SELinux is a set of kernel-space security modules and user-space command-line utilities, and it provides a mechanism for system administrators to have control over *who* can access *what* on the system. SELinux is designed to also protect a system against possible misconfigurations and potentially compromised processes.

SELinux was introduced by the **National Security Agency** (**NSA**) as a collection of **Linux Security Modules** (**LSMs**) with kernel updates. SELinux was eventually released to the open source community in 2000 and into Linux starting with the 2.6 kernel series in 2003.

So, how does SELinux work? We'll look at this next.

Working with SELinux

SELinux uses *security policies* to define various access control levels for applications, processes, and files on a system. A security policy is a set of rules describing what can or cannot be accessed.

SELinux operates with *subjects* and *objects*. When a specific application or process (the *subject*) requests access to a file (the *object*), SELinux checks the required permissions involved in the request and enforces the related access control. The permissions for subjects and objects are stored in a lookup table known as the **Access Vector Cache** (**AVC**). The AVC is generated based on the SELinux *policy database*.

A typical SELinux policy consists of the following resources (files), each reflecting a specific aspect of the security policy:

- **Type enforcement**: The actions that have been granted or denied for the policy (such as, read or write to a file).
- **Interface**: The application interface the policy interacts with (such as logging).
- **File contexts**: The system resources associated with the policy (such as log files).

These policy files are compiled together using SELinux build tools to produce a specific *security policy*. The policy is loaded into the kernel, added to the SELinux policy database, and made active without a system reboot.

When creating SELinux policies, we usually test them in *permissive* mode first, where violations are logged but still allowed. When violations occur, the `audit2allow` utility in the SELinux toolset comes to the rescue. We use the log traces produced by `audit2allow` to create the additional rules required by the policy to account for legitimate access permissions. SELinux violations are logged in `/var/log/messages` and are prefixed with `avc: denied`.

The next section will describe the necessary steps for creating an SELinux security policy.

Creating an SELinux security policy

Let's assume that we have a daemon called `packtd` and that we need to secure it to access `/var/log/messages`. For illustration purposes, the daemon has a straightforward implementation: periodically open the `/var/log/messages` file for writing. Use your favorite text editor (such as `nano`) to add the following content (C code) to a file. Let's name the file `packtd.c`:

```c
1   #include <unistd.h>
2   #include <stdio.h>
3
4   FILE *f;
5   char LOG_FILE[] = "/var/log/messages";
6
7   int main(void)
8   {
9       while (1) {
10          f = fopen(LOG_FILE, "w");
11          sleep(10);
12          fclose(f);
13      }
14  }
```

Figure 9.1 – A simple daemon periodically checking logs

Let's compile and build `packtd.c` to generate the related binary executable (`packtd`):

```
gcc -o packtd packtd.c
```

By default, RHEL/CentOS 8 comes with the `gcc` GNU compiler installed. Otherwise, you may install it with the following command:

```
sudo yum install gcc
```

We are ready to proceed with the steps for creating the `packtd` daemon and the required SELinux security policy:

1. Install the daemon.

2. Generate the policy files.

3. Build the security policy.

4. Verify and adjust the security policy.

Let's start with installing our `packtd` daemon.

Installing the daemon

First, we must create the `systemd` unit file for the `packtd` daemon. You may use your favorite text editor (such as `nano`) to create the related file. We will call this file `packtd.service`:

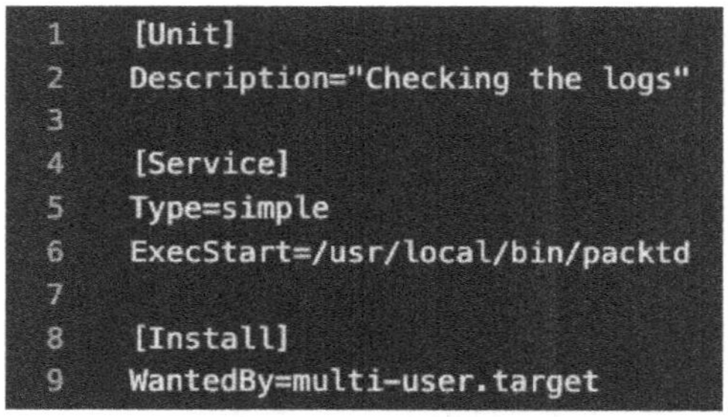

```
1    [Unit]
2    Description="Checking the logs"
3
4    [Service]
5    Type=simple
6    ExecStart=/usr/local/bin/packtd
7
8    [Install]
9    WantedBy=multi-user.target
```

Figure 9.2 – The packtd daemon file

Copy the files we created to their respective locations:

```
sudo cp packtd /usr/local/bin/
sudo cp packtd.service /usr/lib/systemd/system/
```

At this point, we are ready to start our `packtd` daemon:

```
sudo systemctl start packtd
sudo systemctl status packtd
```

The status shows the following output:

```
[packt@jupiter ~]$ sudo systemctl status packtd
● packtd.service - "Checking the logs"
   Loaded: loaded (/usr/lib/systemd/system/packtd.service; disabled; vendor preset
   Active: active (running) since Mon 2020-12-14 19:48:34 PST; 6s ago
 Main PID: 15498 (packtd)
    Tasks: 1 (limit: 23537)
   Memory: 132.0K
   CGroup: /system.slice/packtd.service
           └─15498 /usr/local/bin/packtd
```

Figure 9.3 – The status of the packtd daemon

Let's make sure the `packtd` daemon is not confined or restricted yet by SELinux:

```
ps -efZ | grep packtd | grep -v grep
```

The `-Z` option parameter of `ps` retrieves the SELinux context for processes. The output of the command is as follows:

```
[packt@jupiter ~]$ ps -efZ | grep packtd | grep -v grep
system_u:system_r:unconfined_service_t:s0 root 8523    1  0 16:12 ?        00:00:00
 bash /usr/local/bin/packtd
```

Figure 9.4 – SELinux does not restrict the packtd daemon

The `unconfined_service_t` security attribute suggests that `packtd` is not restricted by SELinux. Indeed, if we tailed `/var/log/messages`, we could see the messages logged by `packtd`:

```
sudo tail -F /var/log/messages
```

Here's an excerpt from the output:

```
[packt@jupiter ~]$ sudo tail -F /var/log/messages
Dec 14 16:30:58 jupiter packtd: Hello from Packt!
Dec 14 16:31:08 jupiter packtd: Hello from Packt!
Dec 14 16:31:18 jupiter packtd: Hello from Packt!
```

Figure 9.5 – The packtd daemon's logging unrestricted

Next, we will generate the security policy files for the `packtd` daemon.

Generating policy files

To build a security policy for `packtd`, we need to generate the related policy files. The SELinux tool for building security policies is `sepolicy`. Also, packaging the final security policy binary requires the `rpm-build` utility. These command-line utilities may not be available by default on your system, so you may have to install the related packages:

```
sudo yum install -y policycoreutils-devel rpm-build
```

The following command generates the policy files for `packtd` (no superuser privileges required):

```
sepolicy generate --init /usr/local/bin/packtd
```

The related output is as follows:

```
[packt@jupiter ~]$ sepolicy generate --init /usr/local/bin/packtd
Created the following files:
/home/packt/packtd.te # Type Enforcement file
/home/packt/packtd.if # Interface file
/home/packt/packtd.fc # File Contexts file
/home/packt/packtd_selinux.spec # Spec file
/home/packt/packtd.sh # Setup Script
```

Figure 9.6 – Generating policy files with sepolicy

Next, we need to rebuild the system policy so that it includes the custom `packtd` policy module.

Building the security policy

We will use the `packtd.sh` build script we created in the previous step here. This command requires superuser privileges since it installs the newly created policy on the system:

```
sudo ./packtd.sh
```

The build takes a relatively short time to complete and yields the following output (excerpt):

```
[packt@jupiter ~]$ sudo ./packtd.sh
Building and Loading Policy
+ make -f /usr/share/selinux/devel/Makefile packtd.pp
Compiling targeted packtd module
Creating targeted packtd.pp policy package
rm tmp/packtd.mod.fc tmp/packtd.mod
+ /usr/sbin/semodule -i packtd.pp
+ sepolicy manpage -p . -d packtd_t
./packtd_selinux.8
+ /sbin/restorecon -F -R -v /usr/local/bin/packtd
++ pwd
```

Figure 9.7 – Building the security policy for packtd

Please note that the build script reinstates the default *SELinux* security context for `packtd` using the `restorecon` command (highlighted in the previous output). Now that we've built the security policy, we're ready to verify the related permissions.

Verifying the security policy

First, we need to restart the `packtd` daemon to account for the policy change:

```
sudo systemctl restart packtd
```

The `packtd` process should now reflect the new SELinux security context:

```
ps -efZ | grep packtd | grep -v grep
```

The output shows a new label (`packtd_t`) for our security context:

```
[packt@jupiter ~]$ ps -efZ | grep packtd | grep -v grep
system_u:system_r:packtd_t:s0    root        15498       1  0 19:48 ?        00:00:00
 /usr/local/bin/packtd
```

Figure 9.8 – The new security policy for packtd

Since SELinux now controls our `packtd` daemon, we should see the related audit traces in `/var/log/messages`, where SELinux logs the system's activity. Let's look at the audit logs for any permission issues. The following command fetches the most recent events for AVC message types using the `ausearch` utility:

```
sudo ausearch -m AVC -ts recent
```

We will immediately notice that `packtd` has no read/write access to `/var/log/messages`:

```
time->Mon Dec 14 19:54:54 2020
type=PROCTITLE msg=audit(1608004494.362:1138): proctitle="/usr/local/bin/packtd"
type=SYSCALL msg=audit(1608004494.362:1138): arch=c000003e syscall=257 success=yes
exit=3 a0=ffffff9c a1=601040 a2=241 a3=1b6 items=0 ppid=1 pid=15498 auid=4294967295
 uid=0 gid=0 euid=0 suid=0 fsuid=0 egid=0 sgid=0 fsgid=0 tty=(none) ses=4294967295
comm="packtd" exe="/usr/local/bin/packtd" subj=system_u:system_r:packtd_t:s0 key=(n
ull)
type=AVC msg=audit(1608004494.362:1138): avc:  denied  { open } for  pid=15498 comm
="packtd" path="/var/log/messages" dev="sda3" ino=1513719 scontext=system_u:system_
r:packtd_t:s0 tcontext=system_u:object_r:var_log_t:s0 tclass=file permissive=1
type=AVC msg=audit(1608004494.362:1138): avc:  denied  { write } for  pid=15498 com
m="packtd" name="messages" dev="sda3" ino=1513719 scontext=system_u:system_r:packtd
_t:s0 tcontext=system_u:object_r:var_log_t:s0 tclass=file permissive=1
```

Figure 9.9 – No read/write access for packtd

To further inquire about the permissions needed by `packtd`, we will feed the output of `ausearch` into `audit2allow`, a tool for generating the required security policy stubs:

```
sudo ausearch -m AVC -ts recent | audit2allow -R
```

The output provides the code macro we're looking for:

```
[packt@jupiter ~]$ sudo ausearch -m AVC -ts recent | audit2allow -R

require {
        type packtd_t;
}

#============= packtd_t ==============
logging_write_generic_logs(packtd_t)
```

Figure 9.10 – Querying the missing permissions for packtd

The `-R` (`--reference`) option of `audit2allow` invokes the stub generation task, which could sometimes yield inaccurate or incomplete results. In such cases, it may take a few iterations to update, rebuild, and verify the related security policies. Let's proceed with the required changes, as suggested previously. We'll edit the *type enforcement* file (`packt.te`) we generated previously and add the lines (copy/paste) exactly, as indicated by the output of `audit2allow`. After saving the file, we need to rebuild the security policy, restart the `packtd` daemon, and verify the audit logs. We're reiterating the last three steps in our overall procedure:

```
sudo ./packtd.sh
sudo systemctl restart packtd
sudo ausearch -m AVC -ts recent | audit2allow -R
```

This time, the SELinux audit should come out clean:

```
[packt@jupiter ~]$ sudo ausearch -m AVC -ts recent | audit2allow -R
<no matches>
Nothing to do
```

Figure 9.11 – No more permission issues for packtd

Sometimes, it may take a little while for `ausearch` to refresh its *recent* buffer. Alternatively, we can specify a starting timestamp to analyze from, such as after we've updated the security policy, using a relatively recent timestamp:

```
sudo ausearch --start 12/14/2020 '22:30:00' | audit2allow -R
```

At this point, we have a basic understanding of SELinux security policy internals. Next, we'll turn to some higher-level operations for managing and controlling SELinux in everyday administration tasks.

Understanding SELinux modes

SELinux is either *enabled* or *disabled* in a system. When enabled, it operates in either of the following modes:

- `Enforcing`: SELinux effectively monitors and controls security policies. In RHEL/CentOS, this mode is enabled by default.

- `Permissive`: Security policies are actively monitored without enforcing access control. Policy violations are logged in `/var/log/messages`.

When SELinux is disabled, security policies are neither monitored nor enforced.

The following command retrieves the current status of SELinux on the system:

```
sestatus
```

The output is as follows:

```
[packt@jupiter ~]$ sestatus
SELinux status:                 enabled
SELinuxfs mount:                /sys/fs/selinux
SELinux root directory:         /etc/selinux
Loaded policy name:             targeted
Current mode:                   enforcing
Mode from config file:          enforcing
Policy MLS status:              enabled
Policy deny_unknown status:     allowed
Memory protection checking:     actual (secure)
Max kernel policy version:      32
```

Figure 9.12 – Getting the current status of SELinux

When SELinux is enabled, the following command retrieves the current mode:

```
getenforce
```

In `permissive mode`, we get the following output:

```
[packt@jupiter ~]$ getenforce
Enforcing
```

Figure 9.13 – Getting the current mode of SELinux

To change from enforcing to `permissive mode`, we can run the following command:

```
sudo setenforce 0
```

The `getenforce` command will display `Permissive` in this case. To switch back into enforcing mode, we can run the following command:

```
sudo setenforce 1
```

The SELinux mode can also be set by editing the `SELINUX` value in `/etc/selinux/config`. The possible values are documented in the configuration file.

> **Important note**
> Manually editing the SELinux configuration file requires a system reboot for the changes to take effect.

With SELinux enabled, a system administrator may choose between the following SELinux policy levels by modifying the `SELINUXTYPE` value in `/etc/selinux/config`: `targeted`, `minimum`, and `mls`. The corresponding values are documented in the configuration file.

> **Important note**
> The default SELinux policy setting is `targeted`, and it's generally recommended not to change this setting, except for `mls`.

With the `targeted` policy in place, only processes that are specifically configured to use SELinux security policies run in a *confined* (or restricted) domain. Such processes usually include system daemons (such as `dhcpd` and `sshd`) and well-known server applications (such as *Apache* and *PostgreSQL*). All other (non-targeted) processes run unrestricted and are usually labeled with the `unconfined_t` domain type.

To completely disable SELinux, we can edit the `/etc/selinux/config` file using a text editor of our choice (such as `sudo nano /etc/selinux/config`) and make the following change:

```
SELINUX=disabled
```

Alternatively, we can run the following command to change the SELinux mode from `enforcing` to `disabled`:

```
sudo sed -i 's/SELINUX=enforcing/SELINUX=disabled/g' /etc/
selinux/config
```

We can retrieve the current configuration with the following command:

```
cat /etc/selinux/config
```

With SELinux `disabled`, we get the following output:

```
[packt@jupiter ~]$ cat /etc/selinux/config

# This file controls the state of SELinux on the system.
# SELINUX= can take one of these three values:
#     enforcing - SELinux security policy is enforced.
#     permissive - SELinux prints warnings instead of enforcing.
#     disabled - No SELinux policy is loaded.
SELINUX=disabled
# SELINUXTYPE= can take one of these three values:
#     targeted - Targeted processes are protected,
#     minimum - Modification of targeted policy. Only selected processes are protec
ted.
#     mls - Multi Level Security protection.
SELINUXTYPE=targeted
```

Figure 9.14 – Disabling SELinux

We need to reboot the system for the changes to take effect:

```
sudo systemctl reboot
```

Next, let's examine how access control decisions are made by introducing SELinux *contexts*.

Understanding SELinux contexts

With SELinux enabled, processes and files are labeled with a *context* containing additional *SELinux-specific* information, such as *user*, *role*, *type*, and *level* (optional). The context data serves for SELinux access control decisions.

SELinux adds the `-Z` option to the `ls`, `ps`, and other commands, thus displaying the security context of filesystem objects, processes, and more.

Let's create an arbitrary file and examine the related SELinux context:

```
touch afile
ls -Z afile
```

The output is as follows:

```
[packt@jupiter ~]$ touch afile
[packt@jupiter ~]$ ls -Z afile
unconfined_u:object_r:user_home_t:s0 afile
```

Figure 9.15 – Displaying the SELinux context of a file

The SELinux context has the following format – a sequence of four fields, separated by a colon (:):

```
USER:ROLE:TYPE:LEVEL
```

We will explain SELinux context fields.

SELinux user

The SELinux *user* is an identity known to the policy that's authorized for a specific set of *roles* and has a particular *level* that's designated by an MLS/MCS *range* (see the *SELinux level* section for more details). Every Linux user account is mapped to a corresponding SELinux user identity using an SELinux policy. This mechanism allows regular Linux users to inherit the policy restrictions associated with SELinux users.

A process owned by a Linux user receives the mapped SELinux user's identity to assume the corresponding SELinux *roles* and *levels.*

The following command displays a list of mappings between Linux accounts and their corresponding SELinux user identities. The command requires superuser privileges. Also, the `semanage` utility is available with the `policycoreutils` package, which you may need to install on your system:

```
sudo semanage login -l
```

The output may slightly differ from system to system:

```
[packt@jupiter ~]$ sudo semanage login -l
[sudo] password for packt:

Login Name              SELinux User            MLS/MCS Range           Service

__default__             unconfined_u            s0-s0:c0.c1023          *
root                    unconfined_u            s0-s0:c0.c1023          *
```

Figure 9.16 – Displaying the SELinux user mappings

For more information on the `semanage` command-line utility, you may refer to the related system reference (`man semanage`, `man semanage-login`).

SELinux roles

SELinux *roles* are part of the RBAC security model, and they are essentially RBAC attributes. In the SELinux context hierarchy, users are authorized for *roles*, and roles are authorized for *types* or *domains*. In the SELinux context terminology, *types* refer to filesystem object types and *domains* refer to process types (see more in the *SELinux type* section).

Take Linux processes, for example. The SELinux *role* serves as an intermediary access layer between *domains* and SELinux *users*. An accessible *role* determines which *domain* (that is, *processes*) can be accessed through that *role*. Ultimately, this mechanism controls which object types can be accessed by the process, thus minimizing the surface for privilege escalation attacks.

SELinux type

The SELinux *type* is an attribute of SELinux *type enforcement* – a *MAC* security construct. For SELinux types, we refer to *domains* as process types and *types* as filesystem object types. SELinux security policies control how specific types can access each other – either with domain-to-type access or domain-to-domain interactions.

SELinux level

The SELinux *level* is an attribute of the *MLS/MCS* schema and an optional field in the SELinux context. A level usually refers to the security clearance of a subject's access control to an object. Levels of clearance include `unclassified`, `confidential`, `secret`, and `top-secret` and are expressed as a *range*. An *MLS range* represents a pair of levels, defined as `low-high` if the levels differ, or just `low` if the levels are identical. For example, a level of `s0-s0` is the same as `s0`. Each level represents a *sensitivity-category* pair, with categories being optional. When a category is specified, the level is defined as `sensitivity:category-set`; otherwise, it's defined as `sensitivity` only.

We are now familiar with SELinux contexts. We'll see them in action, starting with the SELinux contexts for users, next.

SELinux contexts for users

The following command displays the SELinux context associated with the current user:

```
id -Z
```

In our case, the output is as follows:

```
[packt@jupiter ~]$ id -Z
unconfined_u:unconfined_r:unconfined_t:s0-s0:c0.c1023
```

Figure 9.17 – Displaying the current user's SELinux context

In RHEL/CentOS, Linux users are unconfined (unrestricted) by default, with the following context fields:

- unconfined_u: User identity
- unconfined_r: Role
- unconfined_t: Domain affinity
- s0-s0: *MLS* range (the equivalent of s0)
- c0.c1023: Category set, representing all categories (from c0 to c1023)

Next, we'll examine the SELinux context for processes.

SELinux context for processes

The following command displays the SELinux context for the current SSH processes:

```
ps -eZ | grep sshd
```

The command yields the following output:

```
[packt@jupiter ~]$ ps -eZ | grep ssh
system_u:system_r:sshd_t:s0-s0:c0.c1023 1088 ?    00:00:00 sshd
unconfined_u:unconfined_r:unconfined_t:s0-s0:c0.c1023 2598 ? 00:00:00 sshd
```

Figure 9.18 – Displaying the SELinux context for SSH-related processes

From the output, we can infer that the top line refers to the sshd server process, which is running with the system_u user identity, system_r role, and sshd_t domain affinity. The second line refers to the current user's SSH session, hence the unconfined context. System daemons are usually associated with the system_u user and system_r role.

Before concluding this section on SELinux contexts, we'll examine the relatively common scenario of SELinux domain transitions, which is where a process in one domain accesses an object (or *process*) in a different domain.

SELinux domain transitions

Assuming an SELinux-secured process in one domain requests access to an object (or another process) in a different domain, SELinux domain transitions come into play. Unless there's a specific security policy allowing the related domain transition, SELinux would deny access.

An SELinux-protected process transitioning from one domain into another invokes the entrypoint type of the new domain. SELinux evaluates the related *entrypoint permission* and decides if the soliciting process can enter the new domain.

To illustrate a domain transition scenario, we will take the simple case of using the passwd utility when users change their password. The related operation involves the interaction between the passwd process and the /etc/shadow (and possibly /etc/gshadow) file(s). When the user enters (and reenters) the password, passwd would hash and store the user's password in /etc/shadow.

Let's examine the SELinux domain affinities involved:

```
ls -Z /usr/bin/passwd
ls -Z /etc/shadow
```

The corresponding output is as follows:

```
[packt@jupiter ~]$ ls -Z /usr/bin/passwd
system_u:object_r:passwd_exec_t:s0 /usr/bin/passwd
[packt@jupiter ~]$
[packt@jupiter ~]$ ls -Z /etc/shadow
system_u:object_r:shadow_t:s0 /etc/shadow
```

Figure 9.19 – Comparing the domain affinity context

The passwd utility is labeled with the passwd_exec_t type, while /etc/shadow is labeled with shadow_t. There must be a specific security policy chain that allows the related domain to transition from passwd_exec_t to shadow_t; otherwise, passwd will not work as expected.

Let's validate our assumption. We'll use the sesearch tool to query for our assumed security policy:

```
sudo sesearch -s passwd_t -t shadow_t -p write --allow
```

Here's a brief explanation of the preceding command:

- `sesearch`: Searches the SELinux policy database
- `-s passwd_t`: Finds policy rules with `passwd_t` as their source *type* or *role*
- `-t shadow_t`: Finds policy rules with `shadow_t` as their target *type* or *role*
- `-p write`: Finds policy rules with `write` permissions
- `--allow`: Finds policy rules that *allow* the queried permissions (specified with `-p`)

The output of the preceding command is as follows:

```
[packt@jupiter ~]$ sudo sesearch -s passwd_t -t shadow_t -p write --allow | grep pa
sswd_t
allow passwd_t shadow_t:file { append create getattr ioctl link lock map open read
relabelfrom relabelto rename setattr unlink write };
```

Figure 9.20 – Querying SELinux policies

Here, we can see the `append create` permissions, as we correctly assumed.

How did we pick the `passwd_t` source type instead of `passwd_exec_t`? By definition, the *domain* type corresponding to the *executable file* type, `passwd_exec_t`, is `passwd_t`. If we were not sure about *who* has *write* permissions to the `shadow_t` file types, we could have simply excluded the source type (`-s passwd_t`) in the `sesearch` query and parsed the output (for example, using `grep passwd`).

The use of the `sesearch` tool is very convenient when we're querying security policies. There are a handful of similar tools for troubleshooting or managing the SELinux configuration and policies. One of the most notable SELinux command-line utilities is `semanage` for managing SELinux policies. We'll examine it next.

Managing SELinux policies

SELinux provides several utilities for managing security policies and modules, some of which will be briefly described in the *Troubleshooting SELinux issues* section next. Examining each of these tools is beyond the scope of this chapter, but we'll take `semanage` for a quick spin, to reflect on some use cases involving security policy management.

The general syntax of the `semanage` command is as follows:

```
semanage TARGET [OPTIONS]
```

`TARGET` usually denotes a specific namespace for policy definitions (for example, `login`, `user`, `port`, `fcontext`, `boolean`, `permissive`, and so on). Let's look at a few examples to get an idea of how `semanage` works.

Enabling secure binding on custom ports

Let's assume we want to enable SELinux for a custom SSH port instead of the default 22. We can retrieve the current security records (labels) on the SSH port with the following command:

```
sudo semanage port -l | grep ssh
```

For a default configuration, we will get the following output:

```
[packt@jupiter ~]$ sudo semanage port -l | grep ssh
ssh_port_t                        tcp       22
```

Figure 9.21 – Querying the SELinux security label for the SSH port

If we want to enable SSH on a different port (such as 2222), first, we need to configure the related service (sshd) to listen on a different port. We won't go into those details here. Here, we need to enable the secure binding on the new port with the following command:

```
sudo semanage port -a -t ssh_port_t -p tcp 2222
```

Here's a brief explanation of the preceding command:

- -a (--add): Adds a new record (label) for the given type
- -t ssh_port_t: The SELinux type of the object
- -p tcp: The network protocol associated with the port

As a result of the previous command, the new security policy for the ssh_port_t type looks like this:

```
[packt@jupiter ~]$ sudo semanage port -a -t ssh_port_t -p tcp 2222
[packt@jupiter ~]$ sudo semanage port -l | grep ssh
ssh_port_t                        tcp       2222, 22
```

Figure 9.22 – Changing the SELinux security label for the SSH port

We could arguably delete the old security label (for port 22), but that won't really matter if we disable port 22. If we want to delete a port security record, we can do so with the following command:

```
sudo semanage port -d -p tcp 22
```

We used the -d (--delete) option to remove the related security label. To view the local customizations for our semanage port policies, we can invoke the -C (--locallist) option:

```
sudo semanage port -l -C
```

For more information on semanage port, you may refer to the related system reference (man semanage port). Next, we'll look at how to modify security permissions for specific server applications.

Modifying security permissions for targeted services

semanage uses the boolean namespace to toggle specific features of targeted services on and off. A targeted service is a daemon with built-in SELinux protection. In the following example, we want to enable FTP over HTTP connections. By default, this security feature of Apache (httpd) is turned off. Let's query the related httpd security policies:

```
sudo semanage boolean -l | grep httpd | grep ftp
```

We get the following output:

```
[packt@jupiter ~]$ sudo semanage boolean -l | grep httpd | grep ftp
httpd_can_connect_ftp          (off  , off) Allow httpd to act as a FTP client con
necting to the ftp port and ephemeral ports
httpd_enable_ftp_server        (off  , off) Allow httpd to act as a FTP server by
listening on the ftp port.
```

Figure 9.23 – Querying httpd policies related to FTP

As we can see, the related feature – httpd_enable_ftp_server – is turned off by default. The current and persisted states are currently off: (off, off). We can enable it with the following command:

```
sudo semanage boolean -m --on httpd_enable_ftp_server
```

To view the local customizations of the semanage boolean policies, we can invoke the -C (--locallist) option:

```
sudo semanage boolean -l -C
```

The new configuration now looks like this:

```
[packt@jupiter ~]$ sudo semanage boolean -l -C
SELinux boolean                State  Default Description

httpd_enable_ftp_server        (on  ,   on) Allow httpd to act as a FTP server by
listening on the ftp port.
```

Figure 9.24 – Enabling the security policy for FTP over HTTP

In the preceding example, we used the -m (--modify) option with the semanage boolean command to toggle the httpd_enable_ftp_server feature.

For more information on semanage boolean, you may refer to the related system reference (man semanage boolean). Now, let's learn how to modify the security context of specific server applications.

Modifying security contexts for targeted services

In this example, we want to secure SSH keys stored in a custom location on the local system. Since we're targeting a filesystem-related security policy, we will use the fcontext (file context) namespace with semanage.

The following command queries the file context security settings for sshd:

```
sudo semanage fcontext -l | grep sshd
```

Here's a relevant excerpt from the output:

```
/etc/ssh/ssh_host.*_key                        regular file
system_u:object_r:sshd_key_t:s0
/etc/ssh/ssh_host.*_key\.pub                    regular file
system_u:object_r:sshd_key_t:s0
```

Figure 9.25 – The security context of SSH keys

The following command also adds the /etc/ssh/keys/ path to the secure locations associated with the sshd_key_t context type:

```
sudo semanage fcontext -a -t sshd_key_t '/etc/ssh/keys(/.*)?'
```

The '/etc/ssh/keys(/.*)?' regular expression matches any files in the /etc/ssh/keys/ directory, including subdirectories at any nested level. To view the local customizations of the semanage fcontext policies, we can invoke the -C (--locallist) option:

```
sudo semanage fcontext -l -C
```

We should see our new security context:

```
[packt@jupiter ~]$ sudo semanage fcontext -l -C
SELinux fcontext                              type          Context
/etc/ssh/keys(/.*)?                           all files     system_u:objec
t_r:sshd_key_t:s0
```

Figure 9.26 – The modified security context of our SSH keys

We should also initialize the filesystem security context of the /etc/ssh/keys directory (if we've already created it):

```
sudo restorecon -r /etc/ssh/keys
```

restorecon is an SELinux utility for restoring the default security context to a filesystem object. The -r (or -R) option specifies a recursive action on the related path.

For more information on semanage fcontext, you may refer to the related system reference (man semanage fcontext). Next, we'll look at enabling permissive mode for specific server applications.

Enabling permissive mode for targeted services

Earlier in this chapter, we created a custom daemon (packtd) with its security policy. See the related topic in the *Creating an SELinux security policy* section. When we worked on the packtd daemon and tested its functionality, initially, we had to deal with its SELinux policy violations. Eventually, we fixed the required security policy context and everything was fine. During the entire process, we were able to run and test with packtd without having the daemon shut down by SELinux due to non-compliance. Yet, our Linux system runs SELinux in enforcing mode (by default) and is not permissive. See the *Understanding SELinux modes* section for more information on enforcing and permissive modes.

How, then, is it possible that packtd ran unrestricted while violating security policies?

By default, SELinux is permissive to any *untargeted* type in the system. By *untargeted*, we mean a domain (type) that hasn't been forced into a restrictive (or confined) mode yet.

When we built the security policy for our packtd daemon, we let the related SELinux build tools generate the default *type enforcement* file (packt.te) and other resources for our domain. A quick look at the packt.te file shows that our packtd_t type is permissive:

```
cat packt.te
```

Here's the relevant excerpt from the file:

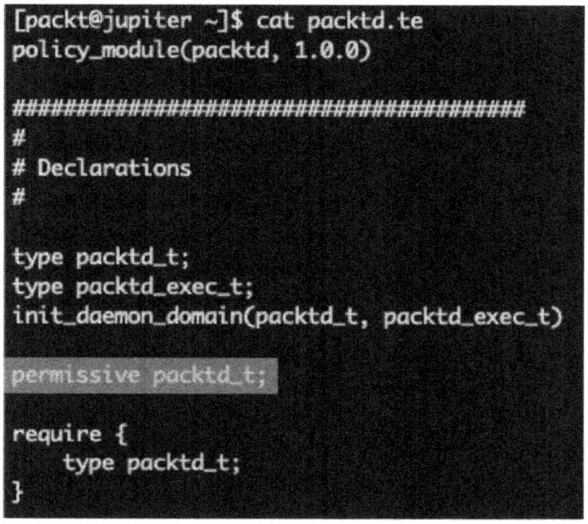

```
[packt@jupiter ~]$ cat packtd.te
policy_module(packtd, 1.0.0)

#####################################
#
# Declarations
#

type packtd_t;
type packtd_exec_t;
init_daemon_domain(packtd_t, packtd_exec_t)

permissive packtd_t;

require {
    type packtd_t;
}
```

Figure 9.27 – The packtd_t domain is permissive

So, the packtd_t domain is permissive by nature. The only way to confine packtd is to remove the permissive line from the packtd.te file and rebuild the related security policy. We will leave that as an exercise to you. The case we wanted to make here was to present a possibly misbehaving – in our case, permissive – domain that we can *catch* by managing permissive types with the semanage permissive command.

To manage permissive mode for individual targets, we can use the semanage command with our permissive namespace. The following command lists all the domains (types) currently in permissive mode:

```
sudo semanage permissive -l
```

In our case, we have the *built-in* packtd_t domain, which is permissive:

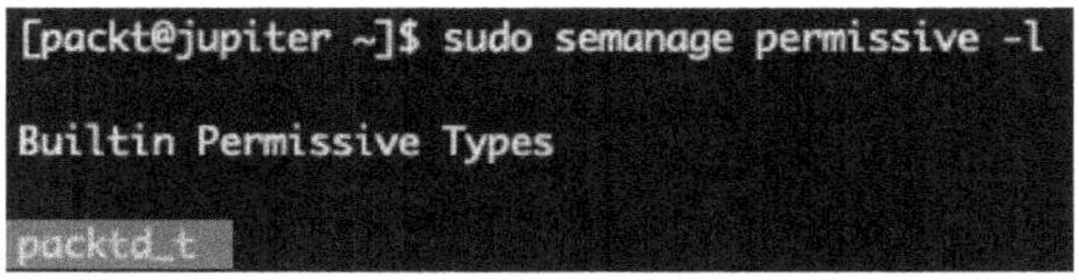

```
[packt@jupiter ~]$ sudo semanage permissive -l

Builtin Permissive Types

packtd_t
```

Figure 9.28 – Displaying permissive types

In general, it is unlikely that a default SELinux configuration would have any permissive types.

We can use the `semanage permissive` command to temporarily place a restricted domain into `permissive mode` while testing or troubleshooting a specific functionality. For example, the following command sets the Apache (`httpd`) daemon in `permissive mode`:

```
sudo semanage permissive -a httpd_t
```

When we query for `permissive` types, we get the following result:

```
[packt@jupiter ~]$ sudo semanage permissive -l

Builtin Permissive Types

packtd_t

Customized Permissive Types

httpd_t
```

Figure 9.29 – Customized permissive types

Domains or types that are made `permissive` with the `semanage permissive` command will show up as `Customized Permissive Types`.

To revert the `httpd_t` domain to the confined (restricted) state, we can invoke the `semanage permissive` command with the `-d` (`--delete`) option:

```
sudo semanage permissive -d httpd_t
```

Note that we cannot confine built-in `permissive types` with the `semanage` command. As we mentioned previously, the `packtd_t` domain is `permissive` by nature and cannot be restricted.

Troubleshooting SELinux issues

Even during our relatively brief journey of exploring SELinux, we used a handful of tools and means to inspect some of the internal workings of security policies and the access control between the subjects (users and processes) and objects (files). SELinux problems usually come down to action being denied, either between specific subjects or between a subject and some objects. SELinux-related issues are not always obvious or easy to troubleshoot, but knowing about the tools that can help is already a good start for tackling these problems.

Here are some of these tools, briefly explained:

- `/var/log/messages`: The log file containing SELinux access control traces and policy violations
- `audit2allow`: Generates SELinux policy rules from the log traces corresponding to denied operations
- `audit2why`: Provides user-friendly translations of SELinux audit messages of policy violations
- `ausearch`: Queries `/var/log/messages` for policy violations
- `ls -Z`: Lists filesystem objects with their corresponding SELinux context
- `ps -Z`: Lists processes with their corresponding SELinux context
- `restorecon`: Restores the default SELinux context for filesystem objects
- `seinfo`: Provides general information about SELinux security policies
- `semanage`: Manages and provides insight into SELinux policies
- `semodule`: Manages SELinux policy modules
- `sepolicy`: Inspects SELinux policies
- `sesearch`: Queries the SELinux policy database

For most of these tools, there is a corresponding system reference (such as `man sesearch`) that provides detailed information about using the tool. Beyond these tools, you can also explore the vast documentation SELinux has to offer. Here's how.

Accessing SELinux documentation

SELinux has extensive documentation, available as an RHEL/CentOS installable package or online at `https://access.redhat.com/documentation/en-us/red_hat_enterprise_linux/8/html/using_selinux/index` (for RHEL/CentOS 8).

The following command installs the SELinux documentation on RHEL/CentOS 8 systems:

```
sudo yum install -y selinux-policy-doc.noarch
```

You can browse a particular SELinux topic with (for example) the following command:

```
man -k selinux | grep httpd
```

SELinux is among the most established and highly customizable security frameworks in the Linux kernel. However, its relatively vast domain and inherent complexity may appear overwhelming for many. Sometimes, even for seasoned system administrators, the choice of a Linux distribution could hang in the balance based on the underlying security module. SELinux is mostly available on RHEL/CentOS platforms. More recent revisions of the Linux kernel are now moving away from SELinux while adopting a relatively lighter and more efficient security framework. The rising star on the horizon is AppArmor.

Introducing AppArmor

AppArmor is an LSM based on the MAC model that confines applications to a limited set of resources. AppArmor uses an ACM based on security profiles that have been loaded into the kernel. Each profile contains a collection of rules for accessing various system resources. AppArmor can be configured to either `enforce` access control or just `complain` about access control violations.

AppArmor proactively protects applications and operating system resources from internal and external threats, including zero-day attacks, by preventing both known and unknown vulnerabilities from being exploited.

AppArmor has been built into the mainline Linux kernel since version 2.6.36 and is currently shipped with *Ubuntu*, *Debian*, *OpenSUSE*, and similar distributions.

In the following sections. we'll use an Ubuntu 20.04 environment to showcase a few practical examples with AppArmor. Most of the related command-line utilities will work the same on any platform with AppArmor installed.

Working with AppArmor

AppArmor command-line utilities usually require superuser privileges.

The following command checks the current status of AppArmor:

```
sudo aa-status
```

Here's an excerpt from the command's output:

Figure 9.30 – Getting the status of AppArmor

The `aa-status` (or `apparmor_status`) command provides a full list of the currently loaded AppArmor profiles (not shown in the preceding excerpt). We'll examine AppArmor profiles next.

Introducing AppArmor profiles

With AppArmor, processes are confined (or restricted) by profiles. AppArmor profiles are loaded upon system start and run either in `enforce mode` or `complain mode`. We'll explain these modes next.

Enforce mode

AppArmor prevents applications running in `enforce mode` from performing restricted actions. Access violations are signaled with log entries in `syslog`. Ubuntu, by default, loads the application profiles in `enforce mode`.

Complain mode

Applications running in `complain mode` can take restricted actions, while AppArmor creates a log entry for the related violation. `complain mode` is ideal for testing AppArmor profiles. Potential errors or access violations can be caught and fixed before switching the profiles to `enforce mode`.

With these introductory notes in mind, let's create a simple application with an AppArmor profile.

Creating a profile

In this section, we'll create a simple application guarded by AppArmor. We hope this exercise will help you get a sensible idea of the inner workings of AppArmor. Let's name this application `appackt`. We'll make it a simple script that creates a file, writes to it, and then deletes the file. The goal is to have AppArmor prevent our app from accessing any other paths in the local system. To try and make some sense of this, think of it as trivial log recycling.

Here's the `appackt` script, and please pardon the thrifty implementation:

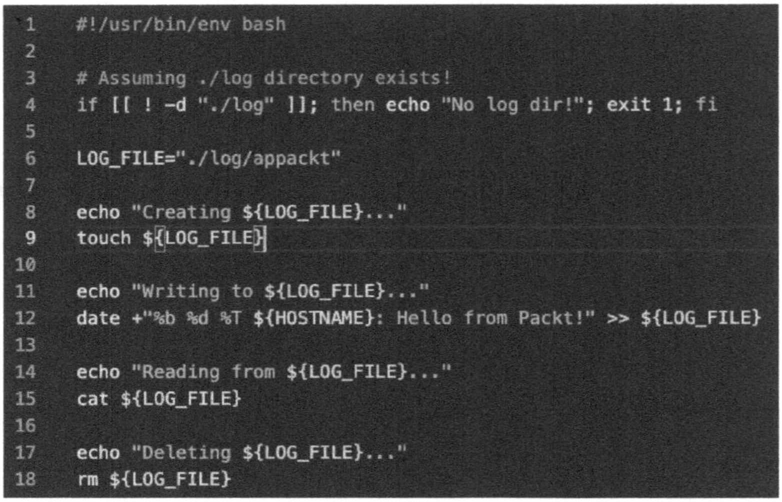

```
 1   #!/usr/bin/env bash
 2
 3   # Assuming ./log directory exists!
 4   if [[ ! -d "./log" ]]; then echo "No log dir!"; exit 1; fi
 5
 6   LOG_FILE="./log/appackt"
 7
 8   echo "Creating ${LOG_FILE}..."
 9   touch ${LOG_FILE}
10
11   echo "Writing to ${LOG_FILE}..."
12   date +"%b %d %T ${HOSTNAME}: Hello from Packt!" >> ${LOG_FILE}
13
14   echo "Reading from ${LOG_FILE}..."
15   cat ${LOG_FILE}
16
17   echo "Deleting ${LOG_FILE}..."
18   rm ${LOG_FILE}
```

Figure 9.31 – The appackt script

We are assuming that the `log` directory already exists at the same location as the script:

```
mkdir ./log
```

Let's make the script executable and run it:

```
chmod a+x appackt
./appackt
```

The output is as follows:

```
packt@neptune:~$ ./appackt
Creating ./log/appackt...
Writing to ./log/appackt...
Reading from ./log/appackt...
Dec 17 05:00:52 neptune: Hello from Packt!
Deleting ./log/appackt...
```

Figure 9.32 – The output of the appackt script

Now, let's work on guarding and enforcing our script with AppArmor. Before we start, we need to install the `apparmor-utils` package – the *AppArmor toolset*:

```
sudo apt-get install -y apparmor-utils
```

We'll use a couple of tools to help create the profile:

- `aa-genprof`: Generates an AppArmor security profile
- `aa-logprof`: Updates an AppArmor security profile

We use `aa-genprof` to monitor our application at runtime and have AppArmor *learn* about it. In the process, we'll be prompted to acknowledge and choose the behavior that's required in specific circumstances.

Once the profile has been created, we'll use the `aa-logprof` utility to make further adjustments while testing in `complain mode`, should any violations occur.

Let's start with `aa-genprof`. We need two terminals: one for the `aa-genprof` monitoring session (in *terminal 1*) and the other for running our script (in *terminal 2*).

We will start with *terminal 1* and run the following command:

```
sudo aa-genprof ./appackt
```

There is a first prompt waiting for us. Next, while the prompt in *terminal 1* is waiting, we will switch to *terminal 2* and run the following command:

```
./appackt
```

Now, we must go back to *terminal 1* and answer the prompts sent by `aa-genprof`, as follows:

Prompt 1 – Waiting to scan

This prompt asks to scan the system log for AppArmor events in order to detect possible complaints (violations).

Answer: S (Scan):

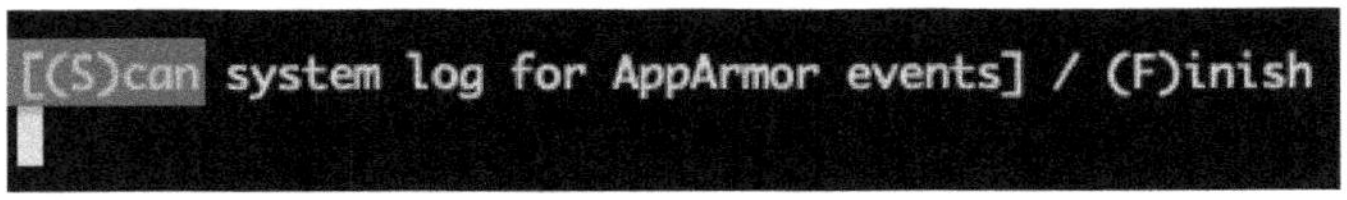

Figure 9.33 – Prompt 1 – Waiting to scan with aa-genprof

Let's look at the next prompt.

Prompt 2 – Execute permissions for /usr/bin/bash

This prompt requests execute permissions for the process (/usr/bin/bash) running our app.

Answer: I (Inherit):

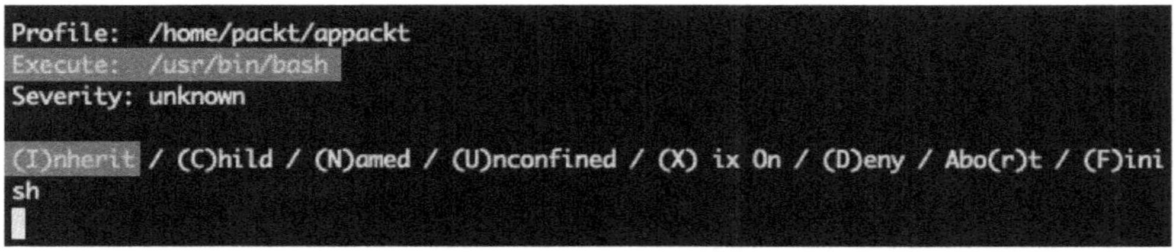

Figure 9.34 – Prompt 2 – Execute permissions for /usr/bin/bash

Let's look at the next prompt.

Prompt 3 – Read/write permissions to /dev/tty

This prompt requests read/write permissions for the app to control the terminal (/dev/tty).

Answer: A (Allow):

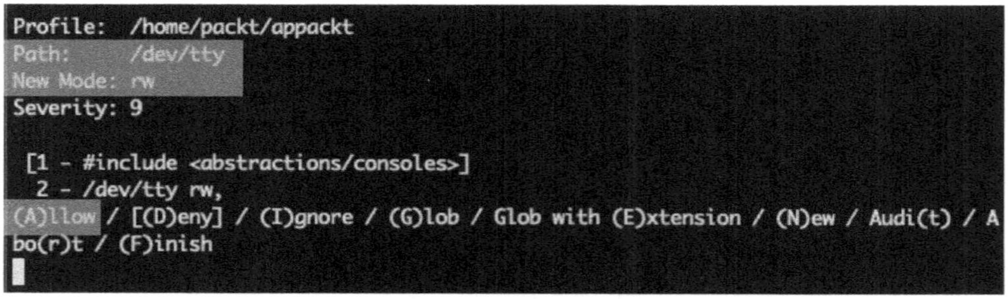

Figure 9.35 – Prompt 3 – Read/write permissions to /dev/tty

Now, let's look at the final prompt.

Prompt 4 – save changes

The prompt is asking to save or review the changes.

Answer: S (Save):

```
Adding #include <abstractions/consoles> to profile.

= Changed Local Profiles =

The following local profiles were changed. Would you like to save them?

[1 - /home/packt/appackt]
(S)ave Changes / Save Selec(t)ed Profile / [(V)iew Changes] / View Changes b/w (C)le
an profiles / Abo(r)t
```

Figure 9.36 – Prompt 4 – Save changes

At this point, we have finished scanning with aa-genprof, and we can answer with F (Finish) to the last prompt. Our app (appackt) is now enforced by AppArmor in complain mode (by default). If we try to run our script, we'll get the following output:

```
packt@neptune:~$ ./appackt
Creating ./log/appackt...
./appackt: line 9: /usr/bin/touch: Permission denied
Writing to ./log/appackt...
./appackt: line 12: ./log/appackt: Permission denied
Reading from ./log/appackt...
./appackt: line 15: /usr/bin/cat: Permission denied
Deleting ./log/appackt...
./appackt: line 18: /usr/bin/rm: Permission denied
```

Figure 9.37 – The first run of appackt with AppArmor confined

As the output suggests, things are not quite right yet. This is where the aa-logprof tool comes to the rescue. For the rest of the steps, we only need one terminal window.

Let's run the aa-logprof command to further tune our appackt security profile:

```
sudo aa-logprof
```

We'll get several prompts again, similar to the previous ones, asking for further permissions needed by our script, namely for the touch, cat, and rm commands. The prompts alternate between Inherit and Allow answers, where appropriate. We won't go into the details here due to space. By now, you should have a general idea about these prompts and their meaning. It's always recommended, though, to ponder upon the permissions asked for and act accordingly.

We may have to run the `aa-logprof` command a couple of times because, with each iteration, new permissions will be discovered and addressed, depending on the child processes that are spawned by our script and so on. Eventually, the `appackt` script will run successfully.

During the iterative process described previously, we may end up with a few *unknown* or orphaned entries in the AppArmor database, which are artifacts of our previous attempts, to secure our application:

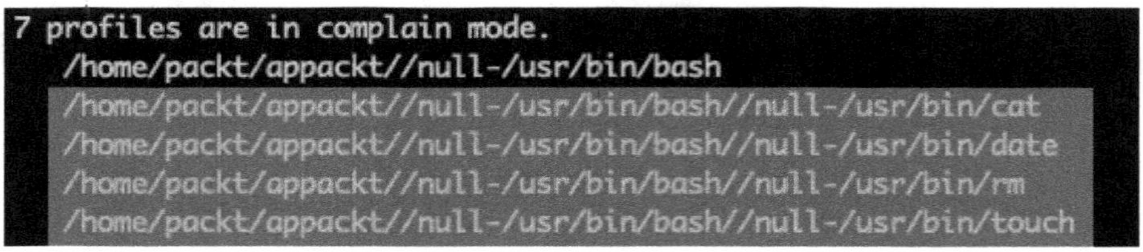

```
7 profiles are in complain mode.
  /home/packt/appackt//null-/usr/bin/bash
  /home/packt/appackt//null-/usr/bin/bash//null-/usr/bin/cat
  /home/packt/appackt//null-/usr/bin/bash//null-/usr/bin/date
  /home/packt/appackt//null-/usr/bin/bash//null-/usr/bin/rm
  /home/packt/appackt//null-/usr/bin/bash//null-/usr/bin/touch
```

Figure 9.38 – Remnants of the iterative process

They will all be named according to the path of or our application (`/home/packt/appackt`). We can clean up these entries with the following command:

```
sudo aa-remove-unknown
```

We can now verify that our app is indeed guarded with AppArmor:

```
sudo aa-status
```

The relevant excerpt from the output is as follows:

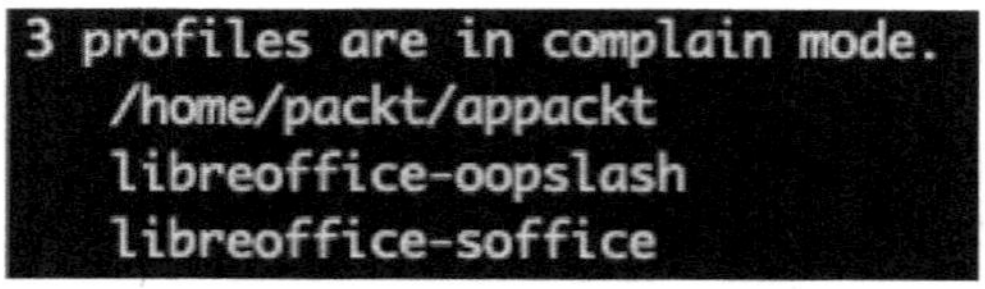

```
3 profiles are in complain mode.
  /home/packt/appackt
  libreoffice-oopslash
  libreoffice-soffice
```

Figure 9.39 – appackt in complain mode

Our application (`/home/packt/appackt`) is shown, as expected, in `complain` mode. The other two are system application-related and are not relevant for us.

Next, we need to validate that our app complies with the security policies enforced by AppArmor. Let's edit the `appackt` script and change the `LOG_FILE` path in line 6 to the following:

```
LOG_FILE="./logs/appackt"
```

We have changed the output directory from `log` to `logs`. Let's create the `logs` directory and run our app:

```
mkdir logs
./appackt
```

The preceding output suggests that `appackt` is attempting to access a path outside the permitted boundaries by AppArmor, thus validating our profile:

```
packt@neptune:~$ mkdir logs
packt@neptune:~$ ./appackt
Creating ./logs/appackt...
touch: cannot touch './logs/appackt': Permission denied
Writing to ./logs/appackt...
./appackt: line 12: ./logs/appackt: Permission denied
Reading from ./logs/appackt...
cat: ./logs/appackt: No such file or directory
Deleting ./logs/appackt...
rm: cannot remove './logs/appackt': No such file or directory
```

Figure 9.40 – appackt acting outside security boundaries

Let's revert the preceding changes and have the `appackt` script act normally. We are now ready to `enforce` our app by changing its profile mode with the following command:

```
sudo aa-enforce /home/packt/appackt
```

The output is as follows:

```
packt@neptune:~$ sudo aa-enforce /home/packt/appackt
Setting /home/packt/appackt to enforce mode.
```

Figure 9.41 – Changing the appackt profile to enforce mode

We can verify that our application is indeed running in `enforce` mode with the following command:

```
sudo aa-status
```

The relevant output is as follows:

```
packt@neptune:~$ sudo aa-status
apparmor module is loaded.
29 profiles are loaded.
27 profiles are in enforce mode.
   /home/packt/appackt
```

Figure 9.42 – appackt running in enforce mode

If we wanted to make further adjustments to our application and then test it with the related changes, we would have to change the profile mode to `complain` and then reiterate the steps described earlier in this section. The following command sets the application profile to `complain` mode:

```
sudo aa-complain /home/packt/appackt
```

AppArmor profiles are plain text files stored in the `/etc/apparmor.d/` directory. Creating or modifying AppArmor profiles usually involves manually editing the corresponding files or the procedure described in this section using the `aa-genprof` and `aa-logprof` tools.

Next, let's look at how to disable or enable AppArmor application profiles.

Disabling and enabling profiles

Sometimes, we may want to disable a problematic application profile while working on a better version. Here's how we do this.

First, we need to locate the application profile we want to disable (for example, `appackt`). The related file is in the `/etc/apparmor.d/` directory, and it's named according to its full path, with dots (`.`) instead of slashes (`/`). In our case, the file is `/etc/apparmor.d/home.packt.appackt`.

To disable the profile, we must run the following commands:

```
sudo ln -s /etc/apparmor.d/home.packt.appackt /etc/apparmor.d/
disable/
sudo apparmor_parser -R /etc/apparmor.d/home.packt.appackt
```

If we run the `aa-status` command, we won't see our `appackt` profile anymore. The related profile is still present in the filesystem, at `/etc/apparmor.d/disable/home.packt.appackt`:

```
packt@neptune:~$ ls -l /etc/apparmor.d/disable/home.packt.appackt
lrwxrwxrwx 1 root root 34 Dec 17 08:19 /etc/apparmor.d/disable/home.packt.appackt ->
/etc/apparmor.d/home.packt.appackt
```

Figure 9.43 – The disabled appackt profile

In this situation, the `appackt` script is not enforced by any restrictions. To reenable the related security profile, we can run the following commands:

```
sudo rm /etc/apparmor.d/disable/home.packt.appackt
sudo apparmor_parser -r /etc/apparmor.d/home.packt.appackt
```

The `appackt` profile should now show up in the `aa-status` output as running in `complain mode`. We can bring it into `enforce mode` with the following:

```
sudo aa-enforce /home/packt/appackt
```

To disable or enable the profile, we used the `apparmor_parser` command, besides the related filesystem operations. This utility assists with loading (`-r`, `--replace`) or unloading (`-R`, `--remove`) security profiles to and from the kernel.

Deleting AppArmor security profiles is functionally equivalent to disabling them. We can also choose to remove the related file from the filesystem altogether. If we delete a profile without removing it from the kernel first (with `apparmor_parser -R`), we can use the `aa-remove-unknown` command to clean up orphaned entries.

Let's conclude our relatively brief study of AppArmor internals with some final thoughts.

Final considerations

Working with AppArmor is relatively easier than SELinux, especially when it comes to generating security policies or switching back and forth between `permissive mode` and `non-permissive mode`. SELinux can only toggle the permissive context for the entire system, while AppArmor does it at the application level. On the other hand, there might be no choice between the two, as some major Linux distributions either support one or the other. AppArmor is a prodigy of Debian, Ubuntu, and, recently, OpenSUSE, while SELinux runs on RHEL/CentOS. Theoretically, you can always try to port the related kernel modules across distros, but that's not a trivial task.

As a final note, we should reiterate that in the big picture of Linux security, SELinux and AppArmor are **ACMs** that act locally on a system, at the application level. When it comes to securing applications and computer systems from the outside world, firewalls come into play. We'll look at firewalls next.

Working with firewalls

Traditionally, a firewall is a network security device that's placed between two networks. It monitors the network traffic and controls access to these networks. Generally speaking, a firewall protects a local network from unwanted intrusion or attacks from the outside. But a firewall can also block unsolicited locally originated traffic targeting the public internet. Technically, a firewall allows or blocks incoming and outgoing network traffic based on specific security rules.

For example, a firewall can block all but a select set of inbound networking protocols (such as SSH and HTTP/HTTPS). It may also block all but approved hosts within the local network from establishing specific outbound connections, such as allowing outbound SMTP connections that originated exclusively from the local email servers.

The following diagram shows a simple firewall deployment regulating traffic between a local network and the internet:

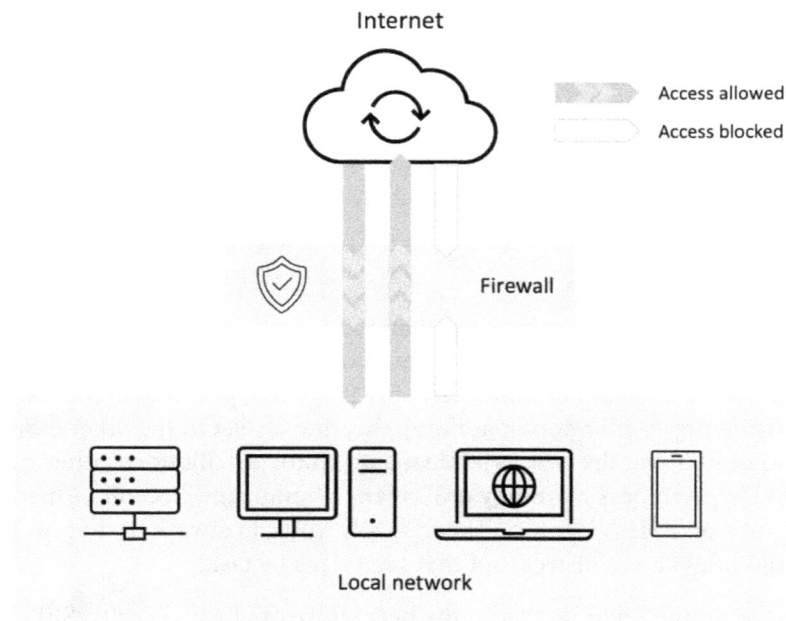

Figure 9.44 – A simple firewall diagram

The outgoing security rules prevent bad actors, such as compromised computers and untrustworthy individuals, directing attacks to the public internet. The resulting protection benefits external networks, but it's ultimately essential for the organization as well. Thwarting hostile actions from the local network avoids them being flagged by **Internet Service Providers (ISPs)** for unruly internet traffic.

Configuring a firewall usually requires a default security policy acting at a global scope, and then configuring specific exceptions to this general rule, based on port numbers (protocols), IP addresses, and other criteria.

In the following sections, we'll explore various firewall implementations and firewall managers. First, let's take a brief look under the hood at how a firewall monitors and controls the network traffic by introducing the Linux firewall chain.

Understanding the firewall chain

At a high level, the TCP/IP stack in the Linux kernel usually performs the following workflows:

- Receives data from an application (process), serializes the data into network packets, and transmits the packets to a network destination, based on the respective IP address and port

- Receives data from the network, deserializes the network packets into application data, and delivers the application data to a process

Ideally, in these workflows, the Linux kernel shouldn't alter the network data in any specific way apart from shaping it due to TCP/IP protocols. However, with distributed and possibly insecure network environments, the data may need further scrutiny. The kernel should provide the necessary hooks to filter and alter the data packets further based on various criteria. This is where firewalls and other network security and intrusion detection tools come into play. They adapt to the kernel's TCP/IP packet filtering interface and perform the required monitoring and control of network packets. The blueprint of the Linux kernel's network packet filtering procedure is also known as the *firewall* or *firewalling chain*:

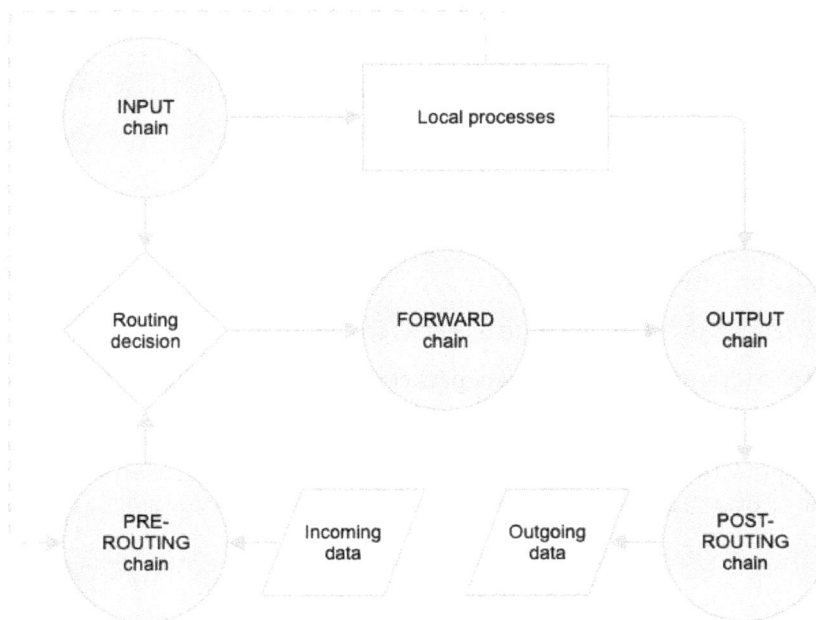

Figure 9.45– The Linux firewall chain

When the *incoming data* enters the firewall packet filtering chain, a *routing decision* is made, depending on the packet's destination. Based on that routing decision, the packet can follow either the **INPUT** chain (for localhost) or the **FORWARD** chain (for a remote host). These chains may alter the incoming data in various ways via the hooks that are implemented by network security tools or firewalls. By default, the kernel won't change the packets traversing the chains.

The **INPUT** chain ultimately feeds the packets into the **local application process** consuming the data. These local applications are usually user space processes, such as network clients (for example, web browsers, SSH, and email clients) or network servers (for example, web and email servers). They may also include kernel space processes, such as the kernel's **Network File System** (**NFS**).

Both the **FORWARD** chain and the **local processes** route the data packets into the **OUTPUT** chain before placing them on the network.

Any of the chains can filter packets based on specific criteria, such as the following:

- The source or destination IP address
- The source or destination port
- The network interface involved in the data transaction

Each chain has a set of security rules that are matched against the input packet. If a rule matches, the kernel routes the data packet to the *target* specified by the rule. Some predefined targets include the following:

- **ACCEPT**: Accepts the data packet for further processing
- **REJECT**: Rejects the data packet
- **DROP**: Ignores the data packet
- **QUEUE**: Passes the data packet to a user space process
- **RETURN**: Stops processing the data packet and passes the data back to the previous chain

For a full list of predefined targets, please refer to the `iptables-extensions` system reference (man `iptables-extensions`).

In the following sections, we'll explore some of the most common network security frameworks and tools based on the kernel's networking stack and firewall chain. We'll start with `netfilter` – the Linux kernel's packet filtering system. Next, we'll look at `iptables` – the traditional interface for configuring `netfilter`. `iptables` is a highly configurable and flexible firewall solution. Then, we'll briefly cover `nftables`, a tool that implements most of the complex functionality of `iptables` wraps it into a relatively easy-to-use command-line interface. Finally, we'll take a step away from the kernel's immediate proximity of packet filtering frameworks and look at *firewall managers* – `firewalld` (RHEL/CentOS) and `ufw` (Debian/Ubuntu) – two user-friendly frontends for configuring Linux firewalls on major Linux distros.

Let's start our journey with `netfilter`.

Introducing netfilter

`netfilter` is a packet filtering framework in the Linux kernel that provides highly customizable handlers (or hooks) to control networking-related operations. These operations include the following:

- Accepting or rejecting packets
- Packet routing and forwarding
- **Network address and port translation (NAT/NAPT)**

Applications that implement the `netfilter` framework use a set of callback functions built around hooks registered with kernel modules that manipulate the networking stack. These callback functions are further mapped to security rules and profiles, which control the behavior of every packet traversing the networking chain.

Firewall applications are first-class citizens of `netfilter` framework implementations. Consequently, a good understanding of the `netfilter` hooks will help Linux power users and administrators create reliable firewall rules and policies.

We'll have a brief look at these `netfilter` hooks next.

netfilter hooks

As packets traverse the various chains in the networking stack, `netfilter` triggers events for the kernel modules that are registered with the corresponding hooks. These events result in notifications in the module or packet filtering application (for example, the firewall) implementing the hooks. Next, the application takes control of the packet based on specific rules.

There are five `netfilter` hooks available for packet filtering applications. Each corresponds to a networking chain, as illustrated in *Figure 9.44*:

- `NF_IP_PRE_ROUTING`: Triggered by incoming traffic upon entering the network stack and before any routing decisions are made about where to send the packet
- `NF_IP_LOCAL_IN`: Triggered after routing an incoming packet when the packet has a localhost destination
- `NF_IP_FORWARD`: Triggered after routing an incoming packet when the packet has a remote host destination
- `NF_IP_LOCAL_OUT`: Triggered by locally initiated outbound traffic entering the network stack
- `NF_IP_POST_ROUTING`: Triggered by outgoing or forwarded traffic, immediately after routing it and just before it exits the network stack

Kernel modules or applications registered with `netfilter` hooks must provide a priority number to determine the order the modules are called in when the hook is triggered. This mechanism allows us to deterministically order multiple modules (or multiple instances of the same module) that have been registered with a specific hook. When a registered module is done processing a packet, it provides a decision to the `netfilter` framework about what should be done with the packet.

The `netfilter` framework's design and implementation is a community-driven collaborative project as part of the **Free and Open-Source Software** (**FOSS**) movement. For a good starting point to the `netfilter` project, you may refer to `http://www.netfilter.org/`.

One of the most well-known implementations of `netfilter` is `iptables` – a widely used firewall management tool that shares a direct interface with the `netfilter` packet filtering framework. A practical examination of `iptables` would further reveal the functional aspects of `netfilter`. Let's explore `iptables` next.

Working with iptables

`iptables` is a relatively low-level Linux firewall solution and command-line utility that uses `netfilter` chains to control network traffic. `iptables` operates with *rules* associated with *chains*. A rule defines the criteria for matching the packets traversing a specific chain. `iptables` uses *tables* to organize rules based on criteria or decision type. `iptables` defines the following tables:

- `filter`: The default table, which is used when we're deciding if packets should be allowed to traverse specific chains (`INPUT`, `FORWARD`, `OUTPUT`).

- `nat`: Used with packets that require a source or destination address/port translation. The table operates on the following chains: PREROUTING, INPUT, OUTPUT, and POSTROUTING.

- `mangle`: Used with specialized packet alterations involving IP headers (such as MSS = Maximum Segment Size or TTL = Time to Live). The table supports the following chains: PREROUTING, INPUT, FORWARD, OUTPUT, and POSTROUTING.

- `raw`: Used when we're disabling connection tracking (NOTRACK) on specific packets, mainly for stateless processing and performance optimization purposes. The table relates to the PREROUTING and OUTPUT chains.

- `security`: Used for **MAC** when packets are subject to SELinux policy constraints. The table interacts with the INPUT, FORWARD, and OUTPUT chains.

The following diagram summarizes the tables with the corresponding chains supported in `iptables`:

Tables ↓ / Chains →	PREROUTING	INPUT	FORWARD	OUTPUT	POSTROUTING
filter		☑	☑	☑	
nat	☑	☑		☑	☑
mangle	☑	☑	☑	☑	☑
raw	☑			☑	
security		☑	☑	☑	

Figure 9.46 – Tables and chains in iptables

The chain traversal order of the packets in the kernel's networking stack is as follows:

- Incoming packets with localhost destination: PREROUTING | INPUT

- Incoming packets with remote host destination: PREROUTING | FORWARD | POSTROUTING

- Locally generated packets (by application processes): OUTPUT | POSTROUTING

Now that we're familiar with some introductory concepts, we can tackle a few practical examples to understand how `iptables` works.

The following examples use an RHEL/CentOS 8 system, but they should work on every major Linux distribution. Please note that starting with RHEL/CentOS 7, the default firewall management application is `firewalld` (discussed later in this chapter). If you want to use `iptables`, first, you need to disable `firewalld`:

```
sudo systemctl stop firewalld
sudo systemctl disable firewalld
sudo systemctl mask firewalld
```

Next, install the `iptables-services` package (on CentOS):

```
sudo yum install iptables-services
```

(On Ubuntu, you must install `iptables` with `sudo apt-get install iptables`).

Now, let's start configuring `iptables`.

Configuring iptables

The `iptables` command requires superuser privileges. First, let's check the current `iptables` configuration. The general syntax for retrieving the *rules* in a *chain* for a specific *table* is as follows:

```
sudo iptables -L [CHAIN] [-t TABLE]
```

The `-L` (`--list`) option lists the rules in a *chain*. The `-t` (`--table`) option specifies a *table*. The CHAIN and TABLE parameters are optional. If the CHAIN option is omitted, *all* chains and their related rules are considered within a table. When no TABLE option is specified, the `filter` table is assumed. Thus, the following command lists all the chains and rules for the `filter` table:

```
sudo iptables -L
```

On a system with a default firewall configuration, the output is as follows:

```
[packt@jupiter ~]$ sudo iptables -L
Chain INPUT (policy ACCEPT)
target     prot opt source               destination

Chain FORWARD (policy ACCEPT)
target     prot opt source               destination

Chain OUTPUT (policy ACCEPT)
target     prot opt source               destination
```

Figure 9.47 – Listing the current configuration in iptables

We can be more specific, for example, by listing all the `INPUT` rules for the `nat` table with the following command:

```
sudo iptables -L INPUT -t nat
```

The `-t` (`--table`) option parameter is only required when `iptables` operations target something other than the default `filter` table.

> **Important note**
> Unless the `-t` (`--table`) option parameter is specified, `iptables` assumes the `filter` table by default.

When you're designing firewall rules from a clean slate, the following steps are generally recommended:

1. Flush any remnants in the current firewall configuration.
2. Set up a default firewall policy.
3. Create firewall rules, making sure the more specific (or restrictive) rules are placed first.
4. Save the configuration.

Let's briefly look at each of the preceding steps by creating a sample firewall configuration using the `filter` table.

Step 1 – Flushing the existing configuration

The following commands flush the rules from the `filter` table's chains (`INPUT`, `FORWARD`, and `OUTPUT`):

```
sudo iptables -F INPUT
sudo iptables -F FORWARD
sudo iptables -F OUTPUT
```

The preceding commands yield no output unless there is an error or you invoke the `iptables` command with the `-v` (`--verbose`) option; for example:

```
sudo iptables -v -F INPUT
```

The output is as follows:

```
[packt@jupiter ~]$ sudo iptables -v -F INPUT
Flushing chain `INPUT'
```

Figure 9.48 – Flushing the INPUT chain in iptables

Next, we'll set up the firewall's default policy.

Step 2 – Setting up a default firewall policy

By default, iptables allows all packets to pass through the networking (firewall) chain. A secure firewall configuration should use DROP as the default target for the relevant chains:

```
sudo iptables -P INPUT DROP
sudo iptables -P FORWARD DROP
sudo iptables -P OUTPUT DROP
```

The -P (--policy) option parameter sets the policy for a specific chain (such as INPUT) to the given target (for example, DROP). The DROP target makes the system gracefully ignore all packets.

At this point, if we were to save our firewall configuration, the system won't be accepting any incoming or outgoing packets. So, we should be careful not to inadvertently drop our access to the system if we used SSH or don't have direct console access.

Next, we'll set up the firewall rules.

Step 3 – Creating firewall rules

Let's create some example firewall rules, such as accepting SSH, DNS, and HTTPS connections.

The following commands enable SSH access from a local network (192.168.0.0/24):

```
sudo iptables -A INPUT -p tcp --dport 22 -m state \
 --state NEW,ESTABLISHED -s 192.168.0.0/24 -j ACCEPT
sudo iptables -A INPUT -p tcp --sport 22 -m state \
 --state ESTABLISHED -s 192.168.0.0/24 -j ACCEPT
```

Let's explain the parameters that were used in the previous code block:

- -A INPUT: Specifies the chain (for example, INPUT) to append the rule to
- -p tcp: The networking protocol (for example, tcp or udp) transporting the packets

- `--dport 22`: The destination port of the packets
- `--sport 22`: The source port of the packets
- `-m state`: The packet property we want to match (for example, `state`)
- `--state NEW,ESTABLISHED`: The state(s) of the packet to match
- `-s 192.168.0.0/24`: The source IP address/mask originating the packets
- `-j ACCEPT`: The target or *what to do* with the packets (such as `ACCEPT`, `DROP`, `REJECT`, and so on)

We used two commands to enable SSH access. The first allows incoming SSH traffic (`--dport 22`) for new and existing connections (`-m state --state NEW,ESTABLISHED`). The second command enables SSH response traffic (`--sport 22`) for existing connections (`-m state –state ESTABLISHED`).

Similarly, the following commands enable HTTPS traffic:

```
sudo iptables -A INPUT -p tcp --dport 443 -m state \
  --state NEW,ESTABLISHED -j ACCEPT
sudo iptables -A INPUT -p tcp --sport 443 -m state \
  --state ESTABLISHED,RELATED -j ACCEPT
```

To enable DNS traffic, we need to use the following commands:

```
sudo iptables -A INPUT -p udp --dport 53 -j ACCEPT
sudo iptables -A INPUT -p udp --sport 53 -j ACCEPT
```

For more information on the `iptables` option parameters, please refer to the following system reference manuals:

- **iptables** (`man iptables`)
- **iptables-extensions** (`man iptables-extensions`).

Now, we're ready to save the `iptables` configuration.

Step 4 – Saving the configuration

To save the current `iptables` configuration, we must run the following command:

```
sudo service iptables save
```

The output is as follows:

```
[packt@jupiter ~]$ sudo service iptables save
[sudo] password for packt:
iptables: Saving firewall rules to /etc/sysconfig/iptables:[  OK  ]
```

Figure 9.49 – Saving the iptables configuration

We can also dump the current configuration to a file (such as `iptables.config`) for later use with the following command:

```
sudo iptables-save -f iptables.config
```

The `-f` (`--file`) option parameter specifies the file to save (backup) the `iptables` configuration in. We can restore the saved `iptables` configuration later with the following command:

```
sudo iptables-restore ./iptables.config
```

Here, we can specify an arbitrary path to our `iptables` backup configuration file.

Exploring more complex rules and topics with `iptables` is beyond the scope of this chapter. The examples we've presented so far, accompanied by the theoretical introduction of `iptables`, should be a good start for everyone to explore more advanced configurations.

On the other hand, the use of `iptables` is generally discouraged, especially with the newly emerging firewall management tools and frameworks that have been shipped with the latest Linux distros, such as `nftables`, `firewalld`, and `ufw`. It is also somewhat accepted that `iptables` has performance and scalability problems.

Next, we'll look at `nftables`, a relatively new framework that was designed and developed by the *Netfilter Project*, built to replace `iptables`.

Introducing nftables

`nftables` is a successor of `iptables`. `nftables` is a firewall management framework that supports packet filtering, **Network Address Translation** (**NAT**), and various packet shaping operations. `nftables` offers notable improvements in terms of features, convenience, and performance over previous packet filtering tools, such as the following:

- Lookup tables instead of linear processing of rules.
- Rules are applied individually instead of them processing a complete ruleset.
- A unified framework for the IPv4 and IPv6 protocols.
- No protocol-specific extensions.

The functional principles behind `nftables` generally follow the design patterns presented in earlier sections about the firewall networking chains; that is, `netfilter` and `iptables`. Just like `iptables`, `nftables` uses tables to store chains. Each chain contains a set of rules for packet filtering actions.

`nftables` is the default packet filtering framework in Debian and RHEL/CentOS 8 Linux distributions, replacing the old `iptables` (and related) tools. The command-line interface for manipulating the `nftables` configuration is `nft`. Yet, some users prefer to use a more user-friendly frontend instead, such as `firewalld`. (`firewalld` recently added backend support for `nftables`.) RHEL/CentOS 8, for example, uses `firewalld` as its default firewall management solution.

In this section, we'll show a few examples of how to use `nftables` and the related command-line utilities to perform simple firewall configuration tasks. For this purpose, we'll take an RHEL/CentOS 8 distribution where we'll disable `firewalld`. Let's have a quick look at the preparatory steps required to run the examples in this section.

Prerequisites for our examples

If you have an RHEL/CentOS 7 system, `nftables` is not installed by default. You can install it with the following command:

```
sudo yum install -y nftables
```

The examples in this section use an RHEL/CentOS 8 distribution. To directly configure `nftables`, we need to disable `firewalld` and potentially `iptables` (if you ran the examples in the related section). The steps for disabling `firewalld` were shown at the beginning of the *Configuring iptables* section.

Also, if you have `iptables` enabled, you need to stop and disable the related service with the following commands:

```
sudo systemctl stop iptables
sudo systemctl disable iptables
```

Next, we need to enable and start `nftables`:

```
sudo systemctl enable nftables
sudo systemctl start nftables
```

We can check the status of `nftables` with the following command:

```
sudo systemctl status nftables
```

A running status of `nftables` should show `active`:

```
[packt@jupiter ~]$ sudo systemctl status nftables
● nftables.service - Netfilter Tables
    Loaded: loaded (/usr/lib/systemd/system/nftables.service; enabled; vendor pre>
    Active: active (exited) since Sun 2020-12-13 18:28:20 PST; 2s ago
      Docs: man:nft(8)
   Process: 22867 ExecStart=/sbin/nft -f /etc/sysconfig/nftables.conf (code=exite>
  Main PID: 22867 (code=exited, status=0/SUCCESS)

Dec 13 18:28:20 jupiter systemd[1]: Starting Netfilter Tables...
Dec 13 18:28:20 jupiter systemd[1]: Started Netfilter Tables.
```

Figure 9.50 – Checking the status of nftables

At this point, we are ready to configure `nftables`. Let's work with a few examples.

Working with nftables

`ntftables` loads its configuration from `/etc/sysconfig/nftables.conf`. We can display the content of the configuration file with the following command:

```
sudo cat /etc/sysconfig/nftables.conf
```

A default `nftables` configuration has no active entries in `nftables.conf`, except for a few comments:

```
[packt@jupiter ~]$ sudo cat /etc/sysconfig/nftables.conf
[sudo] password for packt:
# Uncomment the include statement here to load the default config sample
# in /etc/nftables for nftables service.

#include "/etc/nftables/main.nft"

# To customize, either edit the samples in /etc/nftables, append further
# commands to the end of this file or overwrite it after first service
# start by calling: 'nft list ruleset >/etc/sysconfig/nftables.conf'.
```

Figure 9.51 – The default nftables configuration file

As the comments suggest, to change the `nftables` configuration, we have a few options:

- Directly edit the `nftables.conf` file.
- Manually edit the `/etc/nftables/main.nft` configuration file, then uncomment the related line in `nftables.conf`.
- Use the `nft` command-line utility to edit the rules and then dump the current configuration into `nftables.conf`.

Regardless of the approach taken, we need to reload the updated configuration by restarting the `nftables` service. In this section, we'll use `nft` command-line examples to change the `nftables` configuration. Power users usually write `nft` configuration scripts, but it's best to learn the basic steps first.

The following command displays all the rules in the current configuration:

```
sudo nft list ruleset
```

Your system may already have some default rules set up. You may choose to do a backup of the related configuration (for example, `/etc/sysconfig/nftables.conf` and `/etc/nftables/main.nft`) before proceeding with the next steps.

The following command will flush any preexisting rules:

```
sudo nft flush ruleset
```

At this point, we have an empty configuration. Let's design a simple firewall that accepts SSH, HTTP, and HTTPS traffic, blocking anything else.

Accepting SSH, HTTP, and HTTPS traffic

First, we need to create a *table* and a *chain*. The following command creates a table named `packt_table`:

```
sudo nft add table inet packt_table
```

Next, we'll create a chain called `packt_chain` within `packt_table`:

```
sudo nft add chain inet packt_table packt_chain { type filter
hook input priority 0 \; }
```

Now, we can start adding rules to `packt_chain`. Allow SSH, HTTP, and HTTPS access:

```
sudo nft add rule inet packt_table packt_chain tcp dport {ssh,
http, https} accept
```

Let's also enable ICMP (ping):

```
sudo nft add rule inet packt_table packt_chain ip protocol icmp
accept
```

Finally, we will `reject` everything else:

```
sudo nft add rule inet packt_table packt_chain reject with icmp
type port-unreachable
```

Now, let's have a look at our new configuration:

```
sudo nft list ruleset
```

The output is as follows:

```
[packt@jupiter ~]$ sudo nft list ruleset
table inet packt_table {
        chain packt_chain {
                type filter hook input priority filter; policy accept;
                tcp dport { 22, 80, 443 } accept
                ip protocol icmp accept
                meta nfproto ipv4 reject
        }
}
```

Figure 9.52 – A simple firewall configuration with nftables

The output suggests the following settings for our input chain (`packt_chain`):

- Allow TCP traffic on destination ports 22, 80, and 443 (`tcp dport { 22, 80, 443 } accept`).
- Allow `ping` requests (`ip protocol icmp accept`).
- Reject everything else (`meta nfproto ipv4 reject`).

Next, we will save the current configuration to `/etc/nftables/packt.nft`:

```
sudo nft list ruleset | sudo tee /etc/nftables/packt.nft
```

Finally, we will point the current `nftables` configuration to `/etc/nftables/packt.nft` in the `/etc/sysconfig/nftables.conf` file by adding the following line:

```
include "/etc/nftables/packt.nft"
```

We will use `nano` (or your editor of choice) to make this change:

```
sudo nano /etc/sysconfig/nftables.conf
```

The new `nftables.conf` now contains the reference to our `packt.nft` configuration:

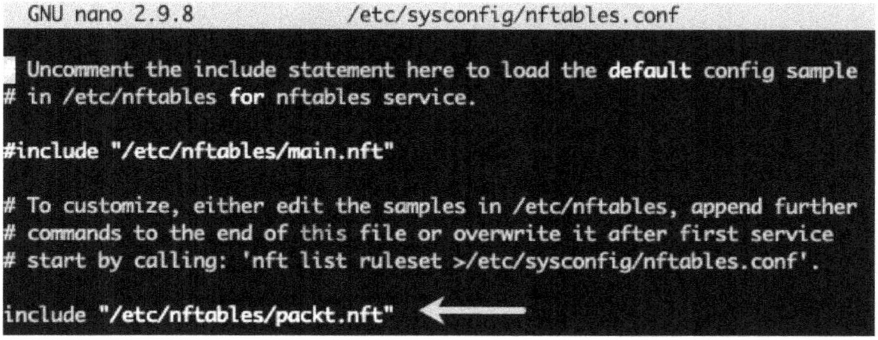

Figure 9.53 – Including the new configuration in nftables

The following command reloads the new `nftables` configuration:

```
sudo systemctl restart nftables
```

After this exercise, you can quickly write a script for configuring `nftables` using the output of the `nft list ruleset` command. As a matter of fact, we just did that with the `/etc/nftables/packt.nft` configuration file.

With that, we will conclude our examination of packet filtering frameworks and the related command-line utilities. They enable power users to have granular control over every functional aspect of the underlying network chains and rules. Yet, some Linux administrators may find the use of such tools overwhelming and turn to relatively simpler firewall management utilities instead.

Next, we'll look at a couple of native Linux firewall management tools that provide a more streamlined and user-friendly command-line interface for configuring and managing firewalls.

Using firewall managers

Firewall managers are command-line utilities with a relatively easy-to-use configuration interface of firewall security rules. Generally, these tools require superuser privileges, and they are a significant asset for Linux system administrators.

In the following sections, we'll present two of the most common firewall managers that are widely used across modern-day Linux distributions:

- `firewalld`: On RHEL/CentOS platforms
- `ufw`: On Ubuntu/Debian

Firewall managers are similar to other network security tools (such as `iptables`, `netfilter`, and `nftables`), with the main difference being that they offer a more streamlined user experience for firewall security. An essential benefit of using a firewall manager is the convenience of not having to restart network daemons when you're operating various security configuration changes.

Let's start with `firewalld`, the default firewall manager for RHEL/CentOS.

Using firewalld

`firewalld` is the default firewall management utility for a variety of Linux distributions, including the following:

- RHEL/CentOS 7 (and newer)
- OpenSUSE 15 (and newer)
- Fedora 18 (and newer)

On CentOS, if `firewalld` is not present, we can install it with the following command:

```
sudo yum install -y firewalld
```

We may also have to enable the `firewalld` daemon at startup with the following command:

```
sudo systemctl enable firewalld
```

Before proceeding, let's make sure `firewalld` is enabled:

```
systemctl status firewalld
```

The status should yield `active (running)`, as shown in the following screenshot:

```
[packt@jupiter ~]$ systemctl status firewalld
● firewalld.service - firewalld - dynamic firewall daemon
   Loaded: loaded (/usr/lib/systemd/system/firewalld.service; enabled; vendor p
   Active: active (running) since Mon 2020-12-07 18:10:45 PST; 1 day 19h ago
     Docs: man:firewalld(1)
 Main PID: 1054 (firewalld)
    Tasks: 2 (limit: 23810)
   Memory: 32.0M
   CGroup: /system.slice/firewalld.service
           └─1054 /usr/libexec/platform-python -s /usr/sbin/firewalld --nofork
```

Figure 9.54 – Making sure firewalld is active

`firewalld` has a set of command-line utilities for different tasks:

- `firewall-cmd`: The primary command-line tool of `firewalld`
- `firewall-offline-cmd`: Used for configuring `firewalld` while it's offline (not running)
- `firewall-config`: A graphical user interface tool for configuring `firewalld`
- `firewall-applet`: A system-tray app for providing essential information about `firewalld` (such as running status, connections, and so on)

In this section, we will look at a few practical examples of using the `firewall-cmd` utility. For any of the other utilities, you may refer to the related system reference manual (such as `man firewall-config`) for more information.

`firewalld` (and `firewalld-cmd`, for that matter) operates with a few key concepts related to monitoring and controlling network packets: *zones*, *rules*, and *targets*.

Zones

Zones are the top organizational units of the `firewalld` configuration. A network packet monitored by `firewalld` belongs to a zone if it matches the network interface or IP address/netmask source associated with the zone. The following command lists the names of the predefined zones:

```
sudo firewall-cmd --get-zones
```

The command yields the following output:

```
[packt@jupiter ~]$ sudo firewall-cmd --get-zones
block dmz drop external home internal libvirt public trusted work
```

Figure 9.55 – The predefined zones in firewalld

For detailed information about all the zones that have currently been configured, we can run the following command:

```
sudo firewall-cmd --list-all-zones
```

Here's an excerpt of the related output:

```
trusted
    target: ACCEPT
    icmp-block-inversion: no
    interfaces:
    sources:
    services:
    ports:
    protocols:
    masquerade: no
    forward-ports:
    source-ports:
    icmp-blocks:
    rich rules:

work
    target: default
    icmp-block-inversion: no
    interfaces:
    sources:
    services: cockpit dhcpv6-client ssh
    ports:
    protocols:
    masquerade: no
    forward-ports:
    source-ports:
    icmp-blocks:
    rich rules:
```

Figure 9.56 – Listing firewalld zones with details

The preceding output illustrates a couple of zones (trusted and work), each with its own attributes, some of which will be explained next. Zones associated with an *interface* and a *source* are known as *active* zones. The following command retrieves the active zones:

```
sudo firewall-cmd --get-active-zones
```

The output, in our case, is as follows:

```
[packt@jupiter ~]$ sudo firewall-cmd --get-active-zones
libvirt
    interfaces: virbr0
public
    interfaces: ens33
```

Figure 9.57 – The firewalld active zones

Interfaces represent the network adapters that are attached to the localhost. Active interfaces are assigned to either the default zone or a user-defined zone. An interface cannot be assigned to multiple zones.

Sources are incoming IP addresses or address ranges, and they can also be assigned to zones. A single source or multiple overlapping IP address ranges cannot be assigned to more than one zone. Doing so would result in undefined behavior, as it would be unclear which rule takes precedence for the related zone.

By default, `firewalld` assigns all network interfaces to the `public` zone without associating any sources with it. Also, by default, `public` is the only active zone and thus the default zone. The following command displays the default zone:

```
sudo firewall-cmd --get-default-zone
```

The default output is as follows:

```
[packt@jupiter ~]$ sudo firewall-cmd --get-default-zone
public
```

Figure 9.58 – Displaying the default zone in firewalld

Specifying a source for a zone is optional. Consequently, for every data packet, there will be a zone with a matching network *interface*. However, there won't necessarily be a zone with a matching *source*. This paradigm would play an essential role in the order in which the matching rules are evaluated. We'll discuss the related topic in the *Rule precedence* section. But first, let's get acquainted with the `firewalld` *rules*.

Rules

The *rules* or `rich` rules that are defined in the `firewalld` configuration represent the configuration settings for controlling the data packets associated with a specific *zone*. Usually, a rule would decide if the packet is accepted or rejected, based on some criteria.

For example, to block the use of ping (ICMP protocol) for the `public` zone, we can add the following `rich` rule:

```
sudo firewall-cmd --zone=public --add-rich-rule='rule protocol
value="icmp" reject'
```

The related output is as follows:

```
[packt@jupiter ~]$ sudo firewall-cmd --zone=public --add-rich-rule='rule protocol
value="icmp" reject'
success
```

Figure 9.59 – Disabling ICMP access with firewalld

We can retrieve the `public` zone information with the following command:

```
sudo firewall-cmd --info-zone=public
```

The `rich` rules attribute reflects the updated configuration:

```
[packt@jupiter ~]$ sudo firewall-cmd --info-zone=public
public (active)
  target: default
  icmp-block-inversion: no
  interfaces: ens33
  sources:
  services: cockpit dhcpv6-client ssh
  ports:
  protocols:
  masquerade: no
  forward-ports:
  source-ports:
  icmp-blocks:
  rich rules:
        rule protocol value="icmp" reject
```

Figure 9.60 – Getting the public zone configuration with firewalld

At this point, our host won't respond anymore to ping (ICMP) requests. We can remove the `rich` rule we just added with the following command:

```
sudo firewall-cmd --zone=public --remove-rich-rule='rule
protocol value="icmp" reject'
```

Alternatively, we can enable ICMP access with the following command:

```
sudo firewall-cmd --zone=public --add-rich-rule='rule protocol
value="icmp" accept'
```

Please note that changes that are made without the `--permanent` option of the `firewall-cmd` utility are transient and won't persist after a system or `firewalld` restart.

When no `rich` rules are defined or matched for a *zone*, `firewalld` uses the zone's *target* to control the packet's behavior. Let's look at *targets* next.

Targets

When a packet matches a specific zone, `firewalld` controls the packet's behavior based on the corresponding zone's `rich` rules. If there are no `rich` rules defined, or none of the `rich` rules match the data packet, the packet's behavior is ultimately determined by the `target` associated with the zone. The possible target values are as follows:

- `ACCEPT`: Accepts the packet
- `REJECT`: Rejects the packet and responds with a reject reply
- `DROP`: Drops the packet without a reply
- `default`: Defers to the default behavior of `firewalld`

Zones, *rules*, and *targets* are the key configuration elements used by `firewalld` when analyzing and handling data packets. Packets are matched using *zones* and then acted upon using either the *rules* or *targets*. Due to the dual nature of *zones* – based on network *interfaces* and IP address/range *sources* – `firewalld` follows a specific order (or precedence) when calculating the matching criteria. We'll look at this next.

Rule precedence

Let's define the terminology first. We'll refer to the zones associated with interfaces as *interface zones*. The zones associates with sources are known as *source zones*. Since zones can have both interfaces and sources assigned to them, a zone can act as either an *interface zone*, a *source zone,* or both.

`firewalld` handles a data packet in the following order:

1. Checks the corresponding *source zone*. There will be, at most, one such zone (since sources can only be associated with a single zone). If there is a match, the packet is handled according to the *rules* or *target* associated with the zone. Otherwise, data packet analysis follows as the next step.

2. Checks the corresponding *interface zone*. Exactly one such zone would (always) exist. If we have a match, the packet is handled according to the zone's *rules* or *target*. Otherwise, the packet validation follows as the next step.

Let's assume the default target of `firewalld` – it accepts ICMP packets and rejects everything else.

The key takeaway from the preceding validation workflow is that **source zones** have precedence over **interface zones**. A typical design pattern for multi-zone `firewalld` configurations defines the following zones:

- **Privileged source zone**: Elevated system access from select IP addresses
- **Restrictive interface zone**: Limited access for everyone else

Let's explore some more potentially useful examples using the `firewall-cmd` utility.

The following command displays the services enabled in the firewall:

```
sudo firewall-cmd --list-services
```

With a default configuration, we get the following output:

```
[packt@jupiter ~]$ sudo firewall-cmd --list-services
cockpit dhcpv6-client ssh
```

Figure 9.61 – Displaying the enabled services in firewalld

The following command enables HTTPS access (port 443):

```
sudo firewall-cmd --zone=public --add-service=https
```

To add a user-defined service or port (for example, 8443), we can run the following command:

```
sudo firewall-cmd --zone=public --add-port=8443/tcp
```

The following command lists the open ports in the firewall:

```
sudo firewall-cmd --list-ports
```

In our case, the output is as follows:

```
[packt@jupiter ~]$ sudo firewall-cmd --list-ports
8443/tcp
```

Figure 9.62 – Displaying the enabled ports in firewalld

Invoking the `firewall-cmd` command without the `--permanent` option results in transient changes that won't persist after a system (or `firewalld`) restart. To reload the previously saved (permanent) configuration of `firewalld`, we can run the following command:

```
sudo firewall-cmd --reload
```

For more information on `firewalld`, refer to the related system reference (man `firewalld`) or `https://www.firewalld.org`.

Using ufw

The **Uncomplicated Firewall (ufw)** is the default firewall manager in Ubuntu. `ufw` provides a relatively simple management framework for `iptables` and `netfilter` and an easy-to-use command-line interface for manipulating the firewall.

Let's look at a few examples of using `ufw`. Please note that the `ufw` command-line utility needs superuser privileges. The following command reports the status of `ufw`:

```
sudo ufw status
```

By default, `ufw` is `inactive` (disabled):

```
packt@neptune:~$ sudo ufw status
Status: inactive
```

Figure 9.63 – Displaying the current status of ufw

We can enable `ufw` with the following command:

```
sudo ufw enable
```

Always be careful when you enable the firewall or perform any changes that may affect your access to the system. By default, when enabled, `ufw` will block all incoming access except ping (ICMP) requests. If you're logged in with SSH, you may get the following prompt while trying to enable `ufw`:

```
packt@neptune:~$ sudo ufw enable
Command may disrupt existing ssh connections. Proceed with operation (y|n)?
```

Figure 9.64 – Enabling ufw could disrupt existing connections

To play it safe, you may want to abort the preceding operation by pressing n (No) and enabling SSH access in the firewall:

```
sudo ufw allow ssh
```

If SSH access is already enabled, the output suggests that the related security rule will not be added:

```
packt@neptune:~$ sudo ufw allow ssh
Skipping adding existing rule
Skipping adding existing rule (v6)
```

Figure 9.65 – Attempting to add an existing rule to ufw

At this point, you can safely enable ufw without fearing that your current or existing SSH connections will be dropped. Upon enabling ufw, we get the following output:

```
packt@neptune:~$ sudo ufw enable
Command may disrupt existing ssh connections. Proceed with operation (y|n)? y
Firewall is active and enabled on system startup
```

Figure 9.66 – Enabling ufw

To check on the detailed status of the firewall, you can run the following command:

```
sudo ufw status verbose
```

The following output suggests that SSH (22/tcp) and HTTP/HTTPS (80,443/tcp) access are enabled:

```
packt@neptune:~$ sudo ufw status verbose
Status: active
Logging: on (low)
Default: deny (incoming), allow (outgoing), disabled (routed)
New profiles: skip

To                            Action      From
--                            ------      ----
80,443/tcp (Nginx Full)       ALLOW IN    Anywhere
22/tcp                        ALLOW IN    Anywhere
80,443/tcp (Nginx Full (v6)) ALLOW IN     Anywhere (v6)
22/tcp (v6)                   ALLOW IN    Anywhere (v6)
```

Figure 9.67 – The detailed status of ufw

As we can see, HTTP/HTTPS access is enabled via the Nginx Full application profile. This rule was automatically added to ufw by the Nginx installation. Please be aware that other client or server applications may also add such rules to ufw. It's always recommended to check your firewall settings to ensure that inadvertent access to the system is not allowed.

We can list the current application security profiles with the following command:

```
sudo ufw app list
```

In our case, the output is as follows:

```
packt@neptune:~$ sudo ufw app list
Available applications:
  CUPS
  Nginx Full
  Nginx HTTP
  Nginx HTTPS
  OpenSSH
```

Figure 9.68 – Listing application security profiles in ufw

To remove a specific service's access (such as HTTP), we can run the following command:

```
sudo ufw deny http
```

The output shows that a new rule has been *added*:

```
packt@neptune:~$ sudo ufw deny http
Rule added
Rule added (v6)
```

Figure 9.69 – Disabling HTTP access in ufw

A subsequent detailed status check would show that access to port 80/tcp has been denied. Yet, the resulting status is somewhat convoluted:

```
packt@neptune:~$ sudo ufw status verbose
Status: active
Logging: on (low)
Default: deny (incoming), allow (outgoing), disabled (routed)
New profiles: skip

To                         Action      From
--                         ------      ----
80,443/tcp (Nginx Full)    ALLOW IN    Anywhere
22/tcp                     ALLOW IN    Anywhere
80/tcp                     DENY IN     Anywhere
80,443/tcp (Nginx Full (v6)) ALLOW IN   Anywhere (v6)
22/tcp (v6)                ALLOW IN    Anywhere (v6)
80/tcp (v6)                DENY IN     Anywhere (v6)
```

Figure 9.70 – Complex rules in ufw

We've only highlighted the IPv4 equivalents of the rules involving HTTP access. In our case, we have two rules controlling HTTP access:

```
80,443/tcp (Nginx Full)  ALLOW IN  Anywhere
80/tcp                    DENY IN   Anywhere
```

By focusing only on HTTP, we can read that the first rule *allows* incoming access to HTTP from anywhere. The second rule *denies* incoming access to HTTP from anywhere. The resulting rule: HTTP is *allowed* from anywhere. Why? Because the *first* rule that matches the criteria wins. Subsequent rules that match the same criteria (that is, *access to 80/tcp from anywhere*) would be discarded.

> **Important note**
> Always put more *specific* (restrictive) rules first. As rules are being added or changed, you may need to delete old entries or rearrange their order to ensure that the rules are appropriately placed and evaluated.

In our case, we need to delete the `Nginx Full` rule. Keep in mind that this rule also enables HTTPS access (`443/tcp`), which we may want to keep. To reinstate the rules in the right order, let's get a `numbered` output of the rule list first:

```
sudo ufw status numbered
```

The output yields the following result:

```
packt@neptune:~$ sudo ufw status numbered
Status: active

        To                 Action      From
        --                 ------      ----
[ 1] Nginx Full            ALLOW IN    Anywhere
[ 2] OpenSSH               ALLOW IN    Anywhere
[ 3] 80/tcp                DENY IN     Anywhere
[ 4] Nginx Full (v6)       ALLOW IN    Anywhere (v6)
[ 5] OpenSSH (v6)          ALLOW IN    Anywhere (v6)
[ 6] 80/tcp (v6)           DENY IN     Anywhere (v6)
```

Figure 9.71 – Numbered list of rules in ufw

The order of the rules is suggested by sequence numbers. We will remove the `Nginx Full` rule next, using the corresponding rule ID (1):

```
sudo ufw delete 1
```

We will get a prompt to approve this operation:

```
packt@neptune:~$ sudo ufw delete 1
Deleting:
 allow 'Nginx Full'
Proceed with operation (y|n)? y
Rule deleted
```

Figure 9.72 – Deleting a rule in ufw

The firewall's status is now as follows:

```
packt@neptune:~$ sudo ufw status numbered
Status: active

     To                        Action      From
     --                        -------     ----
[ 1] OpenSSH                   ALLOW IN    Anywhere
[ 2] 80/tcp                    DENY IN     Anywhere
[ 3] Nginx Full (v6)           ALLOW IN    Anywhere (v6)
[ 4] OpenSSH (v6)              ALLOW IN    Anywhere (v6)
[ 5] 80/tcp (v6)               DENY IN     Anywhere (v6)
```

Figure 9.73 – The firewall's status after removing the Nginx Full application profile in ufw

Similarly, we delete the corresponding IPv6 profile, Nginx Full (v6), with the corresponding ID (3). Please be aware that the rule list has been reindexed upon the previous ufw delete operation:

```
sudo ufw delete 3
```

Now, it's safe to re-add the Nginx HTTPS profile to *only* enable HTTPS access (443/tcp):

```
sudo ufw allow 'Nginx HTTPS'
```

The final status now yields the following output:

```
packt@neptune:~$ sudo ufw status
Status: active

To                    Action      From
--                    ------      ----
OpenSSH               ALLOW       Anywhere
80/tcp                DENY        Anywhere
Nginx HTTPS           ALLOW       Anywhere
OpenSSH (v6)          ALLOW       Anywhere (v6)
80/tcp (v6)           DENY        Anywhere (v6)
Nginx HTTPS (v6)      ALLOW       Anywhere (v6)
```

Figure 9.74 – More specific rules should go first in ufw

As we can see, the more specific (restrictive) rule (80/tcp DENY) goes first (highlighted only for IPv4). We could have even allowed the Nginx Full profile, which would have enabled HTTP access. Still, the corresponding rule (80/tcp ALLOW) would be placed after the more restrictive counterpart and thus discarded.

Alternatively, we could use the insert option to add a specific rule at a given position. For example, the following command places the 80/tcp DENY rule in the second position (as shown in the previous screenshot):

```
sudo ufw insert 2 deny http
```

Let's look at a few more examples of using ufw. The following command enables SSH access (port 22) for all protocols (any) from a specific source address range (192.168.0.0/24):

```
sudo ufw allow from 192.168.0.0/24 to any port 22
```

The following command enables ufw logging:

```
sudo ufw logging on
```

The corresponding log traces are usually in /var/log/syslog:

```
grep -i ufw /var/log/syslog
```

The following log trace indicates a failed attempt (UFW BLOCK) from a source address (SRC=172.16.191.1) to our destination address (DST=172.16.191.4), targeting the HTTP service on port 80 (DPT=80), using the TCP protocol (PROTO=TCP):

```
packt@neptune:~$ grep -i ufw /var/log/syslog
Dec 11 10:34:55 neptune kernel: [114517.668532] [UFW BLOCK] IN=ens33 OUT= MAC=00:
0c:29:27:15:24:16:9d:99:d7:b3:64:08:00 SRC=172.16.191.1 DST=172.16.191.4 LEN=60 T
OS=0x10 PREC=0x00 TTL=64 ID=0 PROTO=TCP SPT=61297 DPT=80 WINDOW=65534 RES=0x00 AC
K PSH FIN URGP=0
```

Figure 9.75 – Analyzing ufw logs

To disable ufw logging, run the following command:

```
sudo ufw logging off
```

The following command reverts ufw to the system's defaults:

```
sudo ufw reset
```

The preceding command results in removing all the rules and disabling ufw.

For more information about ufw, you may wish to explore the *UFW Community Help* at https://help.ubuntu.com/community/UFW or the related system reference (man ufw).

The use of firewall management tools such as ufw and firewalld may have more appeal to some Linux administrators, compared with lower-level packet filtering utilities (for example, netfilter, iptables, and nftables). One of the arguments for choosing one tool over the other, besides platform considerations, is related to scripting and automation capabilities. Some power users may consider the nft command-line utility the tool of choice for designing their firewall rules, due to the granular control provided by nftables. Other users may be inclined to use iptables, especially on older legacy platforms. In the end, it's a matter of choice or preference, as all of these tools are capable of configuring and managing a firewall to roughly the same extent.

Let's wrap up our chapter with some final considerations.

Summary

The relatively vast content of this chapter may appear overwhelming. A key takeaway should be the focus on the *frameworks (modules)*. If we're discussing firewalls, we should look at packet filtering frameworks such as iptables, netfilter, and nftables. For access control, we have security modules such as SELinux and AppArmor. We covered some of the pros and cons of each. The pivoting choice, possibly deciding the Linux distro, is between AppArmor and SELinux. One is perhaps swifter than the other, with the related administration effort hanging in the balance. For example, choosing AppArmor would narrow down the major Linux distributions to Ubuntu, Debian, and OpenSUSE. The distro choice, in turn, would further dictate the available firewall management solutions, and so on.

Mastering the application security frameworks and firewall management tools will help you keep your systems safe with minimal effort. As with any typical Linux system administration task, there are many ways of securing your system. We hope that you will build upon the exploratory knowledge and tools presented in this chapter to make a balanced decision regarding keeping your systems secure.

The next chapter will add a further notch to the safety and protection of your system by introducing disaster recovery, diagnostics, and troubleshooting practices.

Questions

Here's a brief quiz about some of the essential concepts that were covered in this chapter:

1. Enumerate at least a couple of ACMs that are used in Linux.

2. Enumerate the fields of the SELinux security context.

3. What is a *domain* in SELinux?

4. Can you think of a significant difference between SELinux and AppArmor in terms of enforcing security policies?

5. What is the AppArmor command-line utility for retrieving the current application profiles?

6. How do we toggle an AppArmor application profile between the `enforce` and `complain` modes?

7. How many chains can you think of in the Linux kernel networking stack?

8. What is the default firewall management solution in RHEL/CentOS 8? How about Ubuntu?

9. Can you think of a best practice for designing firewall rules?

10. If you had to pick a packet filtering framework, which one would you choose? Why?

Further reading

Please refer to the following for more information about the topics that were covered in this chapter:

*   *Mastering Linux Security and Hardening – Second Edition, Donald A. Tevault, Packt Publishing*

*   *Practical Linux Security Cookbook – Second Edition, Tajinder Kalsi, Packt Publishing*

*   *Practical Linux Security (video), Tajinder Kalsi, Packt Publishing*

*   *Linux Firewalls: Enhancing Security with nftables and Beyond – 4th Edition, Steve Suehring, Addison-Wesley Professional*

10
Disaster Recovery, Diagnostics, and Troubleshooting

In this chapter, you will learn how to do a system backup and restore in a disaster recovery scenario, and how to diagnose and troubleshoot a common array of problems. These are skills that each Linux system administrator needs to have if they wish to be prepared for worst-case scenarios such as power outages, theft, or hardware failure. The world's IT backbone runs on Linux and we need to be prepared for anything that life throws at us.

In this chapter, we're going to cover the following main topics:

- Planning for disaster recovery
- Backing up and restoring the system
- Introducing common Linux diagnostic tools for troubleshooting

Planning for disaster recovery

Managing risks is an important asset for every business or individual. The responsibility of this is tremendous for everyone involved in system administration. For all businesses, managing risks should be part of a wider **risk management strategy**. There are various types of risks in IT, starting from natural hazards directly impacting data centers or business locations, all the way up to cyber security threats. IT's footprint inside a company has exponentially grown in the last decade. Nowadays, there is no activity that does not involve some sort of IT operations being behind it, be it inside small businesses, big corporations, government agencies, or the health or education public sectors, just to give a few examples. Each activity is unique in its own way, so it needs a specific type of assessment. Unfortunately, risk management evolved, mostly with regards to the information security field, into a one-size-fits-all practice, based on checklists that should be implemented by IT management.

A very short introduction to risk management

What is risk management? In a nutshell, risk management is comprised of specific operations that are set to mitigate any possible threat that could impact the overall continuity of a business. The risk management process is crucial for every IT department.

As a basis, a **risk management strategy** should have five distinct steps:

1. **Identifying risk**: Identifying possible threats and vulnerabilities that could impact your ongoing IT operations.

2. **Analyzing risk** : Deciding how big or small it is, based on thorough studies.

3. **Evaluating risk**: This is all about evaluating the impact that it could have on your operations; the immediate action is to respond to the risk based on the impact it has. This calls for real actions being performed at every level of your operations.

4. **Responding to risk**: This will activate your **Disaster Recovery Plans (DRPs)**, combined with strategies for prevention and mitigation.

5. **Monitoring and reviewing risk**: A drastic monitoring and reviewing strategy must be triggered. This will ensure that all the IT teams know how to respond to the risk, and also have the tools and abilities to isolate it and enforce the company's infrastructure.

Risk management frameworks initially arose in the United States due to the **Federal Information Systems Modernization Act (FISMA)** laws, which started in 2002. This was the time when the US **National Institute of Standards and Technology (NIST)** began to create new standards and methods for cyber security assessments among all US government agencies. Therefore, security certifications and compliance are of utmost importance for every Linux distribution provider that sees itself as a worthy competitor in the corporate and governmental space. Similar to the US certification bodies discussed previously, there are other agencies in UK and Russia that develop specific security certifications.

In this respect, all major Linux distributions from Red Hat, SUSE, and Canonical have certifications from NIST, UK's **National Cyber Security Centre (NCSC)**, or Russia's **Federal Service for Technic and Export Control (FSTEC)**.

The risk management framework, according to NIST SP 800-37r2 (the official NIST website: `https://csrc.nist.gov/publications/detail/sp/800-37/rev-2/final`), has seven steps, starting with preparing for the framework's execution, up to monitoring the organization's systems on a daily basis. We will not discuss those steps in detail; instead, we will provide a link at the end of this chapter for NIST's official documentation. In a nutshell, the risk management framework is comprised of several important branches, such as the following:

- **Inventory**: A thorough inventory of all available systems on-premises, and a list of all software solutions.

- **System categorization**: This assesses the impact level for each data type that's used with regard to availability, integrity, and confidentiality.

- **Security control**: This is subject to detailed procedures with regard to hundreds of computer systems' security – a compendium of NIST security controls can be found under SP800-53r4 (the following is a link to the official NIST website: `https://csrc.nist.gov/publications/detail/sp/800-53/rev-4/final`).

- **Risk assessment**: A series of steps that cover threat source identification, vulnerability identification, impact determination, information sharing, risk monitoring, and periodic updates.

- **System security plan**: A report based on every security control and how future actions are assessed, including their implementation and effectiveness.

- **Certification, accreditation, assessment, and authorization**: The process of reviewing security assessments and highlighting security issues and effective resolutions that are detailed in a future plan of action.

- **Plan of action**: This is a tool that's used to track security weaknesses and apply the correct response procedures.

There are many types of risks when it comes to information technology, including hardware failure, software errors, spam and viruses, human error, and natural disasters (fires, floods, earthquakes, hurricanes, and so on). There are also risks of a more criminal nature, including security breaches, whistleblowers, employee dishonesty, corporate espionage, or anything else that could be considered a cybercrime.

Risk assessment is extremely important for any business and should be taken very seriously by IT management. Now that we've tackled some concepts of risk assessment, it is time to explain what it really is. Risk assessment is also known as risk calculation or risk analysis.

Risk calculation

Risk assessment is the action of finding and calculating solutions for possible threats and vulnerabilities. Each of the solutions has certain impacts that affect the business. This impact, at a risk assessment level, has a formula that is useful to know. The following are some basic terms you should know for when you talk about risk impact:

- The **Annual Loss Expectancy (ALE)**, which defines the loss that's expected in 1 year.

- The **Single Loss Expectancy (SLE)**, which represents how much loss is expected at any given time.

- The **Annual Rate of Occurrence (ARO)** is the likeliness of a risky event to occur within 1 year.

- The **risk calculation formula** is SLE x ARO = ALE. There is a monetary value that each element of the formula will provide, so the final result is also expressed as a monetary value.

- The **Mean Time Between Failures (MTBF)** is used to measure the time between anticipated and repairable failures.

- The **Mean Time to Failure (MTTF)** is the average time to an unrepairable failure.

- The **Mean Time to Restore (MTTR)** measures the time needed to repair an affected system.

- The **Recovery Time Objective (RTO)** represents the maximum time that's allocated for downtime.

- The **Recovery Point Objective (RPO)** defines the time when a system needs to restore.

Knowing those terms will help you understand risk assessments so that you can perform a well-documented assessment if or when needed. Risk assessment is based on two major types of actions (or better said, strategies): proactive actions and non-active actions.

The proactive actions are as follows:

- **Risk avoidance**: Based on risk identification and finding a quick solution to avoid its mitigation

- **Risk mitigation**: Based on actions taken to reduce the occurrence of a possible risk

- **Risk transference**: Sharing the risk's possible outcome with an external entity

- **Risk deterrence**: Fighting a possible risk with intimidating actions before it happens

The only non-active action is as follows:

- **Risk acceptance**: Accepting the risk if the other proactive actions could exceed the cost of the harm that's done by the risk.

The strategies described here can be applied to the risk associated with generic, on-premises computing, but nowadays, cloud computing is slowly and surely taking over the world. So, how could these risk strategies apply to cloud computing? In cloud computing, you use the infrastructure of a third party, but with your own data. Even though we will start discussing Linux in the cloud in *Chapter 12*, *Cloud Computing Essentials*, there are some concepts that we will introduce now. As we mentioned earlier, the cloud is taking the infrastructure operations from your on-premises environment to a larger player, such as Amazon, Microsoft, or Google. This could generally be seen as outsourcing. This means that some risks that were a threat when you were running services on-premises are now transferred to third parties.

There are three major cloud paradigms that are now buzz words all over technology media:

- **Software as a Service (SaaS)**: A solution for companies looking to reduce IT costs and rely on software subscriptions. Some examples of SaaS solutions are **Slack**, **Microsoft 365**, **Google Apps**, and **Dropbox**, among others.

- **Platform as a Service (PaaS)**: The way you get apps to your clients using another's infrastructure, runtimes, and dependencies is also known as an application platform. This can be on a public cloud, on a private cloud, or on a hybrid solution. Some examples of PaaS are **Microsoft Azure**, **AWS Lambda**, **Google App Engine**, **SAP Cloud Platform**, **Heroku**, and **Red Hat OpenShift**.

- **Infrastructure as a Service (IaaS)**: These are services that are run online and provide high-level **Application Programming Interfaces (APIs)**. A notable example is **OpenStack**.

Details about all these technologies will be provided in *Chapter 12, Cloud Computing Essentials*, but for this chapter's purpose, we have provided enough information. Major risks regarding cloud computing are concerned with data integration and compatibility. Those are among the risks that you must still overcome, since most of the other risks are no longer your concern they are transferred to the third party managing the infrastructure. Risk calculation can be managed in different ways, depending on the IT scenario a company uses. When you're using the on-premises scenario and you're managing all the components in-house, risk assessments become quite challenging. When you're using the IaaS, PaaS, and SaaS scenarios, risk assessment becomes less challenging as responsibilities are gradually transferred to an external entity.

Risk assessment should always be taken seriously by any individual concerned with the safety of their network and systems or by every IT manager. This is when a disaster recovery plan comes into action. The foundation of a good disaster recovery plan and strategy is having an effective risk assessment.

Designing a disaster recovery plan

A **disaster recovery plan (DRP)** is structured around the steps that should be taken when an incident occurs. In most cases, the disaster recovery plan is part of a business continuity plan. This determines how a company should continue to operate based on a functioning infrastructure.

Every disaster recovery plan needs to start from an **accurate hardware inventory**, followed by a software applications inventory and a separate one for data. The most important part of this is the strategy that's designed to back up all the information that's used.

In terms of the hardware that's used, there must be a clear policy for standardized hardware. This will ensure that faulty hardware can easily be replaced. This kind of policy ensures that everything works and is optimized. Standardized hardware surely has good driver support, and this is very important in the Linux world. Nevertheless, using standardized hardware will tremendously limit practices such as **bring your own device (BYOD)**, since employees only need to use the hardware provided by their employer. Using standardized hardware comes with using specific software applications that have been set up and configured by the company's IT department, with limited input available from the user.

The IT department's responsibility is huge, and they play an important role in designing the IT recovery strategies as part of a disaster recovery plan. Key **tolerances** for downtime and loss of data should be defined based on the minimal acceptable **Recovery Point Objective (RPO)** and **Recovery Time Objective (RTO)**.

Deciding on the **roles** regarding who is responsible for what is another key step for a good disaster recovery plan. This way, the response time for implementing the plan will be dramatically reduced and everyone will know their own responsibilities in case any risks occur. In this case, having a good communication strategy is critical. Enforcing clear procedures for every level of the organizational pyramid will provide clear communication, centralized decisions, and a succession plan for backup personnel.

Disaster recovery plans need to be thoroughly tested at least two times a year to prove their efficiency. Unplanned downtime and outages can negatively impact a business, both on-premises and on any multi-cloud environment. Being prepared for worst-case scenarios is important. Therefore, in the following sections, we will show you some of the best tools and practices for troubleshooting Linux.

Backing up and restoring the system

Disasters can occur at any time. Risk is everywhere. In this respect, backing up your system is of utmost importance and needs to be done regularly. It is always better to practice good prevention than to recover from data loss and learn this the hard way.

Backup and recovery need to be done based on a well-thought-out strategy and need to take the RTO and RPO factors into consideration. RTO should answer basic questions such as how fast to recover lost data and how this will affect the business operations, while RPO should answer questions such as how much of your data you can afford to lose.

There are different types and methods of backup. The following are some examples:

Types of Backup		Methods of Backup	
Full backup	Backing up all the files in the destination target	**Manual backup**	User initiated, not necessary on a reliable schedule
Incremental backup	Backing up all the files that changed from the last backup	**Local automated backup**	Automated backup, on a schedule, more reliable, targeting external drives
Differential backup	Backing up all the files that changed from the previous full backup	**Remote automated backup**	Automated backup, on a schedule, targeting external drives over the network

Figure 10.1 – Backup methods and types

When doing a backup, keep the following rules in mind:

- The **321 rule**, which means that you should always have three copies of your data saved on two separate mediums. One backup should always be kept off-site (on a different geographical location). This is also known as "the rule of three;" it can be adapted to anything, such as 312, 322, 311, or 323.

- **Backup checking** is extremely relevant and is overlooked most of the time. It checks a data's integrity and usefulness.

- **Clear and documented backup strategy and procedures** are beneficial for everyone in the IT team who is using the same practices.

In the next section, we will look at some well-known tools for full Linux system backups, starting with the ones that are integrated inside the operating system, to third-party solutions that are equally suited for both home and enterprise use.

Disk cloning solutions

A good option for a backup is to clone the entire hard drive or several partitions that hold sensitive data. Linux offers a plethora of versatile tools for this job. Among those are the dd command, the ddrescue command, and the **Relax-and-Recover (ReaR)** software tool.

The dd command

One of the most well-known disk backup commands is the dd command. We discussed this previously in *Chapter 6, Working with Disks and Filesystems*. Let's recap how it is used in a backup and restore scenario. The dd command is used to copy block by block, regardless of the filesystem type, from a source to a destination.

Let's learn how to clone an entire disk. We have a system on our network that has a 120 GB SSD drive that we want to back up on a 128 GB USB stick. Using the dd command, we will need to make sure that the source file will fit into the destination. Since our disk sizes are very similar, first, we will run the `fdisk -l` command to make sure that the disk sizes are correct:

```
Disk /dev/sda: 111,81 GiB, 120034123776 bytes, 234441648 sectors
Disk model: ADATA SU650
Units: sectors of 1 * 512 = 512 bytes
Sector size (logical/physical): 512 bytes / 512 bytes
I/O size (minimum/optimal): 512 bytes / 512 bytes
Disklabel type: gpt
Disk identifier: 649F2DAA-BC79-4D52-98DD-E12BA7EFDBC1

Device        Start       End   Sectors   Size Type
/dev/sda1      2048   1050623   1048576    512M EFI System
/dev/sda2   1050624 234440703 233390080  111,3G Linux filesystem
```

Figure 10.2 – Using fdisk -l to verify the source disk's size

In the preceding screenshot, you can see the size of the source disk. The following is a screenshot of the destination disk:

```
Disk /dev/sdb: 114,58 GiB, 123010547712 bytes, 240254976 sectors
Disk model: Ultra
Units: sectors of 1 * 512 = 512 bytes
Sector size (logical/physical): 512 bytes / 512 bytes
I/O size (minimum/optimal): 512 bytes / 512 bytes
Disklabel type: dos
Disk identifier: 0x2eaf4dc0

Device     Boot Start       End   Sectors   Size Id Type
/dev/sdb1       2048 240254975 240252928  114,6G 83 Linux
```

Figure 10.3 – Using fdisk -l to verify the destination disk's size

Now that we know that the sizes are OK and that the source can fit into the destination, we will proceed to cloning the entire disk. We will clone the source disk, /dev/sda, to the destination disk, /dev/sdb:

```
alexandru@asus:~$ sudo dd if=/dev/sda of=/dev/sdb conv=noerror,sync status=progress
63941508608 bytes (64 GB, 60 GiB) copied, 6080 s, 10,5 MB/s_
```

Figure 10.4 – Using dd to clone an entire hard drive

The options shown in the preceding command are as follows:

- if=/dev/sda represents the input file, which in our case is the source hard drive.
- if=/dev/sdb represents the output file, which is the destination USB drive.
- conv=noerror represents the instruction that allows the command to continue ignoring errors.
- sync represents the instruction to fill the input error blocks with zeroes so that the data offset will always be synced.
- status=progress shows statistics about the transfer process.

Please keep in mind that this operation could take a while to finish. On our system, it took 200 minutes to complete. We took the preceding screenshot while the operation was only half complete. In the following section, we will show you how to use ddrescue.

The ddrescue command

The ddrescue command is yet another tool you can use to clone your disk. This tool copies from one device or file to another one, trying to copy only the good and healthy parts for the first time. If your disk is failing, you might want to use ddrescue twice, since the first time it will copy only the good sectors and map the errors to a destination file. The second time, it will copy only the bad sectors, so it is better to add an option for several read attempts just to be sure. On Ubuntu, the ddrescue utility is not installed by default. To install it, use the following apt command:

```
sudo apt install gddrescue
```

We will use ddrescue on the same system we used previously and clone the same drive. The command for this is as follows:

```
sudo ddrescue -n /dev/sda /dev/sdb rescue.map --force
```

The output is as follows:

```
alexandru@asus:~$ sudo ddrescue -n /dev/sda /dev/sdb rescue.map --force
GNU ddrescue 1.23
Press Ctrl-C to interrupt
     ipos:   120034 MB,  non-trimmed:       0 B,   current rate:   21848 kB/s
     opos:   120034 MB,  non-scraped:       0 B,   average rate:   40118 kB/s
non-tried:        0 B,   bad-sector:        0 B,     error rate:       0 B/s
  rescued:   120034 MB,    bad areas:       0,        run time:    49m 51s
pct rescued:  100.00%,   read errors:       0,   remaining time:        n/a
                              time since last successful read:        n/a
Finished
```

Figure 10.5 – Using ddrescue to clone the hard drive

We used the `ddrescue` command with the `--force` option to make sure that everything on the destination will be overwritten. This operation is time-consuming too, so be prepared for a lengthy one. In our case, it took almost 1 hour to finish. Next, we will show you how to use another useful tool: the ReaR utility.

Using Relax-and-Recover (ReaR)

ReaR is a powerful disaster recovery and system migration tool written in Bash. It is used by enterprise-ready distributions such as RHEL and SLES, and can also be installed on Ubuntu. It was designed to be easy to use and set up. It is integrated with the local bootloader, with the `cron` scheduler or monitoring tools such as **Nagios**. For more details on this tool, visit the official website at `http://relax-and-recover.org/about/`.

To install it on Ubuntu, use the following command:

```
sudo apt install rear
```

Once the packages have been installed, you will need to know the location of the main configuration file. It is called `/etc/rear/local.conf`, and all the configuration options should be written inside it. ReaR makes ISO files by default, but it also supports Samba (CIFS), USB, and NFS as backup destinations.

Backing up to a local NFS server using ReaR

As an example, we will show you how to back up to an NFS server. On our network, we already have an NFS server set up on one of our Ubuntu machines (Neptune), meaning that we will be able to use that one for our backup server and the local Asus system as the production machine to be backed up.

First, we must configure the NFS server accordingly. We hope that you still remember how to configure the NFS share (covered in *Chapter 8, Configuring Linux Servers*), but if you don't, here is a short reminder. The configuration file for NFS is /etc/exports and it stores information about the share's location. Before you add any new information about the ReaR backup share's location, add a new directory. When we first set up the NFS server, we used the /home/export/ directory. Inside that directory, we will create a new one for our RearR backups. The command to create the new directory is as follows:

```
sudo mkdir /home/export/rear
```

Now, change the ownership of the directory. If the owner remains root, ReaR will not have permission to write the backup to this location. Change the ownership using the following command:

```
sudo chown -R nobody:nogroup /home/export/rear/
```

Once the directory has been created, open the /etc/exports file with your favorite editor and add a new line for the backup directory. It should look similar to the last one in the following screenshot:

```
  GNU nano 4.8                          /etc/exports
# /etc/exports: the access control list for filesystems which may be exported
#               to NFS clients.  See exports(5).
#
# Example for NFSv2 and NFSv3:
# /srv/homes         hostname1(rw,sync,no_subtree_check) hostname2(ro,sync,no_sub>
#
# Example for NFSv4:
# /srv/nfs4          gss/krb5i(rw,sync,fsid=0,crossmnt,no_subtree_check)
# /srv/nfs4/homes    gss/krb5i(rw,sync,no_subtree_check)
#
/home/export/shares      192.168.0.0/24(rw,sync,no_subtree_check)
/home/export/rear        192.168.0.0/24(rw,sync,no_subtree_check)
```

Figure 10.6 – Adding a new line to the /etc/exports NFS file

Once the new line has been introduced, restart the NFS service and run the `exportfs` command using the `-s` option. The output is shown here:

```
packt@neptune:/home/export/shares$ sudo nano /etc/exports
[sudo] password for packt:
packt@neptune:/home/export/shares$ sudo systemctl restart nfs-kernel-server.serv
ice
packt@neptune:/home/export/shares$ sudo exportfs -s
/home/export/shares  192.168.0.0/24(rw,wdelay,root_squash,no_subtree_check,sec=s
ys,rw,secure,root_squash,no_all_squash)
/home/export/rear  192.168.0.0/24(rw,wdelay,root_squash,no_subtree_check,sec=sys
,rw,secure,root_squash,no_all_squash)
```

Figure 10.7 – Restarting the NFS service and exporting the new shares directory

After setting up the NFS server, go back to the local machine and start editing the ReaR configuration file so that you can use the backup server. Edit the `/etc/rear/local.conf` file and add the lines shown in the following output. Use your own system's IP address, not the one you can see in the following screenshot:

```
  GNU nano 4.8                    /etc/rear/local.conf
# Default is to create Relax-and-Recover rescue media as ISO image
# set OUTPUT to change that
# set BACKUP to activate an automated (backup and) restore of your data
# Possible configuration values can be found in /usr/share/rear/conf/default.co>
#
# This file (local.conf) is intended for manual configuration. For configuration
# through packages and other automated means we recommend creating a new
# file named site.conf next to this file and to leave the local.conf as it is.
# Our packages will never ship with a site.conf.
OUTPUT=ISO
OUTPUT_URL=nfs://192.168.0.244/home/export/rear
BACKUP=NETFS
BACKUP_URL=nfs://192.168.0.244/home/export/rear
```

Figure 10.8 – Editing the /etc/rear/local.conf file

The lines shown here represent the following:

- OUTPUT: The bootable image type, which in our case is ISO

- OUTPUT_URL: The backup target, which can be NFS, CIFS, FTP, RSYNC, or FILE

- BACKUP : The backup method used, which in our case is NETFS, the default ReaR method

- BACKUP_URL: The backup target's location

Now, run the mkbackup command with the -v and -d options:

```
sudo rear -v -d mkbackup
```

The output will be large, so we will not show it to you here. The command will take a significant time to finish. Once it has finished, you can check the NFS directory to view its output. The backup should be in there:

```
packt@neptune:/home/export/rear$ ls -la
total 12
drwxr-xr-x 3 nobody nogroup 4096 dec 13 14:50 .
drwxr-xr-x 4 root    root    4096 dec 13 13:38 ..
drwxr-x--- 2 nobody nogroup 4096 dec 13 15:13 asus
packt@neptune:/home/export/rear$ sudo ls -la asus/
[sudo] password for packt:
total 6224928
drwxr-x--- 2 nobody nogroup       4096 dec 13 15:13 .
drwxr-xr-x 3 nobody nogroup       4096 dec 13 14:50 ..
-rw------- 1 nobody nogroup        202 dec 13 14:52 README
-rw------- 1 nobody nogroup        261 dec 13 14:52 VERSION
-rw------- 1 nobody nogroup   18008493 dec 13 15:13 backup.log
-rw------- 1 nobody nogroup 5944884170 dec 13 15:13 backup.tar.gz
-rw------- 1 nobody nogroup  409520128 dec 13 14:52 rear-asus.iso
-rw------- 1 nobody nogroup    1883329 dec 13 14:52 rear-asus.log
```

Figure 10.9 – Checking the backup on the NFS server

There are several files written on the NFS server. Among those, the one called rear-asus.iso is the actual backup and the one that will be used in case a system restore is needed. There is also a file called backup.tar.gz, which contains all the files from our Asus machine.

> **Important note**
>
> The naming convention of ReaR is as follows. The name will consist of the term `rear-`, followed by the system's `hostname` and the `.iso` extension. Our system's hostname is `asus`, which is why the backup file is called `rear-asus.iso` in our case.

Once the backup has been written on the NFS server, you will be able to restore the system by using a USB disk or DVD with the ISO image that was written on the NFS server.

Backing up to USB using ReaR

There is also the option of directly backing up on the USB disk. Insert a disk into the USB port and format it by using the `rear format /dev/sdb` command. The output is as follows:

```
root@asus:~# rear format -- --efi /dev/sdb
USB device /dev/sdb is not formatted with ext2/3/4 or btrfs filesystem
Type exactly 'Yes' to format /dev/sdb with ext3 filesystem
(default 'No' timeout 300 seconds)
Yes
```

Figure 10.10 – Formatting the USB card with ReaR

Now, we need to modify the `/etc/rear/local.conf` file and adapt it so that it uses the USB as the backup destination. These new lines should look as follows:

```
OUTPUT=USB
# OUTPUT_URL=nfs://192.168.0.244/home/export/rear
BACKUP=NETFS
BACKUP_URL="usb:///dev/disk/by-label/REAR-000"
# BACKUP_URL=nfs://192.168.0.244/home/export/rear
```

Figure 10.11 – Adding new lines inside the /etc/rear/local.conf file

To back up the system on the USB disk, run the following command:

```
rear -v mkbackup
```

The operation will take a considerable amount of time to run, so just be patient. Once it has finished, the ISO and the `tar.gz` files will be on the USB drive.

To recover the system, you will need to boot from the USB drive and select the first option, which says `Recover "hostname"`. Here, the hostname is the hostname of the computer you backed up.

System backup and recovery are two very important tasks that should be indispensable to any Linux system administrator. Knowing how to execute those tasks can save the company, the client's data, time, and money. Minimal downtime and having a quick, effective response should be the most important assets on every **Chief Technology Officer's (CTO's)** table. Backup and recovery strategies should always have a strong foundation in terms of good mitigation practices. In this respect, a strong diagnostics toolset and troubleshooting knowledge will always come in handy for every system administrator. This is why, in the next section, we will show you some of the best diagnostic tools in Linux.

Introducing common Linux diagnostic tools for troubleshooting

The openness of Linux is one of its best assets. This opened the door to an extensive number of solutions that can be used for any task at hand. Hence, many diagnostic tools are available to Linux system administrators. Depending on which part of your system you would like to diagnose, there are several tools available. Troubleshooting is essentially problem solving, based on diagnostics generated by specific tools. To reduce the number of diagnostic tools to cover, we will narrow down the issues to the following categories for this section:

- Boot issues
- General system issues
- Network issues
- Hardware issues

There are specific diagnostic tools for each of these categories. We will start by showing you some of the most widely used ones.

Tools for troubleshooting boot issues

To understand the issues that may affect the boot process, it is important to know how the boot process works. We have not covered this in detail yet, so pay attention to everything that we will tell you.

The boot process

All the major Linux distributions, such as Ubuntu, CentOS, OpenSUSE, Debian, Fedora, and RHEL, use **GRUB2** as their default bootloader and `systemd` as their default *init* system. Until GRUB2 initialization and the `systemd` startup were put in place, the Linux boot process had several more stages.

The boot order is as follows:

1. The **Basic Input Output System (BIOS) Power-On Self-Test (POST)**

2. **Grand Unified Bootloader version 2 (GRUB2)** bootloader initialization

3. GNU/Linux kernel initialization

4. `systemd` init system initialization

BIOS POST is a process specific to hardware initialization and testing, and it is similar for every PC, regardless of whether it is using Linux or Windows. The BIOS is making sure that every hardware component inside the PC is working properly. When the BIOS fails to start, there is usually a hardware problem or incompatibility issues. The BIOS searches for the disk's boot record, such as the **Master Boot Record (MBR)** or **GUID Partition Table (GPT)**, and loads it into memory.

GRUB2 initialization is where Linux starts to kick in. This is the stage when the system loads the kernel into memory. It can choose between several different kernels in case there's more than one operating system available. Once the kernel has been loaded into memory, it takes control over the boot process.

The kernel is a self-extracting archive. Once extracted, it runs into the memory and loads the `init` system, the parent of all the other processes on Linux.

The `init` system, called `systemd`, starts by mounting the filesystems and accessing all the available configuration files.

During the boot process, issues may appear. In the next section, we will tell you what to do if disaster strikes and your bootloader won't start.

Repairing GRUB2

If GRUB2 breaks, you will not be able to access your system. This calls for a GRUB repair. At this stage, a live bootable USB drive will save you. We have an Ubuntu 20.04 live disk, and we will use it for this example. Here are the steps you should follow:

1. Plug it in and boot the system.

2. Open the BIOS, select the bootable disk as the main boot device, and restart it.

3. Select the **Try Ubuntu** option from the window on the screen.

4. Once inside the Ubuntu instance, open a Terminal, and `sudo fdisk -l`, and check your disks and partitions.

5. Select the one that GRUB2 is installed on and use the following command:

```
sudo mount -t ext4 /dev/sda1 /mnt
```

6. Install GRUB2 using the following command:

```
sudo chroot /mnt
grub-install /dev/sda
grub-install -recheck /dev/sda
update-grub
```

7. Unmount the partition using the following commands:

```
exit
sudo unmount /mnt
```

8. Reboot the computer.

Dealing with bootloaders is extremely sensitive. Pay attention to all the details and take care of all the commands you type in. If not, everything could go sideways. In the next section, we will show you some diagnostic tools for general system issues.

Tools for troubleshooting general system issues

System issues can be of different types and complexities. Knowing the tools to deal with them is of utmost importance. In this section, we will cover the default tools provided by the Linux distribution. Basic troubleshooting knowledge is necessary for any Linux system administrator as issues can – and will occur – during regular operations.

What could general system issues mean? Well, basically, these are issues regarding disk space, memory usage, system load, and running processes.

Commands for disk-related issues

Disks, be it HDDs or SSDs, are an important part of the system. They provide the necessary space for your data, files, and software of any type, including the operating system. We will not discuss hardware-related issues as this will be the subject of a future section called *Tools for troubleshooting hardware issues*. Instead, we will cover issues related to disk space. The most common diagnostic tools for this are already installed on any Linux system, and they are represented by the following commands:

- du: A utility that shows disk space utilization for files and directories

- df: A utility that shows the disk usage for directories

Below is an example of using the df utility with the -h option. This shows disk usage in a human readable format, with disk sizes shown using kilobytes, megabytes and gigabytes:

```
packt@neptune:~$ df -h
Filesystem      Size  Used Avail Use% Mounted on
udev            3,8G     0  3,8G   0% /dev
tmpfs           768M  3,4M  765M   1% /run
/dev/nvme0n1p2  468G   17G  429G   4% /
tmpfs           3,8G     0  3,8G   0% /dev/shm
tmpfs           5,0M  4,0K  5,0M   1% /run/lock
tmpfs           3,8G     0  3,8G   0% /sys/fs/cgroup
/dev/loop1       56M   56M     0 100% /snap/core18/1932
/dev/loop3      256M  256M     0 100% /snap/gnome-3-34-1804/36
/dev/loop2      163M  163M     0 100% /snap/gnome-3-28-1804/145
/dev/loop4      218M  218M     0 100% /snap/gnome-3-34-1804/60
/dev/loop5       65M   65M     0 100% /snap/gtk-common-themes/1513
/dev/loop0       56M   56M     0 100% /snap/core18/1885
/dev/loop6       65M   65M     0 100% /snap/gtk-common-themes/1514
/dev/loop7       50M   50M     0 100% /snap/snap-store/467
/dev/nvme0n1p1  511M  7,8M  504M   2% /boot/efi
/dev/loop8       32M   32M     0 100% /snap/snapd/10492
/dev/loop9       52M   52M     0 100% /snap/snap-store/498
/dev/loop10      32M   32M     0 100% /snap/snapd/10238
tmpfs           768M   24K  768M   1% /run/user/125
tmpfs           768M  8,0K  768M   1% /run/user/1000
```

Figure 10.12 – Running the df -h command to view disk space usage

If one of the disks runs out of space, it will be shown in the output. This is not an issue in our case, but the tool is still relevant for finding out which of the available disks is having issues with the free available space.

When a disk is full, or almost full, there are several fixes that can be applied. If you have to delete some of the files, we advise you to delete them from your /home directory. Try not to delete important system files. The following are some ideas for troubleshooting available space issues:

- Delete the unnecessary files using the rm command (optionally using -rf) or the rmdir command.

- Move files to an external drive (or to the cloud) using the rsync command.

- Find which directories use the most space in your /home directory.

The following is an example of using the du utility to find the largest directories inside the /home directory. We are using two pipes to transfer the output of the du command to the sort command and finally to the head command with the option of 5 (because we want to show the five largest directories, not all of them):

```
root@neptune:/home/packt# du -h /home | sort -rh | head -5
6,2G    /home
6,0G    /home/export/rear/asus
6,0G    /home/export/rear
6,0G    /home/export
221M    /home/packt
```

Figure 10.13 – Finding the largest directories in your /home directory

There might be issues with the number of inodes being used, not with the space. You can use the df -i command to see if you've ran out of inodes:

```
packt@neptune:~$ df -i /home
Filesystem        Inodes   IUsed    IFree IUse% Mounted on
/dev/nvme0n1p2 31227904 293448 30934456    1% /
```

Figure 10.14 – Checking if you ran out of inodes

Besides the commands shown here, which are the defaults for every Linux distribution, there are many other open source tools for disk space issues, such as **pydf**, **parted**, **sfdisk**, **iostat**, and the GUI-based **Gparted** application.

Commands for memory usage issues

Memory overload, together with CPU loading and disk usage, are responsible for overall system performance. Checking system load with specific tools is of utmost importance. The default tool for checking RAM statistics in Linux is called `free` and it can be accessed in any major distribution:

```
packt@neptune:~$ free -h
              total        used        free      shared  buff/cache   available
Mem:          7,5Gi       763Mi       289Mi       126Mi       6,5Gi       6,3Gi
Swap:         2,0Gi       6,0Mi       2,0Gi
```

Figure 10.15 – Using the free command in Linux

As shown in the preceding screenshot, using the `free` command (with the `-h` option for human-readable output) shows the following:

- `total`: The total amount of memory
- `used`: The used memory, which is calculated as the total memory minus the buffered, cache, and free memory
- `free`: The free or unused memory
- `shared`: Memory used by `tmpfs`
- `buff/cache`: The memory used by kernel buffers and the page cache
- `available`: The amount of memory available for new applications

This way, you can find specific issues related to higher memory usage. Constantly checking memory usage on servers is important to see if resources are being used efficiently.

Another way to check for memory usage is to use the `top` command, as shown in the following screenshot:

```
top - 17:33:07 up 3 days, 15:57,  1 user,  load average: 0,00, 0,00, 0,00
Tasks: 288 total,   1 running, 287 sleeping,   0 stopped,   0 zombie
%Cpu(s):  0.0 us,  0.0 sy,  0.0 ni,100.0 id,  0.0 wa,  0.0 hi,  0.0 si,  0.0 st
MiB Mem :  7674,1 total,    294,2 free,    757,2 used,   6622,7 buff/cache
MiB Swap:  2048,0 total,   2042,0 free,      6,0 used.   6487,8 avail Mem

  PID USER      PR  NI    VIRT    RES    S memory statistics     MAND
  758 root      20   0  128740  10368   9528 S   0,3   0,1   0:45.30 thermald
  946 mysql     20   0 1775676  88876  17952 S   0,3   1,1   8:47.50 mysqld
19622 root      20   0       0      0      0 S   0,3   0,0   0:13.73 usb-sto+
40934 packt     20   0   16564   4260   3492 R   0,3   0,1   0:00.08 top
    1 root      20   0  171160  13308   8476 S   0,0   0,2   0:12.76 systemd
    2 root      20   0       0      0      0 S   0,0   0,0   0:00.18 kthreadd
    3 root       0 -20       0      0      0 I   0,0   0,0   0:00.00 rcu_gp
    4 root       0 -20       0      0      0 I   0,0   0,0   0:00.00 rcu_par+
    6 root       0 -20       0      0      0 I   0,0   0,0   0:00.00 kworker+
    9 root       0 -20       0      0      0 I   0,0   0,0   0:00.00 mm_perc+
   10 root      20   0       0      0      0 S   0,0   0,0   0:00.61 ksoftir+
   11 root      20   0       0      0      0 I   0,0   0,0   0:57.03 rcu_sch+
   12 root      rt   0       0      0      0 S   0,0   0,0   0:02.04 migrati+
   13 root     -51   0       0      0      0 S   0,0   0,0   0:00.00 idle_in+
   14 root      20   0       0      0      0 S   0,0   0,0   0:00.00 cpuhp/0
   15 root      20   0       0      0      0 S   0,0   0,0   0:00.00 cpuhp/1
   16 root     -51   0       0      0      0 S   0,0   0,0   0:00.00 idle_in+
   17 root      rt   0       0      0      0 S   0,0   0,0   0:02.21 migrati+
```

Figure 10.16 – Using the top command to check memory usage

While using the `top` command, there are several sections available on-screen. The output is dynamic, in the sense that it constantly changes, showing real-time information about the processes running on the system. The memory section shows information about total memory used, as well as free and buffered memory. All the information is shown in megabytes by default so that it's easier to read and understand.

Another command that shows information about memory (and other valuable system information) is `vmstat`:

```
packt@neptune:~$ vmstat
procs -----------memory---------- ---swap-- -----io---- -system-- ------cpu-----
 r  b   swpd   free   buff  cache   si   so    bi    bo   in   cs us sy id wa st
 0  0   6144 291644 433616 6355144    0    0     1     4   12    6  0  0 100  0  0
```

Figure 10.17 – Using vmstat with no options

By default, `vmstat` shows information about processes, memory, swap, disk, and CPU usage. The memory information is shown starting from the second column and contains the following details:

- `swpd`: How much virtual memory is being used
- `free`: How much memory is free
- `buff`: How much memory is being used for buffering
- `cache`: How much memory is being used for caching

The `vmstat` command has several options available. To learn about all the options and what all the columns from the output represent, visit the respective manual pages using the following command:

```
man vmstat
```

The options that can be used with `vmstat` to show different information about memory are `-a` and `-s`. By using `vmstat -a`, the output will show the active and inactive memory:

```
packt@neptune:~$ vmstat -a
procs ----------memory---------- --swap-- -----io---- -system-- ------cpu-----
 r  b   swpd   free  inact active si   so    bi    bo   in   cs us sy id wa st
 1  0   2560 232944 5123284 1210084  0    0     0     4   12    7  0  0 100  0
 0
```

Figure 10.18 – Using vmstat -a to show the active and inactive memory

Using `vmstat -s` will show detailed memory, CPU, and disk statistics. An excerpt of some memory statistics from the output is shown here:

```
packt@neptune:~$ vmstat -s
      7858288 K total memory
       778160 K used memory
      1196848 K active memory
      5123084 K inactive memory
       247664 K free memory
       437184 K buffer memory
      6395280 K swap cache
      2097148 K total swap
         2560 K used swap
      2094588 K free swap
```

Figure 10.19 – Using vmstat with the -s option for memory statistics

All the commands discussed in this section are essential for troubleshooting any memory issues. There might be others you can use, but these are the ones you will find by default on any Linux distribution.

Nevertheless, there is one more that deserves to be mentioned in this section: the `sar` command. This can be installed in Ubuntu through the `sysstat` package. Therefore, install the package using the following command:

```
sudo apt install sysstat
```

Once the package has been installed, to be able to use the `sar` command to show detailed statistics about the system's memory usage, you will need to enable the `sysstat` service. It needs to be active to collect data. By default, the service runs every 10 minutes and saves the logs inside the `/var/log/sysstat/saXX` directory. Every directory is named after the day the service runs on. For example, if we were to run the `sar` command on December 16, the service would look for data inside `/var/log/sysstat/sa16`. We ran the `sar` command on December 16 before starting the service, and the following is the error output:

```
packt@neptune:~$ sar
Cannot open /var/log/sysstat/sa16: No such file or directory
Please check if data collecting is enabled
```

Figure 10.20 – Running the sar command before starting the sysstat service

Thus, to enable data collection, first, we will start and enable the `sysstat` service:

```
packt@neptune:~$ sudo systemctl start sysstat
packt@neptune:~$ sudo systemctl enable sysstat
Synchronizing state of sysstat.service with SysV service script with /lib/systemd/
systemd-sysv-install.
Executing: /lib/systemd/systemd-sysv-install enable sysstat
```

Figure 10.21 – Starting and enabling the sysstat service

> **Important note**
> In the event of a system reboot, the service might not restart by default,
> even though the preceding commands were executed. To overcome this, on
> Ubuntu, you should edit the `/etc/default/sysstat` file and change the
> ENABLED status from false to true.

The service's name is **system activity data collector (sadc)** and it uses the `sysstat` name for the package and service.

With the `sar` command, you can generate different reports in real time. For example, if we want to generate a memory report five times every 2 seconds, we will use the `-r` option, as shown in the following screenshot:

```
packt@neptune:~$ sudo sar -r 2 5
[sudo] password for packt:
Linux 5.4.0-58-generic (neptune)        16.12.2020      _x86_64_        (8 CPU)

22:08:41    kbmemfree   kbavail kbmemused  %memused kbbuffers   kbcached   kbcommit  %commit  kbactive   kbinact  kbdirty
22:08:43       238160   6603272    580400      7,39    438496    5562120    4168672    41,87   1116436   5252800      264
22:08:45       238160   6603272    580400      7,39    438496    5562120    4168672    41,87   1116436   5252800      264
22:08:47       238160   6603272    580392      7,39    438504    5562120    4168672    41,87   1116488   5252800        0
22:08:49       238160   6603272    580392      7,39    438504    5562120    4168672    41,87   1116488   5252800        0
22:08:51       238160   6603272    580392      7,39    438504    5562120    4168672    41,87   1116488   5252800        0
Average:       238160   6603272    580395      7,39    438501    5562120    4168672    41,87   1116467   5252800      106
```

Figure 10.22 – Generating memory statistics in real time with the `sar` command

The output will show one line every 2 seconds, five times in a row, and an average line at the end. It is a powerful tool that can be used for more than just memory statistics. There are options for CPU and disk statistics as well.

Overall, in this section, we covered the most important tools to use for troubleshooting memory issues. In the next section, we will cover tools to use for general system load issues.

Commands for system load issues

Similar to what we covered in the previous sections, in this section, we will discuss system load issues. Some of the tools that were used for other types of issues can be used for system load issues too. For example, the `top` command is one of the most widely used when we're trying to determine the sluggishness of a system. All the other tools, such as `vmstat` and `sar`, can also be used for CPU and system load troubleshooting.

A basic command for troubleshooting system load is `uptime`. Generally the uptime's output shows three values at the end. Those values represent the load averages for 1, 5, and 15 minutes. The load average can give you a fair image about what happens with the system's processes.

If you have a single CPU system, a load average of 1 means that that CPU is under full load. If the number is higher, this means that the load is much higher than the CPU can handle, and this will probably put a lot of stress on your system. Because of this, processes will take longer to execute and the system's overall performance will be affected.

High load average means that there are applications that run multiple simultaneous threads at once. Nevertheless, some load issues are not only the result of an overcrowded CPU – they can be the combined effect of CPU load, disk I/O load, and memory load. In this case, the swiss-army knife for troubleshooting system load issues is the top command. The output of the top command constantly changes in real time, based on the system's load:

```
top - 23:21:13 up 3 days, 21:45,  1 user,  load average: 0,00, 0,00, 0,00
Tasks: 285 total,   1 running, 284 sleeping,   0 stopped,   0 zombie
%Cpu(s):  0,1 us,  0,1 sy,  0,0 ni, 99,8 id,  0,0 wa,  0,0 hi,  0,0 si,  0,0 st
MiB Mem :   7674,1 total,    235,6 free,    756,0 used,   6682,5 buff/cache
MiB Swap:   2048,0 total,   2045,5 free,      2,5 used.   6453,1 avail Mem

  PID USER       PR  NI    VIRT    RES    SHR S  %CPU  %MEM     TIME+ COMMAND
 1153 gdm        20   0 4344124 171236  90852 S   0,7   2,2   2:45.98 gnome-shell
44986 root       20   0       0      0      0 I   0,3   0,0   0:00.18 kworker/u16:+
45129 packt      20   0   16568   4396   3628 R   0,3   0,1   0:00.09 top
    1 root       20   0  171160  13328   8476 S   0,0   0,2   0:14.35 systemd
    2 root       20   0       0      0      0 S   0,0   0,0   0:00.19 kthreadd
    3 root        0 -20       0      0      0 I   0,0   0,0   0:00.00 rcu_gp
    4 root        0 -20       0      0      0 I   0,0   0,0   0:00.00 rcu_par_gp
    6 root        0 -20       0      0      0 I   0,0   0,0   0:00.00 kworker/0:0H+
    9 root        0 -20       0      0      0 I   0,0   0,0   0:00.00 mm_percpu_wq
   10 root       20   0       0      0      0 S   0,0   0,0   0:00.65 ksoftirqd/0
   11 root       20   0       0      0      0 I   0,0   0,0   1:01.04 rcu_sched
   12 root       rt   0       0      0      0 S   0,0   0,0   0:02.18 migration/0
   13 root      -51   0       0      0      0 S   0,0   0,0   0:00.00 idle_inject/0
   14 root       20   0       0      0      0 S   0,0   0,0   0:00.00 cpuhp/0
   15 root       20   0       0      0      0 S   0,0   0,0   0:00.00 cpuhp/1
   16 root      -51   0       0      0      0 S   0,0   0,0   0:00.00 idle_inject/1
   17 root       rt   0       0      0      0 S   0,0   0,0   0:02.35 migration/1
   18 root       20   0       0      0      0 S   0,0   0,0   0:01.05 ksoftirqd/1
   20 root        0 -20       0      0      0 I   0,0   0,0   0:00.00 kworker/1:0H+
```

Figure 10.23 – Running top to troubleshoot system load issues

By default, top sorts processes by how much CPU they use. It runs in interactive mode and sometimes, the output is difficult to see on the screen. You can redirect the output to a file and use the command in batch mode using the -b option. This mode only updates the command a specified number of times. To run top in batch mode, run the following command:

```
top -b -n 1 | tee top-command-output
```

The `top` command could be a little intimidating for unexperienced Linux users. This is why we will explain the output a little bit:

- `us`: User CPU time

- `sy`: System CPU time

- `ni`: Nice CPU time

- `id`: Idle CPU time

- `wa`: Input/output wait time

- `hi`: CPU hardware interrupts time

- `si`: CPU software interrupts time

- `st`: CPU steal time

Another useful tool for troubleshooting CPU usage and hard drive input/output time is `iostat`:

```
packt@neptune:~$ iostat
Linux 5.4.0-58-generic (neptune)        17.12.2020      _x86_64_        (8 CPU)

avg-cpu:  %user   %nice %system %iowait  %steal   %idle
           0,06    0,00    0,07    0,00    0,00   99,87

Device            tps    kB_read/s    kB_wrtn/s    kB_dscd/s    kB_read    kB_wrtn    kB_dscd
loop0            0,00         0,00         0,00         0,00        520          0          0
loop1            0,00         0,00         0,00         0,00        522          0          0
loop10           0,00         0,00         0,00         0,00        340          0          0
loop11           0,00         0,00         0,00         0,00          4          0          0
loop2            0,00         0,00         0,00         0,00       1135          0          0
loop3            0,00         0,00         0,00         0,00       1087          0          0
loop4            0,00         0,00         0,00         0,00       1173          0          0
loop5            0,00         0,00         0,00         0,00       1620          0          0
loop6            0,00         0,00         0,00         0,00       1679          0          0
loop7            0,00         0,00         0,00         0,00        576          0          0
loop8            0,04         0,04         0,00         0,00      13649          0          0
loop9            0,00         0,00         0,00         0,00        579          0          0
nvme0n1          1,16         3,78        29,20      1380,11    1299529   10037549  474382048
sda              0,00         0,01         0,00         0,00       2128          0          0
```

Figure 10.24 – The output of iostat

The CPU statistics are similar to the ones from the output of top. The I/O statistics are shown below the CPU statistic. Here is what each column represents:

- tps: Transfers per second to the device (I/O requests)
- kB_read/s: Amount of data read from the device (in terms of the number of blocks – kilobytes)
- kB_wrtn/s: Amount of data written to the device (in terms of the number of blocks – kilobytes)
- kB_dscd/s: Amount of data discarded for the device (in kilobytes)
- kB_read: Total number of blocks read
- kB_wrtn: Total number of blocks written
- kB_dscd: Total number of blocks discarded

For more details about the iostat command, read the respective manual pages by using the following command:

```
man iostat
```

Besides the iostat command, there is another one that you could use, called iotop. It is not installed by default on Ubuntu, but you can install it. First, search for the package with the following command:

```
packt@neptune:~$ sudo apt search iotop
Sorting... Done
Full Text Search... Done
iotop/focal,now 0.6-24-g733f3f8-1 amd64 [installed]
  simple top-like I/O monitor
```

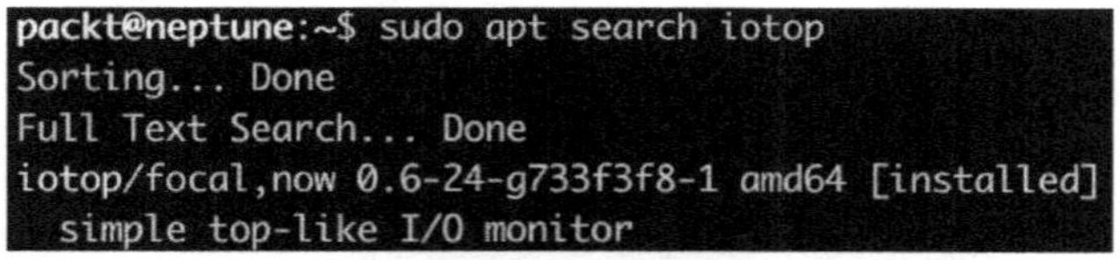

Figure 10.25 – Searching for the iotop package

Then, you can install it with the following command:

```
sudo apt install iotop
```

Once the package has been installed, you will need `sudo` privileges to run it:

```
Total DISK READ:         0.00 B/s | Total DISK WRITE:        14.75 K/s
Current DISK READ:       0.00 B/s | Current DISK WRITE:      22.13 K/s
   TID  PRIO  USER       DISK READ  DISK WRITE  SWAPIN      IO>     COMMAND
   268 be/3 root         0.00 B/s   14.75 K/s   0.00 %    0.10 % [jbd2/nvme0n1p2-]
     1 be/4 root         0.00 B/s    0.00 B/s   0.00 %    0.00 % init splash
     2 be/4 root         0.00 B/s    0.00 B/s   0.00 %    0.00 % [kthreadd]
     3 be/0 root         0.00 B/s    0.00 B/s   0.00 %    0.00 % [rcu_gp]
     4 be/0 root         0.00 B/s    0.00 B/s   0.00 %    0.00 % [rcu_par_gp]
     6 be/0 root         0.00 B/s    0.00 B/s   0.00 %    0.00 % [kworker/0:0H-kblockd]
     9 be/0 root         0.00 B/s    0.00 B/s   0.00 %    0.00 % [mm_percpu_wq]
    10 be/4 root         0.00 B/s    0.00 B/s   0.00 %    0.00 % [ksoftirqd/0]
    11 be/4 root         0.00 B/s    0.00 B/s   0.00 %    0.00 % [rcu_sched]
    12 rt/4 root         0.00 B/s    0.00 B/s   0.00 %    0.00 % [migration/0]
    13 rt/4 root         0.00 B/s    0.00 B/s   0.00 %    0.00 % [idle_inject/0]
    14 be/4 root         0.00 B/s    0.00 B/s   0.00 %    0.00 % [cpuhp/0]
    15 be/4 root         0.00 B/s    0.00 B/s   0.00 %    0.00 % [cpuhp/1]
    16 rt/4 root         0.00 B/s    0.00 B/s   0.00 %    0.00 % [idle_inject/1]
    17 rt/4 root         0.00 B/s    0.00 B/s   0.00 %    0.00 % [migration/1]
    18 be/4 root         0.00 B/s    0.00 B/s   0.00 %    0.00 % [ksoftirqd/1]
```

Figure 10.26 – Running the iotop command

You can also run the `sysstat` service to troubleshoot system load issues, similar to how we used it for troubleshooting memory issues. By default, `sar` will output the CPU statistics for the current day:

```
packt@neptune:~$ sudo sar 2 5
Linux 5.4.0-58-generic (neptune)       17.12.2020      _x86_64_       (8 CPU)

02:16:41        CPU     %user     %nice   %system   %iowait    %steal     %idle
02:16:43        all      0,00      0,00      0,00      0,00      0,00    100,00
02:16:45        all      0,06      0,00      0,00      0,00      0,00     99,94
02:16:47        all      0,00      0,00      0,31      0,00      0,00     99,69
02:16:49        all      0,00      0,00      0,00      0,00      0,00    100,00
02:16:51        all      0,12      0,00      0,00      0,00      0,00     99,88
Average:        all      0,04      0,00      0,06      0,00      0,00     99,90
```

Figure 10.27 – Running sar for CPU load troubleshooting

In the preceding screenshot, `sar` ran five times, every 2 seconds. Our local network servers are not under heavy load at this time, but you can imagine that the output would be different when the command is run on a heavily used server. As we pointed out in the previous section, the `sar` command has several options that could prove useful in finding solutions to potential problems. Run the `man  sar` command to view the manual page containing all the available options.

There are many other tools that could be used for general system troubleshooting. We barely scratched the surface of this subject with the tools shown in this section. We advise you to search for more tools designed for general system troubleshooting if you feel the need to do so. Otherwise, the ones presented here are sufficient for you to generate a viable report about possible system issues.

Network-specific issues will be covered in the next section.

Tools for troubleshooting network issues

Troubleshooting network issues is almost 80% of a system administrator's job – probably even more. The numbers are not backed up by any official studies, but more of a hands-on experience insight. Since most of the server and cloud issues are related to networking, an optimal working network means reduced downtime and happy clients and system administrators.

The tools we will cover in this section are the defaults on all major Linux distributions. All these tools were discussed in *Chapter 7, Networking with Linux, Chapter 8, Configuring Linux Servers*, and *Chapter 9, Securing Linux*, so we will only name them again from a problem-solving standpoint. Let's break down the tools we should use on specific TCP/IP layers. Remember how many layers there are in the TCP/IP model? There are five layers available, and we will start from layer one. As a good practice, troubleshooting a network is best done through the stack, starting from the application layer all the way to the physical layer.

Diagnosing layer 1

Quite often, due to the complexities of a network, issues tend to appear. Networks are essential for everyday living. We use them everywhere, from our wireless smartwatch to our smartphone, to our computer, and up to the cloud. Everything is connected worldwide, to make our lives better and a systems administrator's life a little bit harder. In this interconnected world, things can go sideways quite easily, and network issues need troubleshooting.

One of the basic testing tools and one of the first to be used by most system administrators is the ping command. The name comes from Packet InterNet Groper and it provides a basic connectivity test. Let's do a test on one of our local servers to see if everything is working fine. We will run four tests using the -c option of the ping command. The output is as follows:

```
packt@neptune:~$ ping -c 4 google.com
PING google.com (172.217.22.14) 56(84) bytes of data.
64 bytes from fra16s14-in-f14.1e100.net (172.217.22.14): icmp_seq=1 ttl=111 time=32.3 ms
64 bytes from fra16s14-in-f14.1e100.net (172.217.22.14): icmp_seq=2 ttl=111 time=31.9 ms
64 bytes from fra16s14-in-f14.1e100.net (172.217.22.14): icmp_seq=3 ttl=111 time=32.6 ms
64 bytes from fra16s14-in-f14.1e100.net (172.217.22.14): icmp_seq=4 ttl=111 time=32.0 ms

--- google.com ping statistics ---
4 packets transmitted, 4 received, 0% packet loss, time 3002ms
```

Figure 10.28 – Running a basic test using the ping command

Ping is sending simple ICMP packets to the destination (in our case, it was `google.com`) and is waiting for a response. Once it is received and no packets are lost, this means that everything is working fine. The `ping` command can be used to test connections to local network systems, as well as remote networks. It is the first tool that's used to test and isolate possible problems.

There are times when a simple test with the `ping` command is not enough. In this case, another versatile command is the `ip` command. You can use it to check if there are any issues with the physical layer:

```
packt@neptune:~$ ip link show
1: lo: <LOOPBACK,UP,LOWER_UP> mtu 65536 qdisc noqueue state UNKNOWN mode DEFAULT group de
fault qlen 1000
    link/loopback 00:00:00:00:00:00 brd 00:00:00:00:00:00
2: eno1: <BROADCAST,MULTICAST,UP,LOWER_UP> mtu 1500 qdisc fq_codel state UP mode DEFAULT
group default qlen 1000
    link/ether    ■  ■ ■  ■ ■ brd ff:ff:ff:ff:ff:ff
3: wlp0s20f3: <NO-CARRIER,BROADCAST,MULTICAST,UP> mtu 1500 qdisc noqueue state DOWN mode
DORMANT group default qlen 1000
    link/ether ■ ■■  ■■   ■■ brd ff:ff:ff:ff:ff:ff
```

Figure 10.29 – Showing the state of the physical interfaces with the ip command

In the preceding screenshot, you can see that the Ethernet interface is running well (`state UP`) and that the wireless interface is not working (`state DOWN`). It could be different on any other system and we can bring an interface with the following command. In our case, we will bring the wireless interface up using the following command:

```
ip link set wlp0s20f3 up
```

Once executed, you can check the state of the interface by running the following command:

```
ip link show
```

If you have direct access to a server or system, you can directly check if the wires are connected. If, by any chance, you are using a wireless connection (not recommended), you will need to use the `ip` command.

Another useful tool for layer 1 is `ethtool`. It is not installed by default on Ubuntu, so you will need to install it to use it. Once installed, to check the Ethernet interface, run the command shown in the following screenshot:

```
packt@neptune:~$ ethtool eno1
Settings for eno1:
        Supported ports: [ TP ]
        Supported link modes:    10baseT/Half 10baseT/Full
                                 100baseT/Half 100baseT/Full
                                 1000baseT/Full
        Supported pause frame use: No
        Supports auto-negotiation: Yes
        Supported FEC modes: Not reported
        Advertised link modes:   10baseT/Half 10baseT/Full
                                 100baseT/Half 100baseT/Full
                                 1000baseT/Full
        Advertised pause frame use: No
        Advertised auto-negotiation: Yes
        Advertised FEC modes: Not reported
        Speed: 1000Mb/s
        Duplex: Full
        Port: Twisted Pair
        PHYAD: 1
        Transceiver: internal
        Auto-negotiation: on
        MDI-X: on (auto)
Cannot get wake-on-lan settings: Operation not permitted
        Current message level: 0x00000007 (7)
                               drv probe link
        Link detected: yes
```

Figure 10.30 – Using the ethtool command

By using `ethtool`, we can check if a connection has negotiated the correct speed. In the preceding example, you can see that in our case, the server has correctly negotiated a full 1,000 Mbps full-duplex connection. In the next section, we will show you how to diagnose the layer 2 stack.

Diagnosing layer 2

The second layer in the TCP/IP stack is called the data link layer. It is generally responsible for local area network connectivity. Most of the issues that can occur at this stage happen due to improper IP to MAC address mapping. Some of the tools that can be used in this instance include the `ip` command and the `arp` command. The `arp` command, which comes from the **Address Resolution Protocol (ARP)**, is used to map IP addresses (layer 3) with MAC addresses (layer 2). In Ubuntu, the `arp` command is available through the `net-tools` package. First, proceed and install it using the following command:

```
sudo apt install net-tools
```

To check the entries inside the ARP table, you can use the `arp` command, as shown in the following screenshot:

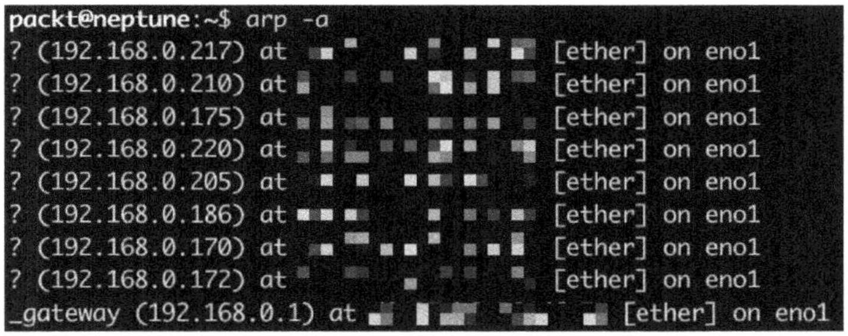

Figure 10.31 – Using the arp command to map the ARP entries

The output of the `arp` command will show all the connected devices, with details about their IP and MAC addresses. Note that the MAC addresses have been blurred for privacy reasons.

Similar to the `arp` command, you can use the `ip neighbor show` command, as shown here:

Figure 10.32 – Listing the ARP entries using the ip command

The `ip` command can be used to delete entries from the ARP list, like so:

```
ip neighbor delete IP dev eno1
```

Here, `IP` is the desired IP you want to delete from the list.

Both the `arp` and `ip` commands have a similar output. They are powerful commands and very useful for troubleshooting possible layer 2 issues. In the next section, we will show you how to diagnose the layer 3 stack.

Diagnosing layer 3

On layer 3, we are working with IP addresses only. We already know about the tools to use here, such as the `ip` command, the `ping` command, the `traceroute` command, and the `nslookup` command. Since we've already covered the `ip` and `ping` commands, we will only discuss how to use `traceroute` and `nslookup` here. The `traceroute` command is not installed by default in Ubuntu. You will have to install it using the following command:

```
sudo apt install traceroute
```

The `nslookup` package is already available in Ubuntu by default. First, to check for the routing table to see the list of gateways for different routes, we can use the `ip route` command:

```
packt@neptune:~$ ip route show
default via 192.168.0.1 dev eno1 proto dhcp metric 100
169.254.0.0/16 dev eno1 scope link metric 1000
192.168.0.0/24 dev eno1 proto kernel scope link src 192.168.0.244 metric 100
```

Figure 10.33 – Showing the routing table using the ip route show command

The `ip route` command is showing the default gateway. An issue would be if it is missing or incorrectly configured.

The `traceroute` tool is used to check the path of traffic from the source to the destination. The following output shows the path of packets traveling from our local gateway to Google's servers:

```
packt@neptune:~$ traceroute google.com
traceroute to google.com (216.58.205.238), 30 hops max, 60 byte packets
 1  _gateway (192.168.0.1)  0.623 ms  0.441 ms  0.694 ms
 2  ▪▪ ▪ ▪▪ .next-gen.ro (▪ ▪ ▪ ▪ )  4.638 ms  4.956 ms  4.929 ms
 3  ▪▪▪ ▪ ▪.next-gen.ro (▪▪ ▪▪     ▪▪)  3.258 ms  3.507 ms  3.730 ms
 4  ▪▪▪▪▪▪▪.next-gen.ro (▪▪  ▪   ▪ ▪)  2.863 ms  4.792 ms  2.810 ms
 5  ▪▪▪▪▪▪▪▪.next-gen.ro (▪▪ ▪ ▪▪ ▪▪)  4.917 ms  5.053 ms  5.270 ms
 6  bucuresti.nxdata.br01.next-gen.ro (▪▪▪▪ ▪▪▪▪)  4 655 ms  3.638 ms  3.816 ms
 7  10.19.141.193 (10.19.141.193)  2.248 ms  1.943 ms  2.211 ms
 8  * * *
 9  10.0.240.186 (10.0.240.186)  28.743 ms  29.263 ms  29.231 ms
10  10.0.240.125 (10.0.240.125)  32.431 ms 57.240.0.110.ap.yournet.ne.jp (110.0.240.57)
29.734 ms 10.0.240.125 (10.0.240.125)  32.637 ms
11  92.87.30.13 (92.87.30.13)  30.908 ms  30.937 ms  29 525 ms
12  * 10.252.43.30 (10.252.43.30)  30 654 ms *
13  74.125.37.98 (74.125.37.98)  28.107 ms 74.125.37.124 (74.125.37.124)  32.775 ms 74.12
5.37.196 (74.125.37.196)  30.686 ms
14  108.170.252.82 (108.170.252.82)  30.470 ms 108.170.252.83 (108.170.252.83)  30.689 ms
    31.106 ms
15  209.85.252.214 (209.85.252.214)  31.104 ms fra15s24-in-f14.1e100.net (216.58.205.238)
    29.002 ms 108.170.238.60 (108.170.238.60)  29.598 ms
```

Figure 10.34 – Using traceroute for path tracing

The `traceroute` tool is used to check the path of traffic from the source to the destination. Packets don't usually have the same route when they're sent and when they return to the source. Packets are sent to gateways to be processed and sent to the destination on a certain route. When packets exceed the local network, their route can be inaccurately represented by the `traceroute` tool, since the packets it relies on could be filtered by many of the gateways on the path (ICMP TTL Exceeded packets are generally filtered).

Similar to traceroute, there is a newer tool called **tracepath**. It is installed by default on Ubuntu and is a replacement for `traceroute`. It is considered much more reliable since it uses UDP ports for tracking compared to `traceroute`, which uses the less reliable ICMP protocol. `Tracepath` can be used with the `-n` option to show the IP address instead of the hostname. The following is an example of this:

```
packt@neptune:~$ tracepath google.com
 1?: [LOCALHOST]                          pmtu 1500
 1:  _gateway                                            0.685ms
 1:  _gateway                                            0.673ms
 2:  ■ ■ ■ ■ ■ ■ next-gen.ro                             2.957ms
 3:  ■ ■ ■ ■ ■ ■ .next-gen.ro                            2.360ms
 4:  ■ ■ ■ ■ ■ ext-gen.ro                               11.696ms
 5:  no reply
 6:  bucuresti.nxdata.br01.next-gen.ro                   3.734ms
 7:  10.19.141.193                                       2.550ms
 8:  no reply
 9:  10.0.200.2                              28.828ms asymm 11
10:  10.0.240.145                            33.099ms
11:  92.87.30.13                             29.735ms asymm 15
^C
packt@neptune:~$ tracepath -n google.com
 1?: [LOCALHOST]                          pmtu 1500
 1:  192.168.0.1                                         1.026ms
 1:  192.168.0.1                                         0.658ms
 2:  ■ ■   ■   ■                                         3.009ms
 3:  ■     ■ ■     ■                                     2.931ms
 4:  ■ ■   ■ ■                                           2.459ms
 5:  no reply
 6:  81.22.150.2                                         3.858ms
 7:  10.19.141.193                                       2.590ms
 8:  no reply
 9:  10.0.200.2                              28.951ms asymm 11
10:  10.0.240.9                              31.620ms
11:  92.87.30.13                             32.359ms asymm 15
^C
```

Figure 10.35 – Using the tracepath command

Checking further network issues could lead to faulty DNS resolution, where a host can only be accessed by the IP address and not by the hostname. To troubleshoot this, even if it is not a layer 3 protocol, you could use the `nslookup` command, combined with the `ping` command:

```
packt@neptune:~$ nslookup google.com
Server:         127.0.0.53
Address:        127.0.0.53#53

Non-authoritative answer:
Name:   google.com
Address: 172.217.22.14
Name:   google.com
Address: 2a00:1450:4001:81a::200e

packt@neptune:~$ ping -c 4 google.com
PING google.com (172.217.22.14) 56(84) bytes of data.
64 bytes from fra16s14-in-f14.1e100.net (172.217.22.14): icmp_seq=1 ttl=111 time=32.2 ms
64 bytes from fra16s14-in-f14.1e100.net (172.217.22.14): icmp_seq=2 ttl=111 time=32.1 ms
64 bytes from fra16s14-in-f14.1e100.net (172.217.22.14): icmp_seq=3 ttl=111 time=32.1 ms
64 bytes from fra16s14-in-f14.1e100.net (172.217.22.14): icmp_seq=4 ttl=111 time=32.0 ms

--- google.com ping statistics ---
4 packets transmitted, 4 received, 0% packet loss, time 3005ms
```

Figure 10.36 – Using nslookup for IP and DNS troubleshooting

The preceding output shows that the nslookup command has the same IP for google.com as the ping command, which means that everything is fine. If a different IP shows up in the output, then you have an issue with your host's configuration. In the next section, we will show you how to diagnose both the layer 4 and layer 5 stacks.

Diagnosing layers 4 and 5

The last two layers, layer 4 (transport) and layer 5 (application), will mainly provide host-to-host communication services for applications. This is why we will cover them in a condensed manner. Two of the most well-known protocols from layer 4 are the **Transmission Control Protocol (TCP)** and the **User Datagram Protocol (UDP)**, and both used and implemented inside every operating system available. TCP and UDP cover all the traffic on the internet. One important tool for troubleshooting layer 4 issues is the ss command. The ss command is a recent replacement for netstat and is used to see the list of all network sockets. As such, a list can have a significant size, so you could use several command options to reduce it. For example, you could use the -t option to see only the TCP sockets, the -u option for UDP sockets, and -x for Unix sockets. Thus, to see TCP and UDP socket information, we will use the ss command, as shown in the following screenshot:

```
packt@neptune:~$ ss -t
State      Recv-Q     Send-Q          Local Address:Port          Peer Address:Port         Process
ESTAB      0          0               192.168.0.244:ssh           192.168.0.217:50281
packt@neptune:~$ ss -u
Recv-Q      Send-Q               Local Address:Port          Peer Address:Port         Process
0           0               192.168.0.244%eno1:bootpc          192.168.0.1:bootps
```

Figure 10.37 – Using the ss command to list TCP and UDP sockets

Furthermore, to see all the listening sockets on your system, you can use the -l option. This one, combined with the -u and -t options, will show you all the UDP and TCP listening sockets on your system. The following is an excerpt from a much longer list:

```
packt@neptune:~$ ss -ltu
Netid  State    Recv-Q  Send-Q   Local Address:Port            Peer Address:Port  Process
udp    UNCONN   0       0        127.0.0.53%lo:domain          0.0.0.0:*
udp    UNCONN   0       0        0.0.0.0:ntp                   0.0.0.0:*
udp    UNCONN   0       0        192.168.0.255:netbios-ns      0.0.0.0:*
udp    UNCONN   0       0        192.168.0.244:netbios-ns      0.0.0.0:*
udp    UNCONN   0       0        0.0.0.0:netbios-ns            0.0.0.0:*
udp    UNCONN   0       0        192.168.0.255:netbios-dgm     0.0.0.0:*
udp    UNCONN   0       0        192.168.0.244:netbios-dgm     0.0.0.0:*
udp    UNCONN   0       0        0.0.0.0:netbios-dgm           0.0.0.0:*
udp    UNCONN   0       0        127.0.0.1:323                 0.0.0.0:*
udp    UNCONN   0       0        0.0.0.0:631                   0.0.0.0:*
tcp    LISTEN   0       50       127.0.0.1:microsoft-ds        0.0.0.0:*
tcp    LISTEN   0       50       192.168.0.244:microsoft-ds    0.0.0.0:*
tcp    LISTEN   0       64       0.0.0.0:nfs                   0.0.0.0:*
tcp    LISTEN   0       4096     0.0.0.0:52101                 0.0.0.0:*
tcp    LISTEN   0       64       0.0.0.0:41001                 0.0.0.0:*
tcp    LISTEN   0       80       127.0.0.1:mysql               0.0.0.0:*
tcp    LISTEN   0       50       127.0.0.1:netbios-ssn         0.0.0.0:*
tcp    LISTEN   0       50       192.168.0.244:netbios-ssn     0.0.0.0:*
tcp    LISTEN   0       4096     0.0.0.0:50347                 0.0.0.0:*
tcp    LISTEN   0       4096     0.0.0.0:sunrpc                0.0.0.0:*
tcp    LISTEN   0       4096     127.0.0.53%lo:domain          0.0.0.0:*
tcp    LISTEN   0       128      0.0.0.0:ssh                   0.0.0.0:*
tcp    LISTEN   0       5        0.0.0.0:ipp                   0.0.0.0:*
tcp    LISTEN   0       4096     0.0.0.0:40955                 0.0.0.0:*
tcp    LISTEN   0       64       [::]:42847                    [::]:*
tcp    LISTEN   0       64       [::]:nfs                      [::]:*
tcp    LISTEN   0       4096     [::]:41603                    [::]:*
tcp    LISTEN   0       4096     [::]:36453                    [::]:*
tcp    LISTEN   0       4096     [::]:sunrpc                   [::]:*
tcp    LISTEN   0       511      *:http                        *:*
tcp    LISTEN   0       4096     [::]:39731                    [::]:*
tcp    LISTEN   0       32       *:ftp                         *:*
tcp    LISTEN   0       128      [::]:ssh                      [::]:*
tcp    LISTEN   0       5        [::]:ipp                      [::]:*
```

Figure 10.38 – A list of listening sockets

The ss command is important for network troubleshooting when you want to verify the available sockets and the ones that are in the LISTEN state. This tool should not be missing from a system administrator's toolbox. Layer 5, the application layer, is comprised of protocols used by applications, and we will remember protocols such as the **Dynamic Host Configuration Protocol (DHCP)**, the **Hypertext Transfer Protocol (HTTP)**, and the **File Transfer Protocol (FTP)**. Since diagnosing layer 5 is mainly an application troubleshooting process, this will not be covered in this section.

In the next section, we will shortly discuss troubleshooting hardware issues.

Tools for troubleshooting hardware issues

The first step in troubleshooting hardware issues is to check your hardware. A very good tool to see details about the system's hardware is the dmidecode command. This command is used to read details about each hardware component in a human-readable format. Each piece of hardware has a specific DMI code, depending on its type. This code is specific to the SMBIOS. The following is a list of every hardware code available. There are 45 codes that are used by the SMBIOS, as follows:

1	System	23	System Reset
2	Base Board	24	Hardware Security
3	Chassis	25	System Power Controls
4	Processor	26	Voltage Probe
5	Memory Controller	27	Cooling Device
6	Memory Module	28	Temperature Probe
7	Cache	29	Electrical Current Probe
8	Port Connector	30	Out-of-band Remote Access
9	System Slots	31	Boot Integrity Services
10	On Board Devices	32	System Boot
11	OEM Strings	33	64-bit Memory Error
12	System Configuration Options	34	Management Device
13	BIOS Language	35	Management Device Component
14	Group Associations	36	Management Device Threshold Data
15	System Event Log	37	Memory Channel
16	Physical Memory Array	38	IPMI Device
17	Memory Device	39	Power Supply
18	32-bit Memory Error	40	Additional Information
19	Memory Array Mapped Address	41	Onboard Device
20	Memory Device Mapped Address	126	Disabled Entries
21	Built-in Pointing Device	127	End-Of-Table
22	Portable Battery	128	OEM specific
		255	OEM specific

To view details about the system's memory, you can use the dmidecode command with the -t option (from TYPE) and code 17, which corresponds to the memory device. An example from our system is as follows:

```
packt@neptune:~$ sudo dmidecode -t 17
Handle 0x0038, DMI type 17, 84 bytes
Memory Device
        Array Handle: 0x0037
        Error Information Handle: Not Provided
        Total Width: 64 bits
        Data Width: 64 bits
        Size: 4096 MB
        Form Factor: SODIMM
        Set: None
        Locator: SODIMM1
        Bank Locator: Memory Channel A
        Type: DDR4
        Type Detail: Synchronous
        Speed: 2400 MT/s
        Rank: 1
        Configured Memory Speed: 2133 MT/s
```

Figure 10.39 – Using dmidecode to view information about memory

To see details about other hardware components, use the command with the specific code. Other quick troubleshooting tools include commands such as lspci, lsblk, and lscpu. The output for each of these commands could be significantly large and will not fit into one screen:

```
packt@neptune:~$ lsblk
NAME        MAJ:MIN RM    SIZE RO TYPE MOUNTPOINT
sda             8:0  1 115,6G  0 disk
└─sda1          8:1  1 115,6G  0 part
nvme0n1       259:0  0   477G  0 disk
├─nvme0n1p1   259:1  0   512M  0 part /boot/efi
└─nvme0n1p2   259:2  0 476,4G  0 part /
```

Figure 10.40 – The output of the lsblk command

The output of the lsblk command shows information about the disks and partitions being used on the system. The lscpu command will show details about the CPU:

```
packt@neptune:~$ lscpu
Architecture:                    x86_64
CPU op-mode(s):                  32-bit, 64-bit
Byte Order:                      Little Endian
Address sizes:                   39 bits physical, 48 bits virtual
CPU(s):                          8
On-line CPU(s) list:             0-7
Thread(s) per core:              2
Core(s) per socket:              4
Socket(s):                       1
NUMA node(s):                    1
Vendor ID:                       GenuineIntel
CPU family:                      6
Model:                           142
Model name:                      Intel(R) Core(TM) i5-10210U CPU @ 1.60GHz
Stepping:                        12
CPU MHz:                         631.463
CPU max MHz:                     4200,0000
CPU min MHz:                     400,0000
BogoMIPS:                        4199.88
Virtualization:                  VT-x
```

Figure 10.41 – The output of lscpu

When you're troubleshooting hardware issues, taking a quick look at the kernel's logs could prove useful. To do this, use the dmesg command like in the following example:

```
dmesg | more
```

As you've seen, hardware troubleshooting is just as important and challenging as all other types of troubleshooting. Solving hardware-related issues is an integral part of any system administrator's job. This involves constantly checking hardware components, replacing the faulty parts with new ones, and making sure that they run smoothly.

Summary

In this chapter, we emphasized the importance of disaster recovery planning, backup and restore strategies, and troubleshooting various system issues. Every system administrator should be able to put their knowledge into practice when disaster strikes. Different types of failures will eventually hit the running servers, so solutions should be given in as soon as possible, to ensure minimal downtime and to prevent data loss.

This chapter represented the culmination of the *Advanced server administration* section of this book. In the next chapter, we will introduce you to cloud computing as a natural step and pinnacle of today's and tomorrow's computing landscape.

Exercises

Before we dive into the cloud section and look at troubleshooting issues on any Linux distribution, let's test everything you already know by now. Troubleshooting is problem solving at its best, and the following questions should test your entire knowledge of basic and advanced Linux administration:

1. Try to draft a DRP for your private network or small business.

2. Back up your entire system using the 321 rule.

3. Find out what your system's top 10 processes use the CPU the most.

4. Find out what your system's top 10 processes use RAM the most.

Further reading

-   Ubuntu 20.04 LTS official documentation: `https://ubuntu.com/server/docs`

-   RHEL 8 official documentation: `https://access.redhat.com/documentation/en-us/red_hat_enterprise_linux/8/`

-   SUSE official documentation: `https://documentation.suse.com/`

Section 3: Cloud Administration

In this section, you will learn about advanced concepts related to cloud computing. By the end of this section, you will be proficient in using specific tools and deploying Linux to the cloud.

This part of the book comprises the following chapters:

- *Chapter 11, Working with Containers and Virtual Machines*
- *Chapter 12, Cloud Computing Essentials*
- *Chapter 13, Deploying to the Cloud with AWS and Azure*
- *Chapter 14, Deploying Applications with Kubernetes*
- *Chapter 15, Automating Workflows with Ansible*

11
Working with Containers and Virtual Machines

In this chapter, you will learn what VMs and containers are. For starters, you will learn how virtualization works in Linux and how to create and use VMs. Once you master that, you will learn about containers and how are they set to change the future of virtualization and application delivery. You will learn about one of the well-known tools for using container – Docker. The topics in this chapter will prepare you for the future of Linux, as it is the foundation of every modern cloud technology. If you wish to remain up to date in a constantly changing landscape, this chapter will be an essential starting point for your journey.

In this chapter, we're going to cover the following main topics:

- Introduction to virtualization on Linux
- Understanding Linux containers
- Working with Docker

Technical requirements

No special technical requirements are needed, just a working installation of Linux on your system. Eighter Ubuntu or CentOS are equally suitable for this chapter's exercises and we will use both for exemplification.

The code for this chapter is available at the following link: `https://github.com/PacktPublishing/Mastering-Linux-Administration`.

Introduction to virtualization on Linux

Virtualization appeared as a need to make more efficient use of computer hardware. It is basically an abstraction layer that takes advantage of the computer's resources. In this section, you will learn about the types of VMs, how they work on Linux, and how to deploy and manage them.

Efficiency in resource usage

The abstraction layer that virtualization uses is a software layer that allows for a more efficient use of all of the computer's components. This allows for better use of all the physical machine's capabilities and resources.

Before going any further into virtualization, let's give you an example. In our testing laboratory, we have several physical machines, in the form of laptops and desktop computers that we use as servers. Each of the systems has significant resources available, more than enough to run the services we need. For instance, our least performant system is a laptop with a dual core Intel i3 (with hyperthreading), 8 GB of DDR3 RAM, and a 120 GB SSD. We also have a fifth generation Intel NUC with the same exact configuration. Those two systems have plenty of resources that could be more efficiently used by using VMs. For running a local web service or any kind of servers on our local network, those resources can be split between several VMs. For example, each physical system could host four different VMs, each using a single CPU core, 2 GB of memory, and 30 GB of storage. This way, one single machine will work as if there were four different ones. This is way more efficient than using different machines for single tasks.

In the following diagram, we are comparing the load on a single computer versus the same load divided between several VMs. This way of using the same hardware resources is more efficient:

Single Computer   Virtual Machines

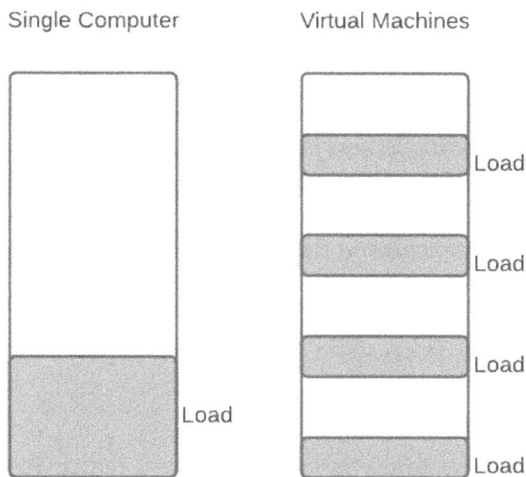

Figure 11.1 – Comparison between single computer use and using multiple VMs

For the purpose of this chapter's exercises, when we will be using Ubuntu, it will be on a tenth generation Intel NUC with a quad core Intel i5 CPU (with hyperthreading), 12 GB of DDR4 RAM, and 512 GB SSD storage. On this particular machine, we could use up to eight VMs, each using one CPU core, 1.5 GB of RAM, and 64 GB of storage. When we will be using CentOS, this will be on a fifth generation Intel NUC with a dual core Intel i3 CPU (with hyperthreading), 8 GB of DDR3 RAM, and 120 GB SSD storage.

Nonetheless, as we will use the hypervisor on top of a host OS, we will have to keep some of the resources for the OS's use, so the number of VMs would be smaller. Following is a sketch of how VMs are working on a host OS:

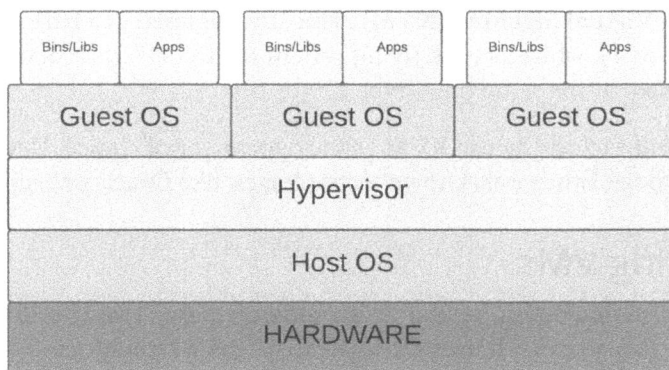

Figure 11.2 – How virtualization works on a host OS

The preceding diagram is the scheme of how virtualization works when used on a host OS. As we will see in the following sections, it is not the only type of virtualization used.

Efficiency is not related solely to the hardware resources used. A significant importance of efficient use of hardware in data centers is related to energy efficiency and the carbon footprint. In this respect, virtualization played a major role for several decades in changing the usage patterns for servers inside data centers. Overall, virtualization and containerization are significant players in the fight against climate change.

In the following sections, we will give you a short introduction to what hypervisors and VMs are.

Introduction to hypervisor

The software layer that virtualization is based on is called a **hypervisor**. The physical resources are divided and used as virtual computers, or better known as **virtual machines**. By using VMs, the limits of physical hardware are overcome by the process of **emulation**. This has a lot of advantages, by enabling the hardware to be of better use.

Hypervisors can be used either on top of an existing OS – **type 2**, or directly on bare metal (hardware) – **type 1**. For each of these types, there are several solutions that can be used, particularly on Linux. For a Linux OS, examples of each type would be as follows:

- Examples of hypervisors that run on top of a host OS (type 2) are Oracle VirtualBox, VMware Workstation/Fusion.
- Examples of hypervisors that run directly on bare metal (type 1) are Citrix Xen Server, VMware ESXi
- **Kernel-based Virtual Machine** (**KVM**) is mostly classified as a bare metal hypervisor (type 1), while its underlying system is a full OS, thus being classified as a host hypervisor at the same time (type 2)

In this section, we will exclusively use KVM as the hypervisor of choice. Nevertheless, we will show you how to use other well-known technologies, like Oracle's VirtualBox.

Understanding VMs

A VM is similar to a standalone computer. It is a software-based emulator that has access to the host computer's resources. It uses the hosts CPU, RAM, storage, networking interface(s), and ports. It is a virtual environment that has the same functions as a physical computer; it is also seen as a virtual computer.

The resources for each VM are managed by the hypervisor. It can relocate resources between existing VMs or create new VMs. The VMs are isolated from each other and from the host computer. As multiple VMs can exist on a single computer, each VM can use different guest OSes. For example, if you use a Windows machine and want to try out Linux, a popular solution would be to create a VM with the Linux distribution that you want to try. The same goes for Mac users, too. The OS installed inside a VM runs similar to an OS installed on bare metal. The user experience could vary from one hypervisor to the next, and so could the resource efficiency and response times. From our experience, running VMs from KVM is much smoother than running from VirtualBox, for example, but use cases could be different from one user to another.

Choosing the hypervisor

In this section, we will show you how to use the hypervisors called VirtualBox and KVM. As an optional solution, we will also cover GNOME Boxes. As both KVM and GNOME Boxes are directly available from Linux repositories, we consider them to be the better solutions for newcomers to Linux. Both KVM and GNOME Boxes share parts of libvirt and qemu code (to be detailed in the following sections), and in this respect, we consider them both as being the same hypervisor, mainly KVM.

In *Chapter 1*, *Installing Linux*, you first encountered the use of a hypervisor to set up a Linux VM. We showed you how to use VMware Fusion and VirtualBox for setting up a Linux VM. The details used then should be sufficient for any user, be it someone who is experienced or a newbie. In this section, we will only give you succinct information about installing VirtualBox on Ubuntu. In *Chapter 1*, *Installing Linux*, we used VirtualBox on macOS.

Using the VirtualBox hypervisor

VirtualBox has been developed by Oracle and is an open source project. It offers cross-platform support for Windows, macOS, and Linux. It only offers support for x86 architectures, and so is usable with both Intel- and AMD-based computers, with support for Apple's arm architecture not yet developed. The most use cases for VirtualBox are on a desktop or laptop computer, with a graphical user interface enabled.

As VirtualBox is not available for download from official Ubuntu repositories, you will need to install it from an external source. Go to the official website and download it from there. The link is `https://www.virtualbox.org/wiki/Linux_Downloads`. From the list shown on the page, click on the link that specifies your host OS. In our case, it is Ubuntu 20.04.1 LTS, so we will click on the link that says **Ubuntu 19.10 / 20.04**. Once the package is downloaded, you will have to install it by double-clicking it. The installation process should only take a few minutes to complete.

We also recommend installing the VirtualBox Extension Pack, available at the following link: `https://www.virtualbox.org/wiki/Downloads`. This package will add support for USB 3.0, PXE, NVMe, and disk encryption. Once you download the package, double-click it in order to install it.

To create a VM in VirtualBox, you will need to click on the **New** button and follow the instructions. Here is a detailed procedure on how to create a first VM:

1. Add a new VM by providing the name, type, version, and destination folder. In the following example, we are creating a VM for openSUSE Leap 15.2. We add the name and click on the Next button.

2. Specify the memory size in the next window. We recommend a minimum of 2 GB/VM.

3. Select the option to create the virtual disk and click the **Create** button.

4. Select the virtual disk type. If you don't plan on using the image with other hypervisors, select the default **VirtualBox Disk Image** (**VDI**).

5. Specify whether the new disk file should by fixed or dynamic. We recommend using the dynamic type, as at this stage, we might not use the entire disk size. If you think that a fixed size is more suited for your use, select that specific type.

6. Select the size of the virtual disk file and its location in the window that follows.

7. The VM is now created. In the main VirtualBox window, click the **Settings** button and, in the new window, select **System** and proceed by selecting the number of CPUs to use. We recommend a minimum of 2 CPU cores to use for a Linux distribution with GUI support. A single core will suffice for a CLI-only server instance, however.

8. Select the openSUSE installation disk and begin the setup process. If you plan to use another distribution, use the image you downloaded for this.

Using VirtualBox is relatively straightforward and easy. Nevertheless, there might be some issues that you will need to overcome, such as issues with secure boot, but nothing that is too taxing.

> **Important note**
>
> The VM created in VirtualBox might not start if you have secure boot enabled. If that is the case, restart the system, open the BIOS, and disable secure boot. Starting the VM after the system boots will work as expected.

VirtualBox offers Guest Additions software, great hardware support, VM groups, and supports a plethora of host OS versions.

The VirtualBox **Guest Additions** software provide several new device drivers (for better peripheral integration and video support) that are not provided by default, together with extra system applications. The guest additions are in the form of an ISO file that is supplied with the package and installed inside the /usr/share/virtualbox/ directory. VirtualBox has several features that will make it a fair candidate for your hypervisor solution, but in our opinion, it still lacks the finesse of KVM.

Using the KVM hypervisor

The KVM hypervisor is an open-source virtualization project available on all major Linux distributions. It is a modern hypervisor that uses specific kernel modules to take advantage of all the benefits that the Linux kernel has to offer, including memory support, scheduler, nested virtualization, GPU passthrough, and so on.

KVM in details – QEMU and libvirt

KVM uses Quick Emulator, or QEMU, as the emulator software for all the hardware components and peripherals. The main management tool and daemon that controls the hypervisor and also the **Application Programming Interface** (**API**) for KVM is called libvirt. The KVM's interface with libvirt, specifically in GNOME, is virt-manager. The CLI for libvirt is called virsh.

The libvirt API provides a common library for managing VMs, being the management layer for VM creation, modification, and provision. It is running in the background as a daemon called libvirtd that manages the connections with the hypervisor at the client's request.

QEMU is both an emulator and a virtualizer. When used as an emulator, QEMU uses **dynamic binary translation** methods to operate. This means that it can use different types of OS on the host machine, even if they are designed for different architectures. Dynamic binary translations are used in software-based virtualization, when hardware is emulated to execute instructions in virtualized environments. This way, QEMU emulates the machine's CPU, using a specific binary translator method called **Tiny Code Generator** (**TCG**), which transforms the binary code for different types of architectures.

When used as virtualizer, QEMU uses what is known as a hardware-based virtualization, where the binary translation is not used, because the instructions are executed directly on the host CPU. The differences between software- and hardware-assisted virtualization is shown in the following diagram:

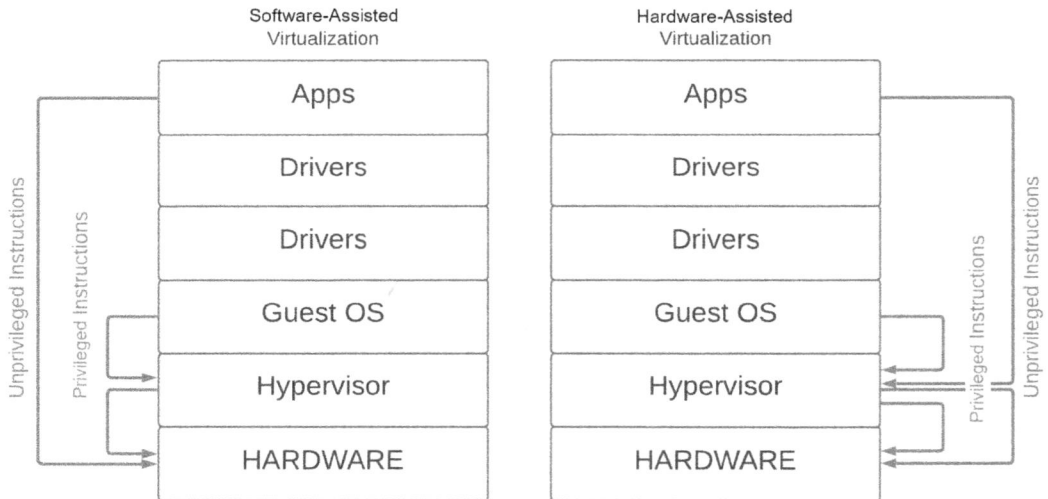

Figure 11.3 – Comparison between software- and hardware-assisted virtualization

As you can see in the diagram, instructions have different paths when using software- and hardware-assisted virtualization. In software-assisted virtualization, when dynamic binary translations are used, the user's unprivileged instructions are sent directly to the hardware, while the guest OS privileged instructions are first sent to the hypervisor before getting to the hardware. In hardware-assisted virtualization, the user's unprivileged instructions are sent to the hypervisor first, and then sent to the hardware, while the privileged instructions from the guest OS have the same path as in the software-assisted virtualization. This ensures a certain level of isolation for the guest OS, thereby achieving better performance and less complexity.

In the following section, we will show you how to install and configure QEMU on a CentOS 8 machine.

Installing QEMU and libvirt

Installing QEMU is a straightforward task. All you need to do is to run the package installer utility of your distribution. In our case, as we are using CentOS, we will use yum as follows:

```
sudo yum install qemu-kvm
```

There is a reasonable chance that the package is already installed on your CentOS 8 distribution. In this case, you will see the following message:

```
[packt@jupiter ~]$ sudo yum install qemu-kvm
Last metadata expiration check: 0:23:00 ago on Fri 01 Jan 2021 11:43:41 AM EET.
Package qemu-kvm-15:4.2.0-34.module_el8.3.0+555+a55c8938.x86_64 is already insta
lled.
Dependencies resolved.
Nothing to do.
Complete!
```

Figure 11.4 – Installing QEMU

We recommend installing the package provided by the distribution, rather than installing from source. This way, you will ensure that the package has all the requisite dependencies installed.

Install the necessary modules with the following command:

```
yum module install virt
```

Besides the qemu-kvm package, you will need to install libvirt together with other necessary packages. Use the yum command in a similar way to before:

```
sudo yum install libvirt libvirt-client virt-manager virt-
install virt-viewer
```

Again, you might come across messages that say the packages are already installed, as in the following output:

```
[packt@jupiter ~]$ sudo yum install libvirt libvirt-client virt-manager virt-ins
tall virt-viewer
[sudo] password for packt:
Last metadata expiration check: 0:34:37 ago on Fri 01 Jan 2021 11:43:41 AM EET.
Package libvirt-6.0.0-28.module_el8.3.0+555+a55c8938.x86_64 is already installed
.
Package libvirt-client-6.0.0-28.module_el8.3.0+555+a55c8938.x86_64 is already in
stalled.
Package virt-manager-2.2.1-3.el8.noarch is already installed.
Package virt-install-2.2.1-3.el8.noarch is already installed.
Package virt-viewer-9.0-4.el8.x86_64 is already installed.
Dependencies resolved.
Nothing to do.
Complete!
```

Figure 11.5 – Installing libvirt and other requisite packages

There is a reason for those outputs that you can see, and this is detailed here.

> **Important note**
>
> The reason why our outputs are showing that the packages are already installed is because we chose a complete virtualization package installation right from the beginning, when we installed CentOS on our machine. If you did not do that, the output will be totally different.
>
> When first installing CentOS, you have the choice of software selection, with options for installing guest agent packages, virtualization client packages, virtualization hypervisor packages, together with container management packages, among others.

Once all the necessary packages are installed, a safe action to take is to check whether your machine is compatible with KVM requirements. To do this, use the `virt-host-validate` command, as a root user, or by using `sudo`. The output is shown as follows:

```
[packt@jupiter ~]$ sudo virt-host-validate
[sudo] password for packt:
setlocale: No such file or directory
  QEMU: Checking for hardware virtualization                         : PASS
  QEMU: Checking if device /dev/kvm exists                           : PASS
  QEMU: Checking if device /dev/kvm is accessible                    : PASS
  QEMU: Checking if device /dev/vhost-net exists                     : PASS
  QEMU: Checking if device /dev/net/tun exists                       : PASS
  QEMU: Checking for cgroup 'cpu' controller support                 : PASS
  QEMU: Checking for cgroup 'cpuacct' controller support             : PASS
  QEMU: Checking for cgroup 'cpuset' controller support              : PASS
  QEMU: Checking for cgroup 'memory' controller support              : PASS
  QEMU: Checking for cgroup 'devices' controller support             : PASS
  QEMU: Checking for cgroup 'blkio' controller support               : PASS
  QEMU: Checking for device assignment IOMMU support                 : PASS
  QEMU: Checking if IOMMU is enabled by kernel                       : WARN (IOMMU appears to be disabled in kernel. Add intel_iommu=on to kernel cmdline arguments)
  QEMU: Checking for secure guest support                            : WARN (Unknown if this platform has Secure Guest support)
```

Figure 11.6 – Checking system compatibility

In the preceding output, a warning about **Input-Output Memory Management Unit** (**IOMMU**) not being activated in the kernel is displayed, together with a warning about support for secure guests. The second warning is due to the fact that secure boot is disabled in the machine's BIOS. As for the first warning, it can easily be fixed by activating IOMMU support at kernel level.

> **Important note**
>
> What is **IOMMU**? It is a memory management unit that manages **direct memory access** (**DMA**) requests to the embedded or DRAM memory of a system. In a nutshell, the IOMMU was designed to virtualize the memory space to allow for a better association between the drivers and the hardware.

To enable IOMMU support for the kernel, go ahead and edit the `/etc/default/grub` file. Find the line that says `GRUB_CMDLINE_LINUX` and add the following text at the end (if you have Intel hardware): `intel_iommu=on`. If you have AMD CPU and motherboard chipsets, add `amd_iommu=on`:

```
GRUB_TIMEOUT=5
GRUB_DISTRIBUTOR="$(sed 's, release .*$,,g' /etc/system-release)"
GRUB_DEFAULT=saved
GRUB_DISABLE_SUBMENU=true
GRUB_TERMINAL_OUTPUT="console"
GRUB_CMDLINE_LINUX="crashkernel=auto resume=/dev/mapper/cl-swap rd.lvm.lv=cl/root rd.lvm.lv=cl/swap rhgb quiet intel_iommu=on"
GRUB_DISABLE_RECOVERY="true"
GRUB_ENABLE_BLSCFG=true
```

Figure 11.7 – Activating kernel IOMMU support

After altering the file, save the changes, refresh the `grub.cfg` file, and reboot the system. To refresh the GRUB2 file, use the following command:

```
[packt@jupiter ~]$ sudo grub2-mkconfig -o /boot/grub2/grub.cfg
Generating grub configuration file ...
done
```

Figure 11.8 – Refreshing the grub.cfg file before restarting

Once the system is on again, you can run the `virt-host-validate` command again. You will see that the warning regarding IOMMU kernel support is no longer present. Please ignore the second warning, as we still have secure boot disabled in the BIOS. In any case, you might not even have this last warning in your output:

```
[packt@jupiter ~]$ sudo virt-host-validate
[sudo] password for packt:
setlocale: No such file or directory
  QEMU: Checking for hardware virtualization                 : PASS
  QEMU: Checking if device /dev/kvm exists                   : PASS
  QEMU: Checking if device /dev/kvm is accessible            : PASS
  QEMU: Checking if device /dev/vhost-net exists             : PASS
  QEMU: Checking if device /dev/net/tun exists               : PASS
  QEMU: Checking for cgroup 'cpu' controller support         : PASS
  QEMU: Checking for cgroup 'cpuacct' controller support     : PASS
  QEMU: Checking for cgroup 'cpuset' controller support      : PASS
  QEMU: Checking for cgroup 'memory' controller support      : PASS
  QEMU: Checking for cgroup 'devices' controller support     : PASS
  QEMU: Checking for cgroup 'blkio' controller support       : PASS
  QEMU: Checking for device assignment IOMMU support         : PASS
  QEMU: Checking if IOMMU is enabled by kernel               : PASS
  QEMU: Checking for secure guest support                    : WARN
(Unknown if this platform has Secure Guest support)
```

Figure 11.9 – Running the virt-host-validate command once more

After seeing that there are no compatibility issues, we can proceed to creating our first VM using the command-line interface.

Creating the first VM using the command line

Before actually creating the first VM, we make sure that the `libvirtd` daemon is actively running. To do this, we will use the `systemctl` utility, as shown in the following code snippet:

```
systemctl start libvirtd; systemctl status libvirtd
```

In order to create a VM, first we will need to download the image file of the guest OS. For our first VM, we will plan to use Debian 10.7 Linux distribution. First, we will download the net-install (smaller) ISO image with the following command:

```
wget https://cdimage.debian.org/debian-cd/current/amd64/iso-cd/
debian-10.7.0-amd64-netinst.iso
```

The download location will be inside /tmp/Downloads:

```
[root@jupiter Downloads]# wget https://cdimage.debian.org/debian-cd/current/amd64/is
o-cd/debian-10.7.0-amd64-netinst.iso
--2021-01-01 14:41:22--  https://cdimage.debian.org/debian-cd/current/amd64/iso-cd/d
ebian-10.7.0-amd64-netinst.iso
Resolving cdimage.debian.org (cdimage.debian.org)... 194.71.11.165, 194.71.11.173, 2
001:6b0:19::165, ...
Connecting to cdimage.debian.org (cdimage.debian.org)|194.71.11.165|:443... connecte
d.
HTTP request sent, awaiting response... 200 OK
Length: 352321536 (336M) [application/x-iso9660-image]
Saving to: 'debian-10.7.0-amd64-netinst.iso'

debian-10.7.0-amd64- 100%[===================>] 336.00M  41.6MB/s    in 9.1s

2021-01-01 14:41:32 (36.8 MB/s) - 'debian-10.7.0-amd64-netinst.iso' saved [352321536
/352321536]
```

Figure 11.10 – Downloading the Debian image for our first VM

Once the Debian image is downloaded, we will use the virt-install command to create the first VM on our host system. For this exercise, we will use a fifth generation Intel NUC system, with an Intel i3 dual core processor, 8 GB of RAM, and 120 GB of storage. The host OS is CentOS 8. We will create one VM that will use a single virtual CPU (vCPU), 2 GB of RAM, and 20 GB of storage.

The virt-install command has the following arguments (mandatory):

- --name: The name of the new VM
- --memory: The amount of RAM used by the VM
- --vcpus: The number of virtual CPUs used by the new VM
- --disk size: The amount of storage used
- --os-type: The OS type, in our case, Linux
- --os-variant: The type of guest OS
- --location: The location of the guest OS ISO file

All these are mandatory, and `-os-variant` could post some problems for you as you may not know what to write. In order to find the OS type, you should use the `osinfo-query os` command. Running it will output a long list of known OSes and their short ID, sorted in alphabetical order. The following is a short excerpt, with Alpine Linux being the first on the list:

```
[root@jupiter Downloads]# osinfo-query os
Short ID            | Name               | Version | ID
--------------------+--------------------+---------+-----------------------------------------
alpinelinux3.5      | Alpine Linux 3.5   | 3.5     | http://alpinelinux.org/alpinelinux/3.5
alpinelinux3.6      | Alpine Linux 3.6   | 3.6     | http://alpinelinux.org/alpinelinux/3.6
alpinelinux3.7      | Alpine Linux 3.7   | 3.7     | http://alpinelinux.org/alpinelinux/3.7
alpinelinux3.8      | Alpine Linux 3.8   | 3.8     | http://alpinelinux.org/alpinelinux/3.8
```

Figure 11.11 – List of known OSes

If you plan on installing Debian 10, too, you will see that the tenth version of the well-known universal OS is not on the list. Nevertheless, using the `debian10` ID will not cause any issues. The command to create a VM is the following (run as a root user):

```
virt-install --name debian-vm1 --memory 2048 --vcpus 1 --disk
size=20 --os-type=Linux --os-variant debian10 --location /tmp/
Downloads/debian-10.7.0-amd64-netinst.iso
```

Since we did not set any `-display` argument inside the `virt-install` command, the output will show two warnings:

```
WARNING  Graphics requested but DISPLAY is not set. Not running virt-viewer.
WARNING  No console to launch for the guest, defaulting to --wait -1

Starting install...
```

Figure 11.12 – Warning that there are no display and console arguments set

The only way to continue the installation is by going into the graphical user interface, starting Virtual Machine Manager, and continuing the Debian installation from there:

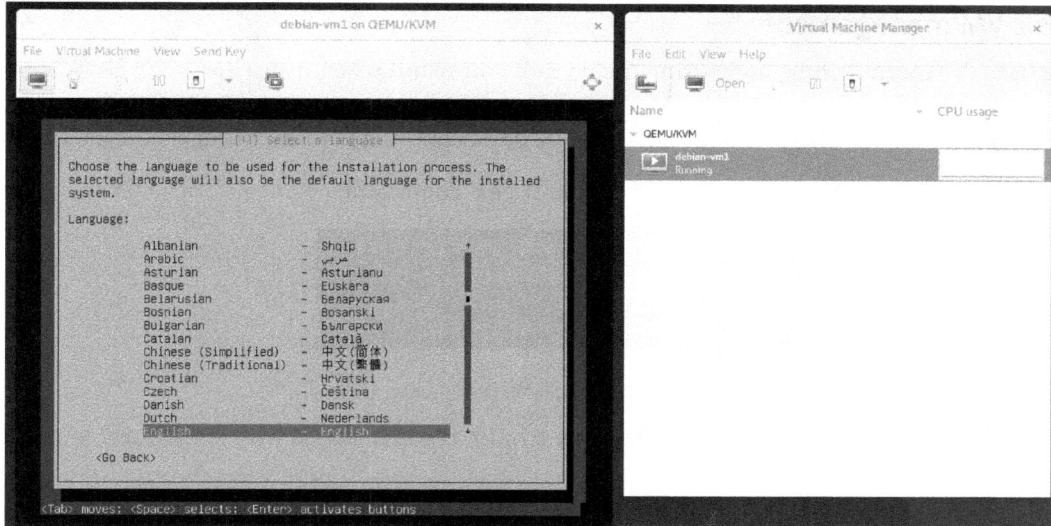

Figure 11.13 – GUI Virtual Machine Manager

To overcome the lack of graphics use, you can use the `–graphics` argument inside the command. The `virt-viewer` package needs to be installed before you can use it. In our case, it is already installed on the system. Now, let's change the `virt-install` command. The new one should appear as follows:

```
virt-install --virt-type=kvm --name Debian-vm --vcpus
1 --memory 2048 --os-variant=debian10 --os-type=Linux
--location=/tmp/debian-10.7.0-amd64-netinst.iso
--network=default --disk size=10 --graphics=vnc
```

By using the command with the `–graphics=vnc` argument, `virt-install` will start `virt-viewer`, which is the default tool for displaying the graphical console using the VNC protocol. Knowing how to create a VM is not sufficient for a system administrator. This is why, in the next section, we will show you some basic VM management tools to use.

Basic VM management

The basic VM tasks can be done using the `virsh` command when using the command-line interface, or Virtual Machine Manager when using a graphical user interface. In the following, we will show you the basic commands to use while inside a CLI.

To list the existing VM guests, use the `virsh list` command:

```
[root@localhost packt]# virsh list
 Id   Name        State
----------------------------
 1    Debian-vm   running
```

Figure 11.14 – Listing all the VMs using virsh

Be aware that listing the VMs cannot be done by just anyone. This is why the following note needs to be considered:

> **Important note**
> When trying to list the existing guest VMs, you will not get a valid output when using a regular user. You will need to be logged in as root or use `sudo` to see the list of VMs.

To change the state of a VM, such as starting, stopping, and pausing, use the following commands:

- **To force stop a VM**: `virsh destroy [vm-name]`:

```
[root@localhost packt]# virsh destroy Debian-vm
Domain Debian-vm destroyed

[root@localhost packt]# virsh list
 Id   Name    State
--------------------
```

Figure 11.15 – Force stopping a VM

- **To reboot a VM**: `virsh reboot [vm-name]`:

```
[root@localhost packt]# virsh reboot Debian-vm
Domain Debian-vm is being rebooted
```

Figure 11.16 – Rebooting a VM

- **To pause (suspend) a VM**: `virsh suspend [vm-name]`:

```
[root@localhost packt]# virsh suspend Debian-vm
Domain Debian-vm suspended

[root@localhost packt]# virsh list
 Id    Name          State
-------------------------------
 2     Debian-vm     paused
```

Figure 11.17 – Suspending a VM

- **To start a VM**: `virsh start [vm-name]`:

```
[root@localhost packt]# virsh start Debian-vm
Domain Debian-vm started

[root@localhost packt]# virsh list
 Id    Name          State
-------------------------------
 3     Debian-vm     running
```

Figure 11.18 – Starting a stopped VM

- **To resume an already suspended (paused) VM**: `virsh resume [vm-name]`:

```
[root@localhost packt]# virsh list
 Id    Name          State
-------------------------------
 3     Debian-vm     paused

[root@localhost packt]# virsh resume Debian-vm
Domain Debian-vm resumed

[root@localhost packt]# virsh list
 Id    Name          State
-------------------------------
 3     Debian-vm     running
```

Figure 11.19 – Resuming a paused VM

- **To completely delete a VM guest**: `virsh undefine [vm-name]`:

```
[root@localhost packt]# virsh list
 Id   Name          State
 --------------------------
 3    Debian-vm     running

[root@localhost packt]# virsh undefine Debian-vm
Domain Debian-vm has been undefined
```

Figure 11.20 – Deleting a VM

The command-line tools for managing VMs are powerful and offer various options. If we consider the fact that, most of the time, a system administrator will be using the CLI rather than the GUI, the ability to use command-line tools is of the utmost importance. For all the options available for `virsh`, please refer to the manual pages using the following command:

```
man virsh
```

Nevertheless, you could still use the GUI tools. All modern Linux distributions that use GNOME as the desktop environment will offer at least two useful tools: The Virtual Machine Manager and GNOME Boxes. The first is simply the GUI interface for `libvirt`, and the latter is a new and simple way to provision VMs for immediate use inside GNOME, based on QEMU/KVM technology. As you already know how to use `virsh`, we will let you explore how to use Virtual Machine Manager by yourself.

In the following section, we will show you how to use GNOME Boxes inside CentOS 8.

Using GNOME Boxes

Boxes is a relatively new tool available in GNOME. It is easy to use and offers a straightforward solution for when you require a virtual environment for testing and/or experimenting. The following is the procedure for creating a new VM using GNOME Boxes:

1. To start Boxes, open the activities menu in GNOME, or hit the Windows logo key on your keyboard. This will bring up the activities window overview, which, by default, has an application dock on the left and a virtual desktop column on the right, with a search box right at the top. Inside the search box, type Boxes and the boxes app icon will appear on screen. Hit *Enter* and the GNOME Boxes app will start. If it is the first time you are using the app, a welcome tutorial will be shown on screen. Feel free to explore it:

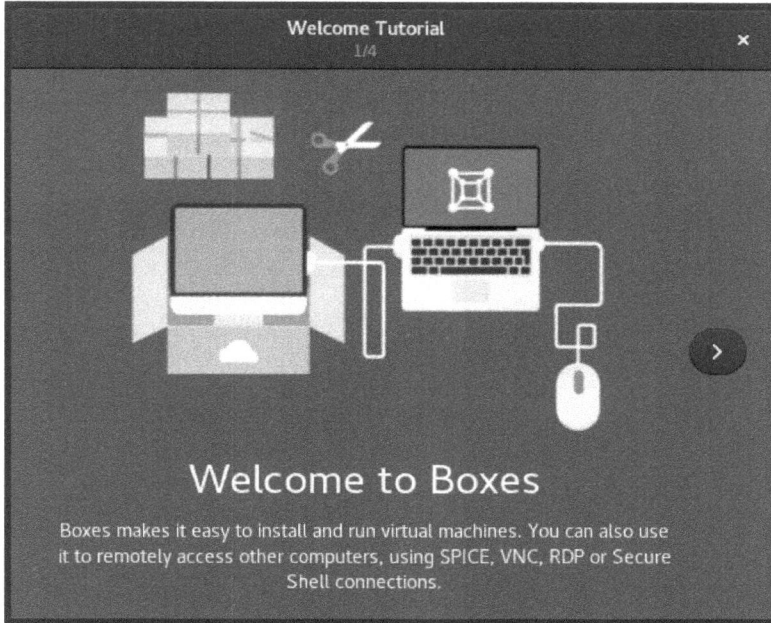

Figure 11.21 – GNOME Boxes tutorial window

2. Once you finish skimming through the tutorial, you will be able to create your first box by clicking on the + button in the upper-left corner of the app. Select **Create a Virtual Machine**:

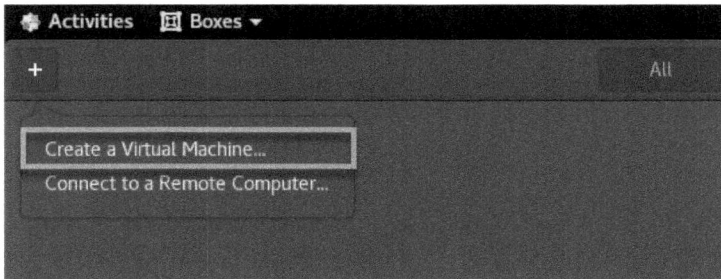

Figure 11.22 – Choosing the option to create a VM

3. A new window will appear, inside which you will be able to select the OS for the guest VM. A short list of distributions is provided upfront, but you also have the option to select another OS. We will choose that and proceed to select openSUSE Leap 15.2 from the relatively long list of distributions:

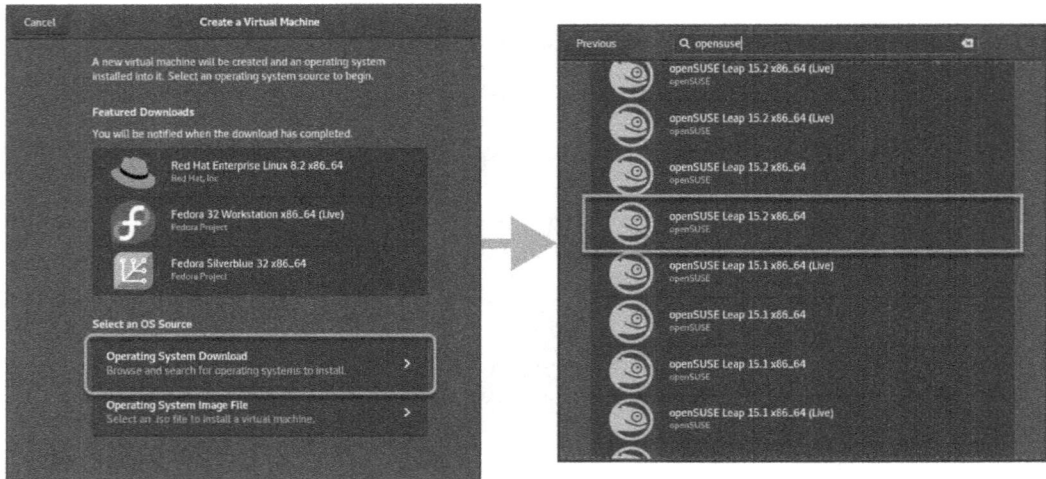

Figure 11.23 – Selecting the guest VM OS

4. Once you click on the desired OS, Boxes will automatically start to download the new VM's OS of your choosing. The download process will not be intrusive at all; simply a small message will pop up from the upper-left corner of the app. The process will take some time to finish, depending on your bandwidth:

Figure 11.24 – New box installation pop-up message

5. Once the download is complete, a new window will appear, with the option of an express installation and setting up a new user. Choose your username and password and proceed by clicking the **Next** button in the upper-right corner:

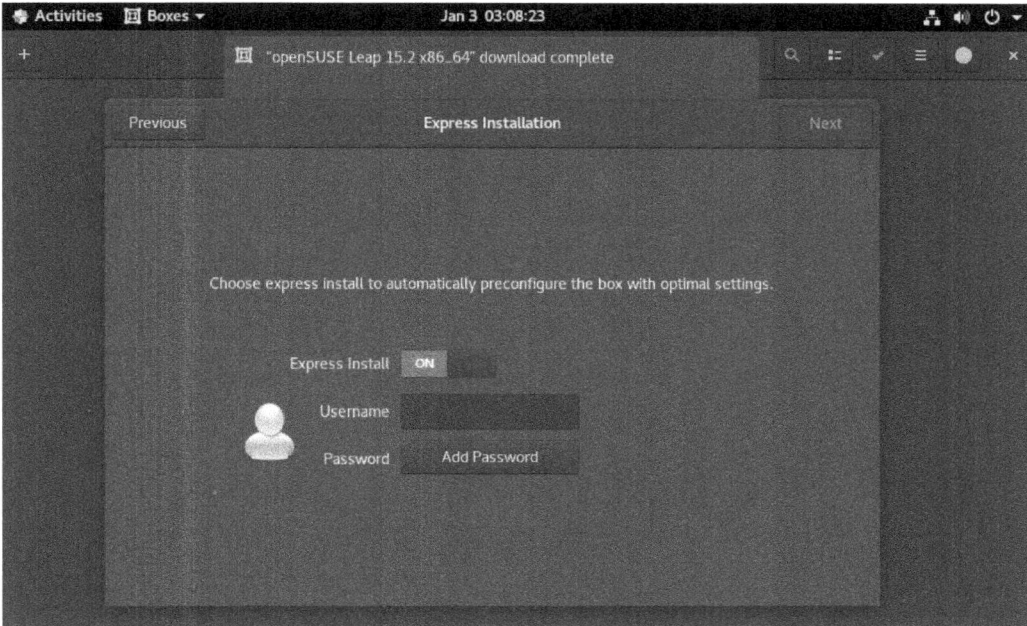

Figure 11.25 – Express installation and user creation

6. Once you have created a username and password, a new window will appear. This time, you will be able to customize the default resource allocation of the VM. You will be shown the amount of memory and disk storage already allocated. If you want to change that, click the **Customize** button. There is no vCPU change option:

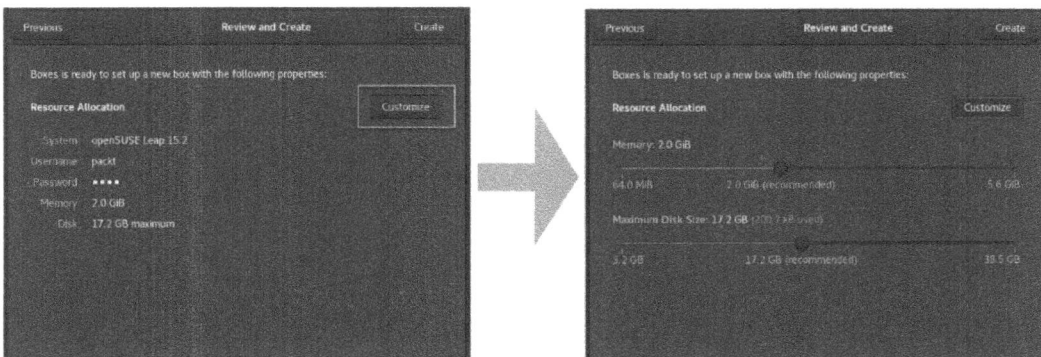

Figure 11.26 – Resource allocation

7. Once you customize the resource allocation, click on the **Create** button in the upper-right corner of the window. The VM creation will start. In the new window, the new VM will appear with a progress sign in front of the name. All you have to do is to click on it to open a new window in which you will be able to interact with the setup process. This might take some time to complete, depending on the resources you provided. Once it is finished, you will be able to use the new VM:

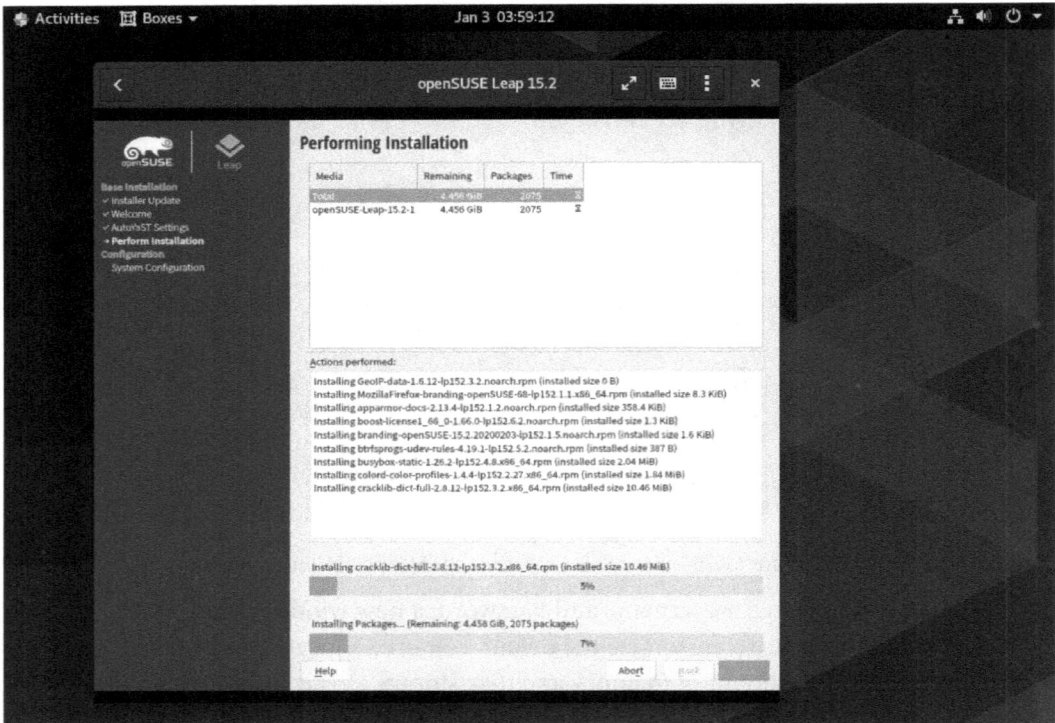

Figure 11.27 – Installing the guest OS

8. After installation, the new VM will be available to use inside Boxes:

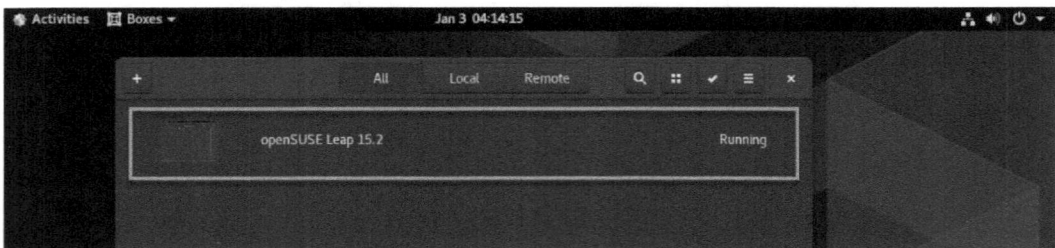

Figure 11.28 – The new VM is ready to use

9. You can now right-click on it and select the **Properties** entry. This way, you will be able to modify resources of the VM, including the number of vCPUs used. In the new **Properties** window, select the **System** tab to modify the number of vCPUs, among other options:

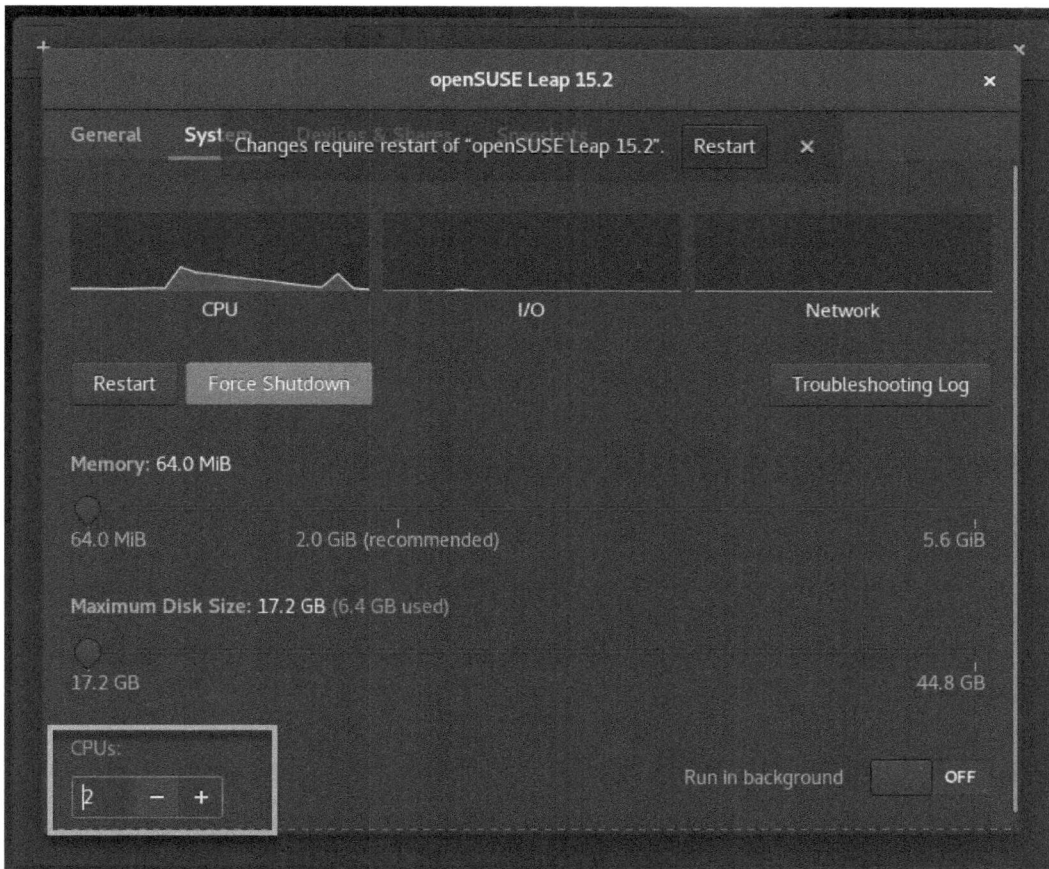

Figure 11.29 – Changing the relevant properties of the VM

10. Restart the VM after you change the number of vCPUs. From now on, the VM is at your disposal, ready to use.

Virtualization is an important part of computing, providing the technology needed to take advantage of the tremendous computing power that modern systems provide. It gives you the ability to get the most out of your investment in hardware technology. Virtualization can be used at several distinct levels:

- **Operating system-level virtualization**: When a single computer can run several different OSes

- **Server-level virtualization**: When a single server can act like many more

- **Network-level virtualization**: When isolated virtual networks can be created from a single original network

Virtualization opens doors to so much potential and administrators, developers, and users are to benefit from it. The optimization of virtualization technology had the advantage of better capabilities and use. Cloud computing benefited from the technology and philosophy of virtualization, and so did the new containers technology. In the following section, we will introduce you to containers – what they are and how they work.

Understanding Linux containers

As has already been demonstrated by now, there are two major types of virtualization: **VM-based** and **container-based**. We discussed VM-based virtualization in the previous section, and now it is time to explain what containers are. At a very basic, conceptual level, containers are similar to VMs. They have similar purposes – allowing an isolated environment to run, only that they are different in so many ways that they can hardly be called similar.

Containers versus VMs

As you already know, a VM emulates the machine's hardware and uses it as if there were several different machines available. By comparison, containers do not replicate the physical machine's hardware. They do not emulate anything.

A container shares the base OS kernel together with shared libraries and binaries needed for certain applications to run. The applications are contained inside the container, isolated from the rest of the system. They also share a network interface with the host, to offer similar connectivity as a VM.

Containers run on top of a **container engine**. Container engines offer OS-level virtualization, used to deploy and test applications by using only the requisite libraries and dependencies. This way, containers make sure that applications can run on any machine by providing the same expected behavior with the one intended by the developer. The following is a visual comparison between containers and VMs:

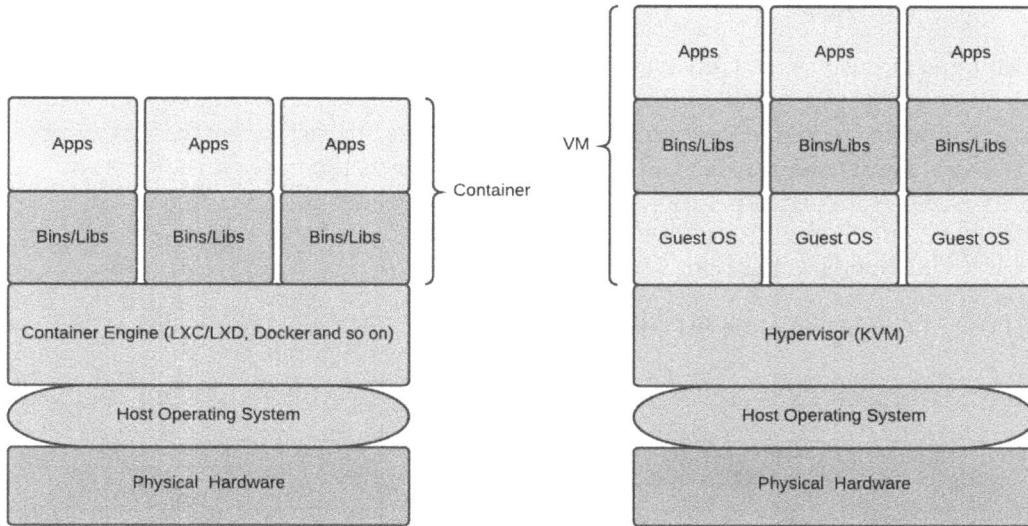

Containers versus Virtual Machines

Figure 11.30 – Containers versus VMs (general scheme)

As you can see, the containers only use the **userspace**, sharing the underlying OS-level architecture.

Historically speaking, containerization has been around for some time now. With the UNIX OS, **chroot** was the tool used for containerization since 1982. In Linux, some of the newest and most frequently used tools are Linux Containers (**LXC/LXD**), introduced in 2008, and **Docker**, introduced in 2013.

Understanding the underlying container technology

One of the earliest forms of containers was introduced 12 years ago, and it was called LXC. The newer form of containers, and the ones that changed the entire container landscape and started all the DevOps hype (more on this later), is called Docker. Containers do not abstract the hardware level like hypervisors do. They use a specific userspace interface that benefits from the kernel's techniques to isolate specific resources. By using Linux containers, you can replicate a default Linux system without using a different kernel, like you would do by using a VM.

Even though LXC is no longer that popular, it is still worth knowing. Docker has taken the crown and center stage in container engine usage. Why this LXC/LXD nomenclature? Well, LXC was the first kid on the container's block, with LXD being a newer, redesigned version of it. We will not use LXC/LXD in our examples, but we will still discuss it for backward compatibility purposes. As of the time of writing this book, there are two supported versions of LXC, version 2.0, with support until June 1, 2021, and version 3.0, with support until June 1, 2023.

According to its developers, LXC uses features to create an isolated environment that is as close as possible to a default Linux installation. Among the kernel technologies that it uses, we could bring up the most important one, which are the backbone of any container inside Linux: kernel **namespaces** and **cgroups**. Besides those, there are still chroots and security profiles for both AppArmor and SELinux. What made LXC appealing when it first appeared were the APIs it uses for multiple programming languages, including Python 3, Go, Ruby, and Haskell.

Let's now explain the basic features that Linux containers use.

Linux namespaces

What are **Linux namespaces**? In a nutshell, namespaces are the ones responsible for the isolation that containers provide. Namespaces are wrapping a global system resource inside an abstraction layer. This process is fooling any app process that is running inside the namespace to believe that the resource it is using is their own. A namespace provides **isolation** at a logical level inside the kernel and also provides **visibility** for any running processes.

To better understand how namespaces work, think of any user on a Linux system and how it can view different system resources and processes. As a user, you can see the global system resources, the running processes, other users, and kernel modules, for example. This amount of transparency could be harmful when wanting to use containers as virtualized environments at the OS level. As it cannot provide the encapsulation and emulation level of a VM, the container engine must overcome this somehow, and the kernel's low-level mechanisms of virtualizing the environment comes in the form of namespaces and cgroups.

There are several types of namespaces inside the Linux kernel, and we will describe them shortly:

- **The mount namespaces**: They restrict visibility for available filesystem mount points within a single namespace so that processes from that namespace have visibility of the filesystem list; processes can have their own root filesystem and different private or shared mounts.

- **The UTS namespaces**: Isolates the system's hostname and domain name.

- **The IPC namespaces**: Allows processes to have their own IPC shared memory, queues, and semaphores.

- **The PID namespaces**: Allows mapping of PID with the possibility of a process with PID 1 (the root of the process tree) to spin off a new tree with its own root process; processes inside a PID namespace only see the processes inside the same PID namespace.

- **The network namespaces**: Abstraction at the network protocol level; processes inside a network namespace have a private network stack with private network interfaces, routing tables, sockets, and iptables rules.

- **The user namespaces**: Allows mapping of UID and GID, including root UID 0 as a non-privileged user.

- **The cgroup namespaces**: A cgroup namespace process can see filesystem paths relative to the root of the namespace.

The preceding namespaces we detailed can be viewed by using the `lsns` command in Linux:

```
packt@neptune:~$ lsns
          NS TYPE    NPROCS    PID USER   COMMAND
4026531835 cgroup       16   6144 packt  /lib/systemd/systemd --user
4026531836 pid          16   6144 packt  /lib/systemd/systemd --user
4026531837 user         16   6144 packt  /lib/systemd/systemd --user
4026531838 uts          16   6144 packt  /lib/systemd/systemd --user
4026531839 ipc          16   6144 packt  /lib/systemd/systemd --user
4026531840 mnt          16   6144 packt  /lib/systemd/systemd --user
4026532008 net          16   6144 packt  /lib/systemd/systemd --user
```

Figure 11.31 – Using lsns to view the available namespaces

In the following section, we will break down **cgroups**, as the second major building block of containers.

Linux cgroups

What are **cgroups**? Their name comes from **control groups**, and they are kernel features that restrict and manage resource allocation to processes. Cgroups control how memory, CPU, I/O, and network are used. They provide a mechanism that determines specific sets of tasks that limit how many resources a process can use. They are based on the concept of **hierarchies**. Every child group will inherit the attributes of its parent group, and multiple cgroups hierarchies can exist at the same time on one system.

Cgroups and namespaces combined are creating the actual isolation that containers are built upon. By using cgroups and namespaces, resources are allocated and managed for each container separately. Compared to VMs, containers are lightweight and run as isolated entities.

Understanding Docker

Docker, similar to LXC/LXD, is based, among other technologies, on kernel namespaces and cgroups. Docker is a platform that is used for developing and shipping applications. The Docker platform provides the underlying infrastructure for containers to operate securely. Docker containers are lightweight entities that run directly on the host's kernel. The platform offers features such as tools to create and manage isolated, containerized applications. Thus, the container is the base unit used for application development, testing, and distribution. When apps are production ready and fit for deployment, they can be shipped as containers or as orchestrated services (we will discuss orchestration in *Chapter 14*, *Deploying Applications with Kubernetes*):

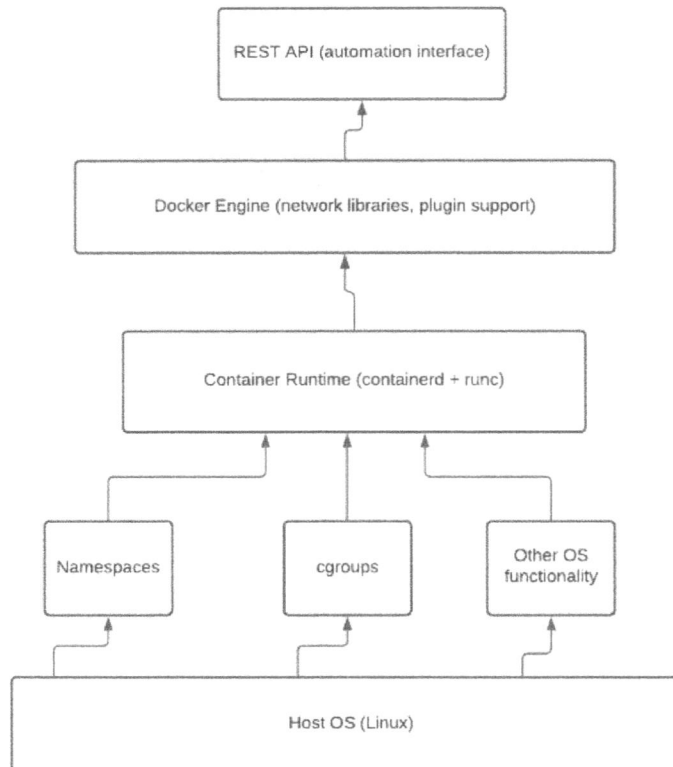

Figure 11.32 – Docker architecture

Let's explain the preceding graph. Docker is using both namespaces and cgroups available in the Linux kernel, and is split into two major components:

- **The container runtime**, which itself is split into **runc** and **containerd**

- **The Docker engine**, which is split into the **dockerd** daemon, the **API** interface, and the **CLI**

Of these components, **containerd** is the one responsible for downloading Docker images and then running them. The **runc** component is responsible for managing namespaces and cgroups for each container. The overseeing authoritative structure that governs the specification for container runtimes is called **Open Container Initiative (OCI)** and defines the open industry standards for containers. The **runc** component follows OCI specifications. According to OCI, the runtime specification defines how to download an image, unpack it, and run it using a specific filesystem bundle. OCI is part of the Linux Foundation. Docker donated **runc** and **containerd** to Cloud Native Foundation, so that more organizations would be able to contribute to both. The following is a diagram showing the details of the Docker architecture, with the core components – the Docker engine and the container runtime, being shown in detail:

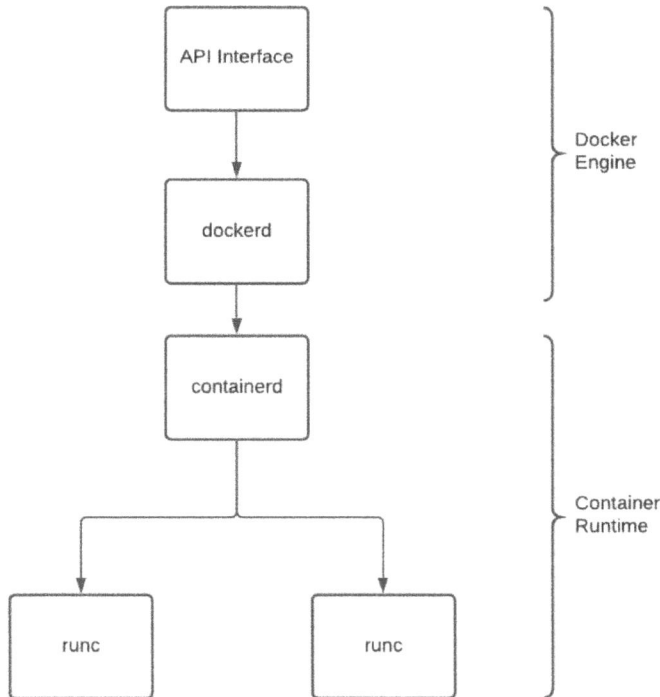

Figure 11.33 – Docker architecture details

The Docker engine is comprised of the API interface and the **dockerd** daemon, while the container runtime has two main components – the **containerd** daemon and **runc** for *namespaces* and *cgroups* management.

Besides the components listed above, in order to run and deploy Docker containers, a number of other components are used. Docker has a client-server architecture and the workflow involves a **host**, or server daemon, a **client**, and a **registry**. The host consists of images and containers (downloaded from the registry), and the client provides the commands needed to manage containers.

The workflow of these components is as follows: the daemon listens for API requests to manage services and objects (such as images, containers, networks, and volumes). The client is the way for users to interact with the daemon through the API. The registries store images, and the Docker Hub is the public registry for anyone to use freely. In addition to this, there are private registries that can be used:

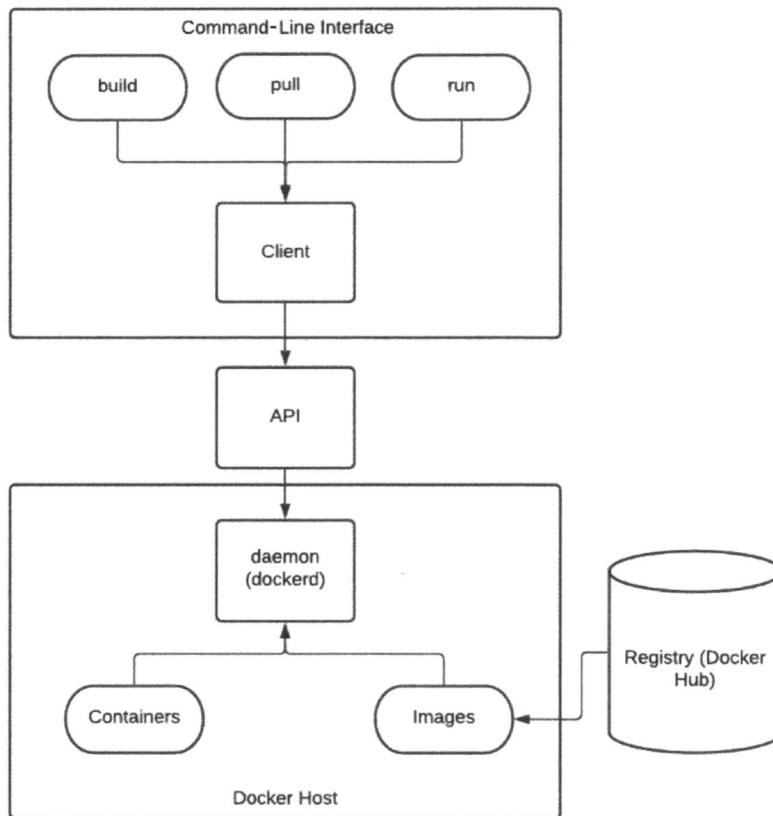

Figure 11.34 – Docker workflow

Docker may seem difficult, even disarming to a beginner. All the different components that work together, all those new typologies, and specific workflows are complicated. Do you feel like you know how Docker works just after reading this section? Of course not. The process of learning Docker has just begun. Having a strong foundation on which to build your Docker knowledge is extremely important. This is why, in the next section, we will show you how to use Docker.

Working with Docker

We will use Ubuntu 20.04.1 LTS for this section's exercises, on an Intel NUC tenth generation machine using a quad core processor, 16 GB of RAM, and a 512 GB SSD drive. The first thing we will do is install Docker Community Edition. But before we do that, let's go into a little detail about how Docker, as an entity, operates.

Which Docker version to choose?

In order for the business to be viable, Docker has two different products available, the Docker **Community Edition** (**CE**) and the Docker **Enterprise Edition** (**EE**). Out of these two, only the EE version is responsible for the revenue of Docker.

The CE edition is available for free and is open source software. The CE is available in two versions – Edge and Stable. The former is the one to deploy the latest features on a monthly release model, while the latter is offering stable software versions, tested, and with a 4-month security update interval.

The EE edition is the enterprise-ready, premium paid-for version. It is fully supported and certified by Red Hat, SUSE, and Canonical for their enterprise-ready Linux distributions. Since 2019, the Docker EE is part of Mirantis, the enterprise container orchestration provider.

For the scope of this section, we will use Docker CE.

Installing Docker CE

Depending on the version of your preferred Linux distribution that you choose, the package available inside the official repository may be out of date. Nevertheless, you have two options: one is to use the official package from the repository, and the other is to download the latest available version from the official Docker website.

As we are using a fresh system, with no prior Docker installation, we will not need to worry about older versions of the software and possible incompatibilities with the new versions.

The procedure to install Docker is as follows:

1. First, add the required packages for the Docker repository:

    ```
    sudo apt install apt-transport-https ca-certificates curl
    software-properties-common gnupg
    ```

2. In order to use the official Docker repository, you will need to add the Docker GPG key. For this, use the following command:

    ```
    packt@neptune:~$ curl -fsSL https://download.docker.com/linux/ubuntu/gpg | sudo
    apt-key add -
    OK
    ```

    Figure 11.35 – Adding the Docker GPG key

3. Set up the repository needed to install the stable version of Docker CE:

    ```
    packt@neptune:~$ sudo add-apt-repository "deb [arch=amd64] https://download.dock
    er.com/linux/ubuntu $(lsb_release -cs) stable"
    ```

    Figure 11.36 – Adding the stable repository

4. When you update the repository list, you should see the official Docker one:

    ```
    packt@neptune:~$ sudo apt update -y
    Hit:1 http://ro.archive.ubuntu.com/ubuntu focal InRelease
    Hit:2 http://security.ubuntu.com/ubuntu focal-security InRelease
    Hit:3 http://ro.archive.ubuntu.com/ubuntu focal-updates InRelease
    Hit:4 https://download.docker.com/linux/ubuntu focal InRelease
    Hit:5 http://ro.archive.ubuntu.com/ubuntu focal-backports InRelease
    ```

    Figure 11.37 – The official Docker repository on the list

5. Install the Docker CE packages, called `docker-ce`, `docker-ce-cli`, and `containerd.io`:

```
packt@neptune:~$ sudo apt install docker-ce docker-ce-cli containerd.io
Reading package lists... Done
Building dependency tree
Reading state information... Done
The following additional packages will be installed:
  docker-ce-rootless-extras git git-man liberror-perl pigz slirp4netns
Suggested packages:
  aufs-tools cgroupfs-mount | cgroup-lite git-daemon-run | git-daemon-sysvinit
  git-doc git-el git-email git-gui gitk gitweb git-cvs git-mediawiki git-svn
The following NEW packages will be installed:
  containerd.io docker-ce docker-ce-cli docker-ce-rootless-extras git git-man
  liberror-perl pigz slirp4netns
0 upgraded, 9 newly installed, 0 to remove and 0 not upgraded.
Need to get 109 MB of archives.
After this operation, 488 MB of additional disk space will be used.
Do you want to continue? [Y/n]
```

Figure 11.38 – Installing Docker CE

6. To verify that you installed the packages from the official Docker repository and not the ones from the Ubuntu repositories, run the following command. If the output shows the source from the `docker.com` website, this means that the source repository is the official Docker one:

```
packt@neptune:~$ apt-cache policy docker-ce
docker-ce:
  Installed: 5:20.10.2~3-0~ubuntu-focal
  Candidate: 5:20.10.2~3-0~ubuntu-focal
  Version table:
 *** 5:20.10.2~3-0~ubuntu-focal 500
        500 https://download.docker.com/linux/ubuntu focal/stable amd64 Packages
        100 /var/lib/dpkg/status
```

Figure 11.39 – Verifying the source repository

7. Check the status of the Docker daemon. It should be started right after installation:

```
packt@neptune:~$ sudo systemctl status docker
[sudo] password for packt:
● docker.service - Docker Application Container Engine
     Loaded: loaded (/lib/systemd/system/docker.service; enabled; vendor preset>
     Active: active (running) since Sat 2021-01-09 14:31:35 EET; 21min ago
TriggeredBy: ● docker.socket
       Docs: https://docs.docker.com
   Main PID: 7638 (dockerd)
      Tasks: 16
     Memory: 44.2M
     CGroup: /system.slice/docker.service
             └─7638 /usr/bin/dockerd -H fd:// --containerd=/run/containerd/cont>
```

Figure 11.40 – Checking the status of the Docker daemon

8. At the time of installation, a Docker group is created. In order to be able to use Docker, your user should be added to the docker group. You can either use your existing user or create a new one. After you add the user, log out and back in again and check whether you were added to the new group:

```
packt@neptune:~$ sudo usermod -aG docker ${USER}
packt@neptune:~$ groups
packt adm cdrom sudo dip plugdev lpadmin lxd sambashare docker
```

Figure 11.41 – Adding a user to the Docker group

9. You have completed the installation of Docker. Now you can enable the Docker daemon to begin at system startup:

```
packt@neptune:~$ sudo systemctl enable docker
Synchronizing state of docker.service with SysV service script with /lib/systemd
/systemd-sysv-install.
Executing: /lib/systemd/systemd-sysv-install enable docker
```

Figure 11.42 – Enabling the Docker daemon

Installing Docker is only the first step. Now let's explore what we can do with it. In the following section, you will learn about the available Docker commands.

Using the Docker commands

Working with Docker means using its **command-line interface** (**CLI**). It has a significant number of sub-commands available. If you want to see them all, you should run the `docker -help` command. There are two main command groups shown. The first group shows the management commands, while the second group shows the regular commands. We will not discuss all the commands in this section. We will only focus on the ones that you will need to get started with Docker.

Before learning anything about the commands, let's first perform a test to see whether the installation is working. We will use the `docker run` command to check whether we can access the Docker Hub and run containers:

```
packt@neptune:~$ docker run hello-world
Unable to find image 'hello-world:latest' locally
latest: Pulling from library/hello-world
0e03bdcc26d7: Pull complete
Digest: sha256:1a523af650137b8accdaed439c17d684df61ee4d74feac151b5b337bd29e7eec
Status: Downloaded newer image for hello-world:latest

Hello from Docker!
This message shows that your installation appears to be working correctly.

To generate this message, Docker took the following steps:
 1. The Docker client contacted the Docker daemon.
 2. The Docker daemon pulled the "hello-world" image from the Docker Hub.
    (amd64)
 3. The Docker daemon created a new container from that image which runs the
    executable that produces the output you are currently reading.
 4. The Docker daemon streamed that output to the Docker client, which sent it
    to your terminal.

To try something more ambitious, you can run an Ubuntu container with:
 $ docker run -it ubuntu bash
```

Figure 11.43 – Running the first docker run command

The preceding screenshot is self-explanatory and a nice touch from the Docker team. It explains what the command did in the background using clear and easy-to-understand sentences. By running the `docker run` command, you both learn about the workflow and about the success of the installation. Also, it is one of the basic Docker commands that you will use relatively often.

Let's now dig deeper and search for other images available on the Docker Hub. Let's search for an Ubuntu image to run containers on. To search for the image, we will use the `docker search` command:

```
docker search ubuntu
```

The output of the command should list all the Ubuntu images available inside the Docker Hub:

```
packt@neptune:~$ docker search ubuntu
NAME                                               DESCRIPTION                                       STARS   OFFICIAL   AUTOMATED
ubuntu                                             Ubuntu is a Debian-based Linux operating sys...   11708   [OK]
dorowu/ubuntu-desktop-lxde-vnc                     Docker image to provide HTML5 VNC interface ..    487                [OK]
websphere-liberty                                  WebSphere Liberty multi-architecture images ..    266     [OK]
rastasheep/ubuntu-sshd                             Dockerized SSH service, built on top of offi...   249                [OK]
consol/ubuntu-xfce-vnc                             Ubuntu container with "headless" VNC session...   229                [OK]
ubuntu-upstart                                     Upstart is an event-based replacement for th...   110     [OK]
neurodebian                                        NeuroDebian provides neuroscience research s...   78      [OK]
1and1internet/ubuntu-16-nginx-php-phpmyadmin-mysql-5   ubuntu-16-nginx-php-phpmyadmin-mysql-5        50                 [OK]
ubuntu-debootstrap                                 debootstrap --variant=minbase --components=m...   44      [OK]
open-liberty                                       Open Liberty multi-architecture images based...   42      [OK]
nuagebec/ubuntu                                    Simple always updated Ubuntu docker images w...   24                 [OK]
i386/ubuntu                                        Ubuntu is a Debian-based Linux operating sys...   24
1and1internet/ubuntu-16-apache-php-5.6             ubuntu-16-apache-php-5.6                          14                 [OK]
1and1internet/ubuntu-16-apache-php-7.0             ubuntu-16-apache-php-7.0                          13                 [OK]
1and1internet/ubuntu-16-nginx-php-phpmyadmin-mariadb-10   ubuntu-16-nginx-php-phpmyadmin-mariadb-10   11                 [OK]
1and1internet/ubuntu-16-nginx-php-5.6-wordpress-4  ubuntu-16-nginx-php-5.6-wordpress-4               8                  [OK]
1and1internet/ubuntu-16-apache-php-7.1             ubuntu-16-apache-php-7.1                          6                  [OK]
darksheer/ubuntu                                   Base Ubuntu Image -- Updated hourly               5                  [OK]
pivotaldata/ubuntu                                 A quick freshening-up of the base Ubuntu doc...   4
1and1internet/ubuntu-16-nginx-php-7.0              ubuntu-16-nginx-php-7.0                           4                  [OK]
pivotaldata/ubuntu16.04-build                      Ubuntu 16.04 image for GPDB compilation           2
smartentry/ubuntu                                  ubuntu with smartentry                            1                  [OK]
pivotaldata/ubuntu-gpdb-dev                        Ubuntu images for GPDB development                1
1and1internet/ubuntu-16-php-7.1                    ubuntu-16-php-7.1                                 1                  [OK]
pivotaldata/ubuntu16.04-test                       Ubuntu 16.04 image for GPDB testing               0
```

Figure 11.44 – Searching for the Ubuntu image

As you can see, the output has five columns. The first column shows the image's name, the second column shows the description, the third column shows the number of stars it has (some sort of popularity), the fourth column shows whether that image is an official one supported by the company behind the distribution/software, and the fifth column shows whether the image has automated scripts.

Once you find the image you are looking for, you can download it onto your system using the `docker pull` command:

```
packt@neptune:~$ docker pull ubuntu
Using default tag: latest
latest: Pulling from library/ubuntu
da7391352a9b: Pull complete
14428a6d4bcd: Pull complete
2c2d948710f2: Pull complete
Digest: sha256:c95a8e48bf88e9849f3e0f723d9f49fa12c5a00cfc6e60d2bc99d87555295e4c
Status: Downloaded newer image for ubuntu:latest
docker.io/library/ubuntu:latest
```

Figure 11.45 – Downloading the image you want with the docker pull command

With this command, the image is downloaded locally onto your computer. Now, containers can be run using this image. To list the images that are already available on your computer, run the `docker images` command:

```
packt@neptune:~$ docker images
REPOSITORY     TAG       IMAGE ID       CREATED         SIZE
ubuntu         latest    f643c72bc252   6 weeks ago     72.9MB
hello-world    latest    bf756fb1ae65   12 months ago   13.3kB
```

Figure 11.46 – A list of the downloaded images

Please note the small size of the Ubuntu Docker image. You may be wondering why is it so small? Because Docker images contain only the base and minimum packages needed to run. This makes the container running on the image extremely efficient in resource usage.

Those few commands we showed you are the most basic ones needed to start using Docker. Now that you know how to download an image, let's show you how to run Docker containers.

Managing Docker containers

We will use the Ubuntu image downloaded earlier. To run it, we will use the `docker run` command with two arguments, `-i` and `-t`, which will give us interactive access to the shell:

```
packt@neptune:~$ docker run -it ubuntu
root@d70e5ff728b1:/#
```

Figure 11.47 – Running a Docker container

You will notice that your command prompt will change. Now it will contain the container ID. The user, by default, is the root user. Basically, you are now inside an Ubuntu image, so you can use it exactly as you would use any Ubuntu command line. You can update the repository, install the requisite applications, remove unnecessary apps, and so on. Any changes that you do make to the container image stay inside the container. To exit the container, simply type `exit`.

You can open a new terminal on your system and check to see how many Docker containers are actively running. Do not close the terminal in which the Ubuntu-based container is currently running.

In the output, you will see the ID of the container that is running in the other terminal. There are also details about the command that run inside the container and the creation time. There are several arguments that you can use with the docker ps command. If you want to see all active and inactive containers, use the docker ps -a command. If you want to see the latest created container, use the docker ps -l command. The following is the output of all three variants of the docker ps command:

```
packt@neptune:~$ docker ps
CONTAINER ID    IMAGE        COMMAND      CREATED         STATUS          PORTS       NAMES
d70e5ff728b1    ubuntu       "/bin/bash"  12 minutes ago  Up 12 minutes               hopeful_ramanujan
packt@neptune:~$ docker ps -a
CONTAINER ID    IMAGE         COMMAND      CREATED         STATUS                PORTS       NAMES
d70e5ff728b1    ubuntu        "/bin/bash"  27 minutes ago  Up 27 minutes                     hopeful_ramanujan
d31545dc6fde    hello-world   "/hello"     6 hours ago     Exited (0) 6 hours ago            serene_kapitsa
packt@neptune:~$ docker ps -l
CONTAINER ID    IMAGE        COMMAND      CREATED         STATUS          PORTS       NAMES
d70e5ff728b1    ubuntu       "/bin/bash"  27 minutes ago  Up 27 minutes               hopeful_ramanujan
```

Figure 11.48 – Listing containers with the docker ps command

In the output, you will also see names assigned to containers, such as hopeful_ramanujan and serene_kapitsa. Those are random names automatically given to containers by the daemon.

When managing containers, such as starting, stopping, or removing a container, you can refer to them by using the container ID or the name assigned by Docker. Let's now show you how to start, stop, and remove a container:

```
packt@neptune:~$ docker ps
CONTAINER ID    IMAGE      COMMAND      CREATED    STATUS      PORTS      NAMES
packt@neptune:~$ docker start hopeful_ramanujan
hopeful_ramanujan
packt@neptune:~$ docker ps
CONTAINER ID    IMAGE      COMMAND          CREATED             STATUS          PORTS
    NAMES
d70e5ff728b1    ubuntu     "/bin/bash"      About an hour ago   Up 3 seconds
    hopeful_ramanujan
packt@neptune:~$ docker stop hopeful_ramanujan
hopeful_ramanujan
packt@neptune:~$ docker ps
CONTAINER ID    IMAGE      COMMAND      CREATED    STATUS      PORTS      NAMES
packt@neptune:~$ docker rm hopeful_ramanujan
hopeful_ramanujan
```

Figure 11.49 – Starting, stopping, and removing a container

In the preceding output, we are first listing the active containers with the `docker ps` command. The result is showing that no containers are active. Then we start the Ubuntu-based Docker container named `hopeful_ramanujan` using the `docker start` command. Immediately after this, we use the `docker ps` command once again to see that the container is listed as active. We then stop the container with the `docker stop` command and run `docker ps` once again to make sure that it is stopped. Then, we remove the container using the `docker rm` command.

Once you remove the container, any changes that you made and that you did not save (commit) will be lost. Let's now show you how to commit changes you made in a container to the Docker image. This means that you will save a specific state of a container as a new Docker image.

Let's say that you would like to develop, test, and deploy your Python application on Ubuntu. The default Docker image of Ubuntu doesn't have Python installed. In the following screenshot, we show you how we started the container and how we checked to see whether Python was installed:

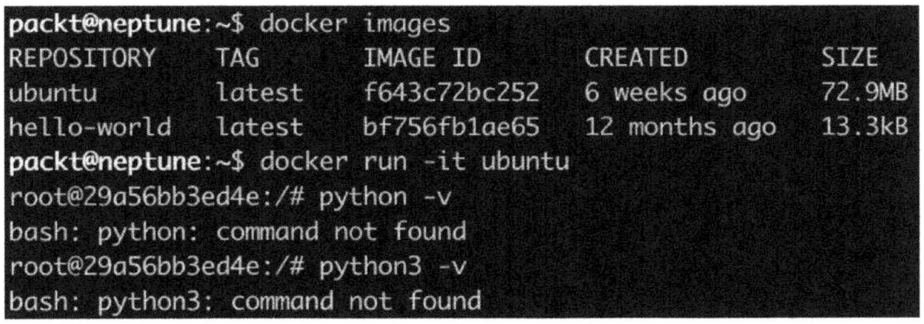

Figure 11.50 – Checking for Python in the container

We check for both Python 2 and Python 3, but neither version is installed on the image. As we want to use the latest version of the programming language, we will use the following command to install Python 3 support:

```
apt install python3
```

Now, with Python 3 installed, we can save the instance of the container to a new Docker image. For this, we will use the `docker commit` command. When using this command, you will save the new image onto your local computer, but there is also the possibility to save it to the Docker Hub, for others to use it too. To save to the Docker Hub, you will need to have an active Docker user created. For now, we will save the new image locally:

```
packt@neptune:~$ docker commit -m "added python 3 to ubuntu" -a "packt user" 47f
75b9281d1 packt/ubuntu-python3
sha256:8c93d9ac6e7591d6d036fceeea999bc215c4b8628c3c5f96d2732855abc25d14
packt@neptune:~$ docker images
REPOSITORY              TAG        IMAGE ID        CREATED          SIZE
packt/ubuntu-python3    latest     8c93d9ac6e75    14 seconds ago   144MB
ubuntu                  latest     f643c72bc252    6 weeks ago      72.9MB
hello-world             latest     bf756fb1ae65    12 months ago    13.3kB
```

Figure 11.51 – A new image committed locally

Notice the increased size of the image we just saved. Installing Python 3 more than doubled the size of the initial Ubuntu image.

By now, you learned how to use extremely basic Docker commands for opening, running, and saving containers. In the final part of this section, we will show you how to use Docker to deploy a very basic application. We will make it so simple that the app to deploy will be a basic static presentation website.

Deploying a containerized application with Docker

For the exercise in this section, we will use a free website template randomly downloaded from the internet (the download link is `https://www.free-css.com/free-css-templates/page262/focus`). We are downloading the file inside our home directory. It is a compressed zip file. We assume you already know how to extract the ZIP file, but we will give you a hint; you can use the `unzip` command:

```
packt@neptune:~$ sudo wget https://www.free-css.com/assets/files/free-css-templa
tes/download/page262/focus.zip
--2021-01-10 23:17:51--  https://www.free-css.com/assets/files/free-css-template
s/download/page262/focus.zip
Resolving www.free-css.com (www.free-css.com)... 217.160.0.242, 2001:8d8:100f:f0
00::28f
Connecting to www.free-css.com (www.free-css.com)|217.160.0.242|:443... connecte
d.
HTTP request sent, awaiting response... 200 OK
Length: 911837 (890K) [application/zip]
Saving to: 'focus.zip'

focus.zip           100%[===================>] 890,47K  4,30MB/s    in 0,2s

2021-01-10 23:17:52 (4,30 MB/s) - 'focus.zip' saved [911837/911837]
```

Figure 11.52 – Downloading a zip file with the website contents

Once the file is downloaded and the archive extracted, you can proceed to creating a **Dockerfile** inside the same directory. As this is the first time you are encountering this type of file, let's explain what it is. A Dockerfile is a text-based file that contains the commands the user will execute when creating an image. This file is used by Docker to automatically build images based on the information the user provides inside the file.

In the following, we will create a Dockerfile inside our present working directory, which is the one where the archive was extracted. The contents of the Dockerfile are shown as follows:

```
GNU nano 4.8                    Dockerfile                 Modified
FROM nginx
COPY . /usr/share/nginx/html_
```

Figure 11.53 – The Dockerfile contents

The file is simple and has only two lines. The first line, using the FROM keyword, specifies the base image that we will use, which will be the official NGINX image available on the Docker Hub. The second line, using the COPY keyword, specifies the location where the contents of our present working directory will be copied inside the new container. The following action is to build the Docker image using the docker build command:

```
packt@neptune:~/focus$ nano Dockerfile
packt@neptune:~/focus$ ls
Dockerfile  css  images  index.html  js
packt@neptune:~/focus$ docker build -t static-website .
Sending build context to Docker daemon  1.921MB
Step 1/2 : FROM nginx
latest: Pulling from library/nginx
6ec7b7d162b2: Pull complete
cb420a90068e: Pull complete
2766c0bf2b07: Pull complete
e05167b6a99d: Pull complete
70ac9d795e79: Pull complete
Digest: sha256:4cf620a5c81390ee209398ecc18e5fb9dd0f5155cd82adcbae532fec94006fb9
Status: Downloaded newer image for nginx:latest
 ---> ae2feff98a0c
Step 2/2 : COPY . /usr/share/nginx/html
 ---> 25437390a7a1
Successfully built 25437390a7a1
Successfully tagged static-website:latest
```

Figure 11.54 – Using the docker build command

The command highlighted above is using the -t parameter to create a tag for the new image. You can verify whether the new image was created using the docker images command:

```
packt@neptune:~/focus$ docker images
REPOSITORY             TAG      IMAGE ID        CREATED          SIZE
static-website         latest   25437390a7a1    12 minutes ago   135MB
packt/ubuntu-python3   latest   8c93d9ac6e75    3 hours ago      144MB
nginx                  latest   ae2feff98a0c    3 weeks ago      133MB
ubuntu                 latest   f643c72bc252    6 weeks ago      72.9MB
hello-world            latest   bf756fb1ae65    12 months ago    13.3kB
```

Figure 11.55 – Verifying whether the new image was created

As the output shows that the new image called static-website was created, you can start a new container using it. As we will need to access the container from the outside, we will need to open specific ports, and we will do that using the -p parameter inside the docker run command. We can either specify a single port or a range of ports. When specifying ports, we will give the ports for the container and for the host, too. We will use the -d parameter to detach the container and run it in the background:

```
packt@neptune:~/focus$ docker run -it -d -p 8080:80 static-website
7af4bcf4b3e898e189c1458cd7c31f4c5c89e5170d8acdcb79a6ef7386641db4
```

Figure 11.56 – Creating a new container using the docker run command

As you can see in the preceding screenshot, we are exposing host port 8080 to port 80 on the container. We could have used both ports 80, but on the host, it might be occupied by other services. You can now access the new containerized application by going to your web browser and typing in the local IP address and port 8080 in the address bar.

Now, we have shown you how to use Docker, how to manage containers, and how to deploy a basic website inside a container. Docker is so much more than that, but this is enough to get you started and make you want to learn more.

Summary

In this chapter, we emphasized the importance of virtualization and containerization.

We showed you the difference between VMs and containers. You learned how to deploy a VM using KVM. We also showed you how to quickly create a VM inside GNOME using Boxes. With those two assets, you are prepared to start your path into virtualization with no fears. We also showed you what containers are, how they work, and why they are so important. Containers are the foundation of the modern DevOps revolution and you are now ready to use them. We also taught you about Docker, the basic commands for a sleek use. You are now ready to start the cloud journey, as this will be the final part of this book.

Virtualization and container technologies are at the heart of cloud technologies. This is why, in the next chapter, we will introduce you to the basics of cloud technologies, OpenStack, AWS, Azure, Ansible, and Kubernetes.

Further reading

For more information on the topics covered in this chapter, you can refer to the following books:

- *Docker Quick Start Guide, Earl Waud, Packt Publishing* (https://www.packtpub.com/product/docker-quick-start-guide/9781789347326)

- *Mastering KVM Virtualization – Second Edition, Vedran Dakic, Humble Devassy Chirammal, Prasad Mukhedkar, Anil Vettathu, Packt Publishing* (https://www.packtpub.com/product/mastering-kvm-virtualization-second-edition/9781838828714)

- *Containerization with LXC, Konstantin Ivanov, Packt Publishing* (https://www.packtpub.com/product/containerization-with-lxc/9781785888946)

12
Cloud Computing Essentials

In this chapter, you will learn the basics of cloud computing and will be presented with the core foundations of cloud infrastructure technologies. You will learn about the *as-a-service* solutions such as **Infrastructure as a Service (IaaS)**, **Platform as a Service (PaaS)**, **Software as a Service (SaaS)**, and **Containers as a Service (CaaS)**. You will be presented with the basics of cloud standards, **Development and Operations (DevOps)**, **continuous integration/continuous deployment (CI/CD)**, and microservices. A base knowledge of the cloud will imply at least a basic introduction to OpenStack, **Amazon Web Services (AWS)**, Azure, and other cloud solutions. By the end of this chapter, we will introduce you to technologies such as Ansible and Kubernetes. This chapter will provide a short and concise theoretical introduction that will be the foundation for the following three cloud-related chapters, which will provide you with important practical knowledge on the many solutions presented here.

In this chapter, we're going to cover the following main topics:

- Introduction to cloud technologies
- Short introduction to OpenStack
- Introducing IaaS solutions
- Introducing PaaS solutions

- Introducing CaaS solutions
- Introducing DevOps
- Introducing cloud management tools

Technical requirements

No special technical requirements are needed as this chapter is a purely theoretical one. All you need is a desire to learn about cloud technologies.

The code for this chapter is available at the following link: `https://github.com/PacktPublishing/Mastering-Linux-Administration`.

Introduction to cloud technologies

The term *cloud computing*, or the simple alternative *cloud*, is not missing from any tech enthusiast's or **Information Technology (IT)** professional's vocabulary these days. You don't even have to be involved in IT at all to hear (or even use) the term *cloud* relatively often. Today's computing landscape is changing at a rapid pace, and the pinnacle of this change is the cloud and the technologies behind it. According to the literature, the term *cloud computing* was used for the first time in 1996, inside a business plan from Compaq (`https://www.technologyreview.com/2011/10/31/257406/who-coined-cloud-computing/`).

Cloud computing is a relatively old concept, even though it was not referred to as this term right from the beginning. Cloud computing is a computing model that was used from the early days of computing. Back in the 1950s, for example, there were mainframe computers that were accessed from different terminals. This model is similar in modern cloud computing, where services are hosted and delivered over the internet to different terminals, from desktop computers to smartphones, tablets, or laptops. This model is based on technologies that are extremely complex and essential to know by anyone who wants to master them.

Why an entire section dedicated to cloud computing and technologies inside a *Mastering Linux* title? Simply because Linux has taken over the cloud in the last decade, in the same way that Linux took over the internet and the high-performance computing landscape. According to the *TOP500* association, the world's top 500 supercomputers all run on Linux (`https://top500.org/lists/top500/2020/11/`). Clouds need to have an operating system to operate on, but it doesn't have to be Linux. Nevertheless, Linux runs on almost 90% of public clouds (`https://www.redhat.com/en/resources/state-of-linux-in-public-cloud-for-enterprises`), mostly because its open source nature appeals to IT professionals inside the public and private sectors alike.

In the following section, we will tackle the subject of cloud standards and why is it good to know about them when planning to deploy or manage a cloud instance.

Understanding the need for cloud computing standards

Before going into more details about cloud computing, let's give you a short introduction about what cloud standards are and their importance in the overall contemporary cloud landscape. You may know that almost every activity in the wider **Information and Communications Technology** (**ICT**) spectrum is governed by some kind of standard or regulation.

Cloud computing is no wild land, and you will be surprised of how many associations, regulatory boards, and organizations are involved in developing standards and regulations for it. Covering all these institutions and standards is out of the scope of this book and chapter, but we will describe some of the most important and relevant ones (in our opinion) so that you can have an idea of their importance in keeping clouds together and web applications running.

International Organization for Standardization/International Electrotechnical Commission

Two of the most widely known standards entities are the **International Organization for Standardization** (**ISO**) and the **International Electrotechnical Commission** (**IEC**), and they currently have 28 published and under-development standards on cloud computing and distributed platforms. They have a joint task group to develop standards for specific cloud core infrastructure, consumer application platforms, and services. Those standards are found under the responsibility of the **Joint Technical Committee 1** (**JTC 1**) **subcommittee 38** (**SC38**), or *ISO/IEC JTC 1/SC 38* for short.

Examples of standards from ISO/IEC include—but are not limited to—cloud computing **service-level agreement** (**SLA**) frameworks (*ISO/IEC 19086-1:2016, ISO/IEC 19086-2:2018, ISO/IEC 19086-3:2017*); cloud computing **service-oriented architecture** (**SOA**) frameworks (*ISO/IEC 18384-1:2016, ISO/IEC 18384-2:2016, ISO/IEC 18384-3:2016*); **Open Virtualization Format** (**OVF**) specifications (*ISO/IEC 17203:2017*); cloud computing **data sharing agreement** (**DSA**) frameworks (*ISO/IEC CD 23751*); **distributed application platforms and services** (**DAPS**) technical principles (*ISO/IEC TR 30102:2012*); and others. To take a closer look at those standards, please go to https://www.iso.org/committee/601355/x/catalogue/. Next on our list of standards development entities is an initiative called the **Cloud Standards Coordination** (**CSC**), created by the **European Commission** (**EC**), together with specialized bodies.

The CSC initiative

Back in 2012, the EC, together with the **European Telecommunications Standards Institute** (**ETSI**), launched the CSC to develop standards and policies for cloud security, interoperability, and portability. The initiative had two phases, with phase 1 starting in 2012 and phase 2 starting in 2015. The final reports of phase 2 (version 2.1.1), were made public, as follows: cloud computing users' needs (*ETSI SR 003 381*); Standards and Open Source (*ETSI SR 003 382*); Interoperability and Security (*ETSI SR 003 391*); and Standards Maturity Assessment (*ETSI SR 003 392*). For more details on each of those standards, access the following link: `http://csc.etsi.org/`. The list continues with one of the most widely known entities in standards development: the **United States** (**US**) **National Institute of Standards and Technology** (**NIST**).

NIST

This will not be the first time you will read about NIST in this book. NIST is the standards development body inside the US Department of Commerce. The main objective of NIST is the standardization of security and interoperability inside US government agencies, so anyone interested in developing for those entities should take a look at the NIST cloud documentation. The NIST document that standardizes cloud computing is called *NIST SP 500-291r2* and can be found at `http://csrc.nist.gov/publications/ nistpubs/800-145/SP800-145.pdf`. We will close our short listing with one of the oldest—if not the oldest—standards development bodies, part of the **United Nations** (**UN**) organization: the **International Telecommunication Union** (**ITU**).

The ITU

The ITU is a body inside the UN, and its main focus is to develop standards for communications, networking, and development. This agency was founded in 1865 and, among other things, it is responsible for global radio frequency spectrum and satellite orbit allocation. It is also responsible for the use of Morse code as a standard means of communication. When it comes to global information infrastructure, internet protocols, next-generation networks, **Internet of Things** (**IoT**), and smart cities, the ITU has a lot of standards and recommendations available. To check them all out, have a look at the following link: `https://www.itu.int/rec/T-REC-Y/en`. To narrow down the document list from the aforementioned link, some specific cloud computing documents can be found using document codes, starting with *Y.3505* up to *Y.3531*. The cloud computing standards were developed by the **Study Group 13** (**SG13**) **Joint Coordination Activity on Cloud Computing** (**JCA-Cloud**) inside the ITU.

Besides the entities described in this section, there are many others, such as the **Cloud Standards Customer Council (CSCC)**, the **Distributed Management Task Force (DMTF)**, and the **Organization for the Advancement of Structured Information Standards (OASIS)**, just to name a few. The main reason for adopting standards for cloud computing is ease of use when it comes to either the **Cloud Service Provider (CSP)** or the client. Both categories need to have an easy access to data, more so for CSPs and application developers, as easy access to data is translated into agility and interoperability. Standardization of application frameworks, network protocols, and **application programming interfaces (APIs)** guarantees success for everyone involved.

Understanding the cloud through APIs

Standards, besides being technically correct, need to be consistent and persistent. According to the literature, there are two main standards groups: ones that are established from practice, and ones that are being regulated. An important part of the cloud standards, from the second category, is the API. APIs are sets of protocols, procedures, and functions: all the bricks needed to build a web-distributed application. Modern APIs emerged at the beginning of the 21st century, firstly as a theory in Roy Fielding's doctoral dissertation. Before the modern APIs, there were SOA standards and the **Simple Object Access Protocol (SOAP)**, based on **Extensible Markup Language (XML)**. Modern APIs are based on a new application architectural style, called **REpresentational State Transfer (REST)**.

REST APIs are based on a series of architectural styles, elements, connectors, and views that are clearly described by Roy Fielding in his thesis. There are six guiding constraints for an API to be RESTful, and those are: uniform user interfaces, client-server clear delineation, stateless operations, cacheable resources, layered system of servers, and code on demand execution. Nevertheless, following those guiding principles is far from following a standard, but REST provides them for developers as a high-level abstraction layer. Unless they are standardized, they will always remain great principles that generate confusion and frustration among developers.

The only organization that managed to standardize REST APIs for the cloud is the DMTF, through the **Cloud Infrastructure Management Interface (CIMI)** model and the RESTful **HyperText Transfer Protocol (HTTP)**-based protocol, in a document coded *DSP0263v2*, which can be downloaded from the following link: `https://www.dmtf.org/standards/cloud`.

There are other specifications that emerge as possible future standards for developers to use when designing REST APIs. Among those, there is the **OpenAPI Specification (OAS)**, an industry standard that provides a language-agnostic description for API development (document available at `http://spec.openapis.org/oas/v3.0.3`), and **GraphQL**, as a query language and server-side runtime, with support for several programming languages such as Python, JavaScript, Scala, Ruby, and **PHP: Hypertext Preprocessor (PHP)**.

REST managed to become the preferred API because it is easier to understand, more lightweight, and simple to write. It is more efficient, uses less bandwidth, supports many data formats, and uses **JavaScript Object Notation (JSON)** as the preferred data format. JSON is easy to read and write and offers better interoperability between applications written in different languages, such as JavaScript, Ruby, Python, Java, and others. By using JSON as the default data format for the API, it makes it friendly, scalable, and platform-agnostic.

APIs are everywhere on the web and in the cloud and are the base for SOA and microservices. For example, microservices use RESTful APIs to communicate between services, by offering an optimized architecture for cloud-distributed resources.

Therefore, if you want to master cloud computing technologies, you should be open to embracing the cloud standards. In the next section, we will discuss the cloud types and architecture.

Knowing the architecture of the cloud

The cloud's architectural design is similar to a building's architectural design. There is one design paradigm that governs the cloud—the one in which the design starts from a blank, clean drawing board where the architects put together different standardized components in order to achieve an architectural design. The final result is based on a certain architectural style. The same happens when designing the cloud's architecture.

The cloud is based on a client-server, layered, stateless network-based architectural style. The REST APIs, SOA, microservices, and web technologies all are the base components that form the foundation of the cloud. The architecture of the cloud has been defined by the US NIST, part of the US Department of Commerce (`https://www.nist.gov/publications/nist-cloud-computing-reference-architecture`).

Some of the technologies behind the cloud have been discussed in *Chapter 11*, *Working with Containers and Virtual Machines*. Indeed, both virtualization and containers are the foundation technologies of cloud computing.

Let's imagine a situation where you would like to have several Linux systems to deploy your apps. What you do first is go to a CSP and request the systems you need. The CSP will create the **virtual machines** (**VMs**) onto its infrastructure, according to your needs, and will put all of them in the same network and share the credentials to access them with you. This way, you will have access to the systems you wanted, in exchange for a subscription fee that is billed either on a daily, monthly, or yearly basis, or based on a resource-consumption basis. Most of the time, those CSP requests are done through a specific web interface, developed by the provider to best suit the needs of their users.

Everything that the cloud uses as technology is based on VMs and containers. Inside the cloud, everything is abstracted and automated.

Describing the cloud types

No matter the type, each cloud has a specific **architecture**, just as we showed you in the previous section. It provides *the blueprints* for the foundation of cloud computing. The cloud architecture is the base for the cloud infrastructure, and the infrastructure is the base for cloud services. See how everything is connected? Let's now see what the infrastructure and services are with regard to the cloud.

There are four main cloud infrastructure types, as follows:

- **Public clouds**: These run on infrastructure owned by the provider and are available mostly off-premises; the largest public cloud providers are AWS, Microsoft Azure, and Google Cloud. A public cloud infrastructure type is depicted in the following diagram:

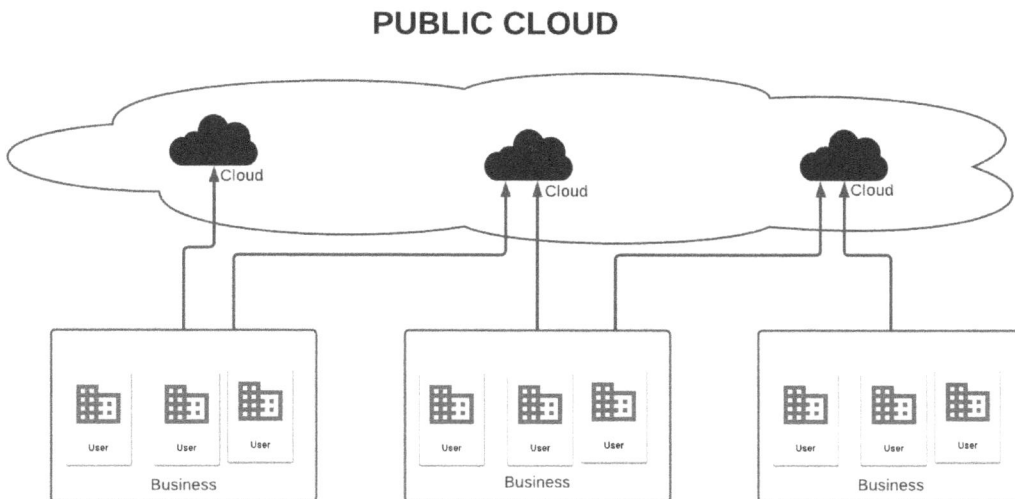

Figure 12.1 – Diagram showing a public cloud type

- **Private clouds**: These run specifically for individuals and groups with isolated access; they are available on both on-premises or off-premises hardware infrastructure. There are managed private clouds available, or dedicated private clouds.

- **Hybrid clouds**: These are both private and public clouds running inside connected environments, with resources available for potential on-demand scaling.

- **Multi-clouds**: These are more than one cloud running from more than one provider.

The different cloud types are depicted in the following diagram:

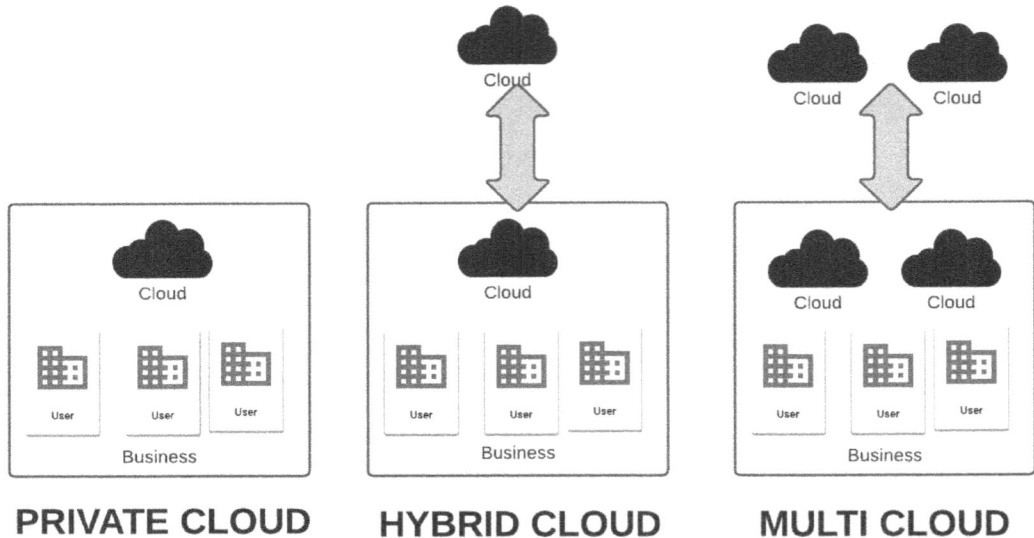

Figure 12.2 – Diagram showing private, hybrid, and multi-cloud types

The preceding diagram shows how private, hybrid, and multi-cloud types would work on a theoretical level. For a better understanding of the concept of a private cloud, we showed the cloud running on-premises, with restricted access outside the business. A hybrid cloud type is shown for a business that uses both private and public clouds, and a multi-cloud type shows how a business could run multiple private, public, or hybrid clouds.

Besides those, there are three main cloud service types:

- **IaaS**: With an IaaS cloud service type, the cloud provider manages all hardware infrastructure such as servers and networking, plus virtualization and storage of data. The infrastructure is owned by the provider and rented by the user; in this case, the user needs to manage the operating system, the runtimes, automation, management solutions, and containers, together with the data and applications. IaaS is the backbone of every cloud computing service, as it provides all the resources.

- **CaaS**: This is considered to be a *subset of IaaS*; it has the same advantages as IaaS, only that the base consists of containers, not VMs, and it is better suited to deploying distributed systems and microservices architectures.

- **PaaS**: With a PaaS cloud service type, the hardware infrastructure, networking, and software platform are managed by the cloud provider; the user manages and owns the data and applications—thus DevOps. Here is a diagrams showing the IaaS, CaaS, and PaaS models:

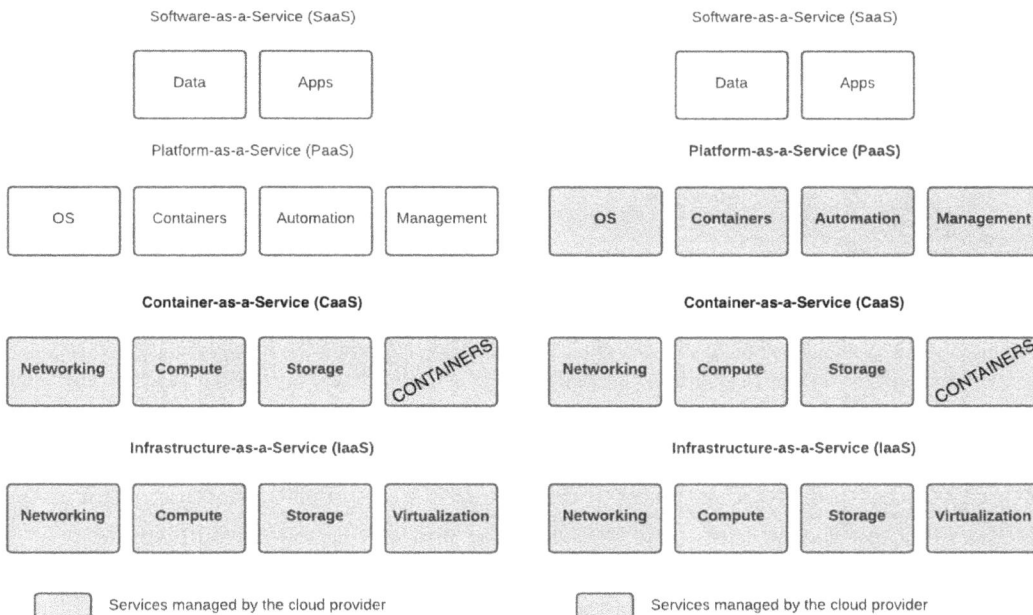

Figure 12.3 – Diagram for IaaS, CaaS, and PaaS

- **SaaS**: With a SaaS cloud service type, the cloud provider manages and owns the hardware, networking, software platform, management, and software applications. This type of service is also known for delivering web apps or mobile apps.

- Besides those types of services, there is another one that we should bring into discussion: **serverless computing** services. Opposed to what the name might suggest, serverless computing still implies the use of servers, but the infrastructure running them is not visible to users, which in most cases are developers. Serverless is similar to SaaS; actually, it would fit right between PaaS and SaaS. It has no infrastructure management, is scalable, offers a faster way to market for app developers, and is efficient when it comes to the use of resources. Here is a diagram showing the components of SaaS and serverless types:

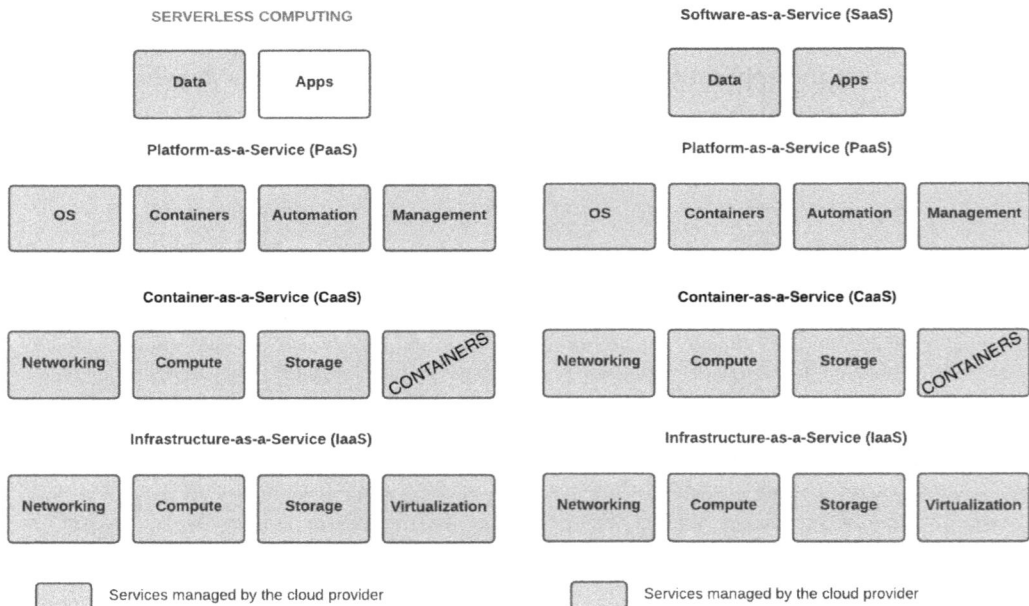

Figure 12.4 – Diagram showing SaaS and serverless types

Now that you know the types of cloud infrastructure and services, you might wonder why you, your business, or anyone you know should migrate to cloud services. First of all, cloud computing is based on an on-demand access to various resources that are hosted and managed by a CSP. This means that the infrastructure is owned or managed by the CSP and the user will be able to access the resources based on a subscription fee. Should you migrate to the cloud? We will discuss the advantages and disadvantages of migrating to the cloud in the next section.

Knowing the key features of cloud computing

Before deciding whether migrating to the cloud would be a good decision, you need to know the advantages and disadvantages of doing this. Cloud computing does provide some essential features, such as the following ones listed:

- **Cost savings**: There are reduced costs generated by the infrastructure setup, which is now managed by the CSP; this puts the user's focus on application development and running the business.

- **Speed**, **agility**, and **resource access**: All the resources are available from any place, just a few clicks away, at any time (dependent on internet connectivity and speed).

- **Reliability**: Resources are hosted in different locations, by providing good quality control, **disaster recovery** (**DR**) policies, and loss prevention; maintenance is done by the CSP, meaning that end users don't need to waste time and money doing this.

Besides the advantages (key features) listed previously, there are possible disadvantages too, such as **performance variations**, **downtime**, and **lack of predictability**. Nevertheless, those are not game stoppers for anyone wanting to migrate to the cloud.

Performance may vary depending on the CSP you choose, but none of the big names out there has any significant performance issues. In most cases, performance is dictated by the local internet speed of the user, so it isn't a CSP problem after all. Downtime could be an issue, but all major providers strive to offer 99.9% uptime. If disaster strikes, issues are solved in a matter of minutes—or in worst-case scenarios, in a matter of hours. There is a lack of predictability with regard to the CSP and its presence on the market, but rest assured that none of the big players will go away anytime soon.

In the next section, we will introduce you to the OpenStack platform.

Short introduction to OpenStack

OpenStack has been governed by the **Open Infrastructure Foundation** (**OpenInfra Foundation**) since October 2020 and offers a set of open source tools that are used for building and administering cloud infrastructures. OpenStack can be used for creating both public and private clouds.

OpenStack is a collection of tools designed to create and manage cloud infrastructures by providing the needed components as a granular model that serves cloud computing-specific services. OpenStack was initially released in 2010 as a joint project between Rackspace and the **National Aeronautics and Space Administration** (**NASA**), is written in Python, and is licensed under the Apache 2.0 license.

Similar to the virtualization technologies, OpenStack uses a software layer based on specific APIs that abstract the virtual resources.

The latest release of OpenStack is called Victoria and was released in October 2020. It offers the founding cloud infrastructure for bare-metal VMs and containers. OpenStack has a modular structure and offers extreme flexibility when it comes to adding features and functionality. The modularity is given by the components it provides, each with specific APIs for accessing the infrastructure resources. Here is a list of some of the OpenStack components, by category:

- **Web interface**:

  a) **Horizon**: A dashboard to manage all OpenStack services

- **Compute**

  a) **Nova**: A compute service that provides resources on-demand

  b) **Zun**: A container service that provides the API for creating and managing containers, using different container technologies

- **Storage**:

  a) **Swift**: An object storage service that is designed for scalability, high availability, and concurrency.

  b) **Cinder**: A block storage service that provides a self-service API for resource management; it can be used with **Logical Volume Manager** (**LVM**).

  c) **Manila**: A service that provides access to shared filesystems.

- **Networking**:

  a) **Neutron**: A component that provides a **software-defined networking** (**SDN**) solution for virtual compute environments

  b) **Octavia**: A component that provides on-demand load balancing solutions for fleets of bare-metal servers, containers, or VMs

  c) **Designate**: A component that provides **Domain Name System** (**DNS**) services

- **Shared services**:

  a) **Keystone**: A component that provides APIs for client authentication, supporting **Lightweight Directory Access Protocol (LDAP)**, **Open Authorization (OAuth)**, OpenID Connect, **Security Assertion Markup Language (SAML)**, and **Structured Query Language (SQL)**

  b) **Barbican**: A key manager service for secure storage and password, certificate, and encryption-key management

  c) **Glance**: An image service that stores and retrieves VM images

- **Orchestration**:

  a) **Heat**: A component for infrastructure resource orchestration

  b) **Senlin**: A clustering service, designed to facilitate orchestration of similar objects inside OpenStack

  c) **Zaqar**: A component providing a cloud messaging service for both web and mobile developers

- **Workload provisioning**:

  a) **Magnum**: This makes container orchestration engines such as Docker Swarm, Kubernetes, and Apache Mesos available on OpenStack.

  b) **Trove**: This provides both relational and non-relational database engines.

  c) **Sahara**: This provides data processing tools such as Hadoop, Spark, and Storm on OpenStack.

The list provided previously is not a comprehensive one. We only listed the components that we thought would be of interest to those new to OpenStack. A full list with more details about each module can be found at `https://www.openstack.org/software/project-navigator/openstack-components#openstack-services`. Here is a diagram showing the OpenStack components and the connections between them:

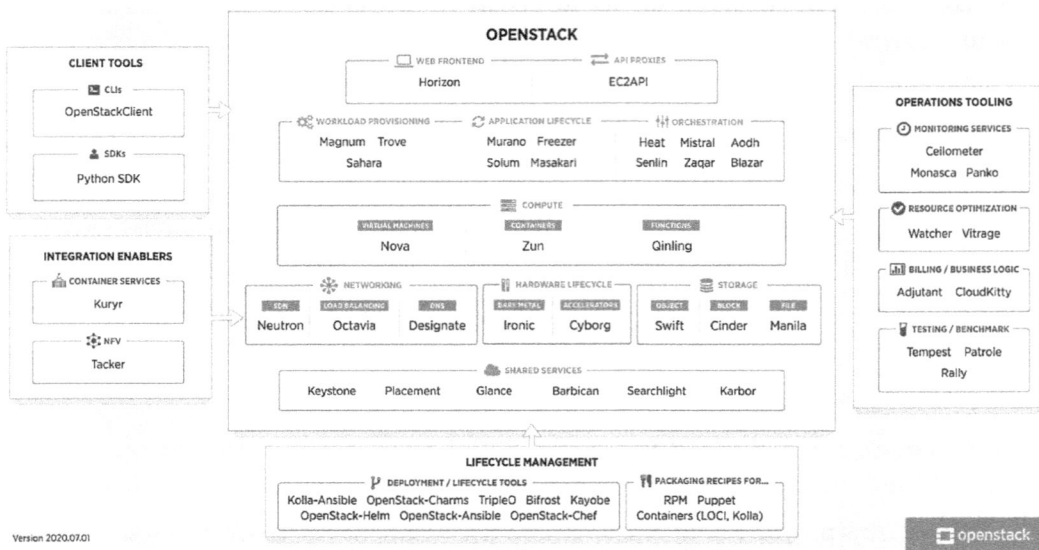

Figure 12.5 – The OpenStack map (image source: https://www.openstack.org/software/)

In the preceding diagram, you can see a map of OpenStack services as provided by the foundation. This is a straightforward way to look over the entire OpenStack services landscape, with all the relations between the provided services, and see how all of them fit together.

Therefore, let's summarize how OpenStack works. It uses a Linux base operating system to run a series of components called *projects*. The projects have scripts that are sent to the operating system to create cloud environments on top of virtualized resources. OpenStack is mainly used to create IaaS solutions. You can see it as a software version of an IT infrastructure generated by the components that create the entire stack. In the next section, we will introduce you to some IaaS solutions.

Introducing IaaS solutions

IaaS is the backbone of cloud computing. It offers on-demand access to resources, such as compute, storage, network, and so on. The CSP uses hypervisors to provide IaaS solutions. In this section, we provide you information about some of the most widely used IaaS solutions available. We will give you details about providers such as **Amazon Elastic Compute Cloud (Amazon EC2)**, Microsoft Azure Virtual Machines, and **Google Compute Engine (GCE)** as the big players, and DigitalOcean as a viable solution.

Amazon EC2

The IaaS solution provided by AWS is called Amazon EC2. It provides a good infrastructure solution for anyone, from low-cost compute instances to a high-power **graphics processing unit (GPU)** for machine learning. AWS was the first provider of IaaS solutions 12 years back and it is doing better than ever, even after the COVID-19 pandemic (`https://www.zdnet.com/article/the-top-cloud-providers-of-2021-aws-microsoft-azure-google-cloud-hybrid-saas/`).

When starting with Amazon EC2, you have several steps to fulfill. First is the option to choose your **Amazon Machine Image (AMI)**, which is basically a preconfigured image of either Linux or Windows. When it comes to Linux, you can choose between the following:

- Amazon Linux 2 (based on CentOS/**Red Hat Enterprise Linux (RHEL)**)
- RHEL 8
- **SUSE Linux Enterprise Server (SLES)** 15 SP2
- Ubuntu Server 20.04 **Long-Term Support (LTS)**

You will need to choose your instance type from a really wide variety. To learn more about EC2 instances, visit `https://aws.amazon.com/ec2/instance-types/` and find out details about each one. EC2, for example, is the only provider that offers Mac instances, based on Mac minis with Intel i7 six-core **central processing units (CPUs)**, with hyperthreading and 32 **gigabytes (GB)** of **random-access memory (RAM)**. To use Linux, you can choose from low-end instances up to high-performance instances, depending on your needs.

Amazon provides an **Elastic Block Store (EBS)** option with **solid-state drive (SSD)** and magnetic mediums available. You can select a custom value, depending on your needs. EC2 is a flexible solution, compared to other options. It has an easy-to-use and straightforward interface, and you will only pay for the time and resources you use. An example on how to deploy on EC2 will be provided to you in *Chapter 13, Deploying to the Cloud with AWS and Azure*.

Microsoft Azure Virtual Machines

Microsoft is the second biggest player on the cloud market, right after Amazon. Azure is the name of their cloud computing offering. Even though it is provided by Microsoft, Linux is the most widely used operating system on Azure (`https://www.zdnet.com/article/microsoft-developer-reveals-linux-is-now-more-used-on-azure-than-windows-server/`).

Azure's IaaS offering is called Virtual Machines and is similar to Amazon's offering; you can choose between many tiers. What is different about Microsoft's offering is the pricing model. They have a pay-as-you-go model, or a reservation-based instance, for 1 to 3 years. Microsoft's interface is totally different from Amazon's, and in our opinion might not be as straightforward as its competitor's, but you will get to know it after some time.

Microsoft offers several types, from economical burstable VMs to powerful memory-optimized instances. The pay-as-you-go model offers a per-hour cost, and this could add to the final bill on those services, so choose with care based on your needs. When it comes to Linux distributions, you can choose from the following: CentOS 6, 7, and 8; Debian 8, 9, and 10; RHEL 6, 7, and 8; SLES for SAP 11, 12, and 15; openSUSE Leap 15; and Ubuntu Server 16, 18, and 20.

Azure has a very powerful SaaS offering too, and this will make it a good option if you use other Azure services. An example on how to deploy to Azure will be provided to you in *Chapter 13*, *Deploying to the Cloud with AWS and Azure*.

Other strong IaaS offerings

DigitalOcean is another important player on the cloud market and offers a strong IaaS solution. DigitalOcean has a very simple and straightforward interface, and it helps you to create a cloud in a very short time. They call their VMs **droplets** and you create one in matter of seconds. All you have to do is to choose the image (Linux distribution); the plan (based on your **virtual CPU** (**vCPU**), memory, and disk space needs); add storage blocks; choose your data center region, authentication method (password or **Secure Shell** (**SSH**) key), and the hostname. You can also assign droplets to certain projects you manage.

DigitalOcean's interface is better-looking and much more user friendly than the other competitors. Following the example of DigitalOcean, other IaaS providers—such as Linode and Hetzner, among others—provide a slim and friendly interface for creating virtual servers.

Linode is another strong competitor on the cloud market, offering powerful solutions. Their VMs are called Linodes. The interface is somewhere between DigitalOcean and Azure, with regard to their ease of use and appearance.

Another strong player, at least on the European market, is **Hetzner**, a German-based cloud provider. They offer a great balance between resources and cost, and similar solutions to the others mentioned in this section. They provide an interface similar to DigitalOcean that is really easy to explore, and the cloud instance will be deployed in a matter of seconds.

Similar to the offerings of DigitalOcean, Linode, and Hetzner, there is a relatively new offering from Amazon (starting in 2017), called **Lightsail**. This service was introduced in order to offer clients an easy way to deploy **virtual private servers** (**VPSs**) or VMs in the cloud. The interface is similar to that seen from the competition, but it comes with the full Amazon infrastructure reliability on top.

Lightsail provides several distributions, together with application bundles. Deploying on AWS, using Lightsail, becomes easier and more straightforward. It is a useful tool to lure in new users wanting a quick and secure solution for delivering their web apps.

There are other solutions available, such as Google's solution, called GCE, which is the IaaS solution from **Google Cloud Platform** (**GCP**). The GCP interface is very similar to the one on the Azure platform.

> **Important note**
> One interesting aspect of using GCP is that when you want to delete a project, the operation is not immediate, and the deletion is scheduled in 1 month's time. This could be seen as a safety net if the deletion was not intentional, and you need to roll back the project.

If you don't like the offerings that major players have for the IaaS platform, you can create your own IaaS using **OpenStack**. In the next section, we will detail some of the PaaS solutions.

Introducing PaaS solutions

PaaS is another form of cloud computing. Compared to IaaS, PaaS provides the hardware layer, together with an application layer. The hardware and software are hosted by the CSP, with no need to manage those from the client side. Clients of PaaS solutions are, in a majority of cases, application developers.

The CSPs that offer PaaS solutions are mostly the same as those that offer IaaS solutions. We have Amazon, Microsoft, and Google as the major PaaS providers.

Amazon Elastic Beanstalk

Amazon offers the **Elastic Beanstalk** service, whose interface is straightforward and simple. You can create a sample application or upload your own, and Beanstalk takes care of the rest, from deployment details to load balancing, scaling, and monitoring. You select the AWS EC2 hardware instances to deploy on. Next, we will discuss another major player's offering: Google App Engine.

Google App Engine

Google's PaaS solution is **Google App Engine**, a fully managed serverless environment that is relatively easy to use, with support for a large number of programming languages. Google App Engine is a scalable solution, with automatic security updates and managed infrastructure and monitoring. It offers solutions to connect to Google Cloud storage solutions and support for all major web programming languages such as Go, Node.js, Python, .NET, or Java. Google offers competitive pricing and an interface similar to the one we saw in their IaaS offering. Another major player with solid offerings is DigitalOcean, and we will discuss this next.

DigitalOcean App Platform

DigitalOcean offers a PaaS solution in the form of **App Platform**. It offers a straightforward interface, a direct connection with your GitHub or GitLab repository, and a fully managed infrastructure. DigitalOcean is on the same level as big players such as Amazon and Google, and with App Platform it manages infrastructure, provisioning, databases, application runtimes and dependencies, and the underlying operating system. It offers support for popular programming languages and frameworks such as Python, Node.js, Django, Go, React, Ruby, and others. DigitalOcean App Platform uses open cloud-native standards, with automatic code analysis, container creation, and orchestration. A distinctive competence of this solution is the free starter tier, for deploying up to three static websites. For prototyping dynamic web apps, there is a basic tier, and for deploying professional apps on the market, there is a professional tier available. DigitalOcean's interface is pleasant and could be attractive to newcomers. Their pricing acts as an advantage too.

Besides ready-to-use solutions from providers listed previously, there are open source PaaS solutions provided by **Cloud Foundry**, **Red Hat OpenShift**, **Heroku**, and others, which we will not detail in this section. Nevertheless, the three mentioned previously are worth at least a short introduction, so here they are, detailed in the following section.

Red Hat OpenShift

Red Hat OpenShift is a container platform for application deployments. Its base is a Linux distribution (RHEL) paired with a container runtime and solutions for networking, registry, and authentication and monitoring. OpenShift was designed to be a viable, hybrid PaaS solution with total Kubernetes integration (Kubernetes will be covered briefly in the following section, *Introducing CaaS solutions*, and in more detail in *Chapter 14*, *Deploying Applications with Kubernetes*). OpenShift took advantage of the CoreOS acquisition (discussed later on) by bringing some unique solutions. The new CoreOS Tectonic container platform is merging with OpenShift to bring to the user the best of both worlds.

Cloud Foundry

Cloud Foundry is a cloud platform designed as an enterprise-ready PaaS solution. It is open source and can be deployed on different infrastructures, from on-premises to IaaS providers such as Google GCP, Amazon AWS, Azure, or OpenStack. It offers various developer frameworks and a choice of Cloud Foundry certified platforms, such as Atos Cloud Foundry, IBM Cloud Foundry, SAP Cloud Platform, SUSE Cloud Application Platform, and VMware Tanzu.

The Heroku platform

Heroku is a Salesforce company, and the platform was developed as an innovative PaaS. It is based on a container system called Dynos, which uses Linux-based containers run by a container management system, designed for scalability and agility. It offers fully managed data services with support for Postgres, Redis, and Apache Kafka, and Heroku Runtime, a component responsible for container orchestration, scaling, and configuration management. Heroku also supports a plethora of programming languages, such as Node.js, Ruby, Python, Go, Scala, Clojure, and Java.

PaaS has many solutions for developers, helping them create and deploy an application by taking away the burden of managing the infrastructure. As you might have learned by now, many of the solutions described in this section rely on the use of containers. This is why, in the next section, we will detail a subset of IaaS called CaaS, where we will introduce you to container orchestration and container-specialized operating systems.

Introducing CaaS solutions

CaaS is a subset of the IaaS cloud service model. It lets customers use individual containers, clusters, and applications on top of a provider-managed infrastructure. CaaS can be used either on-premises or in the cloud, depending on customer's needs. In a CaaS model, the container engines and orchestration are provided and managed by the CSP. The user's interaction with containers can be done either through an API or a web interface. The container orchestration platform used by the provider—mainly **Kubernetes** and **Docker**—is important and is a key differentiator between different solutions.

We covered containers (and VMs) in *Chapter 11*, *Working with Containers and Virtual Machines*, without giving any detailed information about orchestration or container-specialized micro operating systems. We will now provide you with some more details on those subjects.

Introducing the Kubernetes container orchestration solution

Kubernetes is an open source project developed by Google to be used for the automatic deployment and scaling of containerized applications. It was written in the Go programming language. The name *Kubernetes* comes from Greek, and it represents a ship's helmsman or captain. Kubernetes is a tool to automate container management, together with infrastructure abstraction and service monitoring.

Many newcomers confuse Kubernetes with Docker, or vice versa. They are complementary tools, each used for a specific purpose. Docker creates a container (like a box) in which you want to deploy your application, and Kubernetes takes care of the containers (or boxes) once the applications are packed inside and deployed. Kubernetes provides a series of services that are essential to running containers, such as service discovery and load balancing, storage orchestration, automated backups and self-healing, and privacy. The Kubernetes architecture consists of several components that are crucial for any administrator to know. We will break them down for you in the next section.

Introducing the Kubernetes components

When you run Kubernetes, you mainly manage clusters of hosts, which are usually containers running Linux. In short, this means that when you run Kubernetes, you run clusters. Here is a list of the basic components found in Kubernetes:

- *A cluster* is the core of Kubernetes, as its sole purpose is to manage lots of clusters. Each cluster consists of at least a control plane and one or more nodes, each node running containers inside pods.

- *A control plane* consists of processes that control nodes. The components of a control plane are listed here: kube-apiserver is the API server as the frontend of the control plane; etcd is the backing store for all the data inside the cluster; kube-scheduler looks for pods that have no assigned node and connects them to a node to run; kube-controller-manager runs the controller processes, including the node controller, the replication controller, the endpoints controller, and the token controller; cloud-controller-manager is a tool that allows you to link your cluster to your cloud provider's API; it includes the node controller, the route controller, and the service controller.

- *Nodes* are either a VM or a physical machine running services needed for pods. The node components run on every node and are responsible for maintaining the running pods. The components are kube-proxy, which is responsible for network rules on each node, and kubelet, which makes sure that each container is running inside a Pod.

- *Pods* are a collection of different containers running in the cluster. They are the components of the workload.

Kubernetes clusters are extremely complicated to master. Understanding the concepts around it needs a lot of practice and dedication. No matter how complex it is, Kubernetes does not do everything for you. You still have to choose the container runtime (supported runtimes are Docker, containerd, and **Container Runtime Interface (CRI-O)**, CI/CD tools, the storage solution, access control, and app services. Here is a diagram showing the Kubernetes cluster architecture:

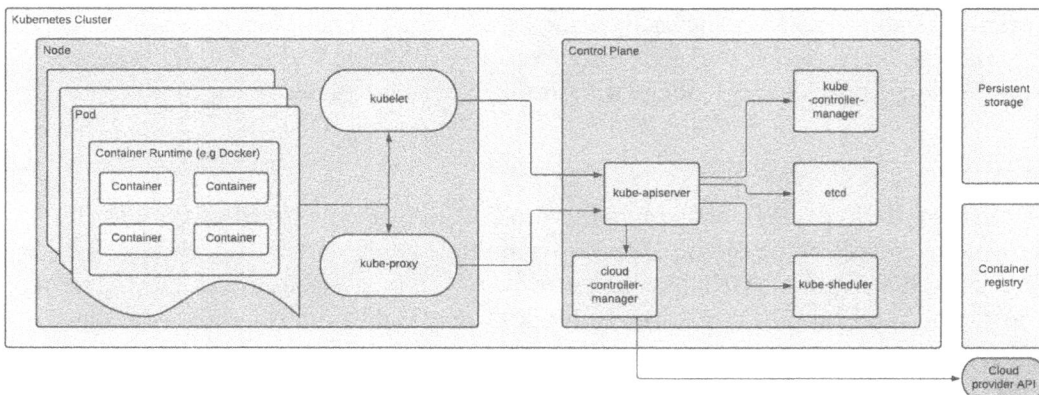

Figure 12.6 – Kubernetes cluster architecture

Managing Kubernetes clusters is out of the scope of this chapter, but you will learn about this in *Chapter 14, Deploying Applications with Kubernetes*. This short introduction was needed for you to understand the concepts and tools that Kubernetes uses.

Besides Kubernetes, there are several other container orchestration tools, such as **Docker Swarm**, **Apache Mesos**, or **Nomad** from HashiCorp. They are extremely powerful tools, used by many people around the world. We will not cover these in detail here, but we thought it would be useful to at least enumerate them at the end of this container orchestration section. In the next section, we will provide you with some information about container solutions in the cloud.

Deploying containers in the cloud

You can use container orchestration solutions in the cloud, and offerings such as Amazon **Elastic Container Service (ECS)**, **Amazon Elastic Kubernetes Service (EKS)**, **Google Kubernetes Engine (GKE)**, and **Azure Kubernetes Service (AKS)** are essential for this.

Amazon ECS

Amazon ECS is a fully managed service for orchestrating containers. It offers an optional, serverless solution (**AWS Fargate**) and is run inside by some of Amazon's key services, which assures that the tool is tested and is secure enough for anyone to use. The container runtime it uses is Docker. Amazon also offers an EKS service for orchestrating Kubernetes applications.

Amazon EKS

Amazon EKS is a tool used for container orchestration. It is based on **Amazon EKS Distro (EKS-D)**, which is a Kubernetes distribution developed by Amazon, based on the original open source Kubernetes. By using EKS-D, you can run Kubernetes either on premises or on Amazon's own EC2 instances, or on VMware vSphere VMs. Another strong solution is provided by Google, with its GKE service, detailed in the next section.

GKE

GKE offers pre-built deployment templates, with Pod auto-scaling based on the CPU and memory usage. Scaling can be done across multiple pools, with enhanced security provided by GKE Sandbox. GKE Sandbox provides an extra layer of security, by protecting the host kernel and running applications. Besides Google and Amazon, Microsoft offers a strong solution for container orchestration, with the AKS. We will present it in the following section.

Microsoft AKS

AKS is a managed service for deploying clusters of containerized applications. As with the other providers, Microsoft offers a fully managed solution by handling resource maintenance and health monitoring. The AKS nodes use Azure VMs to run and support different operating systems, such as Microsoft Windows Server images. It also offers free upgrades to the newest-available Kubernetes images. Among other solutions, AKS offers GPU-enabled nodes, storage volume support, and special development tool integration with Microsoft's own **Visual Studio Code** (**VS Code**).

After seeing some of the solutions available to deploy Kubernetes in the cloud and learning about the main components of Kubernetes and how they work, in the following section we will discuss cloud-specialized minimal operating systems. Kubernetes is an essential part for some of them, therefore in the next section we will introduce you to micro operating systems.

The rise of micro operating systems

When using containers to deploy applications to users, we need the underlying operating system to be as slim as possible. You might remember the small size of the Ubuntu operating system when used with Docker in *Chapter 11*, *Working with Containers and Virtual Machines*. This need for slimmed-down hosts enabled the rise of container-focused, specialized, and minimalist operating systems. These are called **micro operating systems**. The most well-known specialized micro operating systems are CoreOS, Atomic Host (obsolete), RancherOS, and Ubuntu Core (formerly called Ubuntu Snappy). CoreOS was acquired by Red Hat and steadily eliminated Atomic Host from the landscape. Originally, Atomic Host was developed by Fedora. RancherOS was recently acquired by SUSE and still keeps its name and structure. Each of these minimalistic and specialized operating systems is considered to be the future of operating systems, with new immutable architectures and transactional updates.

Short overview of CoreOS (Fedora and RHEL)

CoreOS was the first to introduce a container-specialized Linux distribution, with a small footprint and automatic updates. After the Red Hat acquisition, a new specialized version of Container Linux emerged, with the name of **Red Hat CoreOS**. This new distribution is based on Fedora and RHEL. Red Hat CoreOS is now the base of the **Red Hat OpenShift Container Platform**. This next generation of this platform is based on **CoreOS Tectonic**, a new, fully automated container platform based on Kubernetes. The cluster architecture of CoreOS changed since the introduction of Tectonic and the rise of **Fedora CoreOS** and **Red Hat CoreOS**, and is now based entirely on Kubernetes.

Fedora CoreOS offers what was best from both Atomic Host and Container Linux, by combining automatic update and provisioning tools with support for **Oracle Cloud Infrastructure** (**OCI**) and Docker and **Security-Enhanced Linux** (**SELinux**) security.

Short overview of RancherOS

RancherOS is a specialized operating system for running Docker containers in production. Every process in RancherOS is a Docker container, which makes the system very lightweight and easy on resources. When the system is initialized, there are two main Docker instances that start up. One is called System Docker and is used for running specialized Docker containers with other system services inside. The other instance is called Docker and runs on top of System Docker; this is considered a daemon that manages user's containers, and it is simple and intuitive. Everything runs on the Linux kernel as the foundation. Here is a diagram showing RancherOS's architecture:

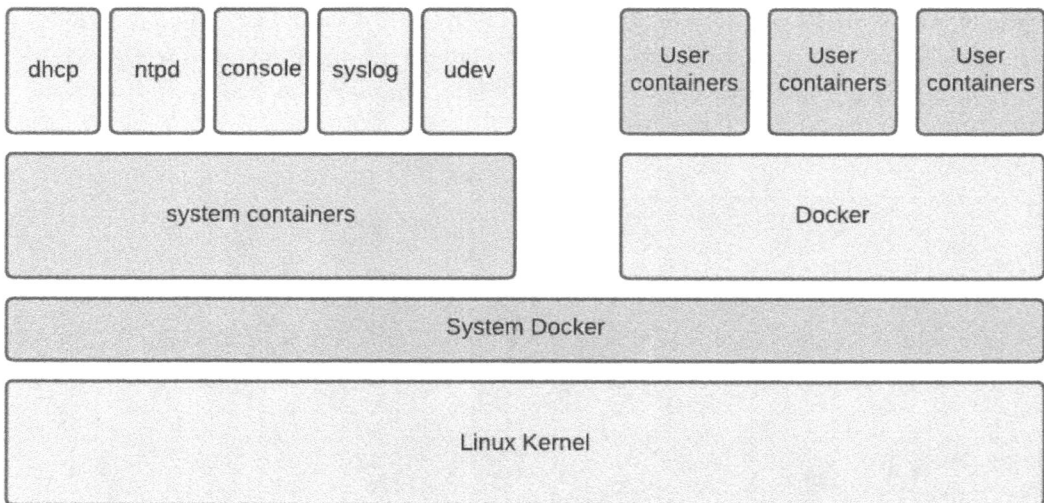

Figure 12.7 – RancherOS architecture

Both **Fedora CoreOS** (**Red Hat CoreOS**) and RancherOS are worthy specialized operating systems to use for containers. There is yet another kid on the block: Ubuntu Core (formerly known as Snappy), which we will discuss briefly in the next section.

Short overview of Ubuntu Core

Compared to the other two solutions presented earlier, **Ubuntu Core** is a different beast. It is not based on a specific container-based architecture (such as Docker, for example), as it was initially developed to run on IoT and is based on Canonical's in-house developed snap packages. It is a stripped-down version of Ubuntu that uses a read-only filesystem, transactional updates, and snap packages. As snaps are self-contained and include every dependency and even the filesystem, the underlying operating system has only the minimal required packages to run. It was designed to deploy and run applications in a secure and reliable way, which is why the entire operating system was developed around snaps. Even the kernel is incapsulated and treated as a snap. There are other types of snaps in Ubuntu Core, besides the kernel snap. There is the gadget snap, which manages system properties; the core snap, which provides the execution environment; and app snaps, which are the packaged applications, daemons, and tools that run on Ubuntu Core.

In this section, dedicated to specialized minimal operating systems for containers, we showed you three different solutions, each using a different technology underneath. All three offerings are very powerful tools that suit their intended purposes very well. In the next section, we will discuss the importance of microservices in cloud computing.

Introducing microservices

A **microservice** is an architectural style used in application delivery. Over time, application delivery evolved from a monolithic model toward a decentralized one, all thanks to the evolution of cloud technologies. Starting with the historical launch of AWS in 2006, followed by the launch of Heroku in 2007 and Vagrant in 2010, application deployment started to change too, in order to take advantage of the new cloud offerings. Applications moved from having a single, large and monolithic code base to a model where each application would benefit from different sets of services. This would make the code base more lightweight and dependent on different services. For example, in the following diagram, we have a comparison between two models, a monolithic application versus a microservices architecture:

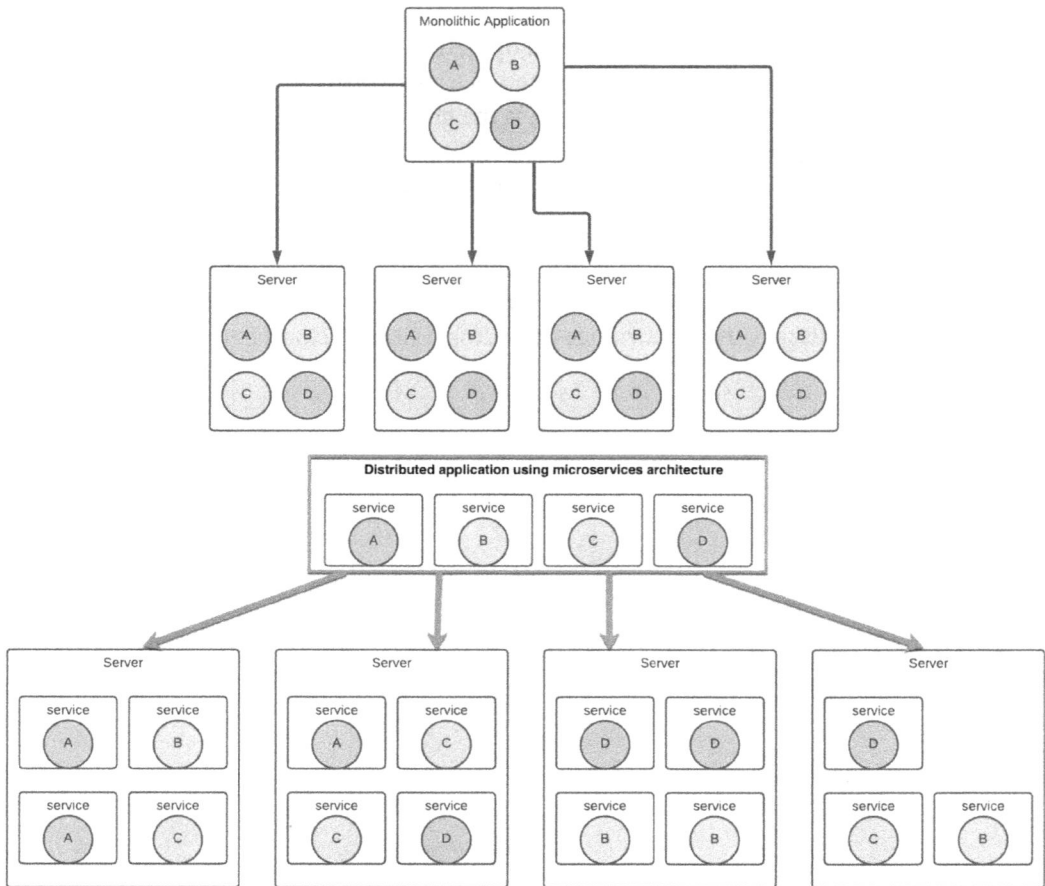

Figure 12.8 – Monolithic versus microservices models for application deployment

On the upper side of the preceding diagram is a monolithic application model, which has all the functionalities (circles with letters from **A** to **D**) inside a single process, represented by the entire app. It is deployed by simply replicating on multiple servers. By comparison, on the lower side, we have an application that has its functionalities (represented by letters from **A** to **D**) separated into different services. Those services are then distributed and scaled across different servers, depending on the user's needs.

A microservices architecture has a modular-based approach. Each module will correspond to a specific service. Services work independently from one another and are connected through REST APIs based on the HTTP protocol. This means that each app functionality can be developed in different languages, depending on which one is better suited. This modular base can also take advantage of new container technologies. Here is another diagram, showing how a microservices architecture works:

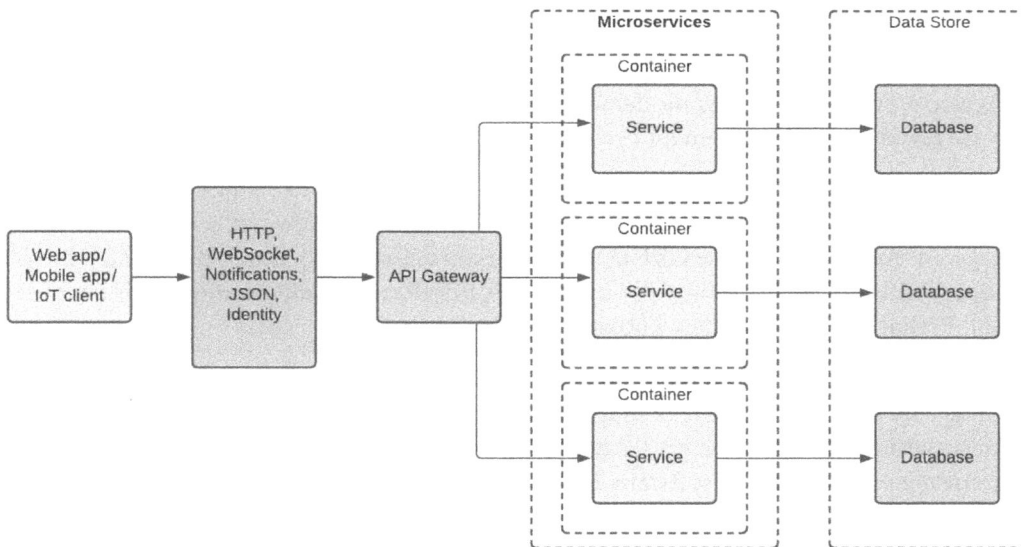

Figure 12.9 – Microservices architecture diagram

A microservices architecture is known for rapidly delivering complex applications. It has no technology or language lock-in; it offers independent scaling and update for each service and component, with no disturbance to other running services; and it has a fail-proof architecture.

The microservices model can be adapted to existing monolithic applications by breaking them down to individual, modular services. There is no need to rewrite the entire application, only splitting the entire code base into smaller parts.

Microservices are optimized for DevOps and CI/CD practices, thanks to their modular approach. In the next section, we will introduce you to DevOps practices and tools.

Introducing DevOps

DevOps is a culture. Its name comes from a combination of development and operations, and it envisions the practices and tools that are used to deliver rapidly. DevOps is about speed, agility, and time. We all know the phrase *time is money*, and this applies very well to the IT sector. The ability to deliver services and applications at a high speed can make the difference between being successful as a business and being irrelevant on the market.

DevOps is a model of cooperation between different teams involved in delivering services and applications. This means that the entire life cycle, from development and testing, up to deployment and management, is done by teams that are equally involved at every stage. The DevOps model assumes that no team is operating in a closed environment, but rather operates in a transparent manner in order to achieve the agility they need to succeed. There is also a different DevOps model whereby security and quality assurance teams are equally involved in the development cycle. It is called DevSecOps.

Crucial for the DevOps model are the automated processes that are created using specific tools. This mindset of agility and speed determined the rise of a new name associated with DevOps, and that is CI/CD. The CI/CD mindset assures that every development step is continuous, with no interruptions. To support this mindset, new automation tools have emerged. Perhaps the most widely known is Jenkins.

Jenkins is probably the most popular open source automation tool. It is a modular tool and can extended with the use of plugins. The ecosystem around the application is quite large, with hundreds of plugins available to choose from. Jenkins is written in Java and was designed to automate software development processes, from building and testing, up to delivery. One of the assets of Jenkins is the ability to create a pipeline through the use of specialized plugins. A pipeline is a tool that adds support for CD as an automated process to the application life cycle. Jenkins can be used either on-premises or in the cloud. It is also a viable solution for use as a SaaS offering.

The DevOps philosophy is not only related to application deployment. Healthy CD and CI are closely tied to the state of infrastructure. This is why configuration and management at the infrastructure level is extremely important. In the next section, you will learn about the cloud infrastructure management and what IaaC is.

Introducing cloud management tools

Today's software development and deployment relies on a plethora of physical systems and VMs. Managing all the related environments for development, testing, and production is a tedious task and involves the use of automated tools. The most widely used solutions for cloud infrastructure management are tools such as Ansible, Puppet, and Chef Infra. All these configuration management tools are powerful and reliable, and we will reserve *Chapter 15*, *Infrastructure and Automation with Ansible*, to teach you how to use only one of them: Ansible. Nevertheless, we will briefly introduce you to all of them in this section.

Ansible is an open source project currently owned by Red Hat. It is considered a simple automation tool, used for diverse actions such as application deployment, configuration management, cloud provisioning, and service orchestration. It was developed in Python and uses the concept of **nodes** to define categories of systems, with a **control node** as the master machine running Ansible, and different **managed nodes** as other machines that are controlled by the master. All the nodes are connected over SSH and controlled through an application called an **Ansible module**. Each module has a specific task to do on the managed nodes, and when the task is completed, they will be removed from that node. The way modules are used is determined by an **Ansible playbook**. The playbook is written in **YAML** (a recursive acronym for **YAML Ain't Markup Language**), a language mostly used for configuration files. Ansible also uses the concept of **inventory**, where lists of the managed nodes are kept. When running commands on nodes, you can apply them based on the lists inside your inventory, based on **patterns**. Ansible will apply the commands on every node or group of nodes available in a certain pattern. Ansible is considered one of the easiest automation tools available. It supports Linux/Unix and Windows for client machines, but the master machine must be Linux/Unix.

Puppet is one of the oldest automation tools available. Puppet's architecture is different from that of Ansible. It uses the concepts of **primary server** and **agents**. Puppet works with infrastructure code written using **domain-specific language** (**DSL**) code specific to Puppet, based on the Ruby programming language. The code is written on the primary server and transferred to the agent, and then translated into commands that are executed on the system you want to manage. Puppet also has an inventory tool called **Facter**, which stores data about the agents, such as hostname, IP address, and operating system. Information stored is sent back to the primary server in the form of a **manifest**, which will then be transformed into a JSON document called a **catalog**. All the manifests are kept inside **modules**, which are tools that are used for specific tasks. Each module contains information in the form of code and data. This data is centralized and managed by a tool called **Hiera**. All the data that Puppet generates is stored inside databases and managed through APIs by every app that needs to manage it. Compared to Ansible, Puppet seems a lot more complex. Puppet's primary server supports only Linux/Unix.

Chef Infra is another automation tool. It uses a client-server architecture. The main components are the **Chef Server**, the **Chef Client**, and the **Chef Workstation**. It uses the concepts of **cookbooks** and **recipes**. The Chef Workstation manages cookbooks that are used for infrastructure administration. The Chef Server is similar to a hub that handles all the configuration data. It is mainly used to upload cookbooks to the Chef Client, which is an application that is installed on every node from the infrastructure that you manage. Chef Infra uses the same Ruby-based code similar to Puppet, called DSL. The server needs to be installed on Linux/Unix, and the client supports Windows too.

All the automation and configuration tools presented in this section use different architectures but do the same thing, which is to provide an abstraction layer that defines the desired state of the infrastructure. Each tool is a different beast, having its own strengths and weaknesses. Chef Infra and Puppet might have a steeper learning curve with their Ruby/DSL-based code, while Ansible could be easier to approach due to its simpler architecture and use of the Python programming language. Nevertheless, you can't go wrong with either one.

Summary

In this chapter, we introduced you to cloud computing by showing you some of the most important concepts, tools, and solutions used. This should be enough for you to start learning about cloud technologies. This subject is very large, complex, and intimidating.

We talked about cloud standards, an important and largely overlooked subject, and about the main cloud types and services. You now have an idea what each as-a-service solution means and what the main differences are between them. You know what the most important solutions are and how they are provided by the main players in this field: Amazon, Google, and Microsoft. We introduced you to container orchestration with Kubernetes, and you know what OpenStack is and how it works. You learned about APIs and minimal container-specialized operating systems and about the DevOps culture, microservices, and infrastructure automation tools. You learned a lot in this chapter, but keep in mind that all these subjects have only scratched the surface of cloud computing. In the next chapter, we will introduce you to the more practical side of cloud deployments. You will learn how to deploy Linux on major clouds such as AWS and Azure.

Further reading

If you want to learn more about cloud technologies, please check out the following titles:

- *OpenStack for Architects - Second Edition, Ben Silverman, Michael Solberg, Packt Publishing* (https://www.packtpub.com/product/openstack-for-architects-second-edition/9781788624510)

- *Learning DevOps, Mikael Krief, Packt Publishing* (https://www.packtpub.com/product/learning-devops/9781838642730)

- *Hands-On Microservices with Kubernetes, Gigi Sayfan, Packt Publishing* (https://www.packtpub.com/product/hands-on-microservices-with-kubernetes/9781789805468)

13
Deploying to the Cloud with AWS and Azure

Recent years have seen a significant shift from on-premises computing platforms to private and public clouds. In an ever-changing and accelerating world, deploying and running applications in a highly scalable, efficient, and secure infrastructure is critical for businesses and organizations everywhere. On the other hand, the cost and expertise required to maintain the equivalent level of security and performance with on-premises computing resources become barely justifiable compared to current public cloud offerings. Businesses and teams, small and large, are adopting public cloud services in increasing numbers, albeit large enterprises are relatively slow to make a move.

One of the best metaphors for cloud computing is application services *on tap*. Do you need more resources for your apps? Just *turn on the tap* and provision the virtual machines or instances in any number you need (scale horizontally). Or perhaps, for some instances, you need more CPUs or memory (scale vertically). When you no longer need resources, just *turn off the tap*.

Public cloud services provide all these functions at relatively low rates, taking away the operations overhead that you might otherwise have with maintaining the on-premises infrastructure accommodating such features.

This chapter will introduce you to **Amazon Web Services** (**AWS**) and Microsoft Azure – two of the major public cloud providers – and offer some practical guidance for deploying your applications in the cloud. In particular, we'll focus on typical cloud management workloads, using both the web administration console and the command-line interface.

At the end of this chapter, you'll know how to use the AWS management console, the AWS CLI, the Azure web portal, and the Azure CLI to manage your cloud resources with the two most popular cloud providers of our time. You'll also learn how to make a prudent decision about creating and launching your resources in the cloud, striking a sensible balance between performance and cost.

We hope that Linux administrators – novice and experienced alike – will find the content in this chapter relevant and refreshing. Our focus is purely practical as we explore the AWS and Azure cloud workloads. We will refrain from comparing the two, as such an endeavor would go beyond this chapter's scope. To make the journey less boring, we'll also steer away from keeping a perfect symmetry between describing AWS and Azure management tasks. We all know AWS blazed the trail into the public cloud realm first. Other major cloud providers followed, adopted, and occasionally improved the underlying paradigms and workflows. As we introduce AWS first, we'll cover more ground on some of the cloud provisioning concepts (such as Regions and **Availability Zones** (**AZs**)), which in many ways are very similar in Azure.

Finally, we'll leave the ultimate choice between using AWS or Azure up to you. We're giving you the map. The road is yours to take.

Here are some of the key topics you'll learn about:

- Working with the AWS console
- Understanding AWS EC2 provisioning types
- Creating and managing AWS EC2 instances
- Working with the AWS CLI
- Introducing the Azure web portal
- Creating virtual machines in Azure
- Managing virtual machines and related cloud resources in Azure
- Working with the Azure CLI

Technical requirements

This chapter requires AWS and Azure accounts if you want to follow along with the practical examples. Both cloud providers provide free subscriptions:

- AWS Free Tier: `https://aws.amazon.com/free`
- Microsoft Azure free account: `https://azure.microsoft.com/en-us/free/`

You will also need a local machine with a Linux distribution of your choice to install and experiment with the AWS CLI and Azure CLI utilities. The web-console-driven management tasks for both AWS and Azure require a modern web browser, and you can access the related portals on any platform. To run the AWS and Azure CLI commands, you need a Linux command-line terminal and intermediate-level proficiency using the shell.

With this, let's start with our first contender, AWS EC2.

Working with AWS EC2

The AWS **Elastic Compute Cloud** (**EC2**) is a scalable computing infrastructure that allows users to lease virtual computing platforms and services to run their cloud applications. AWS EC2 has gained extreme popularity in recent years due to its outstanding performance and scalability combined with relatively cost-effective service plans. This section provides some basic functional knowledge to get you started with deploying and managing AWS EC2 instances running your applications. In particular, we'll introduce you to the following:

- EC2 instance types – differentiating between various provisioning and related pricing tiers
- **Amazon Machine Images** (**AMI**) – the functional unit required to launch an EC2 instance
- Accessing your EC2 instances – using SSH to connect and SCP to transfer files to and from your EC2 instances
- Backing up and restoring your EC2 instances using EBS snapshots
- Working with the AWS CLI

By the end of this section, you'll have a basic understanding of AWS EC2 and how to choose, deploy, and manage your EC2 instances.

Let's start with launching EC2 instances.

Creating AWS EC2 instances

AWS EC2 provides various instance types, each with its provisioning, capacity, pricing, and use case models. Choosing between different EC2 instance types is not always trivial. This section will briefly describe each EC2 instance type, some of the pros and cons of using them, and how to choose the most cost-effective solution. With each instance type, we'll show you how to launch one using the AWS console. Next, we'll look at two essential EC2 deployment features – *AMIs* and *placement groups* – that allow you to be proficient and resourceful when deploying and scaling your EC2 instances.

Let's look at the EC2 instance types next.

Introducing AWS EC2 instance types

We can look at EC2 instance types from two perspectives:

- **Provisioning** – the capacity and computing power of your EC2 instance
- **Pricing** – how much you pay for running your EC2 instance

When you choose an EC2 instance, you'll have to consider both. Let's look at each of these options briefly.

EC2 instance provisioning options

The main differentiating feature of each EC2 instance provisioning type is the computing power expressed with the following:

- CPU or **virtual CPU (vCPU)**
- RAM (memory)
- Storage (disk capacity)

Some EC2 instance types also provide **Graphical Processing Unit (GPU)** or **Field Programmable Gate Array (FPGA)** computing capabilities.

Here's a quick enumeration of the EC2 instance types based on provisioning:

- **General-purpose** – suitable for a wide range of workloads, with a balanced CPU offering, memory, and storage. Instance categories: m4, m5, m5a, m5ad, m5d, m5dn, m5n, m5zn, m6g, m6gd, mac1, t2, t3, t3a, t4g. For example, an m4.large instance has 2 vCPUs and 8 GB memory.

- **Compute-optimized** – ideal for compute-bound applications with high-performance processing. Instance categories: c4, c5, c5a, c5ad, c5d, c5n, c6g, c6gd, c6gn. For example, a c4.large instance has 2 vCPUs and 3.75 GB memory.

- **Memory-optimized** – designed for high-performance workloads with large in-memory datasets. Instance categories: r4, r5, r5a, r5ad, r5b, r5d, 5dn, r5n, r6g, r6gd, u-6tb1, u-12tb1, u-18tb1, u-24tb1, x1, x1e, z1d. For example, an r4.large instance has 2 vCPUs and 15.25 GB memory.

- **Storage-optimized** – useful for running workloads with high, sequential read-write operations with large datasets on local storage. Instance categories: d2, d3, d3en, h1, i3, i3en. For example, a d2.xlarge instance has 4 vCPUs and 30.5 GB memory.

- **Accelerated computing** – good for using hardware acceleration for specific functions, such as graphics processing, floating-point calculations, and data pattern matching. Instance categories: p2, p3, g2, g3, g3s, g4ad, g4dn, f1, inf1. For example, a p2.xlarge instance has 4 vCPUs and 61 GB memory, and 1 accelerator (i.e. co-processor or GPU).

A detailed view of EC2 instance provisioning types is beyond the scope of this chapter. You can explore the related information here: https://docs.aws.amazon.com/AWSEC2/latest/UserGuide/instance-types.html. In addition to provisioning, you'll have to consider the pricing model of your EC2 instances. Let's take a look at the EC2 purchasing options next.

EC2 instance purchasing options

At the time of this writing, the EC2 instance types based on purchasing options are as follows:

- **On-demand instances** – You pay for computing capacity per second, without long-term commitments.

- **Reserved instances** – These provide significant savings compared to on-demand instances if specific instance attributes are set for the long term, such as type and region.

- **Spot instances** – These EC2 instances are available for reuse at a lower price than on-demand instances.

- **Dedicated instances** – These EC2 instances run in a **Virtual Private Cloud** (**VPC**) assigned to a single-payer account.

We'll cover each of the previous EC2 instance types in the following sections. For each of these types, we show an example of launching a corresponding instance. But before creating an instance, we'll look at another key concept regarding EC2 instances – *AZs*.

EC2 Availability Zones

The AWS EC2 service is available in multiple locations around the globe, known as *Regions*. Examples of Regions include *US West Oregon* (us-west-2) and *Asia Pacific Mumbai* (ap-south-1). EC2 defines multiple **AZs** in a region, which are essentially one or more data centers. Regions are entirely isolated from each other in terms of the underlying infrastructure to provide fault tolerance and high availability. If a Region becomes unavailable, only the EC2 instances within the affected Region are unreachable. Other Regions with their EC2 instances will continue to work uninterrupted. Here is a simple illustration of Regions and AZs in AWS:

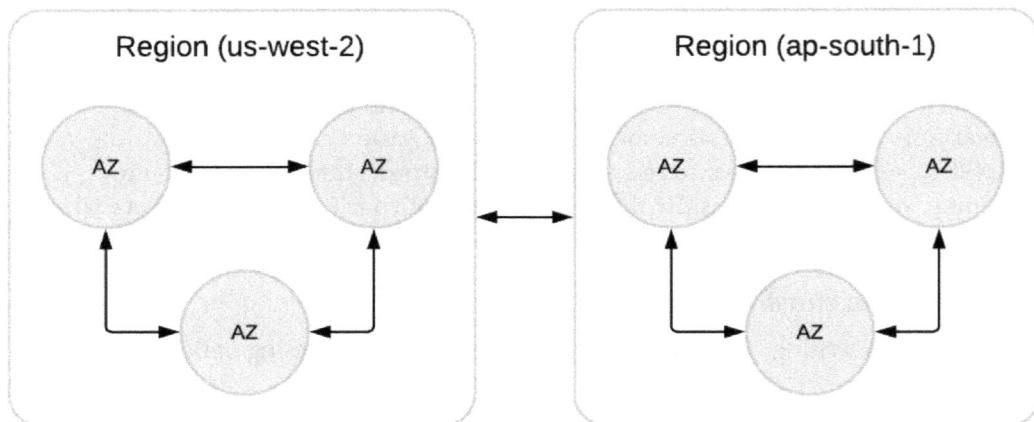

Figure 13.1 – AWS EC2 AZs

Similarly, AZs are connected to each other while providing highly available and fault-tolerant EC2 services within a Region. Launching an EC2 instance creates it in the current Region selected in the AWS console. An AWS EC2 administrator may switch between different Regions when managing EC2 instances. Only instances within the selected Region are visible in the EC2 administration console. An EC2 administrator will usually choose a Region based on the geographical location of the users accessing the EC2 instance.

We now have the preliminary knowledge to launch various EC2 instance types based on pricing. Let's look at on-demand instances first.

EC2 on-demand instances

The AWS EC2 *on-demand instances* use a *pay-as-you-go* pricing model for resource usage per second, without a long-time contract. On-demand instances are best suited for experimenting with uncertain workloads where the resource usage is not fully known (such as during development). The flexibility of the on-demand instances comes with a higher price than reserved instances, for example.

Let's launch an on-demand instance. First, we log into our AWS console at `https://console.aws.amazon.com`. In the upper right corner of the dashboard, we select the Region of preference, in our case, US West (**Oregon**):

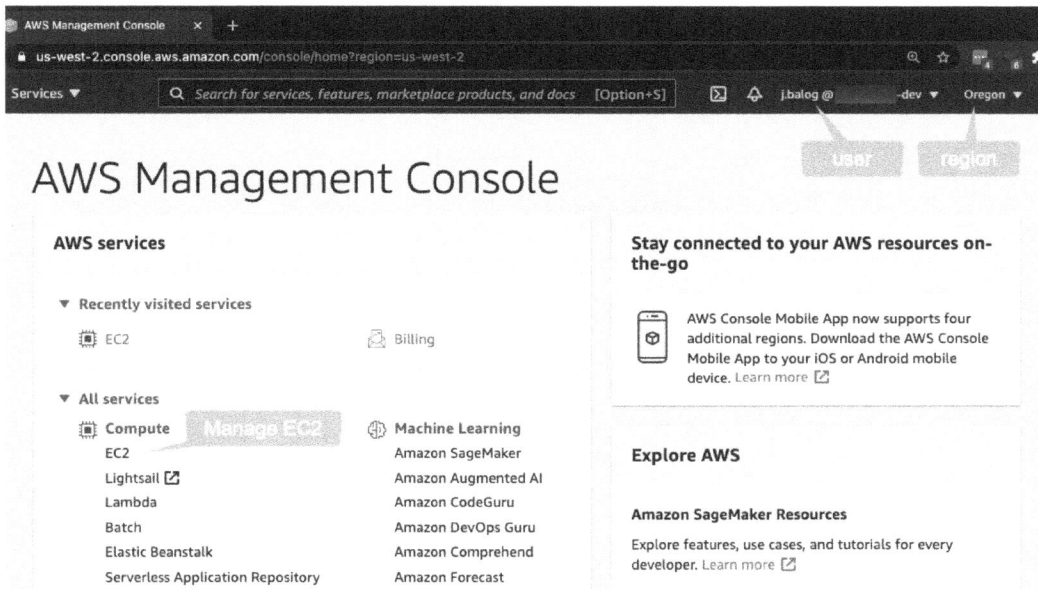

Figure 13.2 – The AWS Management Console

Next, we'll choose the EC2 service. The **Launch instance** button will begin the step-by-step process of creating our on-demand instance:

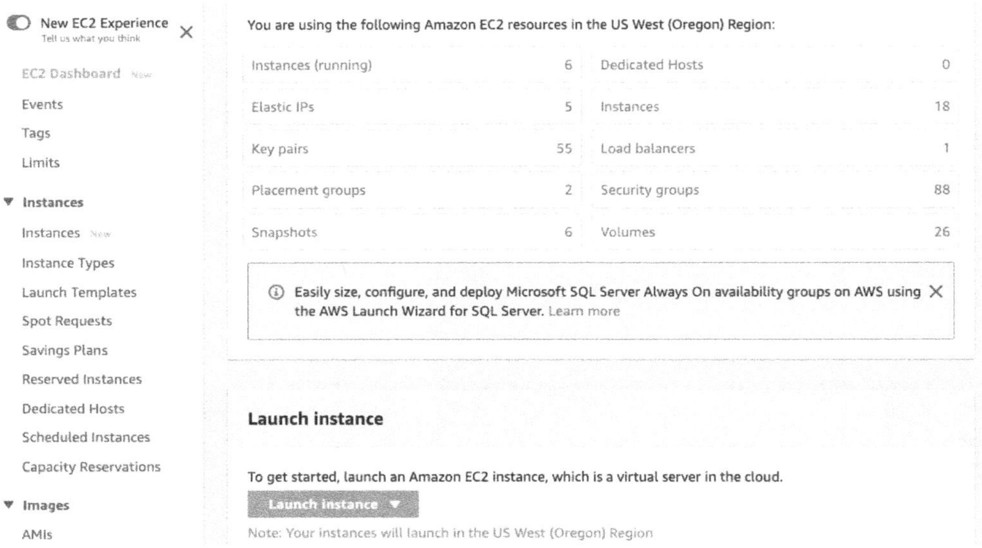

Figure 13.3 – Launching an on-demand EC2 instance

Let's walk through this process together.

Step 1: Choose an Amazon Machine Image (AMI)

On this screen, we choose an AMI with the Linux distribution of our choice. We recommend using a cost-saving **Free tier eligible** AMI. Let's pick the **Amazon Linux 2 AMI** for our on-demand EC2 instance:

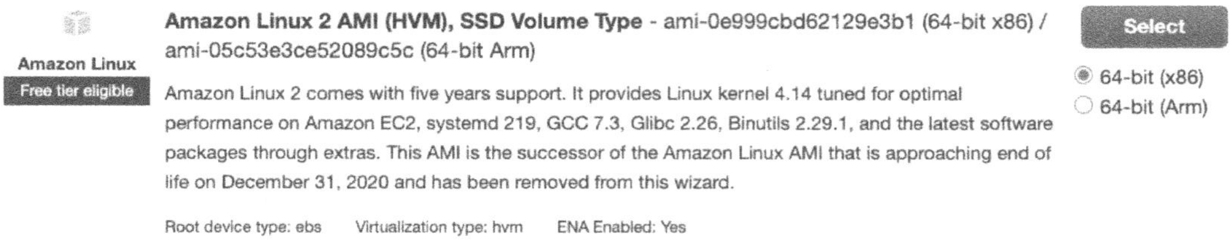

Figure 13.4 – Choosing the AMI for the EC2 instance

Pressing the **Select** button will take us to the next step.

Step 2: Choose an instance type

Here, we select the instance type based on our provisioning needs. We'll choose the `t2.micro` type, which is also **Free tier eligible**, with 1 vCPU and 1 GB memory (RAM):

1. Choose AMI   **2. Choose Instance Type**   3. Configure Instance   4. Add Storage   5. Add Tags   6. Configure Security Group   7. Review

Step 2: Choose an Instance Type

Amazon EC2 provides a wide selection of instance types optimized to fit different use cases. Instances are virtual servers that can run applications. They have varying combinations of CPU, memory, storage, and networking capacity, and give you the flexibility to choose the appropriate mix of resources for your applications. Learn more about instance types and how they can meet your computing needs.

Filter by: All instance families   ⌄   Current generation   ⌄   Show/Hide Columns

Currently selected: t2.micro (- ECUs, 1 vCPUs, 2.5 GHz, -, 1 GiB memory, EBS only)

	Family	Type	vCPUs (i)	Memory (GiB)	Instance Storage (GB) (i)	EBS-Optimized Available (i)	Network Performance (i)	IPv6 Support (i)
	t2	t2.nano	1	0.5	EBS only	-	Low to Moderate	Yes
■	t2	t2.micro [Free tier eligible]	1	1	EBS only	-	Low to Moderate	Yes
	t2	t2.small	1	2	EBS only	-	Low to Moderate	Yes
	t2	t2.medium	2	4	EBS only	-	Low to Moderate	Yes
	t2	t2.large	2	8	EBS only	-	Low to Moderate	Yes

Cancel   Previous   **Review and Launch**   Next: Configure Instance Details

Figure 13.5 – Choosing the EC2 instance type

At this point, we are ready to launch our instance by pressing the **Review and Launch** button. Alternatively, we could follow further configuration steps; otherwise, EC2 will assign some default values. Let's quickly go through these steps.

Step 3: Configure instance details

On this page, we can choose the number of EC2 instances we want to launch, configure the network settings, and select the OS-level shutdown behavior of our instance, to name just a few of the options. For any of the configuration options, you'll get a detailed description by pressing the information button next to it:

1. Choose AMI   2. Choose Instance Type   **3. Configure Instance**   4. Add Storage   5. Add Tags   6. Configure Security Group   7. Review

Step 3: Configure Instance Details

Number of instances (i)	1 Launch into Auto Scaling Group (i)
Purchasing option (i)	☐ Request Spot instances
Network (i)	vpc-dd4c82bb (default) ⬍ C Create new VPC
Subnet (i)	No preference (default subnet in any Availability Zone ⬍ Create new subnet
Auto-assign Public IP (i)	Use subnet setting (Enable) ⬍
Placement group (i)	☐ Add instance to placement group
Capacity Reservation (i)	Open ⬍
Domain join directory (i)	No directory ⬍ C Create new directory
IAM role (i)	None ⬍ C Create new IAM role
CPU options (i)	☐ Specify CPU options
Shutdown behavior (i)	Stop ⬍

Cancel   Previous   **Review and Launch**   Next: Add Storage

Figure 13.6 – Configuring EC2 instance details

In the next step, we'll look at the storage devices attached to our EC2 instance.

Step 4: Add storage

On this page, we choose the amount of storage space we want to have available on our instance, either as local disks or volume mounts:

Step 4: Add Storage

Your instance will be launched with the following storage device settings. You can attach additional EBS volumes and instance store volumes to your instance, or edit the settings of the root volume. You can also attach additional EBS volumes after launching an instance, but not instance store volumes. Learn more about storage options in Amazon EC2.

Volume Type (i)	Device (i)	Snapshot (i)	Size (GiB) (i)	Volume Type (i)	IOPS (i)	Throughput (MB/s) (i)	Delete on Termination (i)	Encryption (i)
Root	/dev/xvda	snap-036485f7aaef475e2	8	General Purpose S ∨	100 / 3000	N/A	☑	Not Encrypte ▼

Add New Volume

Figure 13.7 – Adding storage to an EC2 instance

In the next step, we'll tag our EC2 instance with specific information.

Step 5: Add tags

On this page, we define key-value pairs to label or identify our instance. When we manage a large number of EC2 instances, tags will help with user-friendly finding and filtering operations. For example, if we want to identify our EC2 instance as part of a *Packt* environment, we may create a tag with the following key-value pair:

- **Key**: env
- **Value**: packt

We can add multiple tags (key-value pairs) to a given instance:

Step 5: Add Tags

A tag consists of a case-sensitive key-value pair. For example, you could define a tag with key = Name and value = Webserver.
A copy of a tag can be applied to volumes, instances or both.
Tags will be applied to all instances and volumes. Learn more about tagging your Amazon EC2 resources.

Key (128 characters maximum)	Value (256 characters maximum)	Instances (i)	Volumes (i)	Network Interfaces (i)	
env	packt	☑	☑	☑	⊗

Add another tag   (Up to 50 tags maximum)

Figure 13.8 – Adding tags to an EC2 instance

In the next step, we'll configure security settings related to our instance.

Step 6: Configure security group

On this page, we configure a set of firewall rules controlling the inbound and outbound traffic to and from our EC2 instance:

Step 6: Configure Security Group

A security group is a set of firewall rules that control the traffic for your instance. On this page, you can add rules to allow specific traffic to reach your instance. For example, if you want to set up a web server and allow Internet traffic to reach your instance, add rules that allow unrestricted access to the HTTP and HTTPS ports. You can create a new security group or select from an existing one below. Learn more about Amazon EC2 security groups.

Assign a security group: ◉ Create a **new** security group
                          ○ Select an **existing** security group

Security group name: launch-wizard-20

Description: launch-wizard-20 created 2021-02-16T16:09:01.682-08:00

Type ⓘ	Protocol ⓘ	Port Range ⓘ	Source ⓘ		Description ⓘ	
SSH ⌄	TCP	22	Custom ⌄	0.0.0.0/0	e.g. SSH for Admin Desktop	⊗

Add Rule

Cancel Previous **Review and Launch**

Figure 13.9 – Configure the security group of an EC2 instance

In the last step, we review the configuration settings and launch our EC2 instance.

Step 7: Review

On this page, we have a summary of our EC2 instance configuration. We can edit and change any of the settings before launching the instance:

Step 7: Review Instance Launch

▼ AMI Details   Edit AMI

Amazon Linux 2 AMI (HVM), SSD Volume Type - ami-0e999cbd62129e3b1

Free tier eligible   Amazon Linux 2 comes with five years support. It provides Linux kernel 4.14 tuned for optimal performance on Amazon EC2, systemd 219, GCC 7.3, Glibc 2.26, Binutils 2.29.1, and the latest software packages through extras. This AMI is the successor of the Amazon Linux AMI that is a...

Root Device Type: ebs   Virtualization type: hvm

▼ Instance Type   Edit instance type

Instance Type	ECUs	vCPUs	Memory (GiB)	Instance Storage (GB)	EBS-Optimized Available	Network Performance
t2.micro	-	1	1	EBS only	-	Low to Moderate

▼ Security Groups   Edit security groups

Security group name   launch-wizard-20
Description   launch-wizard-20 created 2021-02-16T16:09:01.682-08:00

Type ⓘ	Protocol ⓘ	Port Range ⓘ	Source ⓘ	Description ⓘ
SSH	TCP	22	0.0.0.0/0	

▶ Instance Details   Edit instance details

Cancel Previous Launch

Figure 13.10 – Review and launch the EC2 instance

When launching the EC2 instance, we are asked to create or select a certificate key pair for remote SSH access into our instance:

Select an existing key pair or create a new key pair   ✕

A key pair consists of a **public key** that AWS stores, and a **private key file** that you store. Together, they allow you to connect to your instance securely. For Windows AMIs, the private key file is required to obtain the password used to log into your instance. For Linux AMIs, the private key file allows you to securely SSH into your instance.

Note: The selected key pair will be added to the set of keys authorized for this instance. Learn more about removing existing key pairs from a public AMI.

```
Create a new key pair                                                                      ⌄
```
Key pair name
```
packt-ec2
```

Download Key Pair

> ••• You have to download the **private key file** (*.pem file) before you can continue. **Store it in a secure and accessible location.** You will not be able to download the file again after it's created.

Cancel   **Launch Instances**

Figure 13.11 – Select or create a certificate key pair for SSH access

Let's create a new certificate key pair and name it `packt-ec2`. Download the related file (`packt-ec2.pem`) to a secure location on your local machine, where you can use it with the `ssh` command to access your EC2 instance:

```
ssh -i aws/packt-ec2.pem ec2-user@EC2_INSTANCE
```

We'll look closer at how to connect via SSH to our EC2 instances later in this chapter.

Pressing the **Launch Instances** button will create and launch our EC2 instance. The next screen will show a **View Instances** button, which will take you to the EC2 dashboard showing your instances in the current region. You may also filter the view based on various instance properties, including tags. For example, filtering by the `env: packt` tag, we'll get a view of the EC2 instance we just created:

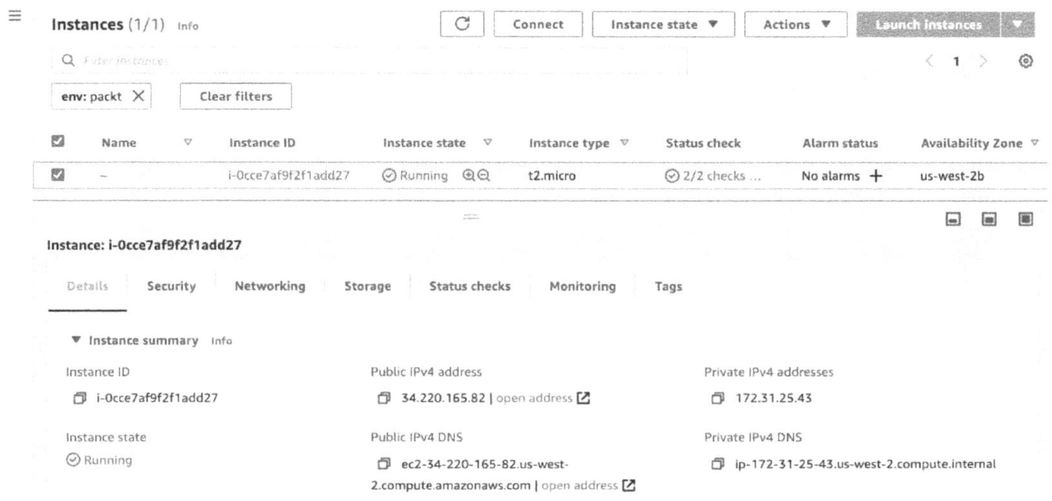

Figure 13.12 – An EC2 instance in the running state

For more information about on-demand instances, please visit https://docs.aws. amazon.com/AWSEC2/latest/UserGuide/ec2-on-demand-instances. html.

Now that we have learned the basics of launching an EC2 instance – namely an *on-demand instance* – let's look at *reserved instances* next.

EC2 reserved instances

With reserved instances, we lease EC2 computing capacity of a specific type for a specific amount of time. The length of time is called a *term* and can either be a 1-year or a 3-year commitment. Here are the main characteristics that need to be set upfront when purchasing reserved instances:

- **Platform** – such as Linux
- **Availability zone** – such as us-west-2
- **Tenancy** – running on **Default** (shared) or **Dedicated** hardware
- **Offering Class** – the options are as follows:

  a) **Standard** – a plain reserved instance with a well-defined set of options

  b) **Convertible** – allowing specific changes, such as modifying the instance type (for example, from t2.large to t2.xlarge)

- **Instance Type** – such as t2.large

- **Term** – such as 1 year

- **Payment option** – **all upfront**, **partial upfront**, or **no upfront**

With each of these options and the different tiers within, your costs depend on the cloud computing resources involved and the duration of the service. For example, if you choose to pay all upfront, you'll get a better discount than otherwise. Choosing from among the options previously mentioned is ultimately an exercise in cost-saving and flexibility.

A close analogy to purchasing reserved instances is a mobile telephone plan: you decide for all the options you want, and then you make a commitment for a certain amount of time. With reserved instances, you get less flexibility in terms of making changes, but with significant savings in cost – sometimes up to 75%, compared to on-demand instances.

To launch a reserved instance, go to your EC2 dashboard in the AWS console and choose **Reserved Instances** under **Instances** in the left panel, then click on the **Purchase Reserved Instances** button. Here is an example of purchasing a reserved EC2 instance:

Purchase Reserved Instances

☐ Only show offerings that reserve capa

Platform Linux/UNIX ⌄			**Tenancy** Default ⌄			**Offering Class** Convertible ⌄				
Instance Type t2.micro ⌄			**Term** 1 months - 1... ⌄			**Payment Option** All Upfront ⌄			Search	

Seller ⌄	Term ⌄	Effective Rate ⌄	Upfront Price ⌄	Hourly Rate ⌄	Payment Option ⌄	Offering Class ⌄	Quantity Available ⌄	Desired Quantity	Normalized units per hour	
AWS	12 months	$0.008	$68.00	$0.000	All Upfront	convertible	Unlimited	1	0.5	Add to Cart

Figure 13.13 – Purchasing a reserved EC2 instance

For more information about EC2 reserved instances, please visit `https://docs.aws.amazon.com/AWSEC2/latest/UserGuide/ec2-reserved-instances.html`.

We have learned that reserved instances are a cost-effective alternative to on-demand EC2 instances. Let's take our journey further and look at yet another way to reduce the costs by using spot instances.

EC2 spot instances

A *spot instance* is an unused instance waiting to be leased. The amount of time a spot instance is vacant and at your disposal depends on the general availability of the requested capacity in EC2, given that the associated costs are not higher than the amount you are willing to pay for your spot instance. AWS advertises spot instances with an up to 90% discount compared to on-demand EC2 pricing.

The major caveat of using spot instances is the potential *no-vacancy* situation when the required capacity is no longer available at the initially agreed-upon rate. In such circumstances, the spot instance will shut down (and perhaps leased elsewhere). AWS EC2 is kind enough to give a 2-minute warning before stopping the spot instance. This time should be used to properly tear down the application workflows running within the instance.

Spot instances are best suited for non-critical tasks, where application processing could be inadvertently interrupted at any moment and resumed later, without considerable damage or data loss. Such jobs may include data analysis, batch processing, and optional tasks.

To launch a spot instance, go to your **EC2** dashboard and choose **Spot Requests** in the left-hand menu. Under **Instances**, click on **Request Spot Instances**:

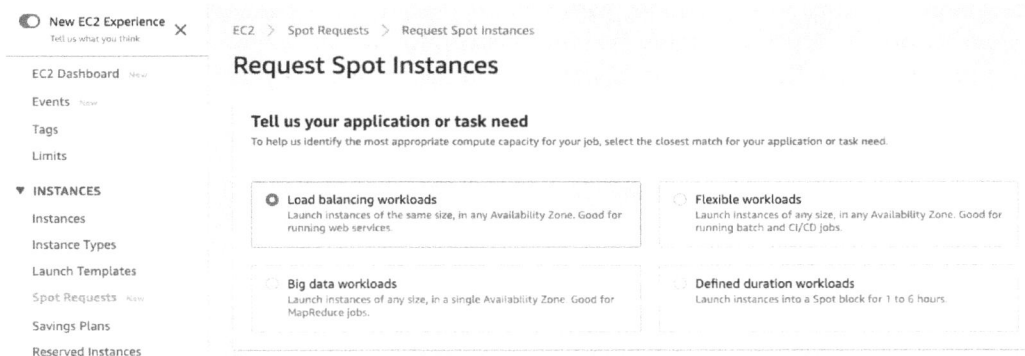

Figure 13.14 – Launching an EC2 spot instance

A detailed explanation of launching a spot instance is beyond the scope of this chapter. The AWS EC2 console does a great job at describing and assisting with the related options. For more information about spot instances, please visit `https://docs.aws.amazon.com/AWSEC2/latest/UserGuide/using-spot-instances.html`.

By default, EC2 instances run on shared hardware, meaning that instances owned by multiple AWS customers share the same machine (or virtual machine). What if you wanted to have a dedicated platform to run your EC2 instances? Let's look at dedicated instances next.

EC2 dedicated instances

Specific businesses require applications to run on dedicated hardware without sharing the platform with anyone. AWS EC2 provides *dedicated hosts* and *dedicated instances* to accommodate this use case. As you may expect, dedicated instances would cost more than other instance types. So, why should we care about leasing such instances?

There are businesses – especially among financial, health, and governmental institutions – required by law to meet strict regulatory requirements for processing sensitive data or to acquire hardware-based licenses for running their applications.

With dedicated instances *without* a dedicated host, EC2 would guarantee that your applications run on a hypervisor exclusively dedicated to you, yet it would not enforce a *fixed* set of machines or hardware. In other words, some of your instances may run on different physical hosts. Choosing **Dedicated host** in addition to **Dedicated** instances would always warrant a fully dedicated environment – hypervisors and hosts – for running your applications exclusively, without sharing the underlying platforms with other AWS customers.

To launch a dedicated instance, you start by following the same steps as launching an on-demand EC2 instance, described earlier in this chapter, in the *EC2 on-demand instances* section. In **Step 3: configure instance details**, for **Tenancy**, you will choose **Dedicated – Run a Dedicated Instance**, as shown in the following screenshot:

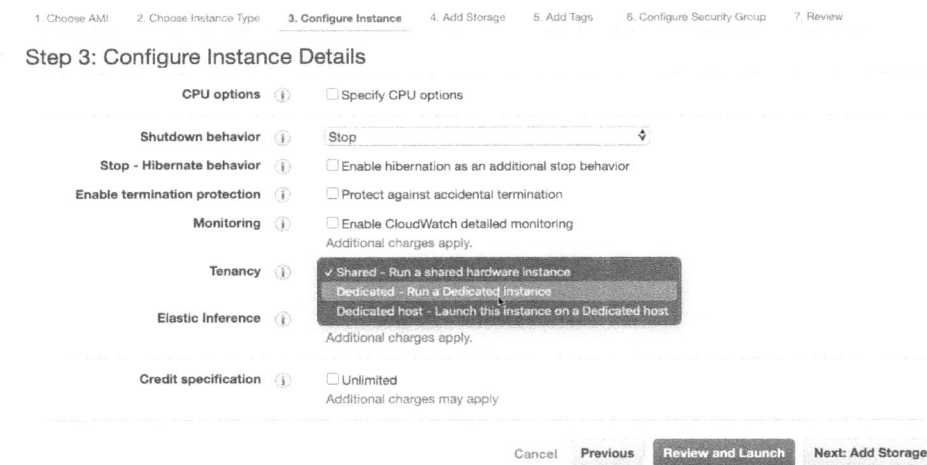

Figure 13.15 – Launching a dedicated EC2 instance

If you want to run your dedicated instance on a dedicated host, you must create a dedicated host first. On the **EC2** dashboard, choose **Dedicated Hosts** in the left menu, under **Instances**:

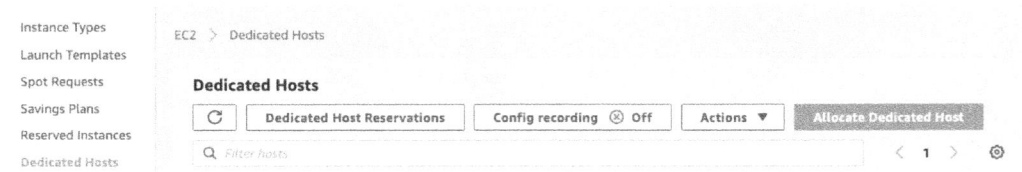

Figure 13.16 – Creating a dedicated EC2 host

Follow the EC2 wizard to allocate your dedicated host according to your preferences. After creating your host, you may launch your dedicated instance as described previously and choose the **Dedicated host – Launch this instance on a Dedicated host** option for **Tenancy** in **Step 3: configure instance details**.

For more information on dedicated hosts, please visit `https://aws.amazon.com/ec2/dedicated-hosts/`. For dedicated instances, see `https://aws.amazon.com/ec2/pricing/dedicated-instances/`.

We conclude our journey through AWS EC2 instance types here. For more information, please visit `https://docs.aws.amazon.com/AWSEC2/latest/UserGuide/Instances.html`.

Next, we'll look at AMIs. For a creative metaphor, an AMI is the egg from which your EC2 instance hatches.

Introducing Amazon Machine Images (AMIs)

An AMI is essentially an EC2 *machine template* that you can launch your instance from. An AMI usually bundles an operating system, but it may also encapsulate any software package or application with specific functionality. Here are the main constituents of an AMI:

- **Root volume template** – the storage volume or *hard drive* containing the image to boot the instance, including operating system files and applications

- **Launch permissions** – specifies *who* can use the AMI – the AWS accounts allowed to launch instances based on the AMI

- **Block device mapping** – a set of additional storage volumes (besides the root volume) for storing additional data, such as logs

Creating an AMI is also known as *registering* the AMI. You can copy the AMI across multiple AZs or share it with other users. AMIs are highly customizable. You can build your AMI by starting with another AMI, modifying it, then launching and saving (or registering) it for a specific use. As with instances, you can assign custom tags to your AMIs for identification purposes or keep them organized, such as versioning (such as *version: 1.0*). When you no longer need an AMI, you may *deregister* it to free up resources.

With a starter AWS EC2 account, you may not have an AMI of your own yet. AWS Marketplace has countless AMIs to choose from, many of them free. You can start with an existing AMI and build your own. The Amazon Linux AMIs are a good starting point. They are free of charge, well maintained, updated regularly, and supported by Amazon. You may also choose from AMIs based on standard Linux distributions, such as RHEL or Ubuntu.

You can create an AMI from a running EC2 instance by selecting your instance and, from the **Actions** menu, choose **Image and Templates**, followed by **Create Image**:

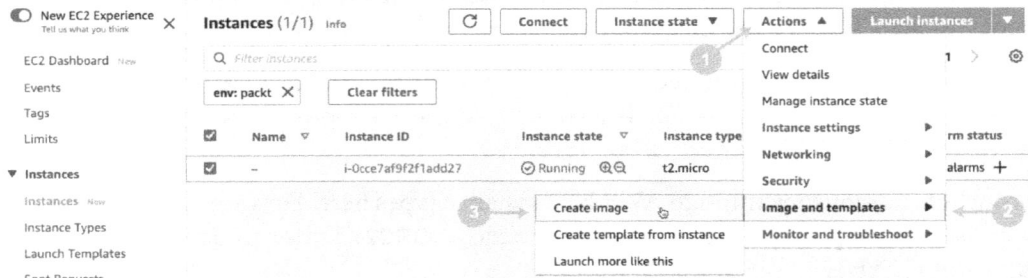

Figure 13.17 – Creating an AMI from an instance

You'll be prompted to name your instance and enter a description together with other options related to your AMI, well documented on the related EC2 dashboard screen, by clicking on the associated information icons.

For more information about AMIs, please visit `https://docs.aws.amazon.com/AWSEC2/latest/UserGuide/AMIs.html`.

Next, we'll look at another significant component of EC2 instances – *placement groups* – controlling how your instances are spread across the EC2 infrastructure for high availability and optimized workloads.

Introducing AWS EC2 placement groups

Placement groups allow you to specify how your EC2 instances are placed across the underlying EC2 hardware or hypervisors, providing strategies to group or separate instances depending on your requirements. Placement groups are offered free of charge.

There are three types of placement groups to choose from:

- Cluster
- Spread
- Partition

Let's quickly go through each of these types and look at their use cases.

Cluster placement groups

With cluster placement groups, instances are placed within a single AZ (data center). They are best suited for low-latency, high-throughput communication between instances, but not with the outside world. Applications with high-performance computing or data replication would greatly benefit from cluster placement, but web servers, for example, not so much.

Spread placement groups

When you launch multiple EC2 instances, there's always a possibility that they may end up running on the same physical machine or hypervisor. This may not be desirable when a single point of failure (such as hardware) would be critical for your applications. Spread placement groups provide hardware isolation between instances. In other words, if you launch multiple instances in a spread placement group, there's a guarantee that they would run on separate physical machines. In the rare case of an EC2 hardware failure, only one of your instances would be affected.

Partition placement groups

Partition placement groups will group your instances in logical formations (partitions) with hardware isolation between the partitions, but not at the instance level. We can view this model as a hybrid between the cluster and spread placement groups. When you launch multiple instances within a partition placement group, EC2 will do its best to distribute the instances between partitions evenly. For example, if you had 4 partitions and 12 instances, EC2 would place 3 instances in each node (partition). We can look at a partition as a computing unit made of multiple instances. In the case of a hardware failure, the isolated partition instances can still communicate with each other but not across partitions. Partition placement groups support up to seven instances in a single logical partition.

To create a placement group, select **Placement Groups** in your EC2 dashboard's left menu, under **Network & Security**, and click on the **Create Placement Group** button. On the next screen, you'll specify a name for the **Placement Group** and a **Placement Strategy**. Optionally, you can add tags (key-value pairs) for organizing or identifying your placement group. When done, click the **Create group** button:

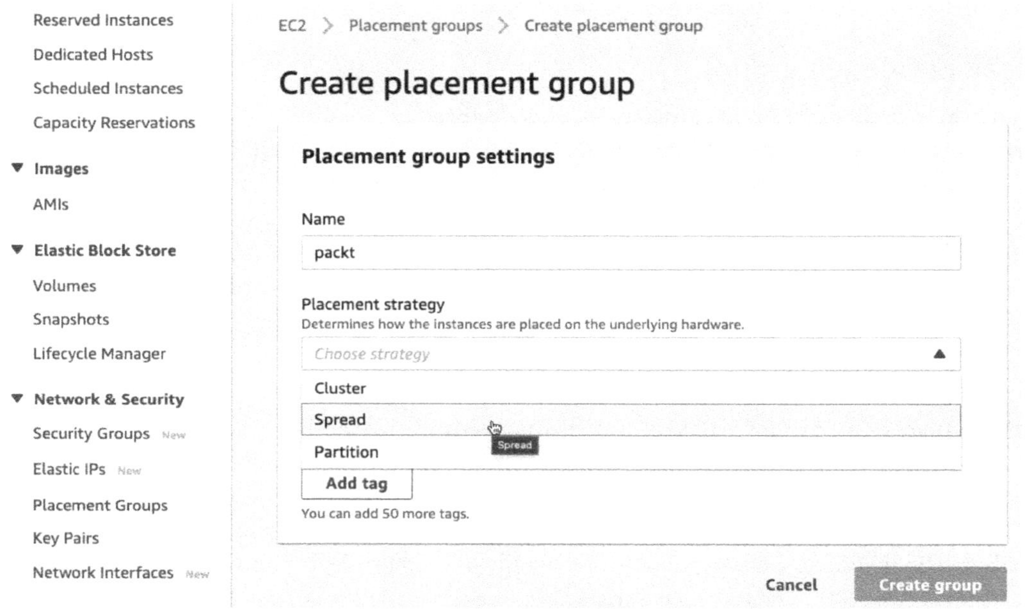

Figure 13.18 – Creating an AMI from an instance

For more information on EC2 placement groups, please visit https://docs.aws. amazon.com/AWSEC2/latest/UserGuide/placement-groups.html.

Now that we are familiar with various EC2 instance types, let's look at how to use our instances.

Using AWS EC2 instances

In this section, we briefly go through some essential operations and management concepts regarding your instances. First, let's look at the life cycle of an EC2 instance.

The life cycle of an EC2 instance

When using or managing EC2 instances, it is important to understand the transitional stages, from launch to running to hibernation, shutdown, or termination. Each of these states affects the billing and the way we access our instances:

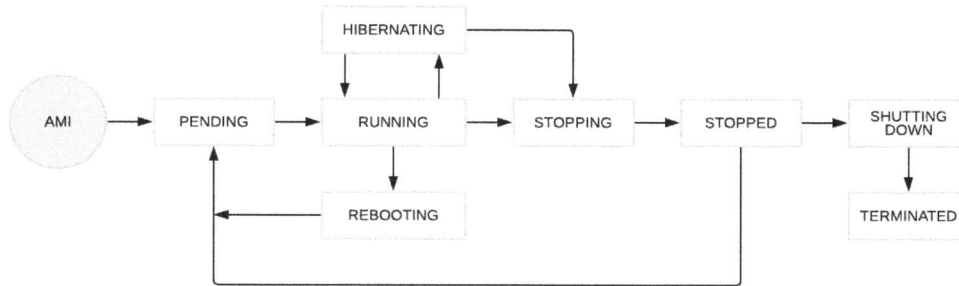

Figure 13.19 – The life cycle of an EC2 instance

The **PENDING** state corresponds to the bootup and initialization phase of our instance. Transitioning from **PENDING** to **RUNNING** is not always immediate, and it may take a while for the applications running within the instance to become responsive. EC2 starts billing our instance in the **RUNNING** state until transitioning to the **STOPPED** state.

In the **RUNNING** state, we can reboot our instance if needed. During the **REBOOTING** state, EC2 always brings back our instance on the same host, whereas stopping and restarting doesn't always guarantee the same host for the instance.

In the **STOPPED** state, we'll no longer be charged for the instance, but there will be costs related to any additional storage (other than the root volume) attached to the instance.

When the instance is no longer needed, we may choose between the **STOPPING** or **HIBERNATING** states. With **HIBERNATING**, we avoid the **PENDING** state's potential latency upon startup. If we no longer use the instance, we may choose to terminate it. Upon termination, there are no more charges related to the instance. When terminating an instance, it may still show up for a little while in the EC2 dashboard before it gets permanently removed.

We can connect to an EC2 instance in a running state using SSH. In the next section, we'll show you how.

Connecting to AWS EC2 instances

EC2 instances, in general, serve the purpose of running a specific application or a group of applications. The related platform's administration and maintenance usually require terminal access. Using the AWS EC2 console and the SSH terminal, we perform administrative tasks on EC2 instances. We call this the *control plane* (or *management plane*) access.

Applications running on EC2 instances may also expose their specific endpoints (ports) for communicating with the outside world. We refer to this as *data plane* access, and EC2 uses security groups to control the related network traffic.

In this section, we briefly look at both control plane and data plane access. In particular, we'll cover the following topics:

- Connecting to an EC2 instance using SSH
- Controlling network traffic with security groups
- Using SCP for file transfer with EC2 instances

First, let's look at how to connect via SSH to our EC2 instance.

Connecting via SSH to an EC2 instance

Using SSH with our EC2 instance allows us to manage it like any on-premises machine on a network. The related SSH command is as follows:

```
ssh -i SSH_KEY ec2-user@EC2_INSTANCE
```

SSH_KEY represents the private key file on our local system that we created and downloaded when launching our instance. See the *EC2 on-demand instances* section (in *Step 7: review*).

ec2-user is the default user assigned by EC2 to our AMI Linux instance. Different AMIs may have different usernames to connect with. You should check with the AMI vendor of your choice about the default username to use with SSH.

EC2_INSTANCE represents the public IP address or DNS name of our EC2 instance. You can find these in the EC2 dashboard for your instance:

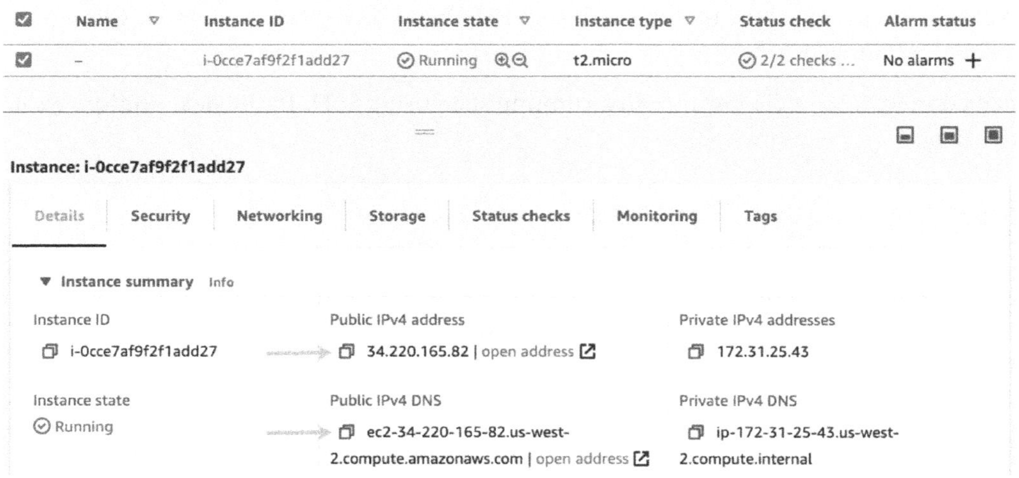

Figure 13.20 – The public IP address and DNS name of an EC2 instance

In our case, the SSH command is as follows:

```
ssh -i aws/packt-ec2.pem ec2-user@34.220.165.82
```

But before we connect, we need to set the right permissions for our private key file so it's not publicly viewable:

```
chmod 400 aws/packt-ec2.pem
```

Failing to do this results in an unprotected key file error while attempting to connect. If you need a refresher on these commands, click the **Connect** button at the top of your **EC2** dashboard with your EC2 instance selected:

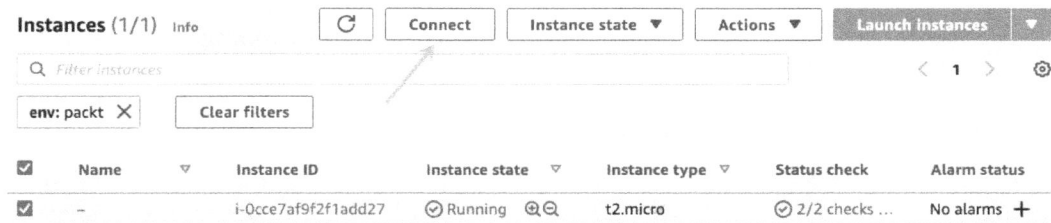

Figure 13.21 – Connecting to your EC2 instance

On the next screen, on the **SSH client** tab, you will have the steps and commands required to connect to your EC2 instance:

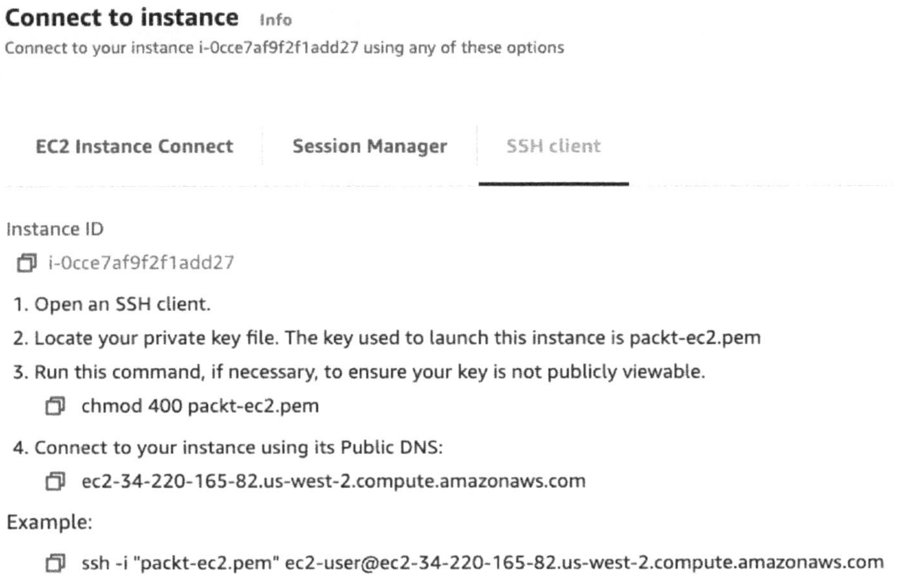

Figure 13.22 – The SSH client commands to connect to your EC2 instance

A successful SSH connection to our EC2 instance yields the following output:

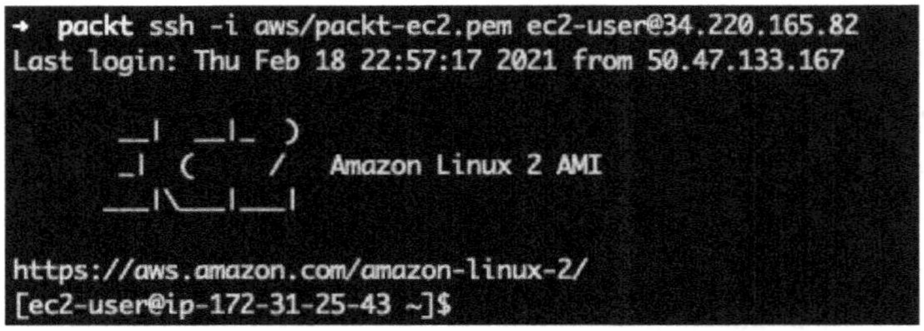

Figure 13.23 – Connecting with SSH to an EC2 instance

At this point, we can interact with our EC2 instance as if it were a standard machine.

Next, let's look at how to control the network access to applications running in EC2 instances.

Controlling network traffic with security groups

Security groups define a set of rules for filtering inbound and outbound network traffic in and out of an EC2 instance. When we create an instance, AWS EC2 automatically creates a default **security group** for it. The related settings are visible in the **Security details** pane with our instance selected on the **Security** tab:

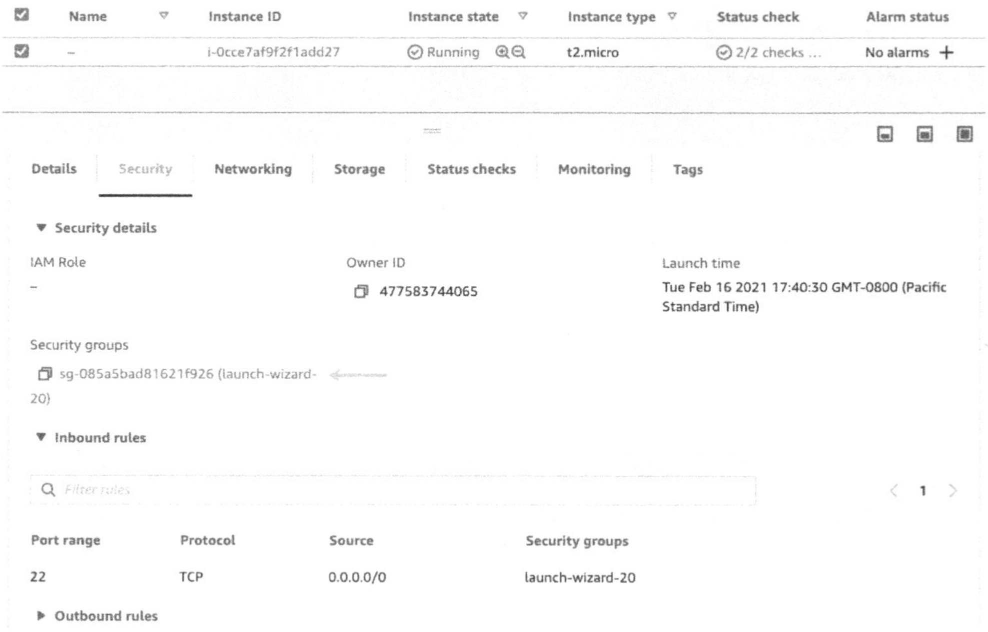

Figure 13.24 – The security settings for our EC2 instance

You may edit the security settings (**Inbound rules** and **Outbound rules**) by clicking the corresponding **security group ID**:

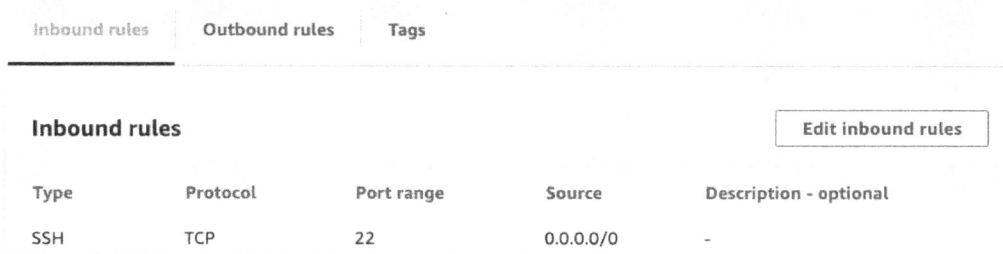

Figure 13.25 – Editing the security settings of your EC2 instance

For example, you may add inbound rules for HTTP and HTTPS connectivity if you run a web server in your instance:

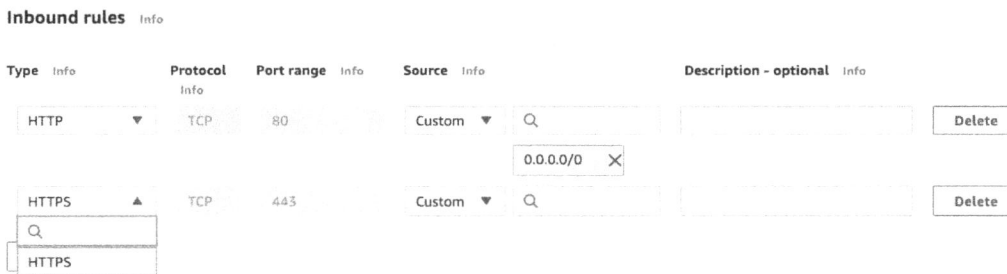

Figure 13.26 – Adding inbound rules for HTTP and HTTPS access to the EC2 instance

Managing the OS platform and applications running in an EC2 instance requires various administrative tasks. There are cases when we have to copy files either to or from the instance. In the next section, we'll show you how.

Using SCP for file transfer

To transfer files to and from an EC2 instance, we use the `scp` utility. `scp` uses the **Secure Copy Protocol (SCP)** to securely transfer files between network hosts.

The following command copies a local file (`README.md`) to our remote EC2 instance:

```
scp -i aws/packt-ec2.pem README.md ec2-user@34.220.165.82:/~
```

The file is copied to the `ec2-user`'s home folder (`/home/ec2-user`) on the EC2 instance. The reverse operation of transferring the `README.md` file from the remote instance to our local directory is as follows:

```
scp -i aws/packt-ec2.pem ec2-user@34.220.165.82:~/README.md  .
```

We should note that the `scp` command invocation is similar to `ssh`, where we specify the private key file (`aws/packt-ec2.pem`) via the `-i` (identity file) parameter.

We'll look next at yet another critical aspect of managing and scaling EC2 instances – storage volumes.

Using EC2 storage volumes

Storage volumes are device mounts within an EC2 instance, providing additional disk capacity (at extra cost). You may need additional storage for large file caches or extensive logging, for example, or you may choose to mount network-attached storage for critical data shared between your EC2 instances.

You can think of EC2 storage volumes as modular hard drives. You mount or unmount them based on your needs.

There are two types of storage volumes provided by EC2:

- Instance store

- **Elastic Block Store (EBS)**

Knowing how to use storage volumes allows you to make better decisions for scaling your applications as they grow. Let's look at the instance store volumes first.

Instance store volumes

Instance store volumes are disks directly (physically) attached to your EC2 instance. Consequently, the maximum size and number of instance store volumes you can connect to your instance are limited by the instance type. For example, a storage-optimized *i3* instance can have up to 8 x 1.9 TB SSD disks attached, while a general-purpose *m5d* instance may only grow up to 4 drives of 900 GB each. See `https://aws.amazon.com/ec2/instance-types/` for more information on instance capacity.

An instance store volume comes at no additional cost if it's the root volume – the volume with the OS platform booting the instance.

Not all EC2 instance types support instance store volumes. For example, the general-purpose *t2* instance types only support EBS storage volumes. On the other hand, if you want to grow your storage beyond the maximum capacity allowed by the instance store, you'll have to use EBS volumes.

The data on instance store volumes only persist with your EC2 instance. If your instance stops or terminates, or there is a failure with it, all of your data is lost. To store and persist critical data with your EC2 instances, you'll have to choose EBS. So, let's look at EBS volumes next.

EBS volumes

EBS volumes are flexible and high-performing network-attached storage devices, serving both the root volume system and additional volume mounts on your EC2 instance. An EBS root volume can only be attached to a single EC2 instance at a time. An EC2 instance can have multiple EBS volumes attached at any time. An EBS volume can also be attached to multiple EC2 instances at a time, using Multi-Attach. For more information on EBS Multi-Attach, see `https://docs.aws.amazon.com/AWSEC2/latest/UserGuide/ebs-volumes-multi.html`.

When you create an EBS volume, it will be automatically replicated within the AZ of your instance to minimize latency and data loss. With EBS, you get live monitoring of drive health and stats via Amazon CloudWatch free of charge. EBS also supports encrypted data storage to meet the latest regulatory standards for data encryption.

EC2 storage volumes are backed by Amazon's **Simple Storage Service (S3)** or **Elastic File System (EFS)** infrastructure. For more information on the EC2 storage types, please visit `https://docs.aws.amazon.com/AWSEC2/latest/UserGuide/Storage.html`.

Now, let's create and configure an EBS storage volume and attach it to our EC2 instance. Here are the steps we'll follow:

1. Create the volume.
2. Attach the volume to an EC2 instance.
3. Format the volume with a filesystem supported by the EC2 instance.
4. Create a volume mount point inside the EC2 instance.

We'll start with the first step, creating the EBS volume.

Creating an EBS volume

In the EC2 dashboard, go to **Volumes** under **Elastic Block Store** in the left navigation pane, and click on the **Create Volume** button at the top:

Volumes > Create Volume

Create Volume

Volume Type	General Purpose SSD (gp2) ▼ ⓘ
Size (GiB)	1 (Min: 1 GiB, Max: 16384 GiB) ⓘ
IOPS	100 / 3000 (Baseline of 3 IOPS per GiB with a minimum of 100 IOPS, burstable to 3000 IOPS) ⓘ
Throughput (MB/s)	Not applicable ⓘ
Availability Zone*	us-west-2b ▼ ⓘ
Snapshot ID	Select a snapshot ▼ C ⓘ
Encryption	☐ Encrypt this volume

Figure 13.27 – Create an EBS volume

Enter the values of your choice for the **Volume Type**, **Size**, and **Availability Zone**. Make sure you choose the AZ where your EC2 instances are. You could also include a **Snapshot ID** if you want to restore the volume from a previous EC2 instance backup (snapshot). We'll look at backup/restore using EBS snapshots later in this chapter.

Press the **Create Volume** button when done. You'll get a **Volume created successfully** message with your new EBS volume ID if all goes well.

Next, we'll attach our volume to an EC2 instance.

Attaching the volume to an EC2 instance

Click on the **Volume ID** or select the volume from the left navigation pane, under **Elastic Block Store** and **Volumes**. Click on the **Actions** button and choose **Attach Volume**:

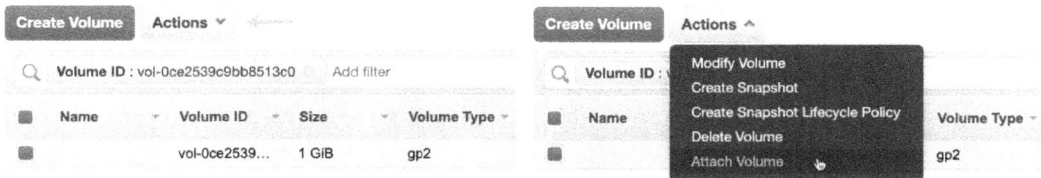

Figure 13.28 – Attach the EBS volume to an EC2 instance

On the next screen, enter your EC2 instance ID (or a name tag to search for it) in the **Instance** field:

Figure 13.29 – Enter the EC2 instance ID to attach the volume

Press the **Attach** button when done. After a few moments, EC2 will initialize your EBS volume, and the **State** changes to **in-use**:

Figure 13.30 – The new EBS volume is ready

The volume device is now ready, but we need to format it with a filesystem to be usable. We'll describe this procedure next.

Formatting the volume

Let's SSH into the EC2 instance where we attached the volume:

```
ssh -i aws/packt-ec2.pem ec2-user@34.220.165.82
```

Next, we retrieve the drives available in our EC2 instance using the lsblk command-line utility to list the block devices:

```
lsblk
```

The output is as follows:

```
[ec2-user@ip-172-31-25-43 ~]$ lsblk
NAME     MAJ:MIN RM SIZE RO TYPE MOUNTPOINT
xvda     202:0    0  8G  0 disk
└─xvda1 202:1    0  8G  0 part /
xvdf     202:80   0  1G  0 disk
```

Figure 13.31 – The local volumes in our EC2 instance

Looking at the size of the volumes, we can immediately tell the one we just added – xvdf with 1G. The other volume (xvda) is the original root volume of our t2.micro instance.

Let's check next if our new EBS volume (xvdf) has a filesystem on it:

```
sudo file -s /dev/xvdf
```

The output is /dev/xvdf: data, meaning that the volume doesn't have a filesystem yet:

```
[ec2-user@ip-172-31-25-43 ~]$ sudo file -s /dev/xvdf
/dev/xvdf: data
```

Figure 13.32 – No filesystem on the new EBS volume

Let's build a filesystem on our volume, using the mkfs (*make filesystem*) command-line utility:

```
sudo mkfs -t xfs /dev/xvdf
```

We invoke the -t (--type) parameter with an xfs filesystem type. XFS is a high-performance journaled filesystem supported by most Linux distributions, installed by default in some of them.

The preceding command yields the following output:

```
[ec2-user@ip-172-31-25-43 ~]$ sudo mkfs -t xfs /dev/xvdf
meta-data=/dev/xvdf              isize=512    agcount=4, agsize=65536 blks
         =                       sectsz=512   attr=2, projid32bit=1
         =                       crc=1        finobt=1, sparse=0
data     =                       bsize=4096   blocks=262144, imaxpct=25
         =                       sunit=0      swidth=0 blks
naming   =version 2              bsize=4096   ascii-ci=0 ftype=1
log      =internal log           bsize=4096   blocks=2560, version=2
         =                       sectsz=512   sunit=0 blks, lazy-count=1
realtime =none                   extsz=4096   blocks=0, rtextents=0
```

Figure 13.33 – Build a new filesystem on the EBS volume

If we check the filesystem using the following command, we should see the filesystem details displayed instead of empty data:

```
sudo file -s /dev/xvdf
```

The output is as follows:

```
[ec2-user@ip-172-31-25-43 ~]$ sudo file -s /dev/xvdf
/dev/xvdf: SGI XFS filesystem data (blksz 4096, inosz 512, v2 dirs)
```

Figure 13.34 – The new filesystem on the EBS volume

The volume drive is now formatted. Let's make it accessible to our local filesystem.

Creating a volume mount point

Let's name our mount point `packt` and create it in the root directory:

```
sudo mkdir /packt
sudo mount /dev/xvdf /packt
```

At this point, the EBS volume is mounted, and when we access the `/packt` directory, we're accessing the EBS volume:

```
[ec2-user@ip-172-31-25-43 ~]$ sudo touch /packt/README.md
[ec2-user@ip-172-31-25-43 ~]$ ls -la /packt
total 0
drwxr-xr-x  2 root root  23 Feb 20 00:10 .
dr-xr-xr-x 19 root root 270 Feb 20 00:07 ..
-rw-r--r--  1 root root   0 Feb 20 00:10 README.md
```

Figure 13.35 – Accessing the EBS volume

EBS volumes may contain critical data we would like to hold on to. Let's look at how to use EBS snapshots for disaster recovery next.

Working with EBS snapshots

When you use EBS, you're likely to run into scenarios where you want to back up your data, either for the long term or to prepare for disaster recovery. We should also note that for specific EC2 instance types (such as general-purpose instances), the root volume – your system – is EBS, and you may want to have a backup if you encounter unexpected failures with your instance. A full backup of an EC2 instance is beyond the scope of this chapter.

Let's look at how to back up an EBS volume using snapshots. Here are the steps:

1. Create a snapshot of the current volume.

2. Attach the snapshot to a new volume.

3. Detach the current volume from the EC2 instance.

4. Attach the new volume to the EC2 instance.

Let's start with the first step.

Creating a snapshot

In your EC2 dashboard, go to **Volumes** under **Elastic Block Store** in the left navigation menu, and select the EBS volume you'd like to back up. In our case, let's create a snapshot of the root volume, the one with 8 GB under **Size**. Next, click on the **Actions** button and choose **Create Snapshot**:

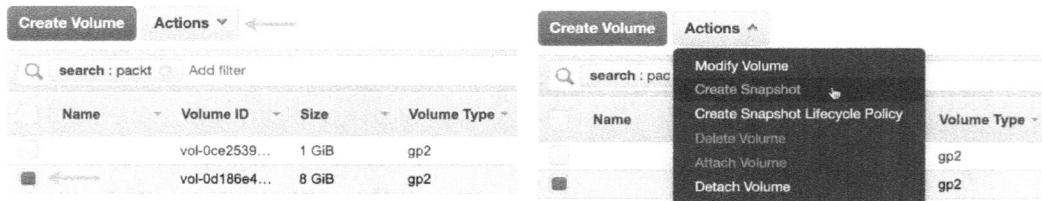

Figure 13.36 – Creating a snapshot of an EBS volume

In the next screen, enter a description for your snapshot (such as `packt-backup`) and click the **Create Snapshot** button:

Volumes > Create Snapshot

Create Snapshot

Volume   vol-0d186e402d2f54fd0 ⓘ

Description   packt-backup   ⓘ

Encrypted   Not Encrypted ⓘ

Figure 13.37 – Describe your EBS snapshot

Upon successfully creating the snapshot, EC2 will show a **Create Snapshot Request Succeeded** message with the corresponding **Snapshot ID**. You can manage your current snapshots in the EC2 dashboard by selecting **Snapshots** under **Elastic Block Store** in the left navigation menu.

A snapshot needs a volume to be consumable. In the next step, we'll create a new volume and attach the snapshot to it.

Attaching the snapshot to a volume

Let's start by locating our snapshot on the **Snapshots** management page and then copy the corresponding **Snapshot ID**. We'll reuse (copy/paste) the snapshot ID in the next step:

▣	Name	▾	Snapshot ID	▴	Size	▾	Description	▾	Status	▾
▣		---→	snap-0ddd84ea954adfb17		8 GiB		packt-backup		◔ completed	

Figure 13.38 – Copy your Snapshot ID

Now, go to your **Volumes**, under **Elastic Block Store** in the left navigation menu, and click **Create Volume**. Paste the snapshot ID previously copied in the **Snapshot ID** field. Make sure the **Availability Zone** of your new EBS volume matches the EC2 instance you attach it to:

Volumes > Create Volume

Create Volume

Volume Type	General Purpose SSD (gp2) ▾ ❶	
Size (GiB)	8	(Min: 1 GiB, Max: 16384 GiB) ❶
IOPS	100 / 3000	(Baseline of 3 IOPS per GiB with a minimum of 100 IOPS, burstable to 3000 IOPS) ❶
Throughput (MB/s)	Not applicable ❶	
Availability Zone*	us-west-2b ←------	▾ ❶
Snapshot ID	snap-0ddd84ea954adfb17 ←------	▾ C ❶

Figure 13.39 – Create a new EBS volume from an existing Snapshot ID

Your new EBS volume will appear with the Snapshot ID it's been created from. Its state is **available**:

Name	Volume ID	Size	Volume Type	Snapshot	Availability Zone ▲	State
■	vol-09a979c...	8 GiB	gp2	snap-0ddd84ea954adfb17	us-west-2b	● available
	vol-0ce2539...	1 GiB	gp2		us-west-2b	○ in-use
	vol-0d186e4...	8 GiB	gp2	snap-036485f7aaef475e2	us-west-2b	○ in-use

Figure 13.40 – The new EBS volume created from a snapshot

We now have a *standalone* volume with the snapshot. To use this volume with a different EC2 instance or restore it later on the same instance, we need to detach the current volume from the instance. In the next section, we'll show you how.

Detaching a volume

In our case, since we're detaching the root volume, we need to stop the EC2 instance. For non-root volumes, we can leave the EC2 instance running during the detach/attach operation.

So, let's stop the EC2 instance first. In the EC2 dashboard, we go to **Instances**, select our EC2 instance, and right-click and choose **Stop Instance**. Next, we'll detach the existing EBS volume from our instance. In the EC2 dashboard, go to **Volumes**, select the current **in-use** volume, right-click, and choose **Detach Volume**. Acknowledge the operation and wait for the volume to be detached.

We're now at the final step of our backup-restore procedure, attaching the new volume containing the snapshot to our EC2 instance.

Attaching a volume

Select the new volume you just created from the EBS snapshot, right-click and select **Attach Volume**:

Name	Volume ID	Size	Volume Type	Snapshot	Availability Zone ▲	State
■	vol-09a979c...	8 GiB	gp2	Modify Volume	s-west-2b	● available
	vol-0ce2539...	1 GiB	gp2	Create Snapshot	s-west-2b	○ in-use
	vol-0d186e4...	8 GiB	gp2	Create Snapshot Lifecycle Policy	s-west-2b	● available
				Delete Volume		
				Attach Volume ⌙		

Figure 13.41 – Attaching the new EBS volume created from a snapshot

In the **Attach Volume** screen, you'll have to specify the ID of your EC2 instance in the **Instance** field, as suggested in *Figure 13.29*. You may also want to make sure the **Device** field has the same device ID as your previous root volume (/dev/xvda), as seen in *Figure 13.31*:

Attach Volume

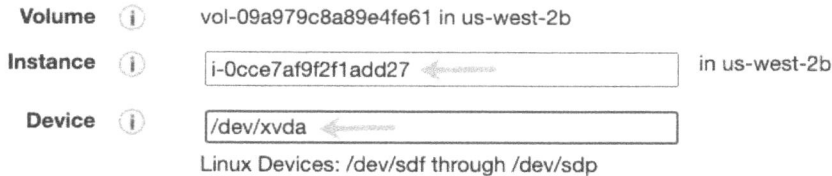

Volume ⓘ	vol-09a979c8a89e4fe61 in us-west-2b	
Instance ⓘ	i-0cce7af9f2f1add27 ◄─────	in us-west-2b
Device ⓘ	/dev/xvda ◄─────	

Linux Devices: /dev/sdf through /dev/sdp

Figure 13.42 – Attaching the new EBS volume as a root volume

With the new volume reattached, we can start our EC2 instance. Go to the **Instances** menu, right-click your EC2 instance, and select **Start Instance**.

> **Important note**
> When you restart your EC2 instance, EC2 may bring your machine up on a different host, and most probably, you'll have a different public IP address.

Once your instance is up and running, make sure you can connect with SSH. In our case, the new public IP address of our EC2 instance has changed, so we'll have to adapt our SSH command accordingly.

You may also want to remove the old EBS volume (if you're not using it anymore), so you're not being charged for it. Go to your **Volumes** page, select the unused volume, right-click, and choose **Delete Volume**.

We conclude our exploration of the AWS EC2 console and related management operations here. For a comprehensive reference on EC2, please refer to the Amazon EC2 documentation at `https://docs.aws.amazon.com/ec2`.

The EC2 management tasks presented so far exclusively used the AWS console. If you're looking to automate your EC2 workloads, you may want to adopt the AWS CLI, a unified tool for managing AWS resources. Let's check it out next.

Working with the AWS CLI

To install the AWS CLI, please visit `https://aws.amazon.com/cli/`. At the time of this writing, the latest release of the AWS CLI is version 2. For the examples in this chapter, we use an Ubuntu machine to install the AWS CLI, following the instructions at `https://docs.aws.amazon.com/cli/latest/userguide/install-cliv2-linux.html`.

We'll start with the download of the AWS CLI v2 package (`awscliv2.zip`):

```
curl "https://awscli.amazonaws.com/awscli-exe-linux-x86_64.zip"
-o "awscliv2.zip"
```

Next, we unzip and install the AWS CLI:

```
unzip awscliv2.zip
sudo ./aws/install
```

We should now have the `aws` command-line utility installed on our system. Let's check the version:

```
aws --version
```

The output is as follows:

```
packt@neptune:~$ aws --version
aws-cli/2.1.27 Python/3.7.3 Linux/5.4.0-65-generic exe/x86_64.ubuntu.20 prompt/off
```

Figure 13.43 – Checking the version of the AWS CLI

You may start exploring the AWS CLI by invoking the help:

```
aws help
```

To manage your AWS EC2 resources using the `aws` utility, first, you need to configure your local environment to establish the required trust with the AWS endpoint.

Configuring the AWS CLI

To configure the local AWS environment on your local machine, run the following command:

```
aws configure
```

The preceding command will prompt you for a few pieces of information, as suggested by the following output:

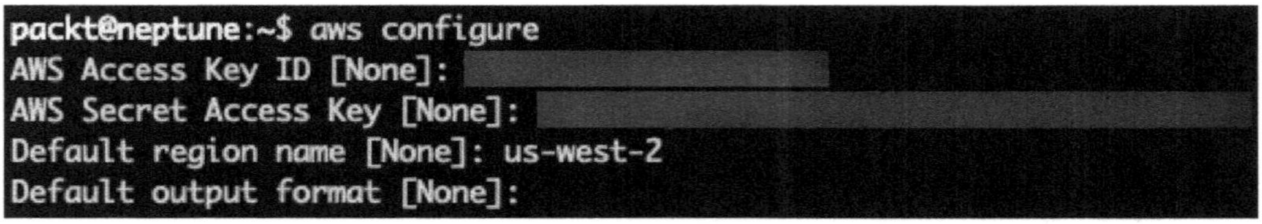

```
packt@neptune:~$ aws configure
AWS Access Key ID [None]:
AWS Secret Access Key [None]:
Default region name [None]: us-west-2
Default output format [None]:
```

Figure 13.44 – Configuring the local AWS CLI environment

The AWS CLI configuration asks for your **AWS Access Key ID** and **AWS Secret Access Key**. You can generate or retrieve these keys by logging into your AWS account. Select the dropdown next to your account name in the upper right corner of the AWS console and choose **My Security Credentials**. If you haven't generated your access key yet, go to the **AWS IAM credentials** tab, under **Access keys for CLI, SDK, & API access**, and click the **Create access key** button. You'll have to store your AWS key ID and secret in a safe place for later reuse:

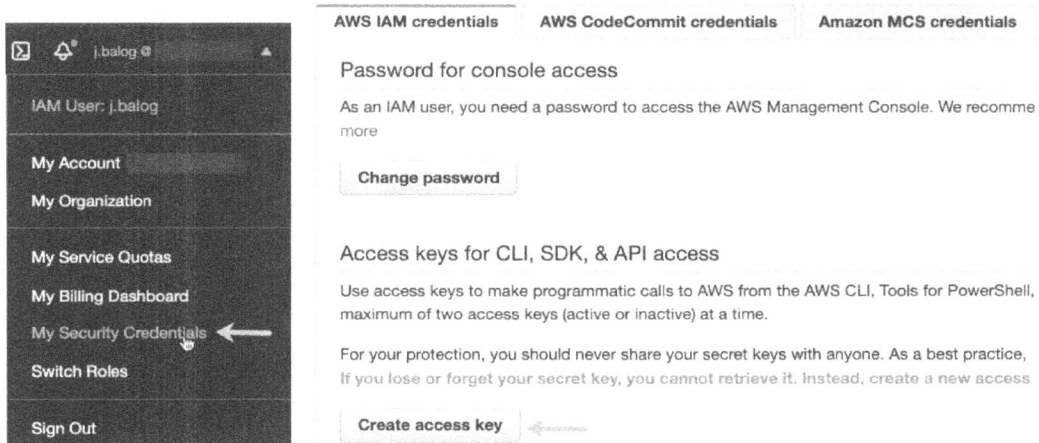

Figure 13.45 – Creating your AWS access key

In the AWS CLI configuration wizard, we also set the **Default region name** as us-west-2. You may want to enter the region of your choice or leave it as the default (None). If you don't have a default region specified, you'll have to enter it every time you invoke the aws command.

At this point, we're ready to use the AWS CLI. Let's start by listing our EC2 instances.

Querying EC2 instances

The following command provides detailed information about EC2 instances:

```
aws ec2 describe-instances
```

The preceding command provides sizeable JSON output, with the details of all the EC2 instances we own in our default region (us-west-2). Alternatively, we can specify the region with the --region parameter:

```
aws ec2 describe-instances --region us-west-2
```

We can get more creative and list only the EC2 instances matching a specific key-value tag, such as env: packt, that we previously tagged our instances with, using the --filters parameter:

```
aws ec2 describe-instances \
  --filters "Name=tag-key,Values=env" \
  --filters "Name=tag-value,Values=packt"
```

The first --filters parameter specifies the key (tag-key=env) and the second points to the value (tag-value=packt).

Combining the aws and jq (*JSON query*) commands, we can extract only the JSON fields we want. For example, the following command lists the InstanceId, ImageId, and BlockDeviceMappings fields of the EC2 instances tagged with env:packt:

```
aws ec2 describe-instances \
  --filters "Name=tag-key,Values=env" \
  --filters "Name=tag-value,Values=packt" | \
  jq '.Reservations[].Instances[] | { InstanceId, ImageId,
BlockDeviceMappings }'
```

If the jq utility is not present on your Linux machine, install it with the following commands:

- **sudo apt install -y jq # on Ubuntu**
- **sudo yum install -y jq # on RHEL/CentOS**

The output of the preceding aws command is as follows:

```
{
  "InstanceId": "i-0cce7af9f2f1add27",
  "ImageId": "ami-0e999cbd62129e3b1",
  "BlockDeviceMappings": [
    {
      "DeviceName": "/dev/sdf",    ←
      "Ebs": {
        "AttachTime": "2021-02-19T22:59:54+00:00",
        "DeleteOnTermination": false,
        "Status": "attached",
        "VolumeId": "vol-0ce2539c9bb8513c0"
      }
    },
    {
      "DeviceName": "/dev/xvda",    ←
      "Ebs": {
        "AttachTime": "2021-02-20T01:59:10+00:00",
        "DeleteOnTermination": false,
        "Status": "attached",
        "VolumeId": "vol-09a979c8a89e4fe61"
      }
    }
  ]
}
```

Figure 13.46 – Querying EC2 instances

We should note the DeviceName properties in the output JSON, reflecting the block devices (/dev/sdf and /dev/xvda) we managed in a previous section when attaching EBS volumes to our instance.

We can filter the output of the aws ec2 describe-instances command by any property. For example, the following command filters our EC2 instances by the AMI image-id:

```
aws ec2 describe-instances \
  --filters "Name=image-id,Values=ami-0e999cbd62129e3b1"
```

Note that the property name used in the filter is a *hyphenated* transformation of the corresponding *camel-cased* JSON property: `image-id` versus `ImageId`. You have to keep this rule in mind when you write your filter queries. Here's an excerpt from the output of the preceding command:

```
packt@neptune:~$ aws ec2 describe-instances \
>    --filters "Name=image-id,Values=ami-0e999cbd62129e3b1"

{
    "Reservations": [
        {
            "Groups": [],
            "Instances": [
                {
                    "AmiLaunchIndex": 0,
                    "ImageId": "ami-0e999cbd62129e3b1",
                    "InstanceId": "i-0cce7af9f2f1add27",
                    "InstanceType": "t2.micro",
                    "KeyName": "packt-ec2",
                    "LaunchTime": "2021-02-20T01:59:28+00:00",
                    "Monitoring": {
                        "State": "disabled"
                    },
                    "Placement": {
                        "AvailabilityZone": "us-west-2b",
                        "GroupName": "",
                        "Tenancy": "default"
                    },
```

Figure 13.47 – Filtering EC2 instances by image ID

Next, let's plan to launch a new EC2 instance of the same AMI type with our current machine and the same security group.

Creating an EC2 instance

The following command retrieves the security groups of the current instance (`i-0cce7af9f2f1add27`):

```
aws ec2 describe-instances \
    --filters "Name=instance-id,Values=i-0cce7af9f2f1add27" \
    --query "Reservations[].Instances[].SecurityGroups[]"
```

The output is as follows:

```
packt@neptune:~$ aws ec2 describe-instances \
>    --filters "Name=instance-id,Values=i-0cce7af9f2f1add27" \
>    --query  "Reservations[].Instances[].SecurityGroups[]"
[
    {
        "GroupName": "launch-wizard-20",
        "GroupId": "sg-085a5bad81621f926"
    }
]
```

Figure 13.48 – Retrieving the security groups of an EC2 instance

To retrieve the `GroupId` directly, we could run the following:

```
aws ec2 describe-instances \
    --filters "Name=instance-id,Values=i-0cce7af9f2f1add27" \
    --query "Reservations[].Instances[].SecurityGroups[].GroupId"
```

The output, in this case, would be as follows:

```
[
    "sg-085a5bad81621f926"
]
```

We used the `--query` parameter to specify the exact JSON path for the field we're looking for (`GroupId`):

```
Reservations[].Instances[].SecurityGroups[].GroupId
```

The use of the `--query` parameter somewhat resembles the piping of the output to the `jq` command, but it's less versatile.

To launch a new instance with the AMI type of our choice and the previous security group ID, we use the `aws ec2 run-instances` command:

```
aws ec2 run-instances \
    --image-id ami-0e999cbd62129e3b1\
    --count 1 \
    --instance-type t1.micro \
    --key-name packt-ec2 \
    --security-group-ids sg-085a5bad81621f926 \
    --placement AvailabilityZone=us-west-2b
```

Here's a brief explanation of the parameters:

- `image-id` – The AMI image ID (`ami-0e999cbd62129e3b1`); we're using the same AMI type (*Amazon Linux*) as with the previous instance we created in the AWS EC2 web console.

- `count` – The number of instances to launch (`1`).

- `instance-type` – The EC2 instance type (`t1.micro`).

- `key-name` – The name of the SSH private key file (`packt-ec2`) to use when connecting to our new instance; we're reusing the SSH key file we created with our first EC2 instance in the AWS console.

- `security-group-ids` – The security groups attached to our instance; we're reusing the security group attached to our current instance (`sg-085a5bad81621f926`).

- `--placement` – The AZ to place our instance in (`AvailabilityZone=us-west-2b`).

Here's an excerpt from the command's output, suggesting that our new instance has been launched, with an `InstanceID` of `i-0e1692c9dfdf07a8d`:

```
packt@neptune:~$ aws ec2 run-instances \
>    --image-id ami-0e999cbd62129e3b1\
>    --count 1 \
>    --instance-type t1.micro \
>    --key-name packt-ec2 \
>    --security-group-ids sg-085a5bad81621f926
{
    "Groups": [],
    "Instances": [
        {
            "AmiLaunchIndex": 0,
            "ImageId": "ami-0e999cbd62129e3b1",
            "InstanceId": "i-0e1692c9dfdf07a8d",    ←
            "InstanceType": "t1.micro",
            "KeyName": "packt-ec2",
            "LaunchTime": "2021-02-22T09:15:17+00:00",
```

Figure 13.49 – Launching a new EC2 instance

Next, let's tag our new instance using the command line.

Tagging an EC2 instance

The following command tags our new instance with the env:packt key-value:

```
aws ec2 create-tags \
  --resources i-0e1692c9dfdf07a8d \
  --tags Key=env,Value=packt
```

Now, we can query our instances based on the aforementioned tag:

```
aws ec2 describe-instances \
  --filters "Name=tag-key,Values=env" \
  --filters "Name=tag-value,Values=packt" \
  --query "Reservations[].Instances[].InstanceId"
```

The output shows our two EC2 instances:

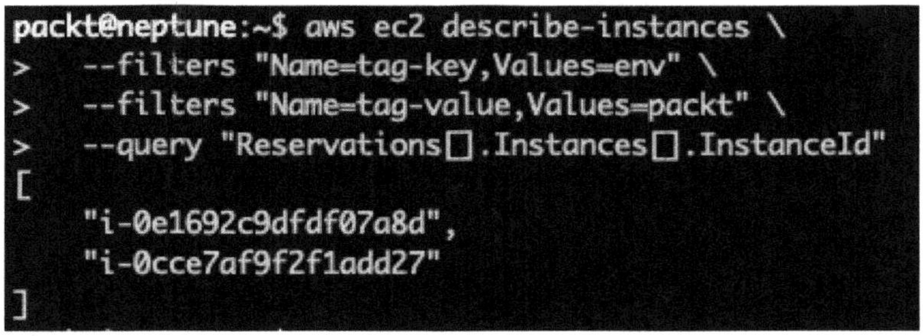

```
packt@neptune:~$ aws ec2 describe-instances \
>     --filters "Name=tag-key,Values=env" \
>     --filters "Name=tag-value,Values=packt" \
>     --query "Reservations[].Instances[].InstanceId"
[
    "i-0e1692c9dfdf07a8d",
    "i-0cce7af9f2f1add27"
]
```

Figure 13.50 – Querying EC2 instance IDs by tag

Let's look at how to add additional storage to our instance.

Adding additional storage to an EC2 instance

First, we need to create a new storage device. The following command creates a general-purpose SSD (gp2) volume type, 8 GB in size, in a US West (us-west-2b) AZ:

```
aws ec2 create-volume \
  --volume-type gp2 \
  --size 8 \
  --availability-zone us-west-2b
```

The output of the command is as follows:

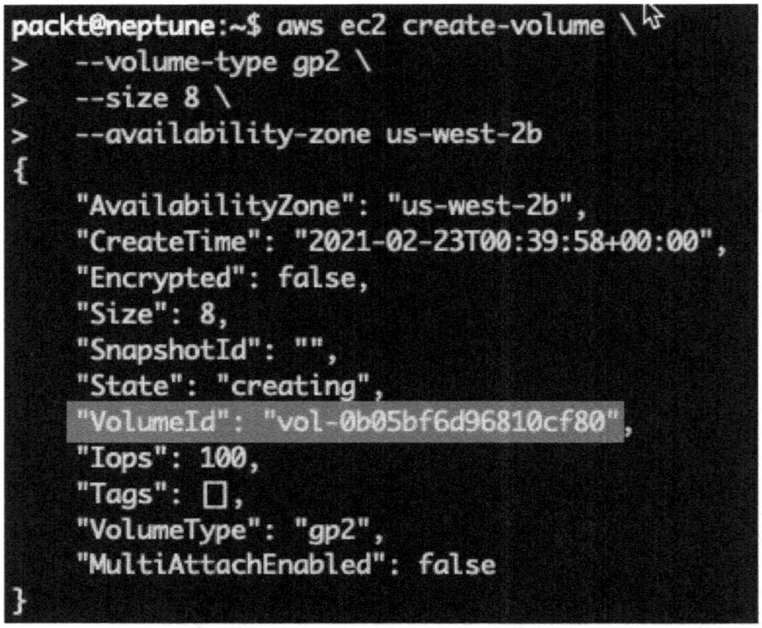

```
packt@neptune:~$ aws ec2 create-volume \
>    --volume-type gp2 \
>    --size 8 \
>    --availability-zone us-west-2b
{
    "AvailabilityZone": "us-west-2b",
    "CreateTime": "2021-02-23T00:39:58+00:00",
    "Encrypted": false,
    "Size": 8,
    "SnapshotId": "",
    "State": "creating",
    "VolumeId": "vol-0b05bf6d96810cf80",
    "Iops": 100,
    "Tags": [],
    "VolumeType": "gp2",
    "MultiAttachEnabled": false
}
```

Figure 13.51 – Creating a new volume

Please note the VolumeId – we'll use it when attaching to the instance.

> **Important note**
>
> Please make sure you create the volume in the same AZ as your instance.
> Otherwise, you won't be able to attach it to the EC2 instance. In our case, the
> AZ is us-west-2b.

The next command attaches the volume to our instance (i-0e1692c9dfdf07a8d)
using the /dev/sdf device identifier:

```
aws ec2 attach-volume \
  --volume-id vol-0b05bf6d96810cf80 \
  --instance-id i-0e1692c9dfdf07a8d \
  --device /dev/sdf
```

The output is as follows:

```
packt@neptune:~$    aws ec2 attach-volume \
>    --volume-id vol-0b05bf6d96810cf80 \
>    --instance-id i-0e1692c9dfdf07a8d \
>    --device /dev/sdf
{
    "AttachTime": "2021-02-23T01:16:07.100000+00:00",
    "Device": "/dev/sdf",
    "InstanceId": "i-0e1692c9dfdf07a8d",
    "State": "attaching",
    "VolumeId": "vol-0b05bf6d96810cf80"
}
```

Figure 13.52 – Attaching a volume to an instance

You'll have to keep in mind that the volume is not initialized with a filesystem, and you'll have to it manually, from within your EC2 instance, as suggested in the *EBS volumes* section earlier in this chapter.

Next, we'll show you how to terminate an EC2 instance.

Terminating an EC2 instance

To terminate an EC2 instance, we'll use the `aws ec2 terminate-instance` command. Note that terminating an instance results in the deletion of the instance. We cannot restart a terminated instance. We could use the `aws ec2 stop-instances` command to stop our instance until later use.

The following command will terminate the instance with the ID `i-0e1692c9dfdf07a8d`:

```
aws ec2 terminate-instances --instance-ids i-0e1692c9dfdf07a8d
```

The output states that our instance is shutting down (from a previously running state):

```
packt@neptune:~$ aws ec2 terminate-instances --instance-ids i-0e1692c9dfdf07a8d
{
    "TerminatingInstances": [
        {
            "CurrentState": {
                "Code": 32,
                "Name": "shutting-down"    <——
            },
            "InstanceId": "i-0e1692c9dfdf07a8d",    <——
            "PreviousState": {
                "Code": 16,
                "Name": "running"
            }
        }
    ]
}
```

Figure 13.53 – Terminating an instance

The instance eventually transitions to the terminated state and will no longer be visible in the AWS EC2 console. The AWS CLI will still list it among the instances until EC2 finally disposes of it. According to AWS, terminated instances can still be visible up to an hour after termination. It's always a good practice to discard the instances in the terminated or shutting-down states when performing queries and management operations via the AWS CLI.

We wrap up here our journey through AWS EC2, and let's just admit that we've only scratched the surface of cloud management workloads in AWS. We learned a few basic concepts about EC2 resources. Next, we looked at typical cloud management tasks, such as launching and managing instances, adding and configuring additional storage, and using EBS snapshots for disaster recovery. Finally, we explored the AWS CLI with hands-on examples of standard operations, including querying and launching EC2 instances, creating and adding additional storage to an instance, and terminating an instance.

The topics covered in this section provide a basic understanding of AWS EC2 cloud resources and help system administrators make better decisions when managing the related workloads. Power users may find the AWS CLI examples a good starting point for automating their cloud management workflows in EC2.

Let's turn our focus now to our next public cloud services contender, Microsoft Azure.

Working with Microsoft Azure

Microsoft Azure, also known as *Azure*, is a public cloud service by Microsoft for building and deploying application services in the cloud. Azure provides a full offering of a highly scalable **IaaS** at relatively low costs, accommodating a wide range of users and business requirements, from small teams to large commercial enterprises, including financial, health, and governmental institutions.

In this section, we'll explore some very basic deployment workflows using Azure, such as the following:

- Creating a Linux virtual machine
- Managing virtual machine sizes
- Adding additional storage to a virtual machine
- Moving a virtual machine between resource groups
- Redeploying a virtual machine
- Working with the Azure CLI

You need an Azure account in order to gain hands-on experience while following the content of this chapter. We encourage you to create a free Azure account, which will provide you access to free popular services for 12 months along with $200 credit for the first 30 days to cover the cost of your resources. Sign up for a free account with Azure at `https://azure.microsoft.com`:

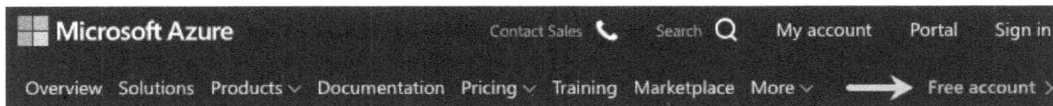

Figure 13.54 – Creating a free Azure account

Once you've created your free Azure account, go to `https://portal.azure.com` to access the Azure portal. You may want to enable the docked view of the portal navigation menu on the left for quick and easy access to your resources. Throughout this chapter, we'll use the docked view for our screen captures. Go to the **Portal settings** cog in the top-right corner and choose **Docked** for the default mode of the portal menu:

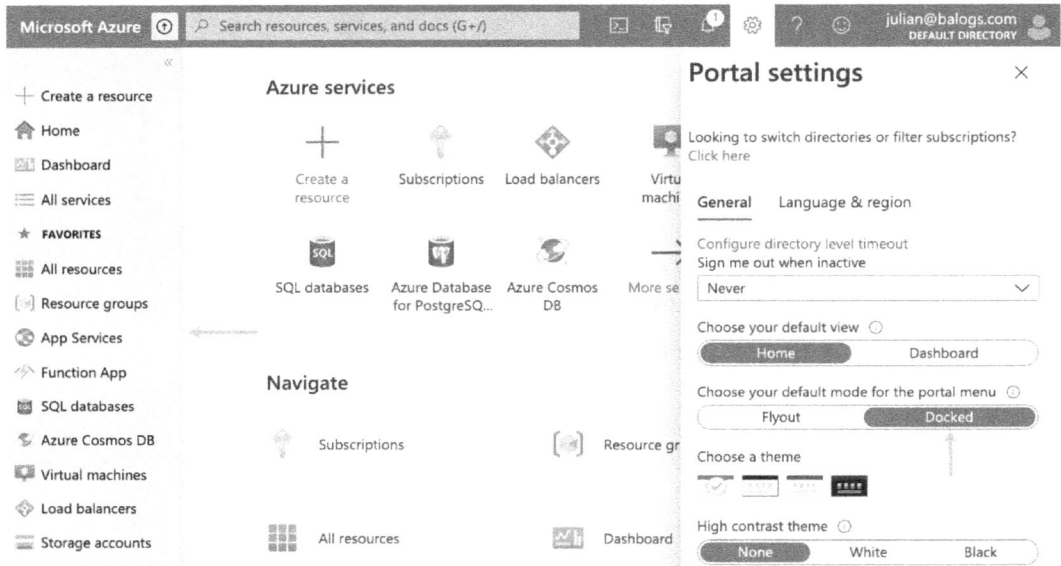

Figure 13.55 – Enable the docked view of the Azure portal menu

Let's create our first resource in Azure – a **Red Hat Enterprise Linux** (**RHEL**) virtual machine.

Creating a virtual machine

We'll follow a step-by-step procedure, guided by the resource wizard in the Azure portal. Here are the steps:

1. Create a compute resource.

2. Create a resource group.

3. Configure the instance details.

4. Configure SSH access.

5. Validate and deploy the virtual machine.

Let's start with the first step, creating a compute resource for the virtual machine.

Creating a compute resource

Start by clicking on the **Create a resource** option in the left navigation menu or under **Azure services** in the main window:

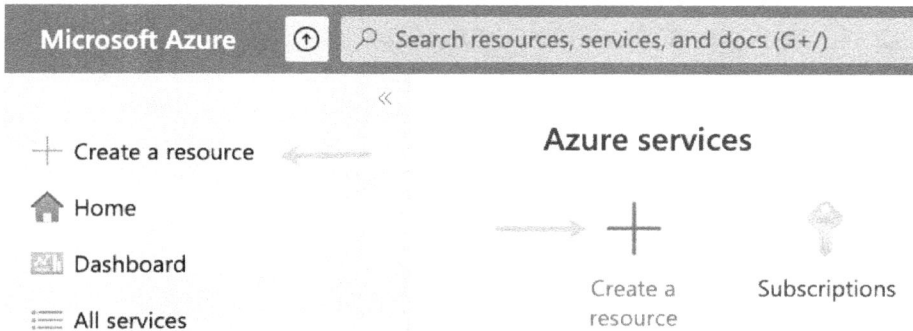

Figure 13.56 – Create a new resource in Azure

The next screen will take us to the Azure Marketplace, where we can search for our resource of choice. You can either search for a relevant keyword or narrow down your selection based on the resource type you're looking for. Let's narrow down our selection by choosing **Compute**, then select **Red Hat Enterprise Linux** from the top available options. You may click on **Learn more** for a detailed description of the image:

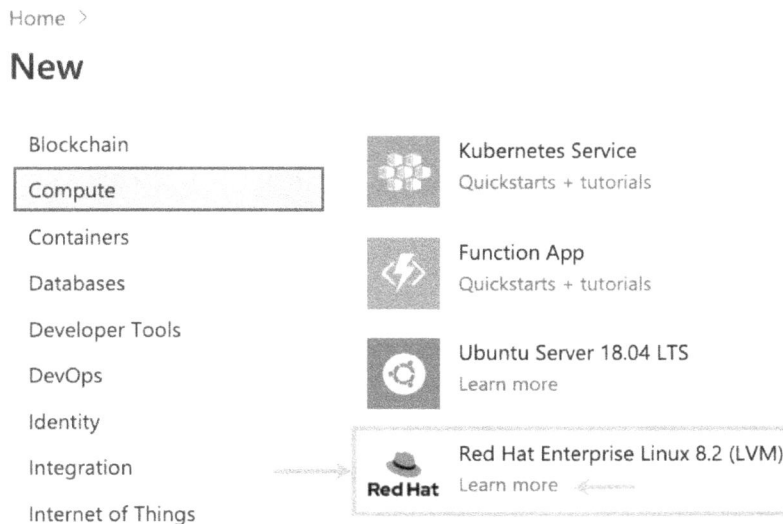

Figure 13.57 – Choosing an RHEL virtual machine

When we select **Red Hat Enterprise Linux**, we're guided through the process of configuring and creating our RHEL virtual machine, starting with a resource group.

Configuring a resource group

First, we need to specify the **Subscription** and **Resource group** associated with the virtual machine. An Azure resource group is a collection of assets related to a specific deployment, including storage, networking interfaces, security groups, and so on. Assuming this is our first virtual machine, we'll create a new resource group and name it packt-demo. If we had a previously created resource group, we could specify it here:

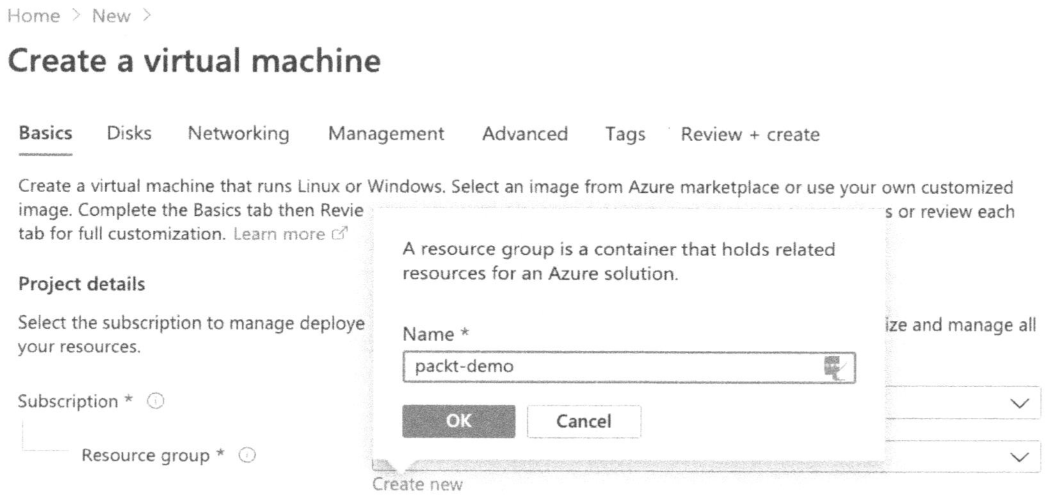

Home > New >

Create a virtual machine

Basics   Disks   Networking   Management   Advanced   Tags   Review + create

Create a virtual machine that runs Linux or Windows. Select an image from Azure marketplace or use your own customized image. Complete the Basics tab then Revie   s or review each tab for full customization. Learn more

Project details

Select the subscription to manage deploye   ize and manage all your resources.

A resource group is a container that holds related resources for an Azure solution.

Subscription *

Name *

packt-demo

Resource group *

OK   Cancel

Create new

Figure 13.58 – Creating a new resource group

Next, we set various properties related to our instance, such as **Virtual machine name**, **Region**, and **Size**.

Configuring instance details

We'll name our virtual machine packt-rhel and place it in the **(US) West US** region, closest to the geographical location where our instance would operate. The size of our machine will directly impact the associated costs:

Subscription * ⓘ	Free Azure Subscription ⌄
Resource group * ⓘ	(New) packt-demo ⌄
	Create new

Instance details

Virtual machine name * ⓘ	packt-rhel ✓
Region * ⓘ	(US) West US ⌄
Availability options ⓘ	No infrastructure redundancy required ⌄
Image * ⓘ	🌰 Red Hat Enterprise Linux 8.2 (LVM) - Gen1 ⌄
	See all images
Size * ⓘ	Standard_A2_v2 - 2 vcpus, 4 GiB memory ($66.43/month) ⌄
	See all sizes

Figure 13.59 – The instance details of the virtual machine

Alternatively, we could browse through different options for **Image** and **Size** by choosing **See all images** or **See all sizes**, respectively. Azure also provides a *pricing calculator* online tool for various resources, at `https://azure.microsoft.com/en-us/pricing/calculator/`.

In the next step, we're asked to configure an SSH key for terminal access to our instance.

Configuring SSH access

In this step, we enable SSH with public key authentication. We'll set the **Username** to `packt` and the **Key pair name** to `packt-rhel`:

Administrator account

Authentication type ⓘ	⦿ SSH public key
	◯ Password
	ⓘ Azure now automatically generates an SSH key pair for you and allows you to store it for future use. It is a fast, simple, and secure way to connect to your virtual machine.
Username * ⓘ	packt ✓
SSH public key source	Generate new key pair ⌄
Key pair name *	packt-rhel ✓

Figure 13.60 – Enabling SSH authentication to the virtual machine

Finally, we set the **Inbound port rules** for our instance to allow SSH access. If, for example, our machine will run a web server application, we can also enable HTTP and HTTPS access:

Inbound port rules

Select which virtual machine network ports are accessible from the public internet. You can specify more limited or granular network access on the Networking tab.

Public inbound ports * ⓘ   ◯ None
                                      ◉ Allow selected ports

Select inbound ports *   ┌───┐
                                      │ SSH (22)   ⌄ │
                                      └───┘

                                      ⚠ **This will allow all IP addresses to access your virtual machine.** This is only
                                      recommended for testing. Use the Advanced controls in the Networking tab
                                      to create rules to limit inbound traffic to known IP addresses.

Figure 13.61 – Enabling SSH access to the virtual machine

At this point, we are ready to create our virtual machine. The wizard can take us further to the additional steps of specifying the *disks* and *networking* configuration associated with our instance. For now, we'll leave them as their defaults and proceed to the last step – reviewing the configuration and deploying the virtual machine.

Validating and deploying the virtual machine

We click the **Review + create** button to initiate the validation process:

┌──────────────────────┐   ┌──────────────────────┐
│ **Review + create** │   < Previous   │   **Next : Disks >**   │
└──────────────────────┘   └──────────────────────┘

Figure 13.62 – Review and create the virtual machine

Next, the deployment wizard validates our virtual machine configuration. In a few moments, if everything goes well, we'll get a **Validation passed** message with the product details and our instance's hourly rate. By clicking **Create**, we agree to the relevant legal terms, and our virtual machine will be deployed shortly:

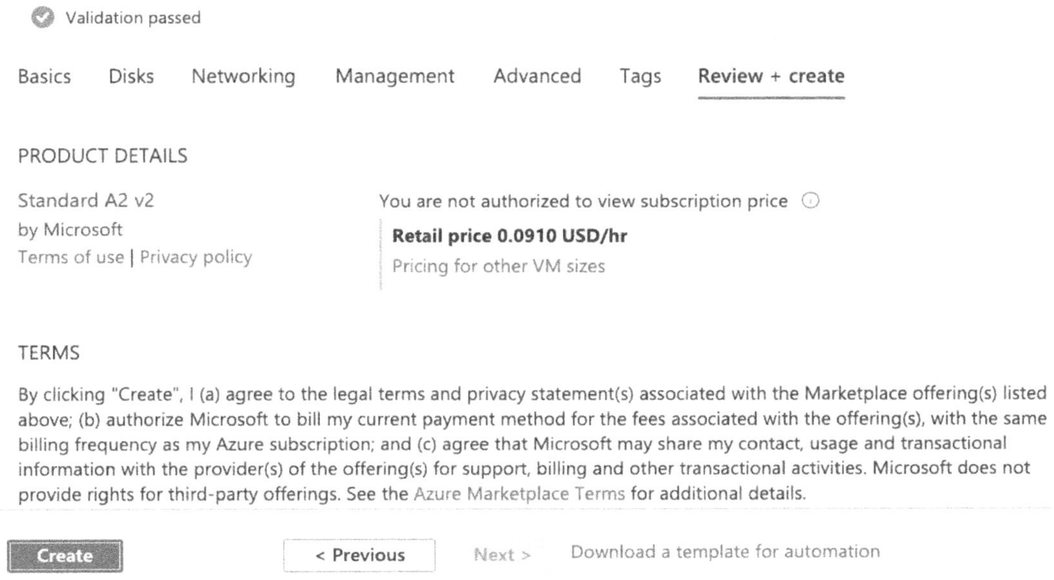

Validation passed

Basics   Disks   Networking   Management   Advanced   Tags   **Review + create**

PRODUCT DETAILS

Standard A2 v2   You are not authorized to view subscription price ⓘ
by Microsoft   **Retail price 0.0910 USD/hr**
Terms of use | Privacy policy   Pricing for other VM sizes

TERMS

By clicking "Create", I (a) agree to the legal terms and privacy statement(s) associated with the Marketplace offering(s) listed above; (b) authorize Microsoft to bill my current payment method for the fees associated with the offering(s), with the same billing frequency as my Azure subscription; and (c) agree that Microsoft may share my contact, usage and transactional information with the provider(s) of the offering(s) for support, billing and other transactional activities. Microsoft does not provide rights for third-party offerings. See the Azure Marketplace Terms for additional details.

[Create]   [< Previous]   Next >   Download a template for automation

Figure 13.63 – Creating the virtual machine

In the process, we'll be prompted to download the SSH private key for accessing our instance:

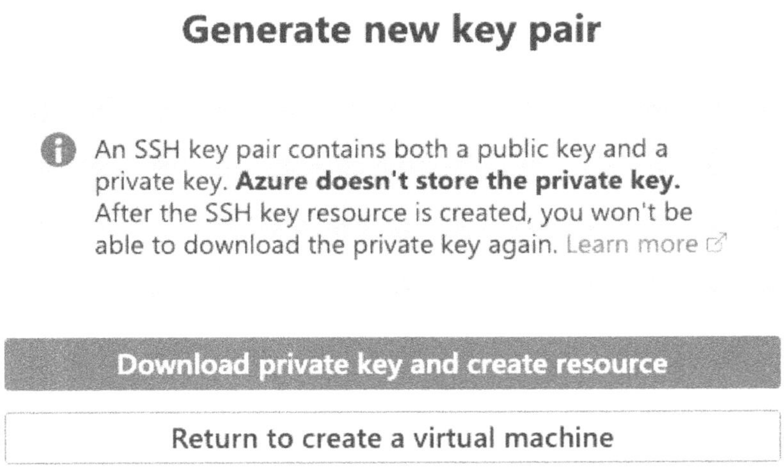

Generate new key pair

ⓘ An SSH key pair contains both a public key and a
   private key. **Azure doesn't store the private key.**
   After the SSH key resource is created, you won't be
   able to download the private key again. Learn more ⌇

[**Download private key and create resource**]

[Return to create a virtual machine]

Figure 13.64 – Downloading the SSH private key for accessing the virtual machine

If the deployment completes successfully, we get a brief pop-up message with **Deployment succeeded** and a **Go to resource** button that will take us to our new virtual machine:

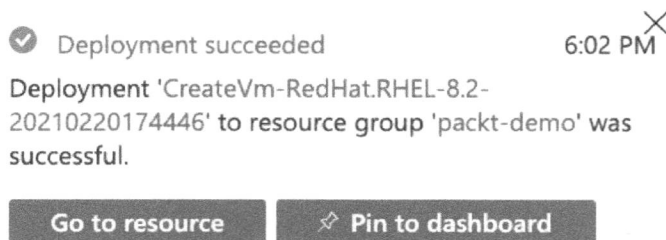

Figure 13.65 – Successfully deploying a virtual machine

We also get a brief report about the deployment details. The related resources are also visible in the **All resources** view from the left navigation menu:

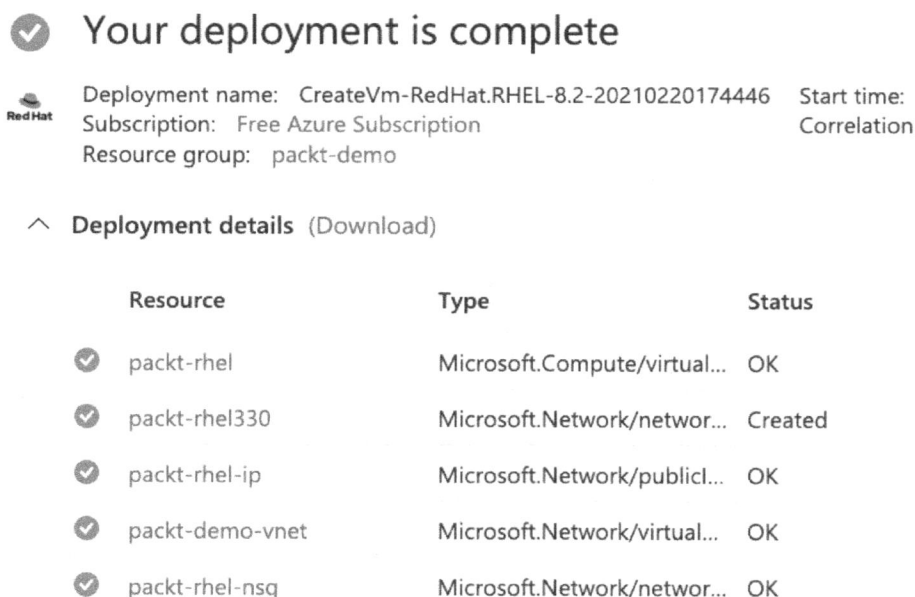

Figure 13.66 – The deployment details

Let's take a quick look at each of the resources created with our virtual machine deployment:

- `packt-rhel` – The virtual machine host
- `packt-rhel330` – The network interface (or network interface card) of the virtual machine

- `packt-rhel-ip` – The IP address of the virtual machine
- `packt-demo-vnet` – The virtual network associated with the resource group (`packt-demo`)
- `packt-rhel-nsg` – The **Network Security Group** (**NSG**) controlling the inbound and outbound access to and from our instance

Azure will create a new set of the resource types mentioned previously with every virtual machine, except the virtual network corresponding to the resource group, when the instance is placed in an existing resource group. We should not forget that we also created a new resource group (`packt-demo`), which is not shown in the deployment report.

Let's try to connect to our newly created instance (`packt-rhel`). Go to **Virtual machines** in the left navigation pane and select the instance (`packt-rhel`):

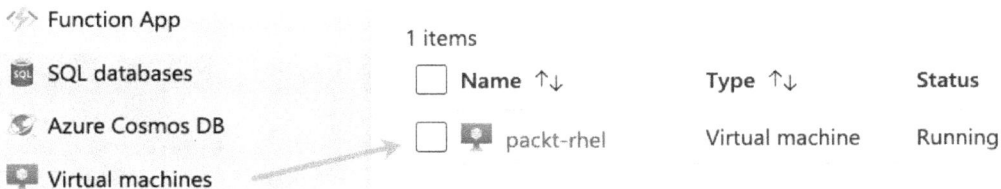

Figure 13.67 – The new instance in the Virtual machines view

In the **Overview** tab, we'll see our virtual machine's essential details, including the **Public IP address** (`104.40.68.161`):

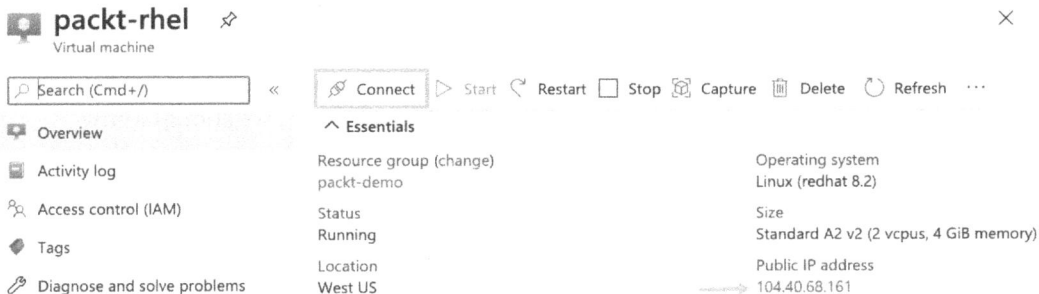

Figure 13.68 – The new instance in the Virtual machines view

Now that we've deployed our virtual machine, we want to make sure we can access it via SSH.

Connecting with SSH to a virtual machine

Before we connect, we need to set the permissions to our SSH private key file, so it's not publicly viewable:

```
chmod 400 azure/packt-rhel.pem
```

Next, we connect to our Azure RHEL instance with the following command:

```
ssh -i azure/packt-rhel.pem packt@104.40.68.161
```

We use the SSH key (`packt-rhel.pem`) and administrator account (`packt`) specified when we created the instance. Alternatively, you can click on the **Connect** button in the virtual machine's **Overview** tab, then click **SSH**. This action will bring up a view where you can see the preceding commands, and copy/paste them into your terminal.

A successful connection to our RHEL instance should yield the following output:

```
→ packt ssh -i azure/packt-rhel.pem packt@104.40.68.161
Activate the web console with: systemctl enable --now cockpit.socket

This system is not registered to Red Hat Insights. See https://cloud.redhat.com/
To register this system, run: insights-client --register
```

Figure 13.69 – Connecting to the RHEL virtual machine

Now that we created our first virtual machine in Azure, let's look at some of the most common management operations performed during a virtual machine's lifetime.

Managing virtual machines

As our applications evolve, so does the computing power and capacity required by the virtual machines hosting the applications. As system administrators, we should know precisely how cloud resources are being utilized. Azure provides the necessary tools for monitoring the health and performance of virtual machines. These tools are available on the **Monitoring** tab of the virtual machine's management page.

A small virtual machine with a relatively low number of virtual CPUs and reduced memory may negatively impact application performance. On the other hand, an oversized instance would yield unnecessary costs. Resizing a virtual machine is a common operation in Azure. Let's see how we can do it.

Changing the size of a virtual machine

Azure makes it relatively easy to resize virtual machines. In the portal, go to **Virtual Machines**, select your instance, and click **Size** under **Settings**:

Settings	VM Size ↑↓	Family ↑↓	vCPUs ↑↓	RAM (GiB) ↑↓	Data disks ↑↓
🖧 Networking	∨ A-Series v2			Best suited for entry level workloads (development or test)	
🖉 Connect	A1_v2	General purpose	1	2	2
🖥 Disks	A2_v2	General purpose	2	4	4
🖳 Size	A2m_v2	General purpose	2	16	4
🛡 Security	A4_v2	General purpose	4	8	8
📶 Advisor recommendations	A4m_v2	General purpose	4	32	8
📄 Extensions					
🔄 Continuous delivery	**Resize**	Prices presented are estimates in your local currency that include only Azure infrastructure costs and any di location. The prices don't include any applicable software costs. Final charges will appear in your local curre views. View Azure pricing calculator.			
🔲 Availability + scaling					

Figure 13.70 – Changing the size of a virtual machine

Our virtual machine (`packt-rhel`) is of **A2_v2** size (2 vCPUs, 4 GB RAM). We can choose to size it up or down. For demo purposes, let's resize to the lower **A1_v2** capacity (1 vCPU, 2 GB RAM). We select the **A1_v2** option and click the **Resize** button. Lowering the size of our instance will also result in cost savings. Azure will stop and restart our virtual machine while resizing. It is always good to stop the machine before changing the size to avoid possible data inconsistencies within the instance.

One of the remarkable features of virtualized workloads in Azure is the ability to scale – including the storage capacity – by adding additional data disks to virtual machines. We can add existing data disks or create new ones.

Let's look at how to add a secondary data disk to our virtual machine next.

Adding additional storage

Azure can add disks to our instance on the fly, without stopping the machine. We can add two types of disks to a virtual machine, *data disks* and *managed disks*.

Let's start with adding a data disk first.

Adding a data disk

To add a new data disk to our virtual machine, go to **Virtual machines** in the left navigation menu and select your instance, click **Disks** under **Settings**, and then click on **Create and attach a new disk**:

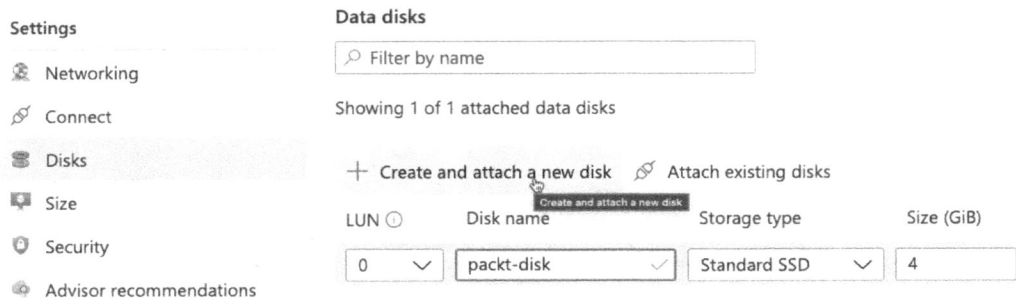

Figure 13.71 – Adding a data disk to a virtual machine

In the disk properties, leave the **Logical Unit Number** (**LUN**) as-is (automatically assigned), specify a **Disk name** (such as packt-disk), **Storage Type**, and **Size** (such as 4 GB). Click **Save** when done:

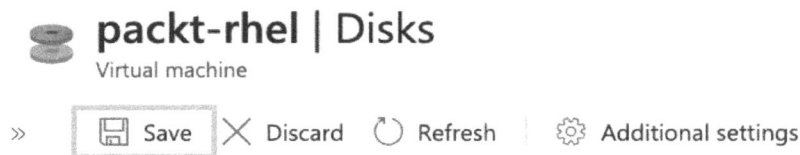

Figure 13.72 – Save the data disk settings

At this point, we have the new disk attached to our virtual machine, but the disk is not initialized with a filesystem yet. We need to follow a similar procedure described earlier in this chapter with AWS storage volumes to initialize the data disk. See the *EBS volumes* section. Let's briefly go through the related commands.

Connect via SSH to our virtual machine:

```
ssh -i azure/packt-rhel.pem packt@104.40.68.161
```

List the current block devices:

```
lsblk
```

Identify the new data disk in the output. Our disk is 4 GB in size, and the related block device is sdc:

```
[packt@packt-rhel ~]$ lsblk
NAME                MAJ:MIN RM  SIZE RO TYPE MOUNTPOINT
sda                     8:0  0   64G  0 disk
├─sda1                  8:1  0  500M  0 part /boot
├─sda2                  8:2  0   63G  0 part
│ ├─rootvg-tmplv      253:0  0    2G  0 lvm  /tmp
│ ├─rootvg-usrlv      253:1  0   10G  0 lvm  /usr
│ ├─rootvg-homelv     253:2  0    1G  0 lvm  /home
│ ├─rootvg-varlv      253:3  0    8G  0 lvm  /var
│ └─rootvg-rootlv     253:4  0    2G  0 lvm  /
├─sda14                8:14  0    4M  0 part
└─sda15                8:15  0  495M  0 part /boot/efi
sdb                    8:16  0   10G  0 disk
└─sdb1                 8:17  0   10G  0 part /mnt
sdc                    8:32  0    4G  0 disk  ←
```

Figure 13.73 – Identify the block device for the new data disk

Verify the block device is empty:

```
sudo file -s /dev/sdc
```

The output is `/dev/sdc: data`, meaning that the data disk doesn't contain a filesystem yet. Next, initialize the volume with an XFS filesystem:

```
sudo mkfs -t xfs /dev/sdc
```

Finally, create a mounting point (`/packt`) and mount the new volume:

```
sudo mkdir /packt
sudo mount /dev/sdc /packt
```

Now we can use the new data disk for regular file storage.

We should note that data disks would only be persisted during the lifetime of the virtual machine. When the virtual machine is paused, stopped, or terminated, the data disk becomes unavailable. When the machine is terminated, the data disk is permanently lost.

For persistent storage, we need to use *managed disks*, similar in behavior to network-attached storage. We attach a managed disk to our virtual machine next.

Adding a managed disk

Let's start by creating a managed disk. In the Azure portal, click **Create a resource** and search for `managed disks` in the Azure Marketplace. Click **Create** on the managed disk resource. On the next screen, enter the **Resource group** containing our virtual machine (`packt-demo`), the managed disk name (`packt-man`), the **Region** for our resource (`(US) West US`), and the size of our disk (**256 GiB, Standard SSD**):

Create a managed disk   ✕

Basics   Encryption   Networking   Advanced   Tags   Review + create

Select the disk type and size needed for your workload. Azure disks are designed for 99.999% availability. Azure managed disks encrypt your data at rest, by default, using Storage Service Encryption. Learn more about disks.

Project details

Select the subscription to manage deployed resources and costs. Use resource groups like folders to organize and manage all your resources.

Subscription * ⓘ | Free Azure Subscription   ⌄ |

    Resource group * ⓘ | packt-demo   ⌄ |
                                     Create new

Disk details

Disk name * ⓘ | packt-man   ✓ |

Region * ⓘ | (US) West US   ⌄ |

Availability zone | None   ⌄ |

Source type ⓘ | None   ⌄ |

Size * ⓘ | **256 GiB**
                                  Standard SSD ⟵ - - - - -
                                  Change size

[Review + create]   < Previous   [Next : Encryption >]

Figure 13.74 – Creating a managed disk

When choosing the disk size, you may also choose between **Standard** (SSD or HDD) and **Premium** storage types. Before choosing **Premium**, make sure your virtual machine supports premium storage disks. For more information about storage types, please visit `https://docs.microsoft.com/en-us/azure/virtual-machines/disks-types`.

The preceding settings will suffice for creating our managed disk. Optionally, you can go through the following steps and specify the encryption and networking options. Click on **Review + create** when done. Azure will validate the deployment of the new resource and prompt us to create the managed disk.

Next, we'll add the managed disk to the virtual machine. The process is very similar to adding a data disk, except we specify **Attach existing disks** to add our managed disk. Note that it may take a while for the managed disk to become available for attaching. Also, Azure will capitalize the name of the resource (for example, PACKT-MAN):

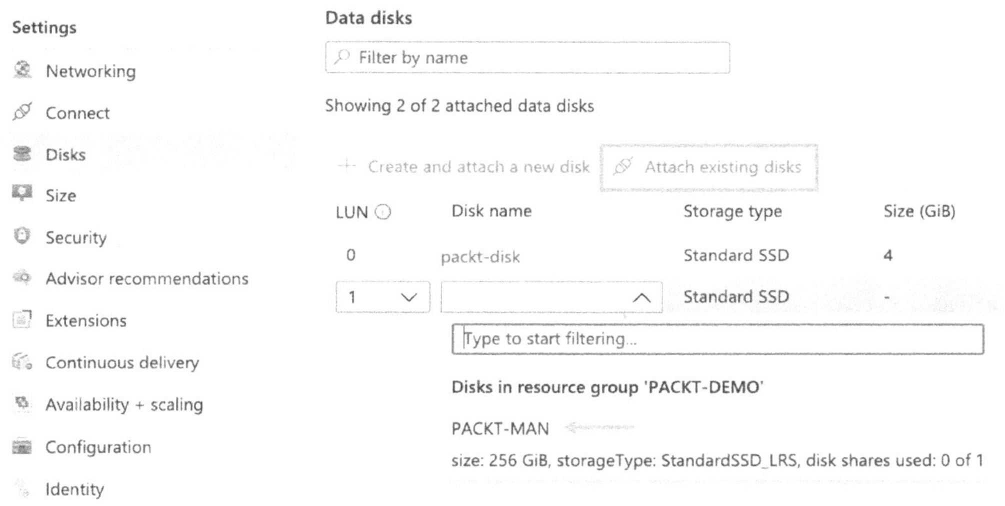

Figure 13.75 – Attaching a managed disk

Select the managed disk (PACKT-MAN) and click **Save** in the top menu bar. The new disk will show up next to the data disk previously added to the virtual machine:

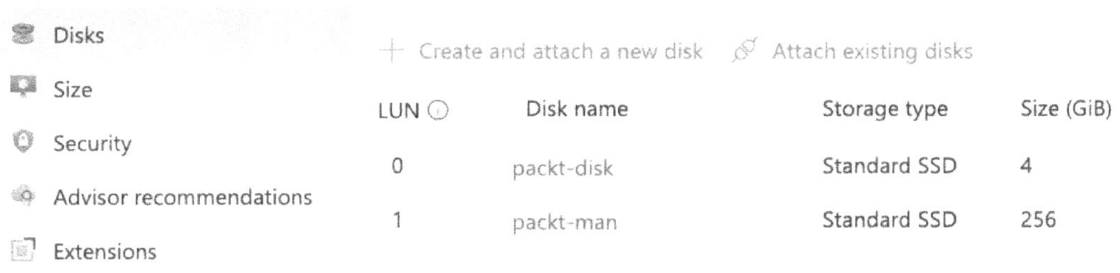

Figure 13.76 – The disks attached to the virtual machine

The remaining steps for initializing the filesystem on the managed storage volume are similar to the data disk, and we won't go over them again.

Another typical operation in the Azure cloud is moving virtual machines across resource groups. Assuming we have a staging and a production environment, at some point, we may want to move a virtual machine from one to the other. In the following section, we'll show you how.

Moving a virtual machine

Moving a virtual machine across resource groups is a relatively straightforward process in Azure. Here are the steps:

1. Create a new resource group.

2. Select the resources to move.

3. Move the resources.

Let's start with the first step and create the new resource group to which we want to move our virtual machine. If your target resource group is already created, you may skip this step.

Creating a new resource group

We'll start by creating a new resource group. Select **Resource groups** in the left navigation pane of the Azure portal, and then click **Create**:

Home > Resource groups >

Create a resource group   ×

Basics   Tags   Review + create

Resource group - A container that holds related resources for an Azure solution. The resource group can include all the resources for the solution, or only those resources that you want to manage as a group. You decide how you want to allocate resources to resource groups based on what makes the most sense for your organization. Learn more ☑

Project details

Subscription * ⓘ | Free Azure Subscription   ⌄ |

          Resource group * ⓘ | packt |

Resource details

Region * ⓘ | (US) West US   ⌄ |

| Review + create | < Previous | Next : Tags > |

Figure 13.77 – Creating a new resource group

We'll name our new resource group `packt` and create it in the same region where our other assets are defined (`(US) West US`). Press **Review + create** when done. As an exercise, you may also create another virtual machine based on Ubuntu (such as `packt-ubuntu`) in the new resource group.

In the next step, we select the resources we want to move.

Selecting the resources to move

We choose **Resource groups** in the left navigation menu and select the resource group we want to move our assets away from (`packt-demo`). Here, we select all the resources we want to move to another resource group. We could, for example, only pick the `packt-rhel` virtual machine while leaving the other related resources in the existing resource group. For consistency, we select all resources related to the `pack-rhel` virtual machine, except the `packt-rhel` SSH key. Moving SSH keys across resource groups is not allowed in Azure, most probably for security reasons.

With the selection done, click on the ellipsis (**...**) in the top-right corner of the resource menu, choose **Move**, and select **Move to another resource group**:

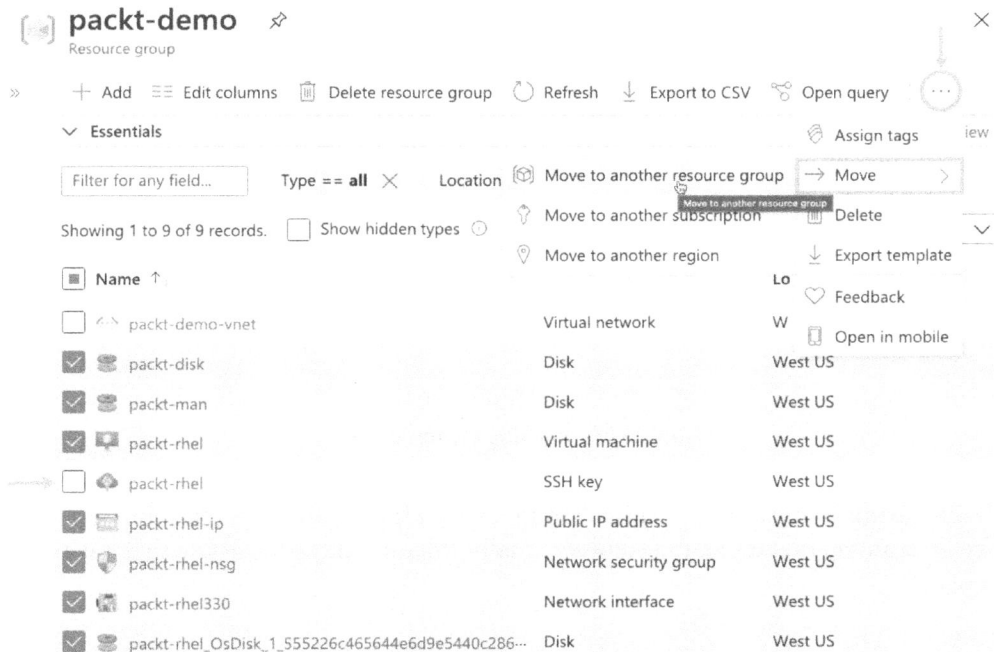

Figure 13.78 – Selecting the resources to move

We're now ready to proceed with the final step.

Moving resources

We enter the name of the target resource group (`packt`). We also have to acknowledge that any automation scripts referencing the resources we're about to move will have to be updated to reflect the new resource group. Before clicking the **OK** button, we can still uncheck items in the list that we don't want to move:

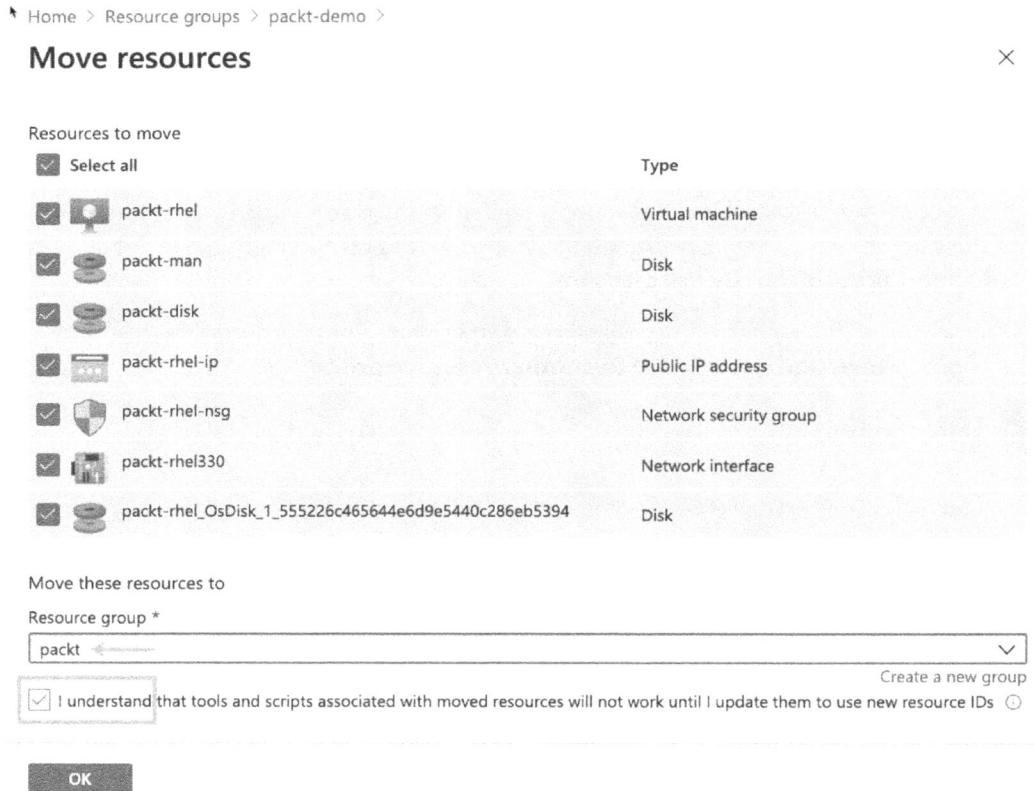

Figure 13.79 – Moving the resources

After a quick validation process, the items we're moving to the new resource group will start disappearing from the old resource group. By clicking the **Refresh** button in the old resource group's top menu bar, the view eventually will only show the assets that have not been moved:

packt-demo 📌
Resource group   ✕

»   + Add   ☰ Edit columns   🗑 Delete resource group   ◯ Refresh   ↓ Export to CSV   ⌘ Open query   · · ·

⌄ Essentials   Refresh   JSON View

Filter for any field...   Type == **all** ✕   Location == **all** ✕   ⊤ Add filter

Showing 1 to 2 of 2 records.   ☐ Show hidden types ⓘ   No grouping   ⌄   List view   ⌄

☐ Name ↑   Type ↑↓   Location ↑↓

☐ ⟋⟍ packt-demo-vnet   Virtual network   West US   · · ·

☐ ☁ packt-rhel   SSH key   West US   · · ·

Figure 13.80 – Refreshing the old resource group

Switching to the new resource group (`packt`), we'll see the resources we just moved placed accordingly:

Home > Resource groups >

packt 📌
Resource group   ✕

»   + Add   ☰ Edit columns   🗑 Delete resource group   ◯ Refresh   ↓ Export to CSV   ⌘ Open query   · · ·

⌄ Essentials   JSON View

Filter for any field...   Type == **all** ✕   Location == **all** ✕   ⊤ Add filter

Showing 1 to 14 of 14 records.   ☐ Show hidden types ⓘ   No grouping   ⌄
List view   ⌄

☐ Name ↑↓   Type ↓   Location ↑↓

☐ ⟋⟍ packt-vnet   Virtual network   West US   · · ·

☐ 🖥 packt-rhel   Virtual machine   West US   · · ·

☐ 🖥 packt-ubuntu   Virtual machine   West US   · · ·

Figure 13.81 – The assets in the new resource group (partial view)

During a virtual machine's lifetime, we may occasionally face problems connecting to our instance, such as in the rare case of an unreachable host or local data center failure. Azure has a convenient feature that allows us to redeploy a virtual machine. The redeployment will place our instance on a new host or in a new data center within the same resource group, making it immediately available. Let's look at how to redeploy a virtual machine next.

Redeploying a virtual machine

To initiate a redeployment using the Azure portal, go to **Virtual machines** in the left navigation menu and select your virtual machine (for example, `packt-rhel`). Next, in the **Settings** blade, select **Redeploy + reapply** under **Support + troubleshooting** and click **Redeploy**:

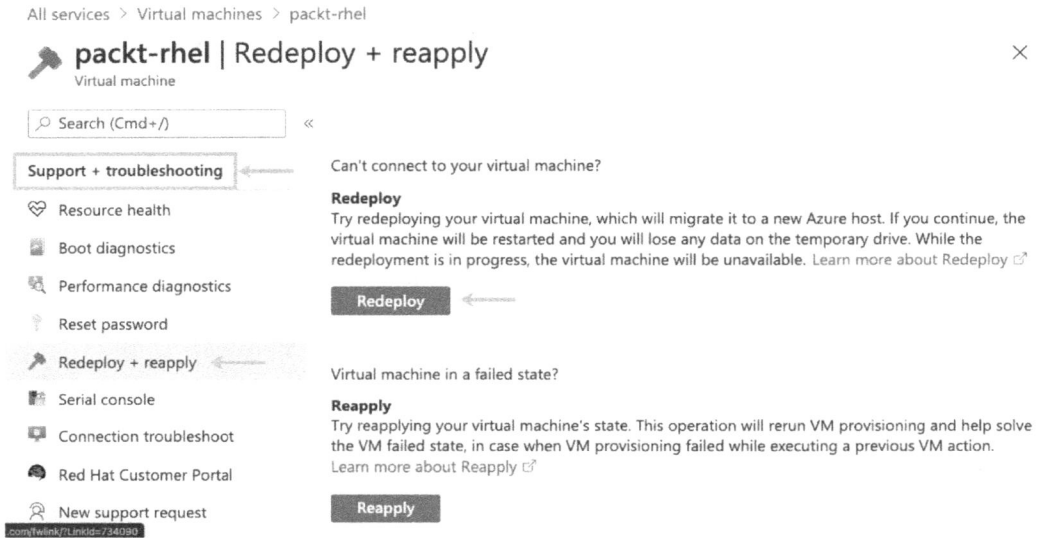

All services > Virtual machines > packt-rhel

packt-rhel | Redeploy + reapply
Virtual machine   ×

| Search (Cmd+/) «

Support + troubleshooting ←------

♡ Resource health

▨ Boot diagnostics

▨ Performance diagnostics

⍦ Reset password

⚒ Redeploy + reapply ←------

▦ Serial console

▱ Connection troubleshoot

● Red Hat Customer Portal

♔ New support request

.com/fwlink/?LinkId=734090

Can't connect to your virtual machine?

Redeploy
Try redeploying your virtual machine, which will migrate it to a new Azure host. If you continue, the virtual machine will be restarted and you will lose any data on the temporary drive. While the redeployment is in progress, the virtual machine will be unavailable. Learn more about Redeploy ⌕

　Redeploy　 ←------

Virtual machine in a failed state?

Reapply
Try reapplying your virtual machine's state. This operation will rerun VM provisioning and help solve the VM failed state, in case when VM provisioning failed while executing a previous VM action. Learn more about Reapply ⌕

　Reapply　

Figure 13.82 – Redeploying a virtual machine

During the redeployment process, Azure will shut down the machine, move it to a new node in the Azure infrastructure, and power the machine back on, with all the configuration options and related resources intact.

So far, we have performed all these management operations in the Azure portal. What if you want to automate your workloads in the cloud using scripting? Azure provides a specialized command-line interface for managing your resources in the cloud. We'll look at the Azure CLI next.

Working with the Azure CLI

First, let's install the Azure CLI on our platform of choice. Follow the instructions here: `https://docs.microsoft.com/en-us/cli/azure/install-azure-cli`. We'll choose the Azure CLI for Linux, and for demo purposes, install it on an Ubuntu machine. The related instructions are captured here: `https://docs.microsoft.com/en-us/cli/azure/install-azure-cli-linux`. From the multiple installation options available, we'll use the following command:

```
curl -sL https://aka.ms/InstallAzureCLIDeb | sudo bash
```

After the installation is complete, we can invoke the Azure CLI with the `az` command:

```
az help
```

The preceding command displays detailed help information about using the `az` utility. Before performing any management operations, we need to authenticate the CLI with our Azure credentials. The following command will set up the local Azure CLI environment accordingly:

```
az login
```

We'll be prompted with a message containing an authentication code and the URL (`https://microsoft.com/devicelogin`) to visit and enter the code:

```
packt@neptune:~$ az login
To sign in, use a web browser to open the page https://microsoft.com/devicelogin
[
  {
    "cloudName": "AzureCloud",
    "homeTenantId": "                    ",
    "id": "                    ",
    "isDefault": true,
    "managedByTenants": [],
    "name": "Free Azure Subscription",
    "state": "Enabled",
    "tenantId": "                    ",
    "user": {
      "name": "julian@balogs.com",
      "type": "user"
    }
  }
]
```

Figure 13.83 – Initializing the Azure CLI environment

At this point, we are ready to use the Azure CLI for management operations. Let's create a new resource group called `packt-dev` in the `westus` region:

```
az group create --name packt-dev --location westus
```

The command yields the following output upon successfully creating the resource group:

```
packt@neptune:~$ az group create --name packt-dev --location westus
{
  "id": "/subscriptions/138189e1-26c3-49f1-a2d9-2caa56d946c7/resourceGroups/packt-dev",
  "location": "westus",
  "managedBy": null,
  "name": "packt-dev",
  "properties": {
    "provisioningState": "Succeeded"
  },
  "tags": null,
  "type": "Microsoft.Resources/resourceGroups"
}
```

Figure 13.84 – Creating a new resource group

Next, we launch an Ubuntu virtual machine named `packt-ubuntu-dev` in the region we just created:

```
az vm create \
  --resource-group packt-dev \
  --name packt-ubuntu-dev \
  --image UbuntuLTS \
  --admin-username packt \
  --generate-ssh-keys
```

Let's quickly go through each of the preceding command-line options:

- `resource-group` – The name of the resource group (`packt-dev`) where we create our virtual machine
- `name` – The name of the virtual machine (`packt-ubuntu-dev`)
- `image` – The Linux distribution to use (`UbuntuLTS`)
- `admin-username` – The username of the machine's administrator account (`packt`)
- `generate-ssh-keys` – Generates a new SSH key pair to access our virtual machine

The command produces the following output:

```
packt@neptune:~$ az vm create \
>    --resource-group packt-dev \
>    --name packt-ubuntu-dev \
>    --image UbuntuLTS \
>    --admin-username packt \
>    --generate-ssh-keys
Command group 'vm' is experimental and under development. Reference and support levels:
https://aka.ms/CLI_refstatus
SSH key files '/home/packt/.ssh/id_rsa' and '/home/packt/.ssh/id_rsa.pub' have been gene
rated under ~/.ssh to allow SSH access to the VM. If using machines without permanent st
orage, back up your keys to a safe location.
{- Finished ..
  "fqdns": "",
  "id": "/subscriptions/138189e1-26c3-49f1-a2d9-2caa56d946c7/resourceGroups/packt-dev/pr
oviders/Microsoft.Compute/virtualMachines/packt-ubuntu-dev",
  "location": "westus",
  "macAddress": "00-0D-3A-35-CD-F4",
  "powerState": "VM running",
  "privateIpAddress": "10.0.0.4",
  "publicIpAddress": "168.62.197.46",
  "resourceGroup": "packt-dev",
  "zones": ""
}
```

Figure 13.85 – Creating a new virtual machine

As the output suggests, the SSH key files have been automatically generated and placed in the local machine's ~/.ssh directory to allow SSH access to the newly created virtual machine. The JSON output also provides the public IP address of the machine:

```
"publicIpAddress": "168.62.197.46"
```

The following command lists all virtual machines:

```
az vm list
```

To get information about a specific virtual machine (packt-ubuntu-dev), we run the following command:

```
az vm show \
   --resource-group packt-dev \
   --name packt-ubuntu-dev
```

To redeploy an existing virtual machine (such as `packt-ubuntu-dev`), we run the following command:

```
az vm redeploy \
  --resource-group packt-dev \
  --name packt-ubuntu-dev
```

The following command deletes a virtual machine (`packt-ubuntu-dev`):

```
az vm delete \
  --resource-group packt-dev \
  --name packt-ubuntu-dev
```

As you have probably noticed, for virtual machine-related commands (`az vm`), we also need to specify the resource group the machine belongs to.

The comprehensive study of the Azure CLI is beyond the scope of this chapter. For detailed information, please visit the Azure CLI online documentation portal: `https://docs.microsoft.com/en-us/cli/azure/`.

This concludes our coverage of public cloud deployments with AWS and Azure. We have covered a vast domain and have merely skimmed the surface of cloud management workloads. We encourage you to build upon this preliminary knowledge and explore more, starting with the AWS and Azure cloud docs. The links are mentioned in the *Further reading* section, together with other valuable resources.

Now, let's look at a summary of what you have learned so far about AWS and Azure.

Summary

AWS and Azure provide a roughly similar set of features for flexible compute capacity, storage, and networking, with pay-as-you-go pricing. They share the essential elements of a public cloud – elasticity, autoscaling, provisioning with self-service, security, and identity access management. This chapter explored both cloud providers strictly from a practical vantage point, focusing on typical deployment and management aspects of everyday cloud administration tasks.

We covered topics such as launching and terminating a new instance or virtual machine. We looked at resizing an instance to accommodate a higher or lower compute capacity and scaling the storage by creating and attaching additional block devices (volumes). Finally, we used CLI tools for scripting various cloud management workloads.

At this point, you should be familiar with the AWS and Azure web administration consoles and CLI tools. You have learned the basics of some typical cloud management tasks and a few essential concepts about provisioning cloud resources. Overall, you've enabled a special skillset of modern-day Linux administrators by engaging in cloud-native administration workflows. Combined with the knowledge you've built so far in the previous chapters, you are assembling a valuable Linux administration toolbelt for on-premises, public, and hybrid cloud systems management.

In the next chapter, we'll take this challenge further and introduce you to managing application deployments using containerized workflows and services with Kubernetes.

Questions

Here is a quick recap of some of the concepts you've learned in this chapter as a quiz:

1. What is an AZ?

2. Between a `t2.small` and a `t2.micro` AWS EC2 instance type, which one yields better performance?

3. You have launched an AWS EC2 instance in the `us-west-1a` AZ and plan to attach an EBS volume created in `us-west-1b`. Would it work?

4. You have two virtual machines and a block device (storage) in Azure, all three within the same resource group. You want to attach the storage to both virtual machines. Would it work? Would it work in AWS?

5. You have a virtual machine in Azure with Standard SSD storage and plan to attach a managed disk with Premium SSD storage. Would it work?

6. What is the SSH command to connect to your AWS EC2 instance or Azure virtual machine?

7. You decided to stop one of your EC2 instances. Later, you restarted the instance and could no longer connect with SSH using the same public IP address. What happened?

8. Why would you redeploy an Azure virtual machine?

9. What is the Azure CLI command for listing your virtual machines? How about the equivalent AWS CLI command?

10. What is the AWS CLI command for launching a new EC2 instance?

11. What is the Azure CLI command for deleting a virtual machine?

Further reading

Here are a few resources to further explore AWS and Azure cloud topics:

- AWS EC2: `https://docs.aws.amazon.com/ec2/index.html`

- Azure: `https://docs.microsoft.com/en-us/azure`

- *AWS for System Administrators, Prashant Lakhera, Packt Publishing* (`https://www.packtpub.com/product/aws-for-system-administrators/9781800201538`)

- *Learning AWS – Second Edition, Aurobindo Sarkar, Amit Shah, Packt Publishing* (`https://www.packtpub.com/product/learning-aws-second-edition/9781787281066`)

- *Learning Microsoft Azure, Geoff Webber-Cross, Packt Publishing* (`https://www.packtpub.com/product/learning-microsoft-azure/9781782173373`)

- *Learning Microsoft Azure: A Hands-On Training [Video], Vijay Saini, Packt Publishing* (`https://www.packtpub.com/product/learning-microsoft-azure-a-hands-on-training-video/9781800203921`)

14
Deploying Applications with Kubernetes

Whether you are a seasoned system administrator managing containerized applications or a DevOps engineer automating app orchestration workflows, Kubernetes could be your platform of choice. This chapter will introduce you to Kubernetes and will guide you through the basic process of building and configuring a Kubernetes cluster. We'll use Kubernetes to run and scale a simple application in a secure and highly available environment. You will also learn how to interact with Kubernetes using the command-line interface.

By the end of this chapter, you'll know how to install, configure, and manage a Kubernetes cluster, either on-premises or using managed services with public cloud providers such as AWS and Azure. We'll also show you how to deploy and scale an application using Kubernetes.

Here's a brief outline of the topics we will cover in this chapter:

- Introducing Kubernetes, architecture and the API object model

- Installing and configuring Kubernetes on desktop, in on-premises virtual machines, and in public cloud environments (AWS and Azure)

- Working with Kubernetes using the `kubectl` command-line tool

- Deploying applications with Kubernetes using imperative and declarative deployment models

Technical requirements

You should be familiar with Linux and the command-line interface in general. A good grasp of TCP/IP networking and Docker containers would go a long way in making your journey easier to learning Kubernetes.

The sections exploring Kubernetes in the cloud require AWS and Azure accounts if you want to follow along with the practical examples. You may sign up for free subscriptions at the following links:

- AWS Free Tier: `https://aws.amazon.com/free`

- Microsoft Azure free account: `https://azure.microsoft.com/en-us/free/`

You will also need a local desktop machine with a Linux distribution of your choice to install and experiment with the CLI tools used in this chapter.

We'll devote a relatively large section to building a Kubernetes cluster using virtual machines. A powerful desktop system with a few CPU cores and at least 8 GB RAM will allow you to replicate the related environment on your desktop. You'll also need a desktop hypervisor, such as *Virtual Box* or *VMware Fusion*. Alternatively, you may choose to explore more lightweight desktop-only environments with Kubernetes, which we'll cover as well.

Now, let's start our journey together to discovering Kubernetes.

Introducing Kubernetes

Kubernetes is an open source *container orchestrator* initially developed by Google. Assuming an application uses containerized microservices, a container orchestration system provides the following features:

- **Elastic orchestration**: Automatically starting and stopping application services (containers) based on specific requirements and conditions – for example, launching multiple web server instances with an increasing number of requests and eventually terminating servers when the number of requests drops below a certain threshold

- **Workload management**: Optimally deploying and distributing application services across the underlying cluster to ensure mandatory dependencies and redundancy – for example, running a web server endpoint on each cluster node for high availability

- **Infrastructure abstraction**: Providing container runtime, networking, and load-balancing capabilities – for example, distributing the load among multiple web server containers and autoconfiguring the underlying network connectivity with a database app container

- **Declarative configuration**: Describing and ensuring the *desired state* of a multi-tiered application – for example, a web server should be ready for serving requests only when the database backend is up and running, and the underlying storage is available

A classic example of workload orchestration is a video-on-demand streaming service. With a popular new TV show in high demand, the number of streaming requests would significantly exceed the average during a regular season. With Kubernetes, we could scale out the number of web servers based on the volume of streaming sessions. We could also control the possible scale-out of some of the middle-tier components, such as database instances (serving the authentication requests) and storage cache (serving the streams). When the TV show goes out of fashion, and the number of requests drops significantly, Kubernetes terminates the surplus instances, automatically reducing the application deployment's footprint and, consequently, the underlying costs.

Here are some key benefits of deploying applications with Kubernetes:

- **Speedy deployment**: Application containers are created and launched relatively fast, using either a *declarative* or *imperative* configuration model (as we'll see later in this chapter).

- **Quick iterations**: Application upgrades are relatively straightforward, with the underlying infrastructure simply seamlessly replacing the related container.

- **Rapid recovery**: If an application crashes or becomes unavailable, Kubernetes automatically restores the application to the desired state by replacing the related container.

- **Reduced operation costs**: The containerized environment and infrastructure abstraction of Kubernetes yields minimal administration and maintenance efforts with relatively low resources for running applications.

Now that we have introduced Kubernetes, let's look at its basic operating principles next.

Understanding the Kubernetes architecture

There are three major concepts at the core of the working model of Kubernetes:

- **Declarative configuration** or **desired state**: This concept describes the overall application state and microservices, deploying the required containers and related resources, including network, storage, and load balancers, to achieve a running functional state of the application.

- **Controllers** or **controller loops**: This concept monitors the desired state of the system and takes corrective action when needed, such as replacing a failed application container or adding additional resources for scale-out workloads.

- **API object model**: This concept represents the actual implementation of the desired state, using various configuration objects and the interaction – **Application Programming Interface (API)** – between these objects.

For a better grasp of the internals of Kubernetes, we need to take a closer look at the Kubernetes object model and the related API.

Introducing the Kubernetes object model

The Kubernetes architecture defines a collection of objects representing the desired state of a system. An *object*, in this context, is a programmatic term to describe the *behavior* of a subsystem. Multiple objects interact with one another via the API, shaping the desired state over time. In other words, the Kubernetes object model is the programmatic representation of the desired state.

So, what are these objects in Kubernetes? We'll briefly enumerate some of the more important ones and further elaborate on each in the following sections:

- API server
- Pods

- Controllers

- Services

- Storage

- Networking

We use these API objects to configure the system's state, using either a *declarative* or *imperative* model. With a *declarative* model, we *describe* the state of the system, usually with a configuration file or manifest (in YAML or JSON format). Such a configuration may include and deploy multiple API objects and regard the system as a whole.

On the other hand, the *imperative* configuration model uses individual commands to configure and deploy specific API objects, usually acting on a single target or subsystem.

Let's look at the API server first – the central piece in the Kubernetes object model.

Introducing the API server

The API server is the core hub of the Kubernetes object model, acting as a management endpoint for the desired state of the system. The API server exposes an HTTP REST interface using JSON payloads. It is accessible *internally*, by other API objects, and *externally*, by configuration and management workflows.

The API server is essentially the gateway of interaction with the Kubernetes cluster, both from the outside and within. A system administrator connects to the API server endpoint to configure and manage a Kubernetes cluster, usually via a CLI. Internally, Kubernetes API objects connect to the API server to provide an update of their state. In return, the API server may further adjust the internal configuration of the API objects toward the desired state.

The API objects are the building blocks of the internal configuration or desired state of a Kubernetes cluster. Let's look at a few of these API objects next.

Introducing Pods

A **Pod** represents the basic *working unit* in Kubernetes, running as a single- or multi-container application. A Pod is also known as the *unit of scheduling* in Kubernetes. In other words, containers within the same Pod are guaranteed to be deployed together on the same cluster node.

A Pod essentially represents a microservice (or a service) within the application's service mesh. Considering the classic example of a web application, we may have the following Pods running in the cluster:

- web server (Nginx)
- authentication (Vault)
- database (PostgreSQL)
- storage (NAS)

Each of these services (or applications) runs within their Pod. Multiple Pods of the same application (for example, web server) make up a *ReplicaSet*. We'll look at ReplicaSets closer in the *Introducing controllers* section next.

One of the essential features of Pods is their *ephemeral* nature. Once a Pod is terminated, it is gone for good. No Pod ever gets redeployed in Kubernetes. Consequently, Pods don't persist any state unless they use persistent storage or a local volume to save their data.

Also, Pods are an *atomic unit* – they are either deployed or not. For a single-container Pod, atomicity is almost a given. For multi-container Pods, atomicity means that a Pod is deployed only when each of the constituent containers is deployed. If any of the containers fail to deploy, the Pod would not be deployed, and hence there's no Pod. If one of the containers within a running multi-container Pod fails, the whole Pod is terminated.

A Pod could be deployed and running, but that doesn't necessarily mean the application or service within the Pod is healthy. Kubernetes uses *probes* to monitor the *health* of an application inside a Pod. For example, a web server Pod can have a probe that checks a specific URL and decides whether it's healthy based on the response.

Kubernetes tracks the state of Pods using controllers. Let's look at controllers next.

Introducing controllers

Controllers in Kubernetes are *control loops* responsible for keeping the system in the desired state or bringing the system closer to the desired state. For example, a controller may detect that a Pod is not responding and request the deployment of a new Pod while terminating the old one.

A controller may also add or remove Pods of a specific type to and from a collection of *Pod replicas*. Such controllers are called **ReplicaSets**, and their responsibility is to accommodate a particular number of Pod replicas based on the current state of the application. For example, suppose an application requires three web server Pods, and one of them becomes unavailable. In that case, the ReplicaSet controller ensures that the failed Pod is deleted, and a new one takes its place.

When deploying applications in Kubernetes, we usually don't use ReplicaSets directly to create Pods. We use the *Deployment* Controller instead. Given the declarative model of Kubernetes, we can define a deployment with one or more ReplicaSets. It's the Deployment Controller's job to create the ReplicaSet with the required number of Pods and manage the ReplicaSet state, in other words, which container image to load and the number of Pods to create.

A Deployment Controller can also manage the transition from one ReplicaSet to another, a functionality used in rollout or upgrade scenarios. Imagine we have a ReplicaSet (v1) with several Pods, all running version 1 of our application, and we want to upgrade them to version 2. Remember, Pods cannot be *regenerated* or *upgraded*. Instead, we'll define a second ReplicaSet (v2), creating the version 2 Pods. The Deployment Controller will tear down the v1 ReplicaSet and bring up v2. Kubernetes performs the rollout seamlessly, with minimal to no disruption of service. The Deployment Controller manages the transition between the v1 and v2 ReplicaSets and even rolls back the transition if needed.

There are many other controller types in Kubernetes, and we encourage you to explore them at https://kubernetes.io/docs/concepts/workloads/controllers/.

As applications scale-out or terminate, the related Pods are deployed or removed. *Services* provide access to the dynamic and transient world of Pods. We'll look at Services next.

Introducing Services

Services provide persistent access to the applications running in Pods. It is the Services' responsibility to ensure that the Pods are accessible by routing the traffic to the corresponding application endpoints. In other words, Services provide the network abstraction for communicating with Pods, such as IP addresses, routing, and DNS resolution. As Pods are deployed or terminated based on the system's desired state, Kubernetes dynamically updates the Service endpoint of the Pods, with minimal to no disruption in terms of accessing the related applications. As users and applications access the Service endpoint's persistent IP address, the service will ensure that the routing information is up to date and traffic is exclusively routed to the running and healthy Pods. Services can also be leveraged to load-balance the application traffic between Pods and scale Pods up or down based on demand.

So far, we have looked at Kubernetes API objects controlling the deployment, access, and life cycle of application services. What about the persistent data that applications require? We'll look at the Kubernetes storage next.

Introducing storage

Kubernetes provides various storage types for applications running within the cluster. The most common are **volumes** and **persistent volumes**. Due to the ephemeral nature of Pods, application data stored within a Pod using volumes is lost when the Pod is terminated. *Persistent volumes* are defined and managed at the Kubernetes cluster level, and they are independent of Pods. Applications (Pods) requiring a persistent state would reserve a persistent volume (of a specific size), using a **PersistentVolumeClaim**. When a Pod using a persistent volume terminates, the new Pod replacing the old one retrieves the current state from the persistent volume and will continue using the underlying storage. For more information on Kubernetes storage types, please refer to `https://kubernetes.io/docs/concepts/storage/`.

Now that we are familiar with the Kubernetes API object model, let's quickly gloss through the architecture of a Kubernetes cluster.

The anatomy of a Kubernetes cluster

A Kubernetes cluster consists of one *control plane node* and one or more *worker nodes*. The following diagram presents a high-level view of the Kubernetes architecture:

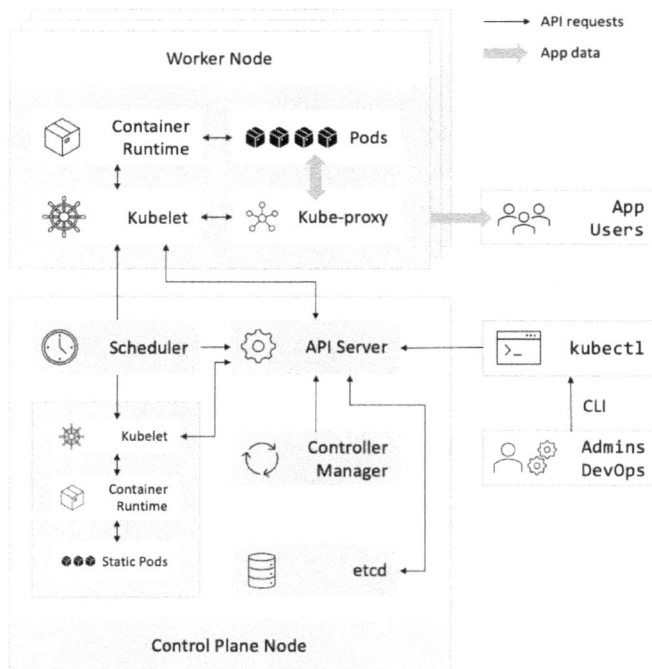

Figure 14.1 – Kubernetes architecture

Let's look at the Kubernetes cluster nodes in some detail next, starting with the Control Plane node.

Introducing the Kubernetes Control Plane

The *Kubernetes Control Plane* provides the essential services for deploying and orchestrating application workloads, and it runs on a dedicated node in the Kubernetes cluster – the *Control Plane node*. The Control Plane node, also known as the *Master node*, implements the core components of a Kubernetes cluster, such as resource scheduling and monitoring. It's also the primary access point for cluster administration. Here are the key subsystems of a Control Plane node:

- **API Server**: The central communication hub between Kubernetes API objects; it also provides the cluster's management endpoint accessible either via CLI or the Kubernetes web administration console (dashboard).

- **Scheduler**: Decides when and which nodes to deploy the Pods on, depending on resource allocation and administrative policies.

- **Controller Manager**: Maintains the control loops, monitoring and shaping the desired state of the system.

- **etcd**, also known as the **cluster store**, is a highly available persisted database, maintaining the state of the Kubernetes cluster and related API objects; the information in `etcd` is stored as key-value pairs.

- **kubectl**: The primary administrative CLI for managing and interacting with the Kubernetes cluster; `kubectl` communicates directly with the API server, and it may connect remotely to a cluster.

A detailed architectural overview of the Kubernetes Control Plane is beyond the scope of this chapter. You may explore the related concepts in more detail at `https://kubernetes.io/docs/concepts/architecture/`.

Next, let's take a brief look at the Kubernetes node – the workhorse of a Kubernetes cluster.

Introducing the Kubernetes nodes

In a Kubernetes cluster, the **nodes** – also referred to as **worker nodes** – run the actual application Pods and maintain their full life cycle. Nodes provide the compute capacity of Kubernetes, and ensure that the workloads are uniformly distributed across the cluster when deploying and running Pods. Nodes can be configured either as physical (bare metal) or virtual machines.

Let's enumerate the key elements of a Kubernetes node:

- **Kubelet**: Processes Control Plane requests (from the Scheduler) to deploy and start application Pods; the Kubelet also monitors the node and Pod state, reporting the related changes to the API server

- **Kube-Proxy**: Dynamically configures the virtual networking environment for the applications running in the Pods; it routes the network traffic, provides load balancing, and maintains the IP addresses of Services and Pods

- **Container Runtime**: Provides the runtime environment for the Pods as application containers; uses the **Container Runtime Interface** (**CRI**) to interact with the underlying container engine (such as **containerd** and **Docker**)

All of the preceding Services run on *each* node in the Kubernetes cluster, including the Control Plane node. These components in the Control Plane are required by special-purpose Pods, providing specific Control Plane services, such as DNS, ingress (load balancing), and dashboard (web console).

For more information on Kubernetes nodes and related architectural concepts, please visit `https://kubernetes.io/docs/concepts/architecture/nodes/`.

Now that we are familiar with some of the key concepts and cluster components, let's get ready to install and configure Kubernetes.

Installing and configuring Kubernetes

Before installing or using Kubernetes, you have to decide on the infrastructure you'll use, on-premises or public cloud. Second, you'll have to choose between an **Infrastructure as a Service** (**IaaS**) or **Platform as a Service** (**PaaS**) model. With IaaS, you'll have to install, configure, manage, and maintain the Kubernetes cluster yourself, either on physical (bare metal) or virtual machines. The related operation efforts are not straightforward and should be considered carefully. If you choose a PaaS solution, available from all major public cloud providers, you'll be limited to only administrative tasks but saved from the burden of maintaining the underlying infrastructure.

In this chapter, we'll cover both IaaS and PaaS deployments of Kubernetes. For IaaS, we'll use a local desktop environment running Ubuntu virtual machines. Then, we'll look at Kubernetes in AWS and Azure.

For the on-premises installation, we may also choose between a lightweight desktop version of Kubernetes or a full-blown cluster with multiple nodes. Let's look at some of the most common desktop versions of Kubernetes next.

Installing Kubernetes on desktop

If you're looking only to experiment with Kubernetes, a desktop version may fit the bill. Desktop flavors of Kubernetes usually deploy a single-node cluster on your local machine. Depending on your platform of choice – Windows, macOS, or Linux – you have plenty of Kubernetes engines to choose from. Here are just a few:

- **Docker Desktop** (macOS, Windows): `https://www.docker.com/products/docker-desktop`
- **minikube** (Linux, macOS, Windows): `https://minikube.sigs.k8s.io/docs/`
- **Microk8s** (Linux, macOS, Windows): `https://microk8s.io/`
- **k3s** (Linux): `https://k3s.io/`

In this section, we'll show you how to install Microk8s, one of the trending Kubernetes desktop engines at the time of writing. Microk8s is available to install via the Snap store. Let's install it on Ubuntu first.

Installing Microk8s on Ubuntu

On Ubuntu 20.04, for example, we can install it with the following command:

```
sudo snap install microk8s --classic
```

A successful installation of Microk8s should yield the following result:

```
packt@neptune:~$ sudo snap install microk8s --classic
[sudo] password for packt:
microk8s (1.20/stable) v1.20.2 from Canonical✓ installed
```

Figure 14.2 – Running Microk8s on Linux

To access the Microk8s CLI without `sudo` permissions, you'll have to add the local user account to the `microk8s` group and also fix the permissions on the `~/.kube` directory with the following command:

```
sudo usermod -aG microk8s $USER
sudo chown -f -R $USER ~/.kube
```

The changes will take effect on the next login, and you can use the `microk8s` command-line utility with non-`sudo` invocations. For example, the following command displays the help for the tool:

```
microk8s help
```

To get the status of the local single-node Microk8s Kubernetes cluster, we run the following command:

```
microk8s status
```

The installation steps of Microk8s on RHEL/CentOS are very similar, with some minor distinctions. Let's look at this next.

Installing Microk8s on RHEL/CentOS

On RHEL/CentOS, we'll have to enable the Snap store first. Snap is available via **Extra Packages for Enterprise Linux (EPEL)**. We install the EPEL repository with the following command:

```
sudo yum install -y epel-release
```

Next, we'll install Snap:

```
sudo yum install -y snapd
```

With Snap installed, we need to make a couple of tweaks to enable the Snap communication socket and the classic Snap support:

```
sudo systemctl enable --now snapd.socket
sudo ln -s /var/lib/snapd/snap /snap
```

Now, we're ready to install Microk8s using `snap`:

```
sudo snap install microk8s --classic
```

Please note that for non-sudo invocations of the **microk8s** CLI, you'll need to fix the required permissions as shown in the *Installing Microk8s on Ubuntu* section.

Kubernetes desktop engines are great for learning and experimenting with the platform, but they are far from matching a real-world production environment. We'll look next at installing a Kubernetes cluster using virtual machines.

Installing Kubernetes on virtual machines

In this section, we'll get closer to a real-world Kubernetes environment – though at a much smaller scale – by deploying a Kubernetes cluster on Ubuntu **Virtual Machines** (**VMs**). You can use any hypervisor, such as Oracle VirtualBox or VMware Fusion, both of them described in *Chapter 1*, *Installing Linux*, of this book.

We'll provision each VM with 2 vCPU cores, 2 GB RAM, and 20 GB disk capacity. You may follow the steps described in the *Installing Ubuntu* section of *Chapter 1*, *Installing Linux*, using your hypervisor of choice.

Before we dive into the Kubernetes cluster installation details, let's take a quick look at our lab environment.

Preparing the lab environment

Here are the specs of our VM environment:

Hypervisor: VMware Fusion

Kubernetes cluster: One **Control Plane** (**CP**) node; three worker nodes

CP node:

- `k8s-cp1:172.16.191.6`

Worker nodes:

- `k8s-n1:172.16.191.8`
- `k8s-n2:172.16.191.9`
- `k8s-n3:172.16.191.10`

VMs: Ubuntu Server 20.04.2, 2 vCPUs, 2 GB RAM, 20 GB disk

User: `packt` (on all nodes), with SSH access enabled

We set the username and hostname settings on each VM node during the Ubuntu Server installation wizard. Also, make sure to enable the OpenSSH server when prompted. Your VM IP addresses will most probably be different from those in the specs, but that shouldn't matter. You may also choose to use static IP addresses for your VMs.

To make hostname resolution simple within the cluster, edit the `/etc/hosts` file on each node and add the related records. For example, we have the following `/etc/hosts` file on the Control Plane node (`k8s-cp1`):

```
packt@k8s-cp1:~$ cat /etc/hosts
127.0.0.1        k8s-cp1 localhost
172.16.191.6     k8s-cp1
172.16.191.8     k8s-n1
172.16.191.9     k8s-n2
172.16.191.10    k8s-n3
```

Figure 14.3 – The /etc/hosts file on the CP node (k8s-cp1)

In production environments, with the firewall enabled on the cluster nodes, we have to make sure that the following rules are configured for accepting network traffic within the cluster (according to `https://kubernetes.io/docs/setup/production-environment/tools/kubeadm/install-kubeadm/`):

Control-plane node(s)

Protocol	Direction	Port Range	Purpose	Used By
TCP	Inbound	6443*	Kubernetes API server	All
TCP	Inbound	2379-2380	etcd server client API	kube-apiserver, etcd
TCP	Inbound	10250	kubelet API	Self, Control plane
TCP	Inbound	10251	kube-scheduler	Self
TCP	Inbound	10252	kube-controller-manager	Self

Worker node(s)

Protocol	Direction	Port Range	Purpose	Used By
TCP	Inbound	10250	kubelet API	Self, Control plane
TCP	Inbound	30000-32767	NodePort Services†	All

Figure 14.4 – The ports used by the Kubernetes cluster nodes

The following sections assume that you have the VMs provisioned and running according to the preceding specs. You may take some initial snapshots of your VMs before proceeding with the next steps. In case anything goes wrong with the installation, you can revert to the initial state and start again.

Here are the steps we'll follow to install the Kubernetes cluster:

- Disable swapping
- Install **containerd**
- Install Kubernetes packages: `kubelet`, `kubeadm`, `kubectl`

We'll have to perform these steps on *each cluster node*. The related commands are also captured in the accompanying chapter source code on GitHub.

Let's start with the first step and disable the memory swap on each node.

Disable swapping

The Kubernetes **kubelet** doesn't work with `swap` enabled on Linux platforms. See `https://github.com/kubernetes/kubernetes/issues/53533`. `swap` is disk space that's used when the memory is full.

To disable the `swap` immediately, we run the following command:

```
sudo swapoff -a
```

To persist the disabled `swap` with system reboots, we need to comment out the `swap`-related entries in `/etc/fstab`. You can do this either manually, by editing `/etc/fstab`, or with the following command:

```
sudo sed -i '/\s*swap\s*/s/^\(.*\)$/# \1/g' /etc/fstab
```

You may want to double-check that *all* swap entries in `/etc/fstab` are disabled:

```
cat /etc/fstab
```

We can see the `swap` mount point commented out in our `/etc/fstab` file:

```
packt@k8s-n2:~$ cat /etc/fstab
# /etc/fstab: static file system information.
#
# Use 'blkid' to print the universally unique identifier for a
# device; this may be used with UUID= as a more robust way to name devices
# that works even if disks are added and removed. See fstab(5).
#
# <file system> <mount point>    <type>  <options>        <dump>  <pass>
# / was on /dev/ubuntu-vg/ubuntu-lv during curtin installation
/dev/disk/by-id/dm-uuid-LVM-QeoPJJysMacrQZRUZifQzWe8L0kvi7ljzEaNHJXlUhvmtt0up4BN3TGdOfW0u1Bb /
  ext4 defaults 0 0
# /boot was on /dev/sda2 during curtin installation
/dev/disk/by-uuid/822fdd70-06c6-43d3-a802-16587ec5276e /boot ext4 defaults 0 0
# /swap.img      none     swap    sw       0        0
```

Figure 14.5 – Disabling swap entries in /etc/fstab

Remember to run the preceding commands on each node in the cluster. Next, we'll look at installing the Kubernetes container runtime.

Installing containerd

`containerd` is the default container runtime in recent versions of Kubernetes. `containerd` implements the **Container Runtime Interface** (**CRI**), required by the Kubernetes container engine abstraction layer. The related installation procedure is not straightforward, and we'll follow the steps described in the official Kubernetes documentation at the time of this writing: `https://kubernetes.io/docs/setup/production-environment/container-runtimes/`. These steps may change at any time, so please make sure to check the latest procedure.

We'll start by installing some `containerd` prerequisites. We enable the `br_netfilter` and `overlay` kernel modules using `modprobe`:

```
sudo modprobe br_netfilter
sudo modprobe overlay
```

We also ensure that these modules are loaded upon system reboots:

```
cat <<EOF | sudo tee /etc/modules-load.d/containerd.conf
br_netfilter
overlay
EOF
```

Next, we apply the CRI required `sysctl` parameters, also persisted across system reboots:

```
cat <<EOF | sudo tee /etc/sysctl.d/99-kubernetes-cri.conf
net.bridge.bridge-nf-call-iptables = 1
net.ipv4.ip_forward = 1
net.bridge.bridge-nf-call-ip6tables = 1
EOF
```

We want the preceding changes to take effect immediately, without a system reboot:

```
sudo sysctl --system
```

Let's make sure the `apt` repository is up to date before installing any new packages:

```
sudo apt-get update
```

Now, we're ready to install `containerd`:

```
sudo apt-get install -y containerd
```

Next, we generate a default `containerd` configuration:

```
sudo mkdir -p /etc/containerd
containerd config default | sudo tee /etc/containerd/config.
toml
```

We need to slightly alter the default `containerd` configuration to use the `systemd` cgroup driver with the container runtime (`runc`). This change is required because the underlying platform (Ubuntu) uses `systemd` as the service manager.

Open the `/etc/containerd/config.toml` file with your editor of choice, such as the following:

```
sudo nano /etc/containerd/config.toml
```

Locate the following section:

```
[plugins."io.containerd.grpc.v1.cri".containerd.runtimes.runc]
```

Then, add the highlighted lines, *adjusting* the appropriate indentation (*very important!*):

```
[plugins."io.containerd.grpc.v1.cri".containerd.runtimes.runc]
   . . .
   [plugins."io.containerd.grpc.v1.cri".containerd.runtimes.
runc.options]
   SystemdCgroup = true
```

Here's the resulting configuration stub:

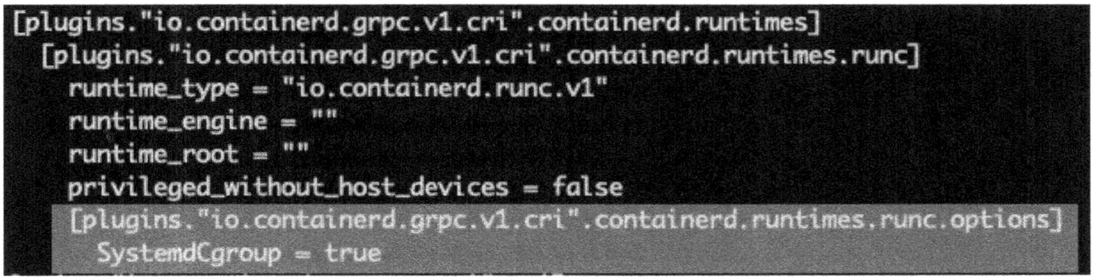

Figure 14.6 – Modifying the containerd configuration

Save the /etc/containerd/config.toml file and restart containerd:

```
sudo systemctl restart containerd
```

With containerd installed and configured, we can proceed with the installation of the Kubernetes packages next.

Installing Kubernetes packages

We'll follow the steps described at https://kubernetes.io/docs/setup/ production-environment/tools/kubeadm/install-kubeadm/. This procedure may also change over time, so please make sure to check out the latest.

Let's start by installing the packages required by the Kubernetes apt repository:

```
sudo apt-get install -y apt-transport-https ca-certificates
curl
```

Next, we download the Google apt repository **GNU Privacy Guard (GNU)** signing key:

```
sudo curl -sfSLo /usr/share/keyrings/kubernetes-archive-
keyring.gpg https://packages.cloud.google.com/apt/doc/apt-key.
gpg
```

The next command adds the Kubernetes apt repository to our system:

```
echo "deb [signed-by=/usr/share/keyrings/kubernetes-archive-
keyring.gpg] https://apt.kubernetes.io/ kubernetes-xenial main"
 | sudo tee /etc/apt/sources.list.d/kubernetes.list
```

Let's read the packages available in the new repository we just added:

```
sudo apt-get update
```

We're now ready to install the Kubernetes packages:

```
sudo apt-get install -y kubelet kubeadm kubectl
```

We want the version of these packages *pinned*, to avoid the inadvertent update via system security patches, and so on. The Kubernetes packages should exclusively be updated using the cluster upgrade procedures. We use the apt-mark hold command to pin the version of the Kubernetes packages, including containerd:

```
sudo apt-mark hold containerd kubelet kubeadm kubectl
```

Finally, ensure that the `containerd` and `kubelet` Services are enabled upon system startup (reboot):

```
sudo systemctl enable containerd
sudo systemctl enable kubelet
```

Now that we have finished installing the Kubernetes packages, let's check the status of the node Services. We retrieve the status of the `containerd` service first:

```
sudo systemctl status containerd
```

`containerd` should be active and running:

```
packt@k8s-cp1:~$ sudo systemctl status containerd
● containerd.service - containerd container runtime
     Loaded: loaded (/lib/systemd/system/containerd.service; enabled; vendor preset: enabled)
     Active: active (running) since Sat 2021-03-13 04:11:52 UTC; 1h 0min ago
       Docs: https://containerd.io
   Main PID: 8747 (containerd)
      Tasks: 11
     Memory: 21.3M
     CGroup: /system.slice/containerd.service
             └─8747 /usr/bin/containerd
```

Figure 14.7 – The running status of containerd

Next, let's check the status of the `kubelet` service:

```
sudo systemctl status kubelet
```

At this time, it should be *no surprise* that the status shows `exited`:

```
packt@k8s-cp1:~$ sudo systemctl status kubelet
● kubelet.service - kubelet: The Kubernetes Node Agent
     Loaded: loaded (/lib/systemd/system/kubelet.service; enabled; vendor preset: enabled)
    Drop-In: /etc/systemd/system/kubelet.service.d
             └─10-kubeadm.conf
     Active: activating (auto-restart) (Result: exit-code) since Sat 2021-03-13 05:16:52 UTC
       Docs: https://kubernetes.io/docs/home/
    Process: 16208 ExecStart=/usr/bin/kubelet $KUBELET_KUBECONFIG_ARGS $KUBELET_CONFIG_ARGS
   Main PID: 16208 (code=exited, status=255/EXCEPTION)

Mar 13 05:16:52 k8s-cp1 systemd[1]: kubelet.service: Main process exited, code=exited, statu
Mar 13 05:16:52 k8s-cp1 systemd[1]: kubelet.service: Failed with result 'exit-code'.
```

Figure 14.8 – The kubelet crashing without cluster configuration

The `kubelet` is looking for the Kubernetes cluster, which is not set up yet. We can see that the `kubelet` attempts to start and activate itself but keeps crashing, as it cannot locate the required configuration.

Next, we'll *bootstrap* (initialize) the Kubernetes cluster using `kubeadm`. Please install the required Kubernetes packages on all cluster nodes following the previous steps before proceeding with the next section.

Introducing kubeadm

`kubeadm` is a helper tool for creating a Kubernetes cluster and essentially has two invocations:

- `kubadm init`: *Bootstrapping* or initializing a Kubernetes cluster
- `kubadm join`: *Adding* a node to a Kubernetes cluster

The default invocation of `kubadm init` – with no parameters – performs the following tasks:

- It runs preliminary (*pre-flight*) *system checks*.
- It creates a *Certificate Authority*.
- It generates `kubeconfig` files.
- It generates *static Pod* manifests.
- It *waits* for the static Pods to start.
- It *taints* the Control Plane node.
- It generates a *bootstrap token*.
- It starts *add-on Pods*.

Let's briefly describe each of these tasks next.

Running preliminary checks

In the very initial phase, `kubadm init` ensures that we have the minimum system resources in terms of CPU and memory, the required user permissions, and a supported CRI-compliant container runtime. If any of these checks fail, `kubeadm init` stops the execution of creating the cluster. If the checks succeed, `kubeadm` creates a **Certificate Authority (CA)** next.

Creating a Certificate Authority

`kubeadm init` creates a self-signed CA used by Kubernetes to generate the certificates required to authenticate and run trusted workloads within the cluster. The CA files are stored in the `/etc/kubernetes/pki/` directory and are distributed on each node upon joining the cluster. After generating the CA, `kubadm` proceeds with the kubeconfig files.

Generating kubeconfig files

The *kubeconfig* files are configuration files used by the Kubernetes cluster components to locate, communicate, and authenticate with the API server. `kubeadm init` creates a default set of kubeconfig files required to bootstrap the cluster. The kubeconfig files are stored in the `/etc/kubernetes/` directory. Next, `kubadm` generates the static Pod manifests.

Generating static Pod manifests

Static Pods are system-specific Pods running exclusively on the Control Plane node and managed by the `kubelet` daemon. Examples of static Pods are the API server, the Controller Manager, Scheduler, and `etcd`. Static Pod manifests are configuration files describing the Control Plane Pods. `kubeadm init` generates the static Pod manifests during the cluster bootstrapping process. The manifest files are stored in the `/etc/kubernetes/manifests/` directory. The `kubelet` service monitors this location and, when it finds a manifest, deploys the corresponding static Pod. After generating the static Pod manifests, `kubeadm` waits for the static Pods to start.

Waiting for the static Pods to start

After the `kubelet` daemon deploys the static Pods, `kubeadm` queries the `kubelet` for the static Pods' state. When the static Pods are up and running, `kubeadm init` proceeds with the next stage, *tainting* the Control Plane node.

Tainting the Control Plane node

Tainting is the process of *excluding* a node from running user Pods. The opposite concept in a Kubernetes environment is *toleration* – controlling the affinity of Pods to specific cluster nodes. `kubeadm init` follows the Kubernetes best practice of tainting the Control Plane to avoid user Pods running on the Control Plane node. The obvious reason is to preserve Control Plane resources exclusively for system-specific workloads. Next, `kubeadm init` generates a bootstrap token.

Generating a bootstrap token

Bootstrap tokens are simple bearer tokens used for joining new nodes to a Kubernetes cluster. `kubeadm init` generates a bootstrap token that can be shared with a trusted node to join the cluster. Finally, `kubeadm init` proceeds with starting the Kubernetes add-on components.

Starting add-on Pods

Kubernetes cluster add-ons are specific Control Plane components (Pods) extending the functionality of the cluster. By default, `kubadm init` creates and deploys the *DNS* and *kube-proxy* add-on Pods.

The stages of a Kubernetes cluster's bootstrapping process are highly customizable. `kubeadm init`, when invoked without additional parameters, runs all the tasks in the preceding order. Alternatively, a system administrator may invoke the `kubeadm` command with different option parameters to control and run any of the stages mentioned.

For more information about `kubeadm`, please refer to the utility's help with the following command:

```
kubeadm help
```

For more information about bootstrapping a Kubernetes cluster using `kubeadm`, you may refer to the official Kubernetes documentation at `https://kubernetes.io/docs/setup/production-environment/tools/kubeadm/`.

In the next section, we'll bootstrap a Kubernetes cluster using `kubeadm` to generate a cluster configuration file and then invoke `kubeadm init` to use this configuration. We'll bootstrap our cluster by creating the Kubernetes Control Plane node next.

Creating a Kubernetes Control Plane node

Here are the steps we'll take to create the Control Plane node:

- Download a manifest file to configure the overlay network for Pods; we'll use the *Calico* networking add-on for Kubernetes.

- Generate a default cluster configuration YAML file using `kubeadm` and adjust it to match our environment.

- Invoke `kubeadm init` using the modified cluster configuration file and the `cri-socket` option parameter pointing to `containerd` as the container engine of choice in our Kubernetes cluster.

- Set the required permissions for the current user to administer the Kubernetes cluster.

- Use kubectl to apply the *Calico* overly network configuration.

Now, let's get to work and configure our Kubernetes Control Plane node. The following commands are performed on the k8s-cp1 host in our VM environment. As the hostname suggests, we choose k8s-cp1 as the Control Plane node of our Kubernetes cluster.

We'll start by downloading the *Calico* manifest for *overlay networking*. The overlay network – also known as **Software Defined Network (SDN)** – is a logical networking layer that accommodates a secure and seamless network communication between the Pods over a physical network that may not be accessible for configuration. Exploring the internals of cluster networking is beyond the scope of this chapter, but we encourage you to read more at https://kubernetes.io/docs/concepts/cluster-administration/networking/. You'll also find references to the Calico networking add-on. To download the related manifest, we run the following command:

```
curl https://docs.projectcalico.org/manifests/calico.yaml -O
```

The command downloads the calico.yaml file in the current directory (/home/packt/) that we'll use with kubectl to configure Pod networking later in the process.

Next, let's open the calico.yaml file using a text editor and look for the following lines (starting with line 3672):

```
# - name: CALICO_IPV4POOL_CIDR
#   value: "192.168.0.0/16"
```

CALICO_IPV4POOL_CIDR points to the network range associated with the Pods. If the related subnet conflicts in any way with your local environment, you'll have to change it here. We'll leave the setting as is.

Next, we'll create a default cluster configuration file using kubeadm. The cluster configuration file describes the settings of the Kubernetes cluster we're building. Let's name this file k8s-config.yaml:

```
kubeadm config print init-defaults | tee k8s-config.yaml
```

You can safely disregard the warning shown at the beginning of the related output regarding the missing Docker runtime. We'll be using `containerd` as the container engine, and we'll make the corresponding adjustment next:

```
packt@k8s-cp1:~$ kubeadm config print init-defaults | tee k8s-config.yaml
W0314 10:21:37.985924  164502 kubelet.go:200] cannot automatically set CgroupDriver when star
ting the Kubelet: cannot execute 'docker info -f {{.CgroupDriver}}': executable file not foun
d in $PATH
apiVersion: kubeadm.k8s.io/v1beta2
bootstrapTokens:
- groups:
  - system:bootstrappers:kubeadm:default-node-token
  token: abcdef.0123456789abcdef
  ttl: 24h0m0s
  usages:
  - signing
  - authentication
```

Figure 14.9 – Creating the cluster configuration file (disregard the Docker warning)

Let's review the `k8s-config.yaml` file we just generated and call out a few changes that we'll have to make. We'll start with the `localAPIEndpoint.advertiseAddress` configuration parameter – the IP address of the API server endpoint. The default value is `1.2.3.4`, and we need to change it to the IP address of the VM running the Control Plane node (`k8s-cp1`), in our case, `172.16.191.6`. Refer to the *Preparing the lab environment* section earlier in this chapter. You'll have to put the IP address matching your environment:

```
localAPIEndpoint:
  advertiseAddress: 1.2.3.4  ⬅
  bindPort: 6443
```

Figure 14.10 – Modifying the advertiseAddress configuration parameter

You can either manually edit the `k8s-config.yaml` file or perform the preceding change with the following command:

```
sed -i 's/ advertiseAddress: 1.2.3.4/ advertiseAddress:
172.16.191.6/' k8s-config.yaml
```

The next change we need to make is pointing the `nodeRegistration.criSocket` configuration parameter to the `containerd` socket (`/run/containerd/containerd.sock`):

```
nodeRegistration:
  criSocket: /var/run/dockershim.sock  ⬅
  name: k8s-cp1
```

Figure 14.11 – Changing the criSocket configuration parameter

The equivalent command is as follows:

```
sed -i 's/  criSocket: \/var\/run\/dockershim.sock/  criSocket:
\/run\/containerd\/containerd.sock/' k8s-config.yaml
```

Next, we change the `kubernetesVersion` parameter to match the version of our Kubernetes environment:

```
imageRepository: k8s.gcr.io
kind: ClusterConfiguration
kubernetesVersion: v1.20.0  ⬅
```

Figure 14.12 – Changing the kubernetesVersion parameter

The default value is `1.20.0`, but our Kubernetes version, according to `kubeadm`, is `1.20.4`:

```
kubeadm version
```

The related output is as follows:

```
packt@k8s-cp1:~$ kubeadm version
kubeadm version: &version.Info{Major:"1", Minor:"20", GitVersion:"v1.20.4", GitCommit:"e87da0
bd6e03ec3fea7933c4b5263d151aafd07c", GitTreeState:"clean", BuildDate:"2021-02-18T16:09:38Z",
GoVersion:"go1.15.8", Compiler:"gc", Platform:"linux/amd64"}
```

Figure 14.13 – Retrieving the current version of Kubernetes

To change the `kubernetesVersion` parameter in `k8s-config.yaml`, we run the following command:

```
sed -i 's/kubernetesVersion: v1.20.0/kubernetesVersion:
v1.20.4/' k8s-config.yaml
```

Our final modification of the cluster configuration file sets the `cgroup` driver of `kubelet` to `systemd`, matching the `cgroup` driver of `containerd`. Please note that `systemd` is the underlying platform's service manager (in Ubuntu), hence the need to yield related service control to the Kubernetes daemons. The corresponding configuration block is not yet present in `k8s-config.yaml`. We can add it manually to the end of the file or with the following command:

```
cat <<EOF | cat >> k8s-config.yaml
---
apiVersion: kubelet.config.k8s.io/v1beta1
kind: KubeletConfiguration
cgroupDriver: systemd
EOF
```

The resulting configuration block is as follows:

```
---
apiVersion: kubelet.config.k8s.io/v1beta1
kind: KubeletConfiguration
cgroupDriver: systemd
```

Figure 14.14 – Pointing the kubelet's cgroup driver to systemd

Now, we're ready to bootstrap the Kubernetes cluster. We invoke the `kubeadm init` command with the `--config` option pointing to the cluster configuration file (`k8s-config.yaml`), and with the `--cri-socket` option parameter pointing to the `containerd` socket:

```
sudo kubeadm init \
    --config=k8s-config.yaml \
    --cri-socket /run/containerd/containerd.sock
```

A failure to specify the `--cri-socket` option results in a pre-flight error pointing to missing Docker runtime due to a current bug in `kubeadm` still assuming that Docker is the container engine. Future versions of `kubeadm` may fix this behavior.

The preceding command takes a couple of minutes to run. A successful bootstrap of the Kubernetes cluster completes with the following output:

```
Your Kubernetes control-plane has initialized successfully!

To start using your cluster, you need to run the following as a regular user:

  mkdir -p $HOME/.kube
  sudo cp -i /etc/kubernetes/admin.conf $HOME/.kube/config     <--- 1
  sudo chown $(id -u):$(id -g) $HOME/.kube/config

Alternatively, if you are the root user, you can run:

  export KUBECONFIG=/etc/kubernetes/admin.conf

You should now deploy a pod network to the cluster.
Run "kubectl apply -f [podnetwork].yaml" with one of the options listed at:
  https://kubernetes.io/docs/concepts/cluster-administration/addons/

Then you can join any number of worker nodes by running the following on each as root:
                                                                              2
kubeadm join 172.16.191.6:6443 --token abcdef.0123456789abcdef \
    --discovery-token-ca-cert-hash sha256:bf5d3a2b9526e98f7403ec4a07cf052b961b3201913c040cd4e
6e28f8818d8c2
```

Figure 14.15 – Successfully bootstrapping the Kubernetes cluster

At this point, our Kubernetes Control Plane node is up and running. In the output, we highlighted the relevant excerpts for the following commands:

- Configuring the current user as the Kubernetes cluster administrator (**1**)
- Joining new nodes to the Kubernetes cluster (**2**)

We recommend taking the time to go over the complete output and identify the related information for each of the kubeadm init tasks, as captured in the *Introducing kubeadm* section earlier in this chapter.

Next, to configure the current user as the Kubernetes cluster administrator, we run the following commands:

```
mkdir -p ~/.kube
sudo cp -i /etc/kubernetes/admin.conf ~/.kube/config
sudo chown $(id -u):$(id -g) ~/.kube/config
```

With our cluster up and running, let's deploy the *Calico* networking manifest to create the Pod network:

```
kubectl apply -f calico.yaml
```

The preceding command creates a collection of resources related to the Pod overlay network. Now, we're ready to take our first peek into the state of our cluster by using the kubectl command to list all the Pods in the system:

```
kubectl get pods --all-namespaces
```

The command yields the following output:

```
packt@k8s-cp1:~$ kubectl get pods --all-namespaces
NAMESPACE     NAME                                         READY   STATUS    RESTARTS   AGE
kube-system   calico-kube-controllers-69496d8b75-vqhhm     1/1     Running   0          6m44s
kube-system   calico-node-l7tz6                            1/1     Running   0          6m44s
kube-system   coredns-74ff55c5b-562j5                      1/1     Running   0          46m
kube-system   coredns-74ff55c5b-nffjj                      1/1     Running   0          46m
kube-system   etcd-k8s-cp1                                 1/1     Running   0          46m
kube-system   kube-apiserver-k8s-cp1                       1/1     Running   0          46m
kube-system   kube-controller-manager-k8s-cp1             1/1     Running   0          46m
kube-system   kube-proxy-7tdx2                             1/1     Running   0          46m
kube-system   kube-scheduler-k8s-cp1                       1/1     Running   0          46m
```

Figure 14.16 – Retrieving the Pods in the Kubernetes cluster

The `--all-namespaces` option retrieves the Pods across all resource groups in the cluster. Kubernetes uses *namespaces* to organize resources. For now, the only Pods running in our cluster are *system Pods*, as we haven't deployed any user Pods yet.

The following command retrieves the current nodes in the cluster:

```
kubectl get nodes
```

The output shows `k8s-cp1` as the only node configured in the Kubernetes cluster, running as Control Plane node:

```
packt@k8s-cp1:~$ kubectl get nodes
NAME       STATUS    ROLES                   AGE    VERSION
k8s-cp1    Ready     control-plane,master    71m    v1.20.4
```

Figure 14.17 – Listing the current nodes in the Kubernetes cluster

You may recall that prior to bootstrapping the Kubernetes cluster, the `kubelet` service was continually crashing (and attempting to restart). With the cluster up and running, the status of the `kubelet` daemon should be *active* and *running*:

```
sudo systemctl status kubelet
```

The output shows the following:

```
packt@k8s-cp1:~$ sudo systemctl status kubelet
● kubelet.service - kubelet: The Kubernetes Node Agent
     Loaded: loaded (/lib/systemd/system/kubelet.service; enabled; vendor preset: enabled)
    Drop-In: /etc/systemd/system/kubelet.service.d
             └─10-kubeadm.conf
     Active: active (running) since Sun 2021-03-14 11:52:38 UTC; 1h 20min ago
       Docs: https://kubernetes.io/docs/home/
   Main PID: 178172 (kubelet)
      Tasks: 15 (limit: 2248)
     Memory: 53.9M
     CGroup: /system.slice/kubelet.service
             └─178172 /usr/bin/kubelet --bootstrap-kubeconfig=/etc/kubernetes/bootstrap-kube
```

Figure 14.18 – A healthy kubelet in the cluster

We encourage you to check out the manifests created in the `/etc/kubernetes/manifests/` directory for each cluster component:

```
ls /etc/kubernetes/manifests/
```

The output shows the configuration files describing the static (system) Pods, corresponding to the *API server*, *Controller Manager*, *Scheduler*, and `etcd`:

```
packt@k8s-cp1:~$ ls /etc/kubernetes/manifests/
etcd.yaml   kube-apiserver.yaml   kube-controller-manager.yaml   kube-scheduler.yaml
```

Figure 14.19 – The static Pod configuration files in /etc/kubernetes/manifests/

You may also look at the `kubeconfig` files in `/etc/kubernetes`:

```
ls /etc/kubernetes/
```

As you may recall from the *Introducing kubeadm* section earlier in this chapter, the `kubeconfig` files are used by the cluster components to communicate and authenticate with the API server.

Next, let's add the worker nodes to our Kubernetes cluster.

Joining a node to a Kubernetes cluster

As previously noted, before adding a node to the Kubernetes cluster, you'll need to run the preliminary steps described in the *Preparing the lab environment* section earlier in this chapter.

To join a node to the cluster, we'll need both the *bootstrap token* and the *discovery token CA certificate hash* generated upon successful bootstrapping of the Kubernetes cluster. The tokens with the related `kubadm join` command were provided in the output at the end of the bootstrapping process with `kubadm init`. Refer to the *Creating a Kubernetes Control plane node* section earlier in this chapter.

Keep in mind that the bootstrap token expires in 24 hours. In case you forgot to copy the command, you can retrieve the related information by running the following commands in the Control Plane node's terminal (on `k8s-cp1`):

- Retrieve the current bootstrap tokens:

    ```
    kubeadm token list
    ```

    The output shows our token (`abcdef.0123456789abcdef`):

```
packt@k8s-cp1:~$ kubeadm token list
TOKEN                      TTL    EXPIRES                USAGES
DESCRIPTION                                              EXTRA GROUPS
abcdef.0123456789abcdef    13h    2021-03-15T11:52:38Z   authentication,signing
<none>                                                   system:bootstrappers:kubead
m:default-node-token
```

Figure 14.20 – Getting the current bootstrap tokens

- Get the CA certificate hash:

```
openssl x509 -pubkey \
    -in /etc/kubernetes/pki/ca.crt | \
    openssl rsa -pubin -outform der 2>/dev/null | \
    openssl dgst -sha256 -hex | sed 's/^.* //'
```

The output is as follows:

```
packt@k8s-cp1:~$ openssl x509 -pubkey -in /etc/kubernetes/pki/ca.crt | \
>     openssl rsa -pubin -outform der 2>/dev/null | \
>     openssl dgst -sha256 -hex | sed 's/^.* //'
bf5d3a2b9526e98f7403ec4a07cf052b961b3201913c040cd4e6e28f8818d8c2
```

Figure 14.21 – Getting the CA certificate hash

You may also generate a new bootstrap token with the following command:

```
kubeadm token create
```

If you choose to generate a new token, you may use the following streamlined command to print out the full `kubeadm join` command with the required parameters:

```
kubeadm token create --print-join-command
```

In the following steps, we'll use our initial tokens as displayed in the output at the end of the bootstrapping process. So, let's switch to the node's command-line terminal (on `k8s-n1`) and run the following command:

1. Make sure to invoke `sudo`, or the command will fail with insufficient permissions:

```
sudo kubeadm join 172.16.191.6:6443 \
    --token abcdef.0123456789abcdef \
    --discovery-token-ca-cert-hash sha256:bf5d3a2b9526e
98f7403ec4a07cf052b961b3201913c040cd4e6e28f8818d8c2
```

The command usually completes in a few seconds, with the following output (excerpt):

```
This node has joined the cluster:
* Certificate signing request was sent to apiserver and a response was received.
* The Kubelet was informed of the new secure connection details.

Run 'kubectl get nodes' on the control-plane to see this node join the cluster.
```

Figure 14.22 – Joining a node to the cluster

2. As the output suggests, we can check the status of the current nodes in the cluster with the following command in the Control Plane node terminal (`k8s-cp1`):

```
kubectl get nodes
```

The output shows our new node (`k8s-n1`) added to the cluster:

```
packt@k8s-cp1:~$ kubectl get nodes
NAME       STATUS   ROLES                   AGE   VERSION
k8s-cp1    Ready    control-plane,master    11h   v1.20.4
k8s-n1     Ready    <none>                  44s   v1.20.4
```

Figure 14.23 – The new node (k8s-n1) added to the cluster

3. We encourage you to repeat the process of joining the other two cluster nodes (`k8s-n2` and `k8s-n3`). During the join, while the Control Plane Pods are being deployed on the new node, you may temporarily see a `NotReady` status for the new node if you query the nodes on the Control Plane node (`k8s-cp1`):

```
kubectl get nodes --watch
```

Invoking the `--watch` option would keep refreshing the output:

```
packt@k8s-cp1:~$ kubectl get nodes --watch
NAME       STATUS     ROLES                   AGE   VERSION
k8s-cp1    Ready      control-plane,master    11h   v1.20.4
k8s-n1     Ready      <none>                  14m   v1.20.4
k8s-n2     NotReady   <none>                  12s   v1.20.4
k8s-n2     Ready      <none>                  20s   v1.20.4
```

Figure 14.24 – The transitory NotReady state while joining a node

In the end, we should have all three nodes showing `Ready` in the output of the `kubectl get nodes` command (on the `k8s-cp1` Control Plane node):

```
packt@k8s-cp1:~$ kubectl get nodes
NAME       STATUS   ROLES                   AGE   VERSION
k8s-cp1    Ready    control-plane,master    11h   v1.20.4
k8s-n1     Ready    <none>                  25m   v1.20.4
k8s-n2     Ready    <none>                  10m   v1.20.4
k8s-n3     Ready    <none>                  32s   v1.20.4
```

Figure 14.25 – The Kubernetes cluster with all nodes running

We completed the installation of our Kubernetes cluster, with a Control Plane node and three worker nodes. We used a local (on-premises) VM environment, but the same process would also apply to a hosted IaaS solution running in a private or public cloud.

Next, we'll look at *managed* Kubernetes services as SaaS offerings from major cloud providers.

Running Kubernetes in the cloud

Managed Kubernetes services are fairly common among public cloud providers. Here are some of the major cloud offerings of Kubernetes at the time of this writing:

- Amazon **Elastic Kubernetes Service (EKS)**: `https://docs.aws.amazon.com/eks/index.html`

- **Azure Kubernetes Service (AKS)**: `https://azure.microsoft.com/en-us/services/kubernetes-service/`

- **Google Kubernetes Engine (GKE)**: `https://cloud.google.com/kubernetes-engine`

In this section, we'll focus on Kubernetes deployments with Amazon EKS and Microsoft AKS. The procedure of deploying a managed Kubernetes cluster is essentially the same, regardless of the cloud provider:

- Authenticate with the cloud provider using a CLI or web management console.

- Deploy a Kubernetes cluster.

- Download the cluster `kubeconfig` file.

- Use `kubectl` to authenticate against the API server and interact with the cluster.

If you want to try out the hands-on examples in this section, you need to have an AWS and an Azure account. Both cloud providers provide free subscriptions:

- AWS Free Tier: `https://aws.amazon.com/free`

- Microsoft Azure free account: `https://azure.microsoft.com/en-us/free/`

Let's start by creating a Kubernetes cluster on Amazon EKS.

Deploying Kubernetes on EKS

We assume you have an AWS account and the AWS CLI installed on your local Linux desktop. See the *Working with the AWS CLI* section in *Chapter 13, Deploying to the Cloud with AWS and Azure*, for the related instructions.

In this section, we'll use an RHEL/CentOS system to interact with Amazon EKS. The operating system or Linux distribution is irrelevant, given the **AWS** and `kubectl` CLI abstractions for communicating with the cluster, and the same steps would apply to any other platform.

> **Important note**
> Before deploying a Kubernetes cluster with EKS, we need to assign the **Amazon EKS Cluster IAM role** to our AWS account. Please follow the related instructions at `https://docs.aws.amazon.com/eks/latest/userguide/service_IAM_role.html#create-service-role`.

With the required EKS Cluster IAM role enabled, we can deploy a Kubernetes cluster in EKS using either of the following methods:

- **AWS web console**
- **AWS CLI**
- `eksctl`

The simplest way to create an EKS cluster is to use the `eksctl` command-line utility – the official CLI of Amazon EKS: `https://docs.aws.amazon.com/eks/latest/userguide/eksctl.html`.

Let's now download and install `eksctl` on our Linux system:

```
curl --silent --location "https://github.com/weaveworks/eksctl/releases/latest/download/eksctl_$(uname -s)_amd64.tar.gz" | tar xz -C /tmp
sudo mv /tmp/eksctl /usr/local/bin/
```

You can explore the capabilities of `eksctl` by invoking the related help:

```
eksctl help
```

`eksctl` specializes in the deployment and teardown of Kubernetes clusters on Amazon EKS while automating the related cloud resource management tasks in AWS. `eksctl` is not a Kubernetes cluster management tool in the sense that `kubectl` is. The two CLI utilities at best complement each other in the end-to-end cluster management workloads.

Next, we install `kubectl` on our system, by observing the Linux installation instructions at `https://kubernetes.io/docs/tasks/tools/install-kubectl-linux/`:

```
curl -LO "https://dl.k8s.io/release/$(curl -L -s https://
dl.k8s.io/release/stable.txt)/bin/linux/amd64/kubectl"
sudo install -o root -g root -m 0755 kubectl /usr/local/bin/
kubectl
```

If `kubectl` is not installed, `eksctl` would still work but would warn that our cluster management capabilities are limited.

To create a Kubernetes cluster on EKS, we run the following command:

```
eksctl create cluster \
    --name k8s-packt \
    --nodes 2 \
    --node-volume-size 20 \
    --region us-west-2 \
    --managed
```

Let's briefly explain each part of the preceding command:

- `eksctl create cluster`: Creates a Kubernetes cluster on Amazon EKS
- `--name k8s-packt`: The name of the cluster (`k8s-packt`)
- `--nodes 2`: The number of nodes in the cluster (the default value is 2)
- `--node-volume-size 20`: The volume size of a node in GB (the default size is 80 GB)
- `--region us-west-2`: The geographical location of the cluster deployment
- `--managed`: Creates an EKS-managed node group in the cluster (for automatic updates, and suchlike)

The command may take up to 15 minutes to complete, sometimes even more. It will create and deploy all the required resources for the cluster, including the **Virtual Private Cloud** (**VPC**), security groups, networks, **Elastic IP** (**EIP**), and the cluster node EC2 instances.

When finished, the command output will show some relevant information regarding our cluster, including the status of the cluster nodes and the location of the `kubeconfig` file:

```
waiting for CloudFormation stack "eksctl-k8s-packt-nodegroup-ng-4ed82a9d"
waiting for the control plane availability...
saved kubeconfig as "/home/packt/.kube/config"
no tasks
all EKS cluster resources for "k8s-packt" have been created
nodegroup "ng-4ed82a9d" has 2 node(s)
node "ip-192-168-11-249.us-west-2.compute.internal" is ready
node "ip-192-168-49-141.us-west-2.compute.internal" is ready
waiting for at least 2 node(s) to become ready in "ng-4ed82a9d"
nodegroup "ng-4ed82a9d" has 2 node(s)
node "ip-192-168-11-249.us-west-2.compute.internal" is ready
node "ip-192-168-49-141.us-west-2.compute.internal" is ready
kubectl command should work with "/home/packt/.kube/config", try 'kubectl get nodes'
EKS cluster "k8s-packt" in "us-west-2" region is ready
```

Figure 14.26 – Creating the Kubernetes cluster on EKS

We can immediately use `kubectl` to interact with our EKS cluster. Make sure that our current `kubeconfig` context is set on the EKS cluster:

```
kubectl config get-contexts
```

The output shows an asterisk (*) for the current `kubeconfig` context, marking our EKS cluster (`k8s-packt.us-west-2.eksctl.io`):

```
[packt@jupiter ~]$ kubectl config get-contexts
CURRENT   NAME                                                    CLUSTER
*  ←      iam-root-account@k8s-packt.us-west-2.eksctl.io          k8s-packt.us-west-2.eksctl.io
```

Figure 14.27 – Checking the current kubeconfig context

The following command is useful when managing multiple Kubernetes clusters from the same `kubectl` terminal. To switch the `kubeconfig` context to our EKS cluster, we run the following command:

```
kubectl config use-context \
    iam-root-account@k8s-packt.us-west-2.eksctl.io
```

Let's take a look at our cluster nodes next:

```
kubectl get nodes
```

The output shows the two worker nodes, while the Control Plane node is hidden from view. That's because the Control Plane is fully managed on EKS:

```
[packt@jupiter ~]$ kubectl get nodes
NAME                                            STATUS   ROLES    AGE   VERSION
ip-192-168-11-249.us-west-2.compute.internal    Ready    <none>   89s   v1.18.9-eks-d1db3c
ip-192-168-49-141.us-west-2.compute.internal    Ready    <none>   90s   v1.18.9-eks-d1db3c
```

Figure 14.28 – Retrieving the cluster nodes on EKS

The following command lists all the Pods in our EKS cluster:

```
kubectl get pods --all-namespaces
```

The output only shows the add-on Pods, as no user Pods are yet deployed. The Control Plane Pods are hidden from view:

```
[packt@jupiter ~]$ kubectl get pods --all-namespaces
NAMESPACE       NAME                         READY   STATUS    RESTARTS   AGE
kube-system     aws-node-dzk5g               1/1     Running   0          2m7s
kube-system     aws-node-zj78h               1/1     Running   0          2m8s
kube-system     coredns-559b5db75d-52fjr     1/1     Running   0          7m9s
kube-system     coredns-559b5db75d-t8gnv     1/1     Running   0          7m9s
kube-system     kube-proxy-d22zl             1/1     Running   0          2m8s
kube-system     kube-proxy-fxggv             1/1     Running   0          2m7s
```

Figure 14.29 – Retrieving the cluster Pods on EKS

To delete an EKS cluster, we invoke `eksctl delete cluster` with the cluster name:

```
eksctl delete cluster --name k8s-packt
```

The command will remove our Kubernetes cluster from EKS and free up all the related cloud resources.

(We won't delete the cluster just yet.)

We can also manage our cluster in the **EKS Clusters** view of the AWS console:

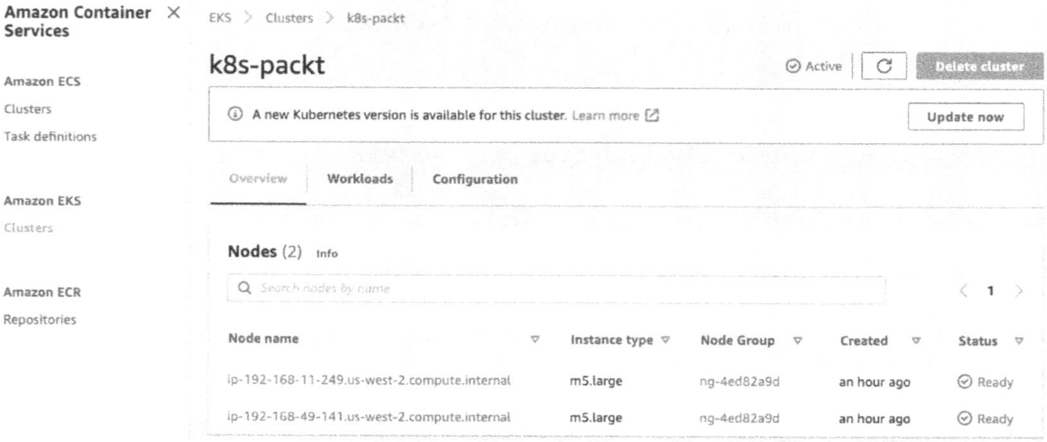

Figure 14.30 – Managing the EKS cluster in the AWS console

Before wrapping up this section, let's look at updating `kubeconfig` with the EKS cluster information. Perhaps we created the EKS cluster on a different machine and would like to consolidate all cluster management under the local `kubectl` environment. Or we may want to use a different user account or **Identity and Access Management (IAM)** role for cluster management. The following section provides a quick and easy way to accomplish this task.

Connecting to an EKS cluster

Suppose we created an EKS cluster using an older system and would like to use our brand-new laptop for cluster management. Let's name the old system *Jupyter* and the new laptop *Neptune*. On Neptune, we already manage our Kubernetes cluster on AKS (which we'll look at in the next section). It would be nice to have both cluster management endpoints on Neptune's `kubeconfig` while choosing between the EKS and AKS cluster contexts.

We'll start by installing the AWS CLI on Neptune. Next, we configure the local AWS environment with the following command:

```
aws configure
```

The related process has been described in the *Working with the AWS CLI* section of *Chapter 13, Deploying to the Cloud with AWS and Azure.* If you are using a different AWS account from the one in the Jupyter environment, make sure that it has the Amazon EKS Cluster role enabled and has access to the cluster resources.

We're now ready to update the local `kubeconfig` on Neptune with the EKS cluster context. Assuming the EKS cluster we created earlier (`k8s-packt`), the command is as follows:

```
aws eks update-kubeconfig \
    --name k8s-packt \
    --region us-west-2
```

Let's check the current `kubeconfig` context:

```
kubectl config get-contexts
```

The output shows that the local `kubeconfig` file includes our Amazon EKS cluster, and that it's also the current (default) context:

```
packt@neptune:~$ kubectl config get-contexts
CURRENT   NAME                                                        CLUSTER
*   ━━━▶  arn:aws:eks:us-west-2:106842557074:cluster/k8s-packt        arn:aws:eks:us-west-2
          k8s-packt-aks                                               k8s-packt
```

Figure 14.31 – Updating kubeconfig with the EKS cluster context

Now, we can manage both clusters on EKS and AKS in the same `kubectl` environment. The EKS cluster's context name bears the **Amazon Resource Name** (**ARN**) of the cluster. Let's rename it to something more user-friendly (`k8s-packt-eks`):

```
kubectl config rename-context \
    arn:aws:eks:us-west-2:106842557074:cluster/k8s-packt \
    k8s-packt-eks
```

If we query the local `kubeconfig` file, the output shows the updated EKS cluster context:

```
packt@neptune:~$ kubectl config get-contexts
CURRENT    NAME             CLUSTER
           k8s-packt-aks    k8s-packt
*  ──────> k8s-packt-eks    arn:aws:eks:us-west-2:106842557074:cluster/k8s-packt
```

Figure 14.32 – Renaming the EKS cluster context

In our would-be scenario, we've consolidated the `kubeconfig` environment on a new machine (Neptune) for managing the EKS and AKS clusters. We assumed that Neptune already had the AKS cluster management configured. Let's turn back in time and create a Kubernetes cluster on AKS.

Deploying Kubernetes on AKS

In this section, we assume you have an Azure account and the Azure CLI installed on your local Linux desktop. Refer to the *Working with the Azure CLI* section in *Chapter 13, Deploying to the Cloud with AWS and Azure*, for the related instructions.

We'll use an Ubuntu machine with the Azure CLI installed and configured with a free (trial) Azure account. At the terminal, begin by logging in to your Azure account:

```
az login
```

Follow the authentication steps as described in the preceding section. Following a successful login, we'll proceed by creating a resource group (`k8s-packt`) for our Kubernetes service:

```
az group create --name "k8s-packt" --location westus
```

We specified our local geographical location (`westus`) for the resource. You may use your region of choice. If the resource group is created successfully, we should see `"provisioningState": "Succeeded"` in the command output:

```
packt@neptune:~$ az group create --name "k8s-packt" --location westus
{
  "id": "/subscriptions/138189e1-26c3-49f1-a2d9-2caa56d946c7/resourceGroups/k8s-packt",
  "location": "westus",
  "managedBy": null,
  "name": "k8s-packt",
  "properties": {
    "provisioningState": "Succeeded"
  },
  "tags": null,
  "type": "Microsoft.Resources/resourceGroups"
}
```

Figure 14.33 – Creating a resource on the Kubernetes cluster

Next, we query the Kubernetes versions available for our location:

```
az aks get-versions --location westus -o table
```

The output shows the Kubernetes versions available, and the respective upgrades supported by Azure:

```
packt@neptune:~$ az aks get-versions --location westus -o table
KubernetesVersion    Upgrades
-------------------  -------------------------
1.20.2(preview)      None available
1.19.7               1.20.2(preview)
1.19.6               1.19.7, 1.20.2(preview)
1.18.14              1.19.6, 1.19.7
1.18.10              1.18.14, 1.19.6, 1.19.7
1.17.16              1.18.10, 1.18.14
1.17.13              1.17.16, 1.18.10, 1.18.14
```

Figure 14.34 – Getting the Kubernetes versions for our location

We can choose a specific version for our Kubernetes cluster or let Azure pick the default value. The following command creates a cluster named k8s-packt, with two nodes using the default Kubernetes version:

```
az aks create \
    --resource-group "k8s-packt" \
    --name "k8s-packt" \
    --generate-ssh-keys \
    --node-count 2
```

Here's a brief explanation of the command and related parameters:

- `az aks create`: Creates a Kubernetes cluster in AKS
- `--resource-group`: The resource group the Kubernetes cluster belongs to (`k8s-packt`)
- `--name`: The name of the Kubernetes cluster (`k8s-packt`)
- `--generate-ssh-keys`: Creates an SSH key pair for accessing the cluster nodes
- `--node-count`: The number of nodes in the cluster (2)

Optionally, we could specify the Kubernetes version with the `--kubernetes-version` parameter.

The command may take a few minutes to complete. We should see `"provisioningState"`: `"Succeeded"` in the related JSON output with a successful deployment. Among other relevant data, we can also see the Kubernetes version of our cluster (`"kubernetesVersion"`: `"1.18.14"`).

The following steps require the `kubectl` command-line utility. We assume you have it installed on your system. If not, you can either follow the instructions at `https://kubernetes.io/docs/tasks/tools/#kubectl` or use the Azure CLI to install `kubectl`:

```
az aks install-cli
```

`kubectl` requires access to the API server endpoint of our AKS cluster. To make this possible, we have to download the related `kubeconfig` to our local system. The following command retrieves the cluster credentials and merges the remote `kubeconfig` into our local environment. `kubectl` will then use the local `kubeconfig` to connect to the remote AKS cluster using certificate-based user authentication:

```
az aks get-credentials \
    --resource-group "k8s-packt" \
    --name "k8s-packt"
```

The command output suggests that the remote `kubeconfig` of the `k8s-packt` cluster has been merged with the current user's configuration in `/home/packt/.kube/config`:

```
packt@neptune:~$ az aks get-credentials \
>       --resource-group "k8s-packt" \
>       --name "k8s-packt"
Merged "k8s-packt" as current context in /home/packt/.kube/config
```

Figure 14.35 – Merging the remote kubeconfig with the local environment

Now, let's retrieve the current user's `kubeconfig` contexts:

```
kubectl config get-contexts
```

As expected, the output shows `k8s-packt` as the only Kubernetes cluster we manage. The asterisk (*) marks it as the current `kubeconfig` context:

```
packt@neptune:~$ kubectl config get-contexts
CURRENT    NAME         CLUSTER      AUTHINFO                          NAMESPACE
*  ⟵       k8s-packt    k8s-packt    clusterUser_k8s-packt_k8s-packt
```

Figure 14.36 – Retrieving the current kubeconfig contexts

In case we are managing multiple Kubernetes clusters, the preceding list would show various entries. With multiple configurations to choose from, we can switch the current `kubeconfig` context to the `k8s-packt` cluster with the following command:

```
kubectl config use-context k8s-packt
```

Now, all `kubectl` commands are directed to the `k8s-packt` cluster in AKS. Let's retrieve the cluster nodes:

```
kubectl get nodes
```

The output shows the two worker nodes of our AKS cluster:

```
packt@neptune:~$ kubectl get nodes
NAME                                STATUS   ROLES   AGE    VERSION
aks-nodepool1-88379849-vmss000000   Ready    agent   135m   v1.18.14
aks-nodepool1-88379849-vmss000001   Ready    agent   135m   v1.18.14
```

Figure 14.37 – Getting the nodes of the AKS cluster

We should note that the Control Plane node is not listed. That's because Azure hides the related information for the apparent reason that the Control Plane is exclusively managed in the cloud, and we should not tamper with it. The same is true for Pods. With no user Pods deployed yet, the following command would only show the add-on Pods of our AKS cluster:

```
kubectl get pods --all-namespaces
```

There are no Control Plane Pods listed in the output:

```
packt@neptune:~$ kubectl get pods --all-namespaces
NAMESPACE     NAME                                    READY   STATUS    RESTARTS   AGE
kube-system   coredns-748cdb7bf4-5l9pg                1/1     Running   0          145m
kube-system   coredns-748cdb7bf4-8gqst                1/1     Running   0          145m
kube-system   coredns-autoscaler-868b684fd4-ws2zk     1/1     Running   0          145m
kube-system   kube-proxy-741c2                        1/1     Running   0          145m
kube-system   kube-proxy-c85b6                        1/1     Running   0          145m
kube-system   metrics-server-58fdc875d5-sqzxt         1/1     Running   0          145m
kube-system   tunnelfront-7948cfb9c-ctxcn             1/1     Running   0          145m
```

Figure 14.38 – Getting the Pods of the AKS cluster

You may compare the preceding output with the corresponding results of our local cluster described in the *Installing Kubernetes on virtual machines* section earlier in this chapter.

To delete the AKS cluster (`k8s-packt`) and free up the related cloud resources, we can run the following command:

```
az aks delete \
    --resource-group "k8s-packt" \
    --name "k8s-packt"
```

(We won't delete the cluster just yet.)

So far, we have exclusively used the Azure CLI and `kubectl` to interact with AKS. We can also use the Azure web portal to manage our `k8s-packt` cluster in the **Kubernetes services** view:

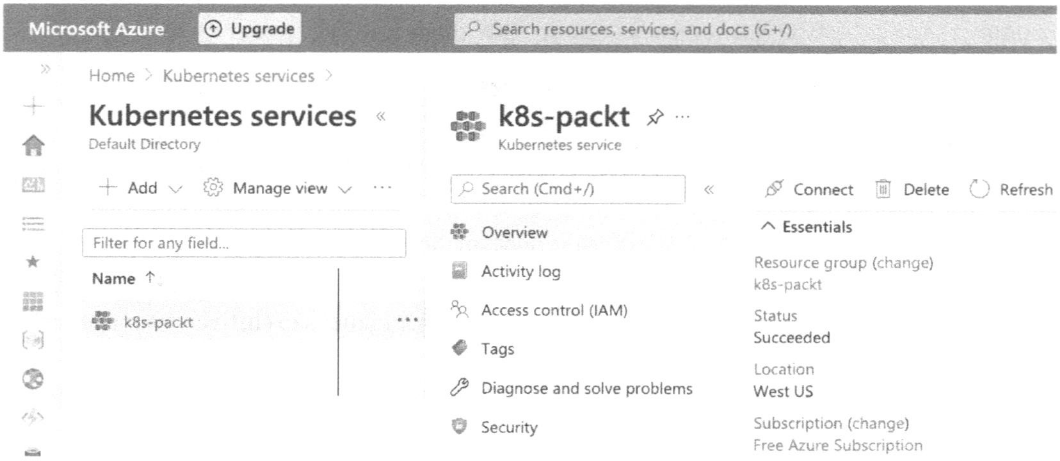

Figure 14.39 – Managing the AKS cluster in the Azure portal

Now, let's consider an imaginary scenario where we'd like to take our current AKS cluster management context to a different machine to manage multiple Kubernetes clusters, possibly running in other clouds or on-premises.

We'll refer to the example of Jupyter and Neptune machines from the *Connecting to an EKS cluster* section earlier in this chapter. Suppose we initially created the AKS cluster using Neptune, but plan to use Jupyter as our consolidated cluster management terminal. Jupyter already has an EKS cluster configured. Let's take a look at how to connect to the AKS cluster on Jupyter and manage both Kubernetes environments in EKS and AKS.

Connecting to an AKS cluster

First, we need to download and install the Azure CLI on Jupyter. With the Azure CLI in place, we use `az login` to authenticate with Azure. We described the related steps in the *Working with the Azure CLI* section in *Chapter 13*, *Deploying to the Cloudwith AWS and Azure*.

Next, we retrieve the access credentials of our AKS cluster (`k8s-packt`):

```
az aks get-credentials \
   --name k8s-packt \
   --resource-group k8s-packt
```

The preceding command merges the AKS cluster's configuration with our local `kubeconfig`:

```
kubectl config get-contexts
```

The output shows both clusters, `k8s-packt` in AKS and `k8s-packt-eks` in EKS. The AKS cluster context is currently active, as indicated by the asterisk (`*`):

```
[packt@jupiter ~]$ kubectl config get-contexts
CURRENT   NAME            CLUSTER                        AUTHINFO
*  ──▶    k8s-packt       k8s-packt                      clusterUser_k8s-packt_k8s-packt
          k8s-packt-eks   k8s-packt.us-west-2.eksctl.io  iam-root-account@k8s-packt.us-west-2.eksctl.io
```

Figure 14.40 – Updating kubeconfig with the AKS cluster context

For consistency, we rename the AKS cluster context to `k8s-packt-aks` to better suggest the cloud environment:

```
kubectl config rename-context k8s-packt k8s-packt-aks
```

The updated `kubeconfig` context yields the following output:

```
[packt@jupiter ~]$ kubectl config get-contexts
CURRENT   NAME            CLUSTER                      AUTHINFO
*         k8s-packt-aks   k8s-packt                    clusterUser_k8s-packt_k8s-packt
          k8s-packt-eks   k8s-packt.us-west-2.eksctl.io   iam-root-account@k8s-packt.us-west-2.eksctl.io
```

Figure 14.41 – The AKS and EKS cluster contexts side by side

With our relatively brief coverage of EKS and AKS, we should note that we just scratched the surface of deploying and managing Kubernetes clusters in the cloud. Yet, here we are, at a significant milestone, where we deployed our first Kubernetes clusters in the cloud and on-premises. You became familiar with the Kubernetes cluster architecture and its major system components. We introduced the `kubeadm` and `kubectl` CLI tools to interact with the cluster. Now, it's time to put this knowledge to some good use and look closer to working with Kubernetes.

In the next section, we'll explore the `kubectl` CLI to a certain depth and use it to create and manage Kubernetes resources. Then, we'll look at deploying and scaling applications using the imperative and declarative deployment models in Kubernetes.

Working with Kubernetes

In this section, we'll use real-world examples of interacting with a Kubernetes cluster. Since we'll be using the `kubectl` CLI to a considerable extent, we're going to deep dive into some of its more common usage patterns. Then, we turn our focus to deploying applications to a Kubernetes cluster. We'll be using the on-premises environment we built in the *Installing Kubernetes on virtual machines* section.

Let's start by taking a closer look at `kubectl` and its usage.

Using kubectl

`kubectl` is the primary tool for managing a Kubernetes cluster and its resources. `kubectl` communicates with the cluster's API server endpoint using the Kubernetes REST API. The general syntax of the `kubectl` command is as follows:

```
kubectl [command] [TYPE] [NAME] [flags]
```

In general, `kubectl` commands execute **CRUD operations** – *Create*, *Read*, *Update*, and *Delete* – against Kubernetes *resources*, such as Pods, deployments, and Services.

One of the essential features of kubectl is the command output format, either in YAML, JSON, or plain text. The output format is handy when creating or editing application deployment manifests. We can capture the YAML output of a kubectl command (such as create a resource) to a file. Later, we can reuse the manifest file to perform the same operation (or sequence of operations) in a *declarative* way. This brings us to the two basic deployment paradigms of Kubernetes:

- **Imperative deployments**: Invoking a single or multiple kubectl commands to operate on specific resources
- **Declarative deployments**: Using *manifest files* and *deploying* them using the kubectl apply command, usually targeting a set of resources with a single invocation

We'll look at these two deployment models more closely in the *Deploying applications* section later in this chapter. For now, let's get back to exploring the kubectl command further.

Here's a shortlist with some of the most common kubectl commands:

- create, apply: Creates resources imperatively/declaratively
- get: Reads resources
- edit, set: Updates resources or specific features of objects
- delete: Deletes resources
- run: Starts a Pod
- exec: Executes a command in a Pod container
- describe: Displays detailed information about resources
- explain: Provides resource-related documentation
- logs: Shows the logs in Pod containers

A couple of frequently used option parameters of the kubectl command are also worth mentioning:

- --dry-run: Runs the command without modifying the system state while still providing the output as if it executed normally
- --output: Specifies various formats for the command output: yaml, json, and wide (additional information in plain text)

In the following sections, we'll look at multiple examples of using the `kubectl` command. Always keep in mind the general pattern of the command:

	operation	resource		options
kubectl	[command]	[TYPE]	[NAME]	[flags]
-------	---------	------	------	-------
kubectl	get	pods	packt-web	--output=wide
kubectl	create	service	packt-svc	--dry-run

Figure 14.42 – The general usage pattern of kubectl

We recommend that you check out the complete `kubectl` command reference at `https://kubernetes.io/docs/reference/kubectl/overview/`. While you are becoming proficient with `kubectl`, you may also want to keep the related cheat sheet to hand: `https://kubernetes.io/docs/reference/kubectl/cheatsheet/`.

Now, let's prepare our `kubectl` environment to interact with the Kubernetes cluster we built earlier with virtual machines. You may skip the next section if you prefer to use `kubectl` on the Control Plane node.

Connecting to a Kubernetes cluster

In this section, we configure the `kubectl` CLI running locally on our Linux desktop to control a remote Kubernetes cluster. As you may recall, our consolidated `kubeconfig` also includes the EKS and AKS cluster contexts. We want to add (merge) yet another cluster configuration to our environment. This time, we connect to an on-premises Kubernetes Control Plane, and we'll use `kubectl` to update the `kubeconfig`.

Here are the steps we'll be taking:

- Make a backup of the current `kubeconfig`
- Copy the remote `kubeconfig` to a temporary location
- Merge the current and new `kubeconfig` to a temporary file using `kubectl`
- Replace the existing `kubeconfig` with the updated file
- Clean up the temporary files

Let's begin by making a backup of the current `kubeconfig` on the local Linux environment:

```
cp ~/.kube/config ~/.kube/config.old
```

Next, we copy the `kubeconfig` from the Control Plane node (`k8s-cp1`, `172.16.191.6`) to a temporary location (`/tmp/config.cp`):

```
scp packt@172.16.191.6:~/.kube/config /tmp/config.cp
```

Now, let's merge the current `kubeconfig` (`.kube/config`) with the one we just copied (`/tmp/config.cp`) to a temporary file (`/tmp/config.new`):

```
KUBECONFIG=~/.kube/config:/tmp/config.cp \
    kubectl config view --flatten > /tmp/config.new
```

Finally, we replace the current `kubeconfig` with the new one:

```
mv /tmp/config.new ~/.kube/config
```

Optionally, we can clean up the temporary files created in the process:

```
rm ~/.kube/config.old /tmp/config.cp
```

Let's get a view of the current `kubeconfig` contexts:

```
kubectl config get-contexts
```

The output shows our new Kubernetes cluster, with the related security principal (`kubernetes-admin@kubernetes`) and cluster name (`kubernetes`):

```
packt@neptune:~$ kubectl config get-contexts
CURRENT   NAME                           CLUSTER
          k8s-packt-aks                  k8s-packt
*         k8s-packt-eks                  arn:aws:eks:us-west-2:106842557074:cluster/k8s-packt
  ──────► kubernetes-admin@kubernetes    kubernetes
```

Figure 14.43 – The new kubeconfig contexts, including the on-premises Kubernetes cluster

We can also see that the current context is set to our AWS EKS cluster (`k8s-packt-eks`). For consistency, let's change the on-premises cluster's context name to `k8s-packt` and make it the default context in our `kubectl` environment:

```
kubectl config rename-context \
    kubernetes-admin@kubernetes \
    k8s-packt
kubectl config use-context k8s-packt
```

The current `kubectl` context becomes `k8s-packt`, and we're now interacting with our on-premises Kubernetes cluster (`kubernetes`):

```
packt@neptune:~$ kubectl config get-contexts
CURRENT   NAME              CLUSTER
*  ──────▶ k8s-packt        kubernetes
          k8s-packt-aks     k8s-packt
          k8s-packt-eks     arn:aws:eks:us-west-2:106842557074:cluster/k8s-packt
```

Figure 14.44 – The current context set to the on-premises Kubernetes cluster

Next, we look at some of the most common `kubectl` commands used with everyday Kubernetes administration tasks.

Working with kubectl

One of the first commands we run when connected to a Kubernetes cluster is the following:

```
kubectl cluster-info
```

The command shows the IP address and port of the API server listening on the Control Plane node, among other information:

```
Kubernetes control plane is running at
https://172.16.191.6:6443
```

The `cluster-info` command can also help to debug and diagnose cluster-related issues:

```
kubectl cluster-info dump
```

To get a detailed view of the cluster nodes, we run the following command:

```
kubectl get nodes --output=wide
```

The `--output=wide` (or `-o wide`) flag yields detailed information about cluster nodes. The output in the following illustration is cropped due to space constraints:

```
packt@neptune:~$ kubectl get nodes -o wide
NAME      STATUS   ROLES                  AGE     VERSION   INTERNAL-IP
k8s-cp1   Ready    control-plane,master   5d11h   v1.20.4   172.16.191.6
k8s-n1    Ready    <none>                 4d23h   v1.20.4   172.16.191.8
k8s-n2    Ready    <none>                 4d23h   v1.20.4   172.16.191.9
k8s-n3    Ready    <none>                 4d23h   v1.20.4   172.16.191.10
```

Figure 14.45 – Getting detailed information about cluster nodes (cropped)

The following command retrieves the Pods running in the default namespace:

```
kubectl get pods
```

As of now, we don't have any user Pods running, and the command returns the following:

```
No resources found in default namespace.
```

To list all the Pods, we append the --all-namespaces flag to the preceding command:

```
kubectl --get pods --all-namespace
```

The output shows all Pods running in the system. Since these are exclusively system Pods, they are associated with the kube-system namespace:

```
packt@neptune:~$ kubectl get pods --all-namespaces
NAMESPACE      NAME                                         READY   STATUS    RESTARTS   AGE
kube-system    calico-kube-controllers-69496d8b75-vqhhm     1/1     Running   0          5d10h
kube-system    calico-node-4bqh7                            1/1     Running   0          4d23h
kube-system    calico-node-8vpqf                            1/1     Running   0          4d23h
kube-system    calico-node-l7tz6                            1/1     Running   0          5d10h
kube-system    calico-node-vsqg4                            1/1     Running   0          5d
kube-system    coredns-74ff55c5b-562j5                      1/1     Running   0          5d11h
kube-system    coredns-74ff55c5b-nffjj                      1/1     Running   0          5d11h
kube-system    etcd-k8s-cp1                                 1/1     Running   0          5d11h
kube-system    kube-apiserver-k8s-cp1                       1/1     Running   5          5d11h
kube-system    kube-controller-manager-k8s-cp1              1/1     Running   7          5d11h
kube-system    kube-proxy-7tdx2                             1/1     Running   0          5d11h
kube-system    kube-proxy-8bdzg                             1/1     Running   0          5d
kube-system    kube-proxy-bnnhc                             1/1     Running   0          4d23h
kube-system    kube-proxy-gvjl5                             1/1     Running   0          4d23h
kube-system    kube-scheduler-k8s-cp1                       1/1     Running   6          5d11h
```

Figure 14.46 – Getting all Pods in the system

We would get the same output if we specified kube-system with the --namespace flag:

```
kubectl get pods --namespace kube-system
```

For a comprehensive view of all resources running in the system, we run the following command:

```
kubectl get all --all-namespaces
```

So far, we have only mentioned some of the more common object types, such as nodes, Pods, and Services. There are many others, and we can view them with the following command:

```
kubectl api-resources
```

The output includes the name of the API object types (such as nodes), their short name or alias (such as no), and whether they can be organized into namespaces (such as false):

```
packt@neptune:~$ kubectl api-resources | more
NAME                      SHORTNAMES      APIVERSION      NAMESPACED
bindings                                  v1              true
componentstatuses         cs              v1              false
configmaps                cm              v1              true
endpoints                 ep              v1              true
events                    ev              v1              true
limitranges               limits          v1              true
namespaces                ns              v1              false
nodes                     no              v1              false
```

Figure 14.47 – Getting all API object types

Suppose you want to find out more about specific API objects, such as nodes. Here's where the explain command comes in handy:

```
kubectl explain nodes
```

The output provides detailed documentation about the nodes API object type, including the related API fields. One of the API fields is spec, describing the implementation details and behavior of an object. You may view the related documentation with the following command:

```
kubectl explain nodes.spec
```

We encourage you to use the explain command to learn about the various Kubernetes API object types in a cluster. Please note that the explain command provides *documentation* about *resource types*. It should not be confused with the describe command, which shows detailed information about the *resources* in the system.

The following commands display cluster node-related information about *all* nodes, and then node k8s-n1 in particular:

```
kubectl describe nodes
kubectl describe nodes k8s-n1
```

For every `kubectl` command, you can invoke `--help` (or `-h`) to get context-specific help. Here are a few examples:

```
kubectl --help
kubectl config -h
kubectl get pods -h
```

The `kubectl` CLI is relatively rich in commands, and becoming proficient with it may take some time. Occasionally, you may find yourself looking for a specific command or remembering its correct spelling or use. The `auto-complete` bash for `kubectl` comes to the rescue. We'll show you how to enable this next.

Enabling kubectl autocompletion

With `kubectl` autocompletion, you'll get context-sensitive suggestions when you hit the *Tab* key twice (*Tab* + *Tab*) while typing `kubectl` commands.

The `kubectl` autocompletion feature depends on **bash-completion**. Most Linux platforms have **bash-completion** enabled by default. Otherwise, you'll have to install the related package manually. On Ubuntu, for example, you install it with the following command:

```
sudo apt-get install -y bash-completion
```

Next, you need to source the `kubectl` autocompletion in your shell (or similar) profile:

```
echo "source <(kubectl completion bash)" >> ~/.bashrc
```

The changes will take effect on your next login to the terminal or immediately if you source the `bash` profile:

```
source ~/.bashrc
```

With the `kubectl` autocomplete active, you'll get context-sensitive suggestions when you hit *Tab* + *Tab* while typing the command. For example, the following sequence provides all the available resources when you try to create one:

```
kubectl create [Tab] [Tab]
```

The `kubectl` autocompletion reaches every part of the syntax: command, resource (type, name), and flags.

Now that we know more about using the `kubectl` command, it's time to turn our attention to deploying applications in Kubernetes.

Deploying applications

When we introduced the `kubectl` command and its usage pattern at the beginning of the *Using kubectl* section, we touched upon the two ways of creating application resources in Kubernetes: *imperative* and *declarative*. As a quick refresher, with imperative deployments, we follow a *sequence* of `kubectl` commands to create the required resources and get to the cluster's desired state, such as running the application. Declarative deployments accomplish the same, usually with a single `kubectl apply` command using a *manifest* file describing multiple resources.

We'll look at both of these models closely in this section while deploying a simple web application. Let's start with the imperative model first.

Working with imperative deployments

Let's begin by creating a *deployment* first. We'll name our deployment `packt`, based on a demo Nginx container we're pulling from the public Docker registry (`docker.io/nginxdemos/hello`):

```
kubectl create deployment packt --image=nginxdemos/hello
```

The command output shows that our deployment was created successfully:

```
deployment.apps/packt created
```

We just created a deployment with a ReplicaSet containing a single Pod running a web server application. We should note that our application is managed by the *Controller Manager* within an app deployment stack (`deployment.apps`). Alternatively, we could just deploy a simple application Pod (`packt-web`) with the following command:

```
kubectl run packt-web --image=nginxdemos/hello
```

The output suggests that our application Pod, in this case, a *standalone* or *bare* Pod (`pod/packt-web`), is not part of a deployment:

```
pod/packt-web created
```

We'll see later in this section that this Pod is not part of a ReplicaSet and so is not managed by the Controller Manager. Let's look at the state of our system by querying the Pods for detailed information:

```
kubectl get pods -o wide
```

Let's analyze the output:

```
packt@neptune:~$ kubectl get pods -o wide
NAME                      READY   STATUS    RESTARTS   AGE    IP               NODE
packt-5dc77bb9bf-bnzsc    1/1     Running   0          29m    192.168.111.193  k8s-n2
packt-web                 1/1     Running   0          3m37s  192.168.215.66   k8s-n1
```

Figure 14.48 – Getting the application Pods with detailed information

We can see that our Pods are up and running, and that Kubernetes deployed them on separate nodes:

- `packt-5dc77bb9bf-bnzsc`: On cluster node `k8s-n2`

- `packt-web`: On cluster node `k8s-n1`

Running the Pods on different nodes is due to internal load balancing and resource distribution in the Kubernetes cluster.

The application Pod managed by the controller is `packt-5dc77bb9bf-bnzsc`. Kubernetes generates a unique name for our managed Pod by appending a *Pod template hash* (`5dc77bb9bf`) and a *Pod ID* (`bnzsc`) to the name of the deployment (`packt`). The Pod template hash and Pod ID are unique within a ReplicaSet.

In contrast, the standalone Pod (`packt-web`) is left as-is since it's not part of an application deployment. Let's *describe* both Pods to obtain more information about them. We'll start with the managed Pod first. Don't forget to use the `kubectl` autocompletion (with *Tab + Tab*):

```
kubectl describe pod packt-5dc77bb9bf-bnzsc
```

The related output is relatively large. Here are some relevant snippets:

- The node where the Pod runs:

  ```
  Node: k8s-n2/172.16.191.9
  ```

- Pod status:

  ```
  Status: Running
  ```

- The pod's internal IP address:

  ```
  IP: 192.168.111.193
  ```

- The ReplicaSet controlling the Pod:

  ```
  Controlled By: ReplicaSet/packt-5dc77bb9bf
  ```

- Pod container image:

```
Image: docker.io/nginxdemos/hello
```

In contrast, the same command for the standalone Pod (packt-web) would be slightly different without featuring the Controlled By field:

```
kubectl describe pod packt-web
```

Here are the corresponding excerpts:

- The node where the Pod runs:

```
Node: k8s-n1/172.16.191.8
```

- Pod status:

```
Status: Running
```

- The pod's internal IP address:

```
IP: 192.168.215.66
```

- Pod container image:

```
Image: docker.io/nginxdemos/hello
```

You can also venture out to any of the cluster nodes where our Pods are running and take a closer look at the related containers. Let's take node k8s-n1 (172.15.191.8), for example, where our standalone Pod (packt-web) is running. We'll SSH into the node's terminal first:

```
ssh packt@172.16.191.8
```

Then we'll use the containerd runtime to query the containers in the system:

```
sudo crictl --runtime-endpoint unix:///run/containerd/
containerd.sock ps
```

The output shows the following:

```
packt@k8s-n1:~$ sudo crictl --runtime-endpoint unix:///run/containerd/containerd.sock ps
CONTAINER ID      IMAGE           CREATED            STATE                 NAME
01c9bcbc9701c     aedf47d433f18   About an hour ago  Running   ------>     packt-web
42d2568413aa8     50b52cdadbcf0   5 days ago         Running               calico-node
e4623a0f654dd     c29e6c5830670   5 days ago         Running               kube-proxy
```

Figure 14.49 – Getting the containers running on a cluster node

Next, we'll show you how to access processes running inside the Pods. Let's switch back to our local `kubectl` environment and run the following command to access the shell in the container running the `packt-web` Pod:

```
kubectl exec -it packt-web -- /bin/sh
```

The command takes us *inside* the container to an *interactive* shell prompt. Here we can run commands as if we were logged in to the `packt-web` host using the terminal. The interactive session is produced using the `-it` options – *interactive terminal* – or `--interactive --tty`.

Let's run a few commands, starting with the process explorer:

```
ps aux
```

Here's a relevant excerpt from the output, showing the processes running inside the `packt-web` container:

```
/ # ps aux
PID   USER      TIME   COMMAND
   1 root       0:00 nginx: master process nginx -g daemon off;
   6 nginx      0:00 nginx: worker process
```

Figure 14.50 – The processes running inside the packt-web container

We can also retrieve the IP address with the following command:

```
ifconfig | grep 'inet addr:' | cut -d: -f2 | awk '{print $1}' |
grep -v '127.0.0.1'
```

The output shows the pod's IP address:

```
192.168.215.66
```

We can also retrieve the hostname with the following command:

```
hostname
```

The output shows the Pod name:

```
packt-web
```

Let's leave the container shell with the `exit` command or by typing *Ctrl + D*. With the `kubectl exec` command, we can run any process inside a Pod, assuming that the related process exists.

We'll experiment next by testing the packt-web application Pod using curl. We should note that, at this time, the only way to access the web server endpoint of packt-web is via its internal IP address. Previously, we used the kubectl get pods -o wide and describe commands to retrieve detailed information regarding Pods, including the pod's IP address. You can also use the following one-liner to retrieve the pod's IP:

```
kubectl get pods packt-web -o jsonpath='{.status.podIP}{"\n"}'
```

In our case, the command returns 192.168.215.66. We used the -o jsonpath output option to specify the JSON query for a specific field, {.status.podIP}. Remember that the pod's IP is only accessible within the *Pod network* (192.168.0.0/16) inside the cluster. (You may look back at the *Creating a Kubernetes Control Plane node* section, where we configured the Calico networking manifest with the Pod network subnet.)

Consequently, we need to probe the packt-web endpoint using a curl command that has originated within the Pod network. An easy way to accomplish such a task is to run a test *Pod* with the curl utility installed. The following command runs a Pod named test, based on the curlimages/curl Docker image:

```
kubectl run test --image=curlimages/curl sleep 600
```

We keep the container *artificially* alive with the sleep command due to the Docker *entrypoint* of the corresponding image, which simply runs a curl command and then exits. Without sleep, the Pod would keep coming up and crashing. With the sleep command, we delay the execution of the curl entrypoint to avoid the exit.

Now, we can run a simple curl command using the test Pod targeting the packt-web web server endpoint:

```
kubectl exec test -- curl http://192.168.215.66
```

We'll get an HTTP response and a corresponding *access log trace* (from the Nginx server running in the Pod) accounting for the request. To view the logs on the packt-web Pod, we run the following command:

```
kubectl logs packt-web
```

The output is as follows:

```
192.168.57.200 - - [21/Mar/2021:04:52:16 +0000] "GET /
HTTP/1.1" 200 7232 "-" "curl/7.75.0-DEV" "-"
```

The logs in the `packt-web` Pod are produced by Nginx and redirected to `stdout` and `stderr`. We can easily verify this with the following command:

```
kubectl exec packt-web -- ls -la /var/log/nginx
```

The output shows the related symlinks:

```
access.log -> /dev/stdout
error.log -> /dev/stderr
```

When you're done using the `test` Pod, you can delete it with the following command:

```
kubectl delete pods test
```

Now, let's rewind to the command we used previously to create the `packt` deployment. Don't run it. Here it is just as a refresher:

```
kubectl create deployment packt --image=nginxdemos/hello
```

The command carried out the following sequence:

1. It created a deployment (`packt`).
2. The deployment created a *ReplicaSet* (`packt-5dc77bb9bf`).
3. The ReplicaSet created the Pod (`packt-5dc77bb9bf-bnzsc`).

We can verify that with the following commands:

```
kubectl get deployments -l app=packt
kubectl get replicasets -l app=packt
kubectl get pods -l app=packt
```

In the preceding commands, we used the `--label-columns` (`-l`) flag to filter results by the `app=packt` label, denoting the `packt` deployment's resources.

We encourage you to take a closer look at each of these resources using the `kubectl describe` command. Don't forget to use the `kubectl` autocomplete feature when typing in the commands:

```
kubectl describe deployment packt | more
kubectl describe replicaset packt | more
kubectl describe pod packt-5dc77bb9bf-bnzsc | more
```

The `kubectl describe` command could be very resourceful when troubleshooting application or Pod deployments. Look inside the *Events* section in the related output for clues on Pods failing to start, errors if any, and possibly understand what went wrong.

Now that we have deployed our first application inside a Kubernetes cluster, let's look at how to expose the related endpoint to the world.

Exposing deployments as services

So far, we have deployed an application (`packt`) with a single Pod (`packt-5dc77bb9bf-bnzsc`) running an Nginx web server listening on port `80`. As we explained earlier, at this time, we can only access the Pod within the Pod network, which is internal to the cluster. In this section, we'll show you how to *expose* the application (or deployment) to be accessible from the outside world. Kubernetes uses the *Service* API object, consisting of a *proxy* and a *selector* routing the network traffic to application Pods in a deployment.

The following command creates a service for our deployment (`packt`):

```
kubectl expose deployment packt \
    --port=80 \
    --target-port=80 \
    --type=NodePort
```

Here's a brief explanation of the preceding command flags:

- `--port=80`: Exposes the service on port `80` externally within the cluster
- `--target-port=80`: Maps to port `80` internally in the application Pod
- `--type=NodePort`: Makes the service available outside the cluster

The output shows the service (`packt`) we just created for exposing our application:

```
service/packt exposed
```

Without the `--type=NodePort` flag, the service type would be `ClusterIP` by default, and the Service endpoint would only be accessible within the cluster. Let's take a closer look at our service (`packt`):

```
kubectl get service packt
```

The output shows the cluster IP assigned to the service (10.103.172.205) and the ports the service is listening on for TCP traffic (80:32081/TCP):

- port 80: Within the cluster

- port 32081: Outside the cluster, on any of the nodes

We should note that the cluster IP is only accessible within the cluster and not from the outside:

```
packt@neptune:~$ kubectl get service packt
NAME     TYPE       CLUSTER-IP       EXTERNAL-IP    PORT(S)        AGE
packt    NodePort   10.103.172.205   <none>         80:32081/TCP   18m
```

Figure 14.51 – The service exposing the packt deployment

Also, EXTERNAL-IP (<none>) should not be mistaken for the cluster node's IP address where our service is accessible. The external IP is usually a load balancer IP address configured by a cloud provider hosting the Kubernetes cluster (configurable via the --external-ip flag).

We should now be able to access our application outside the cluster by pointing a browser to any of the cluster nodes on port 32081. To get a list of our cluster nodes with their respective IP address and hostname, we can run the following command:

```
kubectl get nodes -o jsonpath='{range .items[*]}{.status.
addresses[*].address}{"\n"}'
```

The output is as follows:

```
172.16.191.6 k8s-cp1
172.16.191.8 k8s-n1
172.16.191.9 k8s-n2
172.16.191.10 k8s-n3
```

Let's choose the Control Plane node (172.16.191.6/k8s-cp1) and enter the following address in a browser: http://172.16.191.6:32081.

The web request from the browser is directed to the Service endpoint (`packt`), which routes the related network packets to the application Pod (`packt-5dc77bb9bf-bnzsc`). The `packt` web application responds with a simple **Nginx Hello World** web page, displaying the pod's internal IP address (`192.168.111.193`) and name (**packt-5dc77bb9bf-bnzsc**):

Figure 14.52 – Accessing the packt application service

To verify that the information on the web page is accurate, you may run the following `kubectl` command, retrieving similar information:

```
kubectl get pod packt-5dc77bb9bf-bnzsc -o jsonpath='{.status.
podIP}{"\n"}{.metadata.name}{"\n"}'
```

Suppose we have high traffic targeting our application, and we'd like to scale out the ReplicaSet controlling our Pods. We'll show you how to accomplish this task in the next section.

Scaling application deployments

Currently, we have a *single* Pod in the `packt` deployment. To get the related details, we run the following command:

```
kubectl describe deployment packt
```

The relevant excerpt in the output is as follows:

```
Replicas:          1 current / 1 desired
```

Let's *scale up* our `packt` deployment to *10* `replicas` with the following command:

```
kubectl scale deployment packt --replicas=10
```

The command output is as follows:

```
deployment.apps/packt scaled
```

If we list the Pods of the `packt` deployment, we'll see 10 Pods running:

```
kubectl get pods -l app=packt
```

The output is as follows:

```
packt@neptune:~$ kubectl get pods -l app=packt
NAME                       READY   STATUS    RESTARTS   AGE
packt-5dc77bb9bf-6ggtf     1/1     Running   0          7m36s
packt-5dc77bb9bf-bnzsc     1/1     Running   0          17h
packt-5dc77bb9bf-brqpb     1/1     Running   0          7m36s
packt-5dc77bb9bf-csc89     1/1     Running   0          7m36s
packt-5dc77bb9bf-gv4lk     1/1     Running   0          7m36s
packt-5dc77bb9bf-hzsf8     1/1     Running   0          7m36s
packt-5dc77bb9bf-j4l5h     1/1     Running   0          7m36s
packt-5dc77bb9bf-qtxmf     1/1     Running   0          7m36s
packt-5dc77bb9bf-vp5hj     1/1     Running   0          7m36s
packt-5dc77bb9bf-zfxcj     1/1     Running   0          7m36s
```

Figure 14.53 – Scaling up the deployment replicas

Incoming requests to our application Service endpoint (`http://172.16.191.6:32081`) will be load balanced between the Pods. To illustrate this behavior, we can either use `curl` or a *text-based browser* at the command line to avoid the caching-related optimizations of a modern desktop browser. For a better illustration, we'll use *Lynx*, a simple text-based browser. On our Ubuntu desktop, we install it with the following command:

```
sudo apt-get install -y lynx
```

Next, we point Lynx to our application endpoint:

```
lynx 172.16.191.6:32081
```

Refreshing the page with *Ctrl+R* every few seconds, we observe that the server address and name changes, based on the current Pod processing the request:

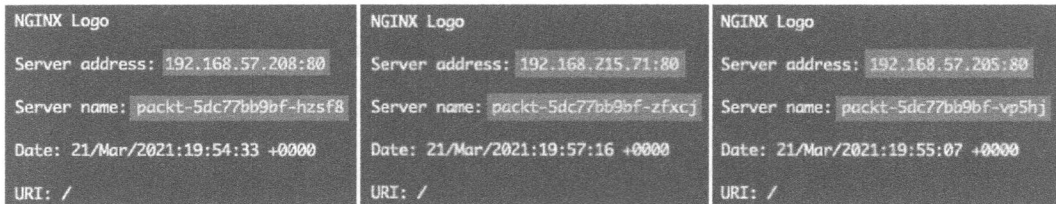

Figure 14.54 – Load balancing requests across Pods

You can exit the Lynx browser by typing *Q* and then *Enter*.

We can scale back our deployment (`packt`) to three replicas (or any other non-zero positive number) with the following command:

```
kubectl scale deployment packt --replicas=3
```

If we query the `packt` application Pods, we can see the surplus Pods terminating until only three Pods are remaining:

```
kubectl get pods -l app=packt
```

The output is as follows:

Figure 14.55 – Scaling back to three Pods

Before concluding our imperative deployments, let's clean up all the resources we have created thus far:

```
kubectl delete service packt
kubectl delete deployment packt
kubectl delete pod packt-web
```

The following command should reflect a clean slate:

```
kubectl get all
```

The output is as follows:

```
packt@neptune:~$ kubectl get all
NAME                 TYPE        CLUSTER-IP   EXTERNAL-IP   PORT(S)   AGE
service/kubernetes   ClusterIP   10.96.0.1    <none>        443/TCP   7d8h
```

Figure 14.56 – The cluster in a default state

In the next section, we'll look at how to deploy resources and applications declaratively in the Kubernetes cluster.

Working with declarative deployments

At the heart of a declarative deployment is a manifest file. Manifest files are generally in YAML format and authoring them usually involves a mix of autogenerated code and manual editing. The manifest is then deployed using the kubectl apply command:

```
kubectl apply -f MANIFEST
```

Deploying resources declaratively in Kubernetes involves the following stages:

- Creating a manifest file
- Updating the manifest
- Validating the manifest
- Deploying the manifest
- Iterating between the preceding stages

To illustrate the declarative model, we follow the example of deploying a simple Hello World web application to the cluster. The result will be similar to our previous approach of using the imperative method.

So, let's start by creating a manifest for our deployment.

Creating a manifest

When we created our packt deployment *imperatively*, we used the following command. (Don't run it just yet!):

```
kubectl create deployment packt --image=nginxdemos/hello
```

The following command will *simulate* the same process without changing the system state:

```
kubectl create deployment packt --image=nginxdemos/hello \
    --dry-run=client --output=yaml
```

We used the following additional options (flags):

- --dry-run=client: Runs the command in the local kubectl environment (*client*) without modifying the system state

- --output=yaml: Formats the command output as YAML

We can use the previous command's output to analyze the changes to be made to the system. Then we can redirect it to a file (packt.yaml) serving as a *draft* of our deployment manifest:

```
kubectl create deployment packt --image=nginxdemos/hello \
    --dry-run=client --output=yaml > packt.yaml
```

We created our first manifest file, packt.yaml. From here, we can edit the file to accommodate more complex configurations. For now, we'll leave the manifest as-is and proceed with the next stage in our declarative deployment workflow.

Validating a manifest

Before deploying a manifest, we recommend validating the deployment, especially if you edited the file manually. Editing mistakes can happen, particularly when working with complex YAML files with multiple indentation levels.

The following command validates the packt.yaml deployment manifest:

```
kubectl apply -f packt.yaml --dry-run=client
```

A successful validation yields the following output:

```
deployment.apps/packt created (dry run)
```

If there are any errors, we should edit the manifest file and correct them prior to deployment. Our manifest looks good, so let's go ahead and deploy it.

Deploying a manifest

To deploy the packt.yaml manifest, we use the following command:

```
kubectl apply -f packt.yaml
```

A successful deployment shows the following message:

```
deployment.apps/packt created
```

We can check the deployed resources with the following command:

```
kubectl get all -l app=packt
```

The output shows that the `packt` deployment resources created declaratively are up and running:

```
packt@neptune:~$ kubectl get all -l app=packt
NAME                          READY   STATUS    RESTARTS   AGE
pod/packt-6c89655654-k54k5    1/1     Running   0          3m21s

NAME                     READY   UP-TO-DATE   AVAILABLE   AGE
deployment.apps/packt    1/1     1            1           3m21s

NAME                              DESIRED   CURRENT   READY   AGE
replicaset.apps/packt-6c89655654 1         1         1       3m21s
```

Figure 14.57 – The deployment resources created declaratively

Next, we want to expose our deployment using a service. We'll repeat the preceding workflow by creating, validating, and deploying the service manifest (`packt-svc.yaml`). For brevity, we simply enumerate the related commands.

Create the manifest file (`packt-svc.yaml`) for the service exposing our deployment (`packt`):

```
kubectl expose deployment packt \
    --port=80 \
    --target-port=80 \
    --type=NodePort \
    --dry-run=client --output=yaml > packt-svc.yaml
```

We explained the preceding command previously in the *Exposing deployments as services* section. Next, we'll validate the service deployment manifest:

```
kubectl apply -f packt-svc.yaml --dry-run=client
```

If the validation is successful, we deploy the service manifest:

```
kubectl apply -f packt-svc.yaml
```

Let's get the current status of the `packt` resources:

```
kubectl get all -l app=packt
```

The output shows all of the `packt` application resources, including the Service endpoint (`service/packt`) listening on port `31168`:

```
packt@neptune:~$ kubectl get all -l app=packt
NAME                          READY    STATUS     RESTARTS    AGE
pod/packt-6c89655654-k54k5    1/1      Running    0           40m

NAME             TYPE        CLUSTER-IP      EXTERNAL-IP          PORT(S)        AGE
service/packt    NodePort    10.106.46.48    <none>    ━━━▶     80:31168/TCP   13m

NAME                     READY    UP-TO-DATE    AVAILABLE    AGE
deployment.apps/packt    1/1      1             1            40m

NAME                                DESIRED    CURRENT    READY    AGE
replicaset.apps/packt-6c89655654    1          1          1        40m
```

Figure 14.58 – The packt application resources deployed

Using a browser, `curl`, or Lynx, we can access our application by targeting any of the cluster nodes on port `31168`. Let's use the Control Plane node (`k8s-cp1`, `172.16.191.6`) by pointing our browser to `http://172.16.191:31168`:

Figure 14.59 – Accessing the packt application endpoint

If we want to change the existing configuration of a resource in our application deployment, we can update the related manifest and redeploy. In the next section, we'll modify the deployment to accommodate a scale-out scenario.

Updating a manifest

Suppose our application is taking a high number of requests, and we'd like to add more Pods to our deployment to handle the traffic. We need to change the `spec.replicas` configuration setting in the `packt.yaml` manifest.

Using your editor of choice, edit the `packt.yaml` file and locate the following configuration section:

```
spec:
  replicas: 1
```

Change the value from 1 to 10 for additional application Pods in the ReplicaSet controlled by the `packt` deployment. The configuration becomes the following:

```
spec:
  replicas: 10
```

Save the manifest file and redeploy with the following command:

```
kubectl -f apply packt.yaml
```

The output suggests that the `packt` deployment has been reconfigured:

```
deployment.apps/packt configured
```

If we query the `packt` resources in the cluster, we should see the new Pods up and running:

```
kubectl get all -l app=packt
```

The output displays the application resources of our `packt` deployment, including the additional Pods deployed in the cluster:

```
packt@neptune:~$ kubectl get all -l app=packt
NAME                          READY   STATUS    RESTARTS   AGE
pod/packt-6c89655654-5rktw    1/1     Running   0          3m28s
pod/packt-6c89655654-6bcw2    1/1     Running   0          3m28s
pod/packt-6c89655654-9c5rd    1/1     Running   0          3m28s
pod/packt-6c89655654-fhxtc    1/1     Running   0          3m28s
pod/packt-6c89655654-k54k5    1/1     Running   0          68m
pod/packt-6c89655654-nqnjb    1/1     Running   0          3m28s
pod/packt-6c89655654-pfvtn    1/1     Running   0          3m28s
pod/packt-6c89655654-psl7z    1/1     Running   0          3m28s
pod/packt-6c89655654-sl9nn    1/1     Running   0          3m28s
pod/packt-6c89655654-t57cs    1/1     Running   0          3m28s


NAME            TYPE       CLUSTER-IP     EXTERNAL-IP   PORT(S)        AGE
service/packt   NodePort   10.106.46.48   <none>        80:31168/TCP   41m


NAME                    READY   UP-TO-DATE   AVAILABLE   AGE
deployment.apps/packt   10/10   10           10          68m


NAME                                DESIRED   CURRENT   READY   AGE
replicaset.apps/packt-6c89655654    10        10        10      68m
```

Figure 14.60 – The additional Pods added for application scale-out

We encourage you to test with the scale-out environment and verify the load balancing workload described in the *Scaling application deployments* section earlier in this chapter.

Let's scale back our deployment to three Pods, but this time by updating the related manifest *on the fly* with the following command:

```
kubectl edit deployment packt
```

The command will open our default editor in the system (**vi**) to make the desired change:

```
spec:
  progressDeadlineSeconds: 600
  replicas: 3    ⬅
```

Figure 14.61 – Making deployment changes on the fly

After saving and exiting the editor, we'll get a message suggesting that our deployment (`packt`) has been updated:

```
deployment.apps/packt edited
```

Please note that the modifications made on the fly with `kubectl edit` *would not be reflected* in the deployment manifest (`packt.yaml`). Nevertheless, the related configuration changes are persisted in the cluster (`etcd`).

We can verify our updated deployment with the help of the following command:

```
kubectl get deployment packt
```

The output now shows only three Pods running in our deployment:

NAME	READY	UP-TO-DATE	AVAILABLE	AGE
packt	3/3	3	3	101m

Before wrapping up, let's clean up our resources once again with the following commands to bring the cluster back to the default state:

```
kubectl delete service packt
kubectl delete deployment packt
```

We have reached the end of our journey here, but we trust that you'll take it to the next level and further explore the exciting domain of application deployment and scaling with Kubernetes. It's been a relatively long chapter, and we barely skimmed the surface of the related field. We encourage you to explore some of the resources captured in the *Further reading* section and strengthen your knowledge regarding some of the key areas of Kubernetes environments, such as networking, security, and scale. Let's now summarize briefly what we have learned in this chapter.

Summary

We began this chapter with a high-level overview of the Kubernetes architecture and API object model, introducing the most common cluster resources, such as Pods, deployments, and Services. Next, we took on the relatively challenging task of building an on-premises Kubernetes cluster from scratch using virtual machines. As we became more familiar with Kubernetes internals in general, we moved to the cloud, working with the EKS and AKS managed cluster services from AWS and Azure. We explored various CLI tools for managing Kubernetes cluster resources in the cloud and on-premises. At the highpoint of our journey, we focused on deploying and scaling applications in Kubernetes using imperative and declarative deployment scenarios.

We believe that novice Linux administrators will benefit greatly from the material covered in this chapter and become more knowledgeable in managing resources across hybrid clouds and on-premises distributed environments, deploying applications at scale, and working with CLI tools. We believe that the structured information in this chapter will also help seasoned system administrators refresh some of their knowledge and skills in the areas covered.

We'll stay within the application deployment realm, and in the next section, we'll look at Ansible, a platform for accelerating application delivery on-premises and in the cloud.

Questions

Here are a few questions for refreshing or pondering upon some of the concepts you've learned in this chapter:

1. Enumerate some of the essential services of a Kubernetes Control Plane node. How do the worker nodes differ?

2. What command did we use to bootstrap a Kubernetes cluster?

3. What is the preferred CLI for managing an EKS cluster? How about the CLI for managing AKS clusters? How are these CLI tools different from kubectl?

4. What is the difference between imperative and declarative deployments in Kubernetes?

5. What is the kubectl command for deploying a Pod? How about for creating a deployment?

6. What is the kubectl command to access the shell within a Pod container?

7. What is the kubectl command to query all resources related to a deployment?

8. You exposed a deployment using a *ClusterIP* service type. Can you access the service outside the Kubernetes cluster? Why?

9. How do you scale out a deployment in Kubernetes? Can you think of the different ways (commands) in which to accomplish the task?

10. How do you delete all resources related to a deployment in Kubernetes?

Further reading

The following resources may help you to consolidate further your knowledge of Kubernetes:

- Kubernetes documentation online: https://kubernetes.io/docs/home/

- The kubectl cheat sheet: https://kubernetes.io/docs/reference/kubectl/cheatsheet/

- Getting started with Amazon EKS: https://docs.aws.amazon.com/eks/latest/userguide/getting-started.html

- **Azure Kubernetes Service** (**AKS**): https://docs.microsoft.com/en-us/azure/aks/

- *Kubernetes and Docker: The Container Masterclass [Video], Cerulean Canvas, Packt Publishing* (https://www.packtpub.com/product/kubernetes-and-docker-the-container-masterclass-video/9781801075084)

- *Mastering Kubernetes – Third Edition, Gigi Sayfan, Packt Publishing* (https://www.packtpub.com/product/mastering-kubernetes-third-edition/9781839211256)

15
Automating Workflows with Ansible

If your day-to-day system administration or development work involves tedious and repetitive operations, **Ansible** could help you run your tasks while saving you precious time. Ansible is a tool for automating software provisioning, configuration management, and application deployment workflows. Initially developed by Michael DeHaan in 2012, Ansible was acquired by Red Hat in 2015 and is now maintained as an open source project.

In this chapter, you'll learn about the fundamental concepts of Ansible, along with a variety of hands-on examples. In particular, we'll explore the following topics:

- Introducing Ansible
- Installing Ansible
- Working with Ansible

Technical requirements

First, you should be familiar with the Linux command-line Terminal in general. Intermediate knowledge of Linux will help you understand some of the intricacies of the practical illustrations used throughout this chapter. You should also be proficient in using a Linux-based text editor. For the hands-on examples, we recommend setting up a lab environment similar to the one we're using. You'll find the related instructions in the README file included with the complementary resources in this book's GitHub repository for this chapter.

If you don't configure a lab environment, you will still benefit from the detailed explanations associated with the practical examples in this chapter.

Now, let's start our journey by covering the introductory concepts surrounding Ansible.

Introducing Ansible

In the opening paragraph of this chapter, we captured one of the essential aspects of Ansible – it's a tool for automating workflows. Almost any Linux system administration task can be automated using Ansible. Using the Ansible CLI, we can invoke simple commands to change the **desired state** of a system. Usually, with Ansible, we execute tasks on a remote host or a group of hosts.

Let's use the classic illustration of package management. Suppose you're managing an infrastructure, which includes a group of web servers, and you plan to install the latest version of a web server application (Nginx or Apache) on all of them. One way to accomplish this task is to SSH into each host and run the related shell commands to install the latest web server package. If you have a lot of machines, this will be a big task. You could argue that you can write a script to automate this job. This is possible, but then you'd have yet another job on your hand; that is, maintaining the script, fixing possible bugs, and, with your infrastructure growing, adding new features.

At some point, you may want to manage users or databases or configure network settings on multiple hosts. Soon, you'll be looking at a Swiss Army knife tool, with capabilities that you'd rather get for free instead of writing them yourself. Here's where Ansible comes in handy. With its myriad of modules – for almost any system administration task you can imagine – Ansible can remotely configure, run, or deploy your management jobs of choice, with minimal effort and in a very secure and efficient way.

We'll consolidate these preliminary thoughts with a brief look at the Ansible architecture.

Understanding the Ansible architecture

The core Ansible framework is written in Python. Let's mention upfront that Ansible has an **agentless** architecture. In other words, Ansible runs on a **control node** that executes commands against remote hosts, without the need for a remote endpoint or service to be installed on the managed host to communicate with the control node. At a minimum, the only requirement for Ansible communication is SSH connectivity to the managed host. Yet, the number of Ansible operations would be relatively limited to only running scripts and raw SSH commands if the host didn't have a Python framework installed. The vast majority of server OS platforms already have Python installed by default.

Ansible can manage a fleet of remote hosts from a single control node using secure SSH connections. The following diagram shows the **logical layout** of a managed infrastructure using Ansible:

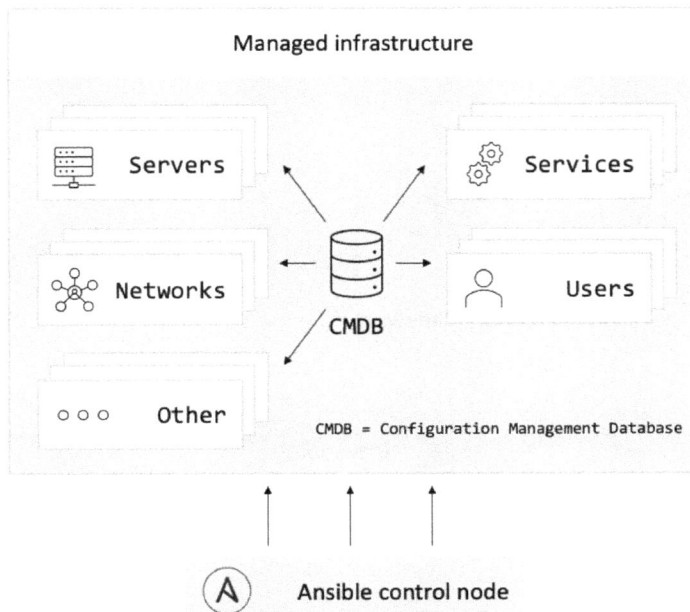

Figure 15.1 – The logical layout of a managed infrastructure using Ansible

Production-grade enterprise environments usually include a **Configuration Management Database (CMDB)** for organizing their **Information Technology (IT)** infrastructure assets. Examples of IT infrastructure assets are servers, networks, services, and users. Although not directly part of the Ansible architecture, the CMDB *describes* the assets and their relationship within a managed infrastructure and can be leveraged to build the Ansible **inventory**.

The inventory is local storage on the Ansible control node – usually an **INI** or **YAML** file – that describes the managed **hosts** or **groups** of hosts. The inventory is either inferred from the CMDB or manually created by the system administrator.

Now, let's have a closer look at the high-level Ansible architecture shown in the following diagram:

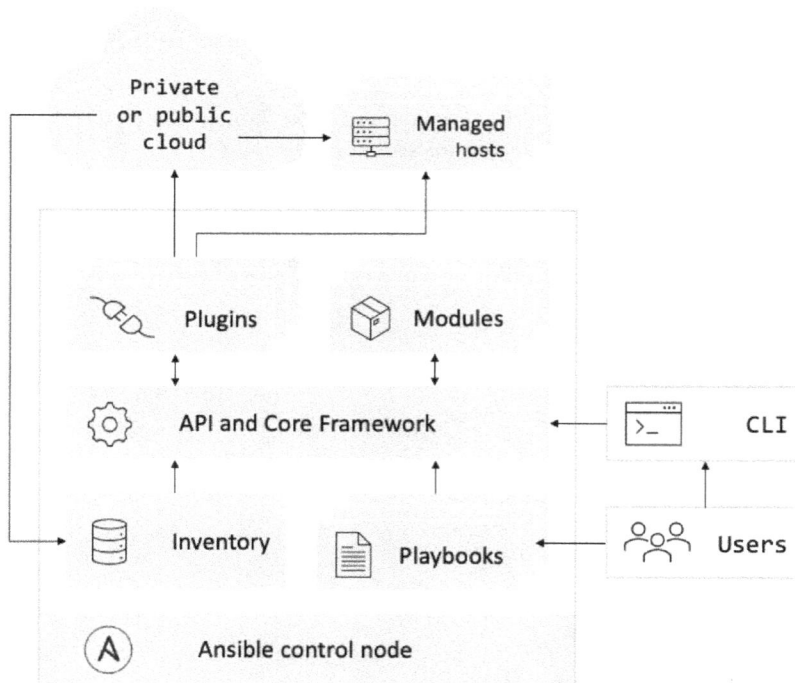

Figure 15.2 – The Ansible architecture

The preceding diagram shows the Ansible control node interacting with the managed hosts in a private or public cloud infrastructure. Here are some brief descriptions of the blocks featured in the architectural view:

- **API and core framework**: The main libraries encapsulating Ansible's core functionality; the Ansible core framework is written in Python.

- **Plugins**: Additional libraries extending the core framework's functionality; examples include **connection plugins** (such as cloud connectors), **test plugins** (verifying specific response data), **callback plugins** (responding to events), and many more.

- **Modules**: These encapsulate specific functions running on the managed hosts; examples include the **user** module (managing users), the **package** module (managing software packages), and so on.

- **Inventory**: The INI or YAML file describing the hosts and groups of hosts targeted by Ansible commands and playbooks.

- **Playbooks**: The Ansible execution files describing a set of tasks that target the managed hosts.

- **Private or public clouds**: The managed infrastructure, hosted on-premises or in various cloud environments (for example, VMware, AWS, and Azure).

- **Managed hosts**: The servers targeted by Ansible commands and playbooks.

- **CLI**: The Ansible CLI tools, such as **ansible**, **ansible-playbook**, **ansible-doc**, and so on.

- **Users**: The administrators, power users, and automated user processes running Ansible commands or playbooks.

Now that we have a basic understanding of the Ansible architecture, let's look at what makes Ansible a great tool for automating management workflows. We'll introduce the concept of **configuration management** next.

Introducing configuration management

If we look back at the old days, system administrators usually managed a relatively low number of servers, running everyday administrative tasks by using a remote shell on each host. Relatively simple operations, such as copying files, updating software packages, and managing users, could easily be scripted and reused regularly. With the recent surge in apps and services, driven by the vast expansion of the internet, modern-day on-premises and cloud-based IT infrastructures – sustaining the related platforms – have grown significantly. The sheer amount of configuration changes involved would by far exceed the capacity of a single admin running and maintaining a handful of scripts. Here's where configuration management comes to the rescue.

With configuration management, the managed hosts and assets are grouped into logical categories, based on specific criteria, as suggested in *Figure 15.1*. Managing assets other than hosts ultimately comes down to performing specific tasks on the servers hosting those assets. The configuration management manifest is the Ansible inventory file. Thus, Ansible becomes the configuration management endpoint.

With Ansible, we can run single one-off commands to carry out specific tasks, but a far more efficient configuration management workflow can be achieved via playbooks. With Ansible playbooks, we can run multiple tasks that target various subsystems of a target platform against any number of hosts. Scheduling ansible-playbook runs for regular maintenance and configuration management tasks is a common practice in IT infrastructure automation.

Running Ansible tasks repeatedly (or on a scheduled basis) against a specific target raises the concern of unwanted changes in the desired state due to repetitive operations. This issue brings us to one of the essential aspects of configuration management – the **idempotency** of configuration changes. We'll look at what idempotent changes are next.

Explaining idempotent operations

In configuration management, an operation is idempotent when running it multiple times yields the same result as running it once. In this sense, Ansible is an idempotent configuration management tool.

Suppose we have an Ansible task creating a user. When the task runs for the first time, it creates the user. Running it for a second time – when the user has already been created – would result in a **no-operation (no-op)**. *Without* idempotency, subsequent runs of the same task would produce errors due to attempting to create a user that already exists.

We should note that Ansible is not the only configuration management tool on the market. We have **Chef**, **Puppet**, and **SaltStack**, to name a few. Most of these platforms have been acquired by larger enterprises, such as SaltStack, being owned by VMware, and some may argue that Ansible's success could be attributed to Red Hat open sourcing the project. Ansible appears to be the most successful configuration management platform of our days. The industry consensus is that Ansible provides a user-friendly experience, high scalability, and affordable licensing tiers in enterprise-grade deployments.

With the introductory concepts covered, let's roll up our sleeves and install Ansible on a Linux platform of your choice.

Installing Ansible

In this section, we'll show you how to install Ansible on a control node. On Linux, we can install Ansible in a couple of ways:

- Using the platform-specific package manager (for example, `apt` on Ubuntu/Debian and `yum` on RHEL/CentOS)
- Using `pip`, the Python package manager

The Ansible community recommends `pip` for installing Ansible since it provides the most recent stable version of Ansible. In this section, we'll use both methods on Ubuntu and RHEL/CentOS. For a complete Ansible installation guide for all major OS platforms, please follow the online documentation at `https://docs.ansible.com/ansible/latest/installation_guide/intro_installation.html`.

On the control node, Ansible requires Python, so before installing Ansible, we need to make sure we have Python installed on our system.

> **Important note**
> Python 2 is no longer supported as of January 1, 2020. Please use Python 3 instead.

Let's start by installing Ansible on Ubuntu.

Installing Ansible on Ubuntu

With Ubuntu 20.04, we have Python 3 installed by default. Let's check the Python 3 version we have by using the following command:

```
python3 --version
```

On our Ubuntu 20.04 machine, the Python version is as follows:

```
Python 3.8.5
```

If you don't have Python 3 installed, you can install it with the following command:

```
sudo apt-get install -y python3
```

With Python 3 installed, we can proceed with installing Ansible. To get the most up to date Ansible packages using `apt`, we need to add the Ansible **Personal Package Archives (PPA)** repository to our system.

Let's start by updating the current `apt` repository:

```
sudo apt-get update
```

Next, we must add the Ansible PPA:

```
sudo apt-get install -y software-properties-common
sudo apt-add-repository -y --update ppa:ansible/ansible
```

Now, we can install the Ansible package with the following command:

```
sudo apt-get install -y ansible
```

With Ansible installed, we can check its current version:

```
ansible --version
```

In our case, the relevant excerpt in the output of the previous command is as follows:

```
ansible 2.9.6
```

Next, we'll look at how to install Ansible on a RHEL/CentOS system.

Installing Ansible on RHEL/CentOS

First, we need to make sure Python is installed on our system. With a minimal **Red Hat Enterprise Linux** (**RHEL**)/CentOS 8 distro, Python is not installed by default. Currently, the latest version of Python that's available on RHEL/CentOS 8 is Python 3.8. Let's install it with the following command:

```
sudo yum install -y python38
```

With Python 3 installed, we can check its current version by running the following command:

```
python --version
```

In our case, the output is as follows:

```
Python 3.8.3
```

On RHEL/CentOS, Ansible is available via the **Extra Packages for Enterprise Linux** (**EPEL**) repository. Let's enable the EPEL repository:

```
sudo yum install -y epel-release
```

Now, we can install Ansible:

```
sudo yum install -y ansible
```

With Ansible installed, we can check its current version:

```
ansible --version
```

In our case, the relevant excerpt of the preceding output is as follows:

```
ansible 2.9.18
```

Next, we'll look at how to install Ansible using `pip`.

Installing Ansible using pip

Before we install Ansible with `pip`, we need to make sure Python is installed on the system. We assume Python 3 is installed based on the steps presented in the previous sections. Next, we should remove any existing version of Ansible that's been installed with the platform-specific package manager (for example, `apt` or `yum`).

To uninstall Ansible on Ubuntu, run the following command:

```
sudo apt-get remove -y ansible
```

On RHEL/CentOS, we can remove Ansible with the following command:

```
sudo yum remove -y ansible
```

Next, we must make sure `pip` is installed. The following command should provide the current version of `pip`:

```
python3 -m pip --version
```

In our case, the output shows the following:

```
pip 21.0.1 from /home/packt/.local/lib/python3.8/site-packages/
pip (python 3.8)
```

If `pip` is not installed, we must download the `pip` installer first:

```
curl -s https://bootstrap.pypa.io/get-pip.py -o get-pip.py
```

We're now ready to install `pip` and Ansible with the following commands:

```
python3 get-pip.py --user
python3 -m pip install --user ansible
```

Please note that, with the previous commands, we only installed `pip` and Ansible for the current user. If you wish to install Ansible *globally* on the system, the equivalent commands are as follows:

```
sudo python3 get-pip.py
sudo python3 -m pip install ansible
```

After the installation completes, you may have to log out and log back into your Terminal again before using Ansible. You can check the Ansible version you have installed with the following command:

```
ansible --version
```

In our case, the output shows the following:

```
ansible 2.10.7
```

As you can see, you can get the most recent version of Ansible by using `pip`. Therefore, it is the recommended method of installing Ansible.

With Ansible installed on our control node, let's look at some practical examples of using Ansible.

Working with Ansible

Starting with this section, we'll use the Ansible CLI tools extensively to perform various configuration management tasks. To showcase our practical examples, we'll work with a custom lab environment, and we highly encourage you to reproduce it for a complete configuration management experience.

Here's the high-level outline of this section:

- Setting up the lab environment
- Configuring Ansible
- Using Ansible ad hoc commands
- Using Ansible playbooks

Let's start with an overview of the lab environment.

Setting up the lab environment

Our lab uses VMware Fusion as a desktop hypervisor for the virtual environment, but any other hypervisor will do. *Chapter 1*, *Installing Linux*, describes the process of creating Linux VMs with VMware Fusion and Oracle VM VirtualBox in detail. We deployed the following virtual machines to mimic a real-world configuration management infrastructure:

- `Neptune`: The Ansible control node (Ubuntu).
- `web1`: The web server (Ubuntu).
- `web2`: The web server (CentOS).
- `db1`: The database server (Ubuntu).
- `db2`: The database server (CentOS).
- The `neptune`, `web1`, and `db1` VMs run on Ubuntu 20.04 LTS Server, while `web2` and `db2` have RHEL/CentOS 8 installed. All VMs have the default server components installed. On each host, we created a default admin user called `packt` with SSH access enabled. We described the related installation process in the first chapter of this book for both the Ubuntu and RHEL/CentOS server platforms.

Now, let's briefly describe the setup for these VMs, starting with the managed hosts first.

Setting up the managed hosts

There are a couple of key requirements for the managed hosts to fully enable configuration management access from the Ansible control node:

- They must be running OpenSSH server.
- They must have Python installed.

We assume you have OpenSSH enabled on your hosts. For installing Python, you may follow the related steps described in the *Installing Ansible* section.

> **Important note**
> The managed hosts don't require Ansible to be installed on the system.

To set the hostname on each VM, you may run the following command (for example, for the `web1` hostname):

```
sudo hostnamectl set-hostname web1
```

We also want to disable the sudo login password on our managed hosts to facilitate unattended privilege escalation when running automated scripts. If we don't make this change, remotely executing Ansible commands will require a password.

To disable the sudo login password, edit the sudo configuration with the following command:

```
sudo visudo
```

Add the following line and save the configuration file. Replace packt with your username if it's different:

```
packt ALL=(ALL) NOPASSWD:ALL
```

You'll have to make this change on all managed hosts.

Next, we'll look at the initial setup for the Ansible control node.

Setting up the Ansible control node

The Ansible control node (neptune) interacts with the managed hosts (web1, web2, db1, and db2) using Ansible commands and playbooks. For convenience, our examples will reference the managed hosts by their hostnames instead of their IP addresses. To easily accomplish this, we added the following entries to the /etc/hosts file on the Ansible control node (neptune):

```
127.0.0.1 neptune localhost
172.16.191.12 web1
172.16.191.13 db1
172.16.191.14 web2
172.16.191.15 db2
```

You'll have to match the hostnames and IP addresses according to your VM environment.

Next, we must install Ansible. Depending on your platform of choice, use the related procedure described in the *Installing Ansible* section earlier in this chapter. In our case, we followed the steps in the *Installing Ansible with pip* section to benefit from the latest Ansible release – version 2.10.7, at the time of writing.

Finally, we'll set up SSH key-based authentication between the Ansible control node and the managed hosts.

Setting up SSH key-based authentication

Ansible uses SSH communication with the managed hosts. The SSH key authentication mechanism enables remote SSH access without the need to enter user passwords. To enable SSH key-based authentication, run the following commands on the Ansible control host (`neptune`).

Use the following command to generate a secure key pair and follow the default prompts:

```
ssh-keygen
```

With the key pair generated, copy the related public key to each managed host. You'll have to target one host at a time and authenticate with the remote `packt` user's password. Accept the SSH key exchange when prompted:

```
ssh-copy-id -i ~/.ssh/id_rsa.pub packt@web1
ssh-copy-id -i ~/.ssh/id_rsa.pub packt@web2
ssh-copy-id -i ~/.ssh/id_rsa.pub packt@db1
ssh-copy-id -i ~/.ssh/id_rsa.pub packt@db2
```

Now, you should be able to SSH into any of the managed hosts from the Ansible control node (`neptune`) without being prompted for a password. For example, to access `web1`, you can test this with the following command:

```
ssh packt@web1
```

The command will take you to the remote server's (`web1`) Terminal. Make sure you go back to the Ansible control node's Terminal (on `neptune`) before following the next steps.

We're now ready to configure Ansible on the control node.

Configuring Ansible

This section explores some of the basic configuration concepts of Ansible that are related to the Ansible **configuration file** and **inventory**. Using a configuration file and the parameters within, we can change the *behavior* of Ansible, such as privilege escalation, connection timeout, and default inventory file path. The inventory *defines* the managed hosts, acting as the configuration management database of Ansible.

Let's look at the Ansible configuration file first.

Creating an Ansible configuration file

The following command provides some helpful information about our Ansible environment, including the current configuration file:

```
ansible --version
```

Here's the complete output of the preceding command:

```
packt@neptune:~$ ansible --version
ansible 2.10.7
  config file = /etc/ansible/ansible.cfg
  configured module search path = ['/home/packt/.ansible/plugins/modules', '/usr/share/ansible/plugins/modules']
  ansible python module location = /home/packt/.local/lib/python3.8/site-packages/ansible
  executable location = /home/packt/.local/bin/ansible
  python version = 3.8.5 (default, Jan 27 2021, 15:41:15) [GCC 9.3.0]
```

Figure 15.3 – The default Ansible configuration settings

A default Ansible installation will set the configuration file path to `/etc/ansible/ansible.cfg`. As you can probably guess, the default configuration file has a global scope, which means that it's used by default when we run Ansible tasks.

What if there are multiple users on the same control host running Ansible tasks? Our instinct suggests that each user may have their own set of configuration parameters. Ansible resolves this problem by looking into the user's home directory for the `~/.ansible.cfg` file. Let's verify this behavior by creating a dummy configuration file in our user's (`packt`) home directory:

```
touch ~/.ansible.cfg
```

A new invocation of the `ansible --version` command now yields the following config file path:

```
config file = /home/packt/.ansible.cfg
```

In other words, `~/.ansible.cfg` takes *precedence* over the global `/etc/ansible/ansible.cfg` configuration file.

Now, suppose our user (`packt`) creates multiple Ansible projects, some managing on-premises hosts and others interacting with public cloud resources. Again, we may need a different set of Ansible configuration parameters (such as a connection timeout and inventory file). Ansible accommodates this scenario by looking for the `./ansible.cfg` file in the current folder.

Let's create a dummy `ansible.cfg` file in a new `~/ansible/` directory:

```
mkdir ~/ansible
touch ~/ansible/ansible.cfg
```

Switching to the `~/ansible` directory and invoking the `ansible --version` command shows the following config file:

```
config file = /home/packt/ansible/ansible.cfg
```

We could have named our project directory anything, not necessarily `/home/packt/ansible`. Ansible prioritizes the `./ansible.cfg` file over the `~/.ansible.cfg` configuration file in the user's home directory.

Finally, we may want the ultimate flexibility of a configuration file that doesn't depend on the directory or location originating from our Ansible commands. Such a feature could be helpful while testing ad hoc configurations without altering the main configuration file. For this purpose, Ansible reads the `ANSIBLE_CONFIG` environment variable for the path of the configuration file.

Assuming we are in the `./ansible` project folder, where we already have our local `ansible.cfg` file defined, let's create a dummy test configuration file named `test.cfg`:

```
cd ~/ansible
touch test.cfg
```

Now, let's verify that Ansible will read the configuration from `test.cfg` instead of `ansible.cfg` when the `ANSIBLE_CONFIG` environment variable is set:

```
ANSIBLE_CONFIG=test.cfg ansible --version
```

The output shows the following:

```
config file = /home/packt/ansible/test.cfg
```

We should note that the configuration file should always have the `.cfg` extension. Otherwise, Ansible will discard it.

Here's a list summarizing the order of precedence for Ansible configuration files, from low to high priority:

- `/etc/ansible/ansible.cfg`
- The `~/.ansible.cfg` file in the user's home directory
- The `./ansible.cfg` file in the local directory
- The `ANSIBLE_CONFIG` environment variable

In our examples, we'll rely on the `ansible.cfg` configuration file in a local project directory (`~/ansible`). Let's create this configuration file and leave it empty for now:

```
mkdir ~/ansible
cd ~/ansible
touch ansible.cfg
```

For the rest of this chapter, we'll run our Ansible commands from the `~/ansible` folder unless we specify otherwise.

Unless we specifically define (override) configuration parameters in our configuration file, Ansible will assume the system defaults. One of the attributes that we'll add to the config file is the inventory file path. But first, we'll need to create the inventory. The following section will show you how.

Creating an Ansible inventory

The Ansible inventory is a regular **INI** or **YAML** file describing the managed hosts. In its simplest form, the inventory could be a flat list of hostnames or IP addresses, but Ansible can also organize the **hosts** into **groups**. Ansible inventory files are either **static** or **dynamic**, depending on whether they are created and updated manually or dynamically. For now, we'll use a static inventory.

In our demo environment with two web servers (`web1`, and `web2`) and two database servers (`db1`, and `db2`), we can define the following inventory (in INI format):

```
[webservers]
web1
web2

[databases]
db1
db2
```

We classified our hosts into a couple of groups, featured in bracketed names; that is, [webservers] and [databases]. Groups are logical arrangements of hosts based on specific criteria. Hosts can be part of multiple groups, for example, by adding the [ubuntu] and [centos] groups, as follows:

```
[ubuntu]
web1
db1

[centos]
web2
db2
```

Group names are case sensitive, should always start with a letter, and should not contain hyphens (-) or spaces.

Ansible has two default groups:

- **all**: Every host in the inventory
- **ungrouped**: Every host in **all** that is not a member of another group

We can also define groups based on specific patterns. For example, the following group includes a range of hostnames starting with web and ending with a number in the range of 1 – 2:

```
[webservers]
web[1:2]
```

Patterns are helpful when we're managing a large number of hosts. For example, the following pattern includes all the hosts within a range of IP addresses:

```
[all_servers]
172.16.191.[11:15]
```

Ranges are defined as [START:END] and include all values from START to END. Examples of ranges are [1:10], [01:10], and [a-g].

Groups can also be nested. In other words, a group may contain other groups. This nesting is described with the :children suffix. For example, we can define a [platforms] group that includes the [ubuntu] and [centos] groups:

```
[platforms:children]
ubuntu
centos
```

Let's name our inventory file hosts. Please note that we are in the ~/ansible directory. Using a Linux editor of your choice, add the following content to the hosts file:

```
[webservers]
web1
web2

[databases]
db1
db2

[ubuntu]
web1
db1

[centos]
web2
db2

[platforms:children]
ubuntu
centos
```

Figure 15.4 – The inventory file in INI format

The hosts file is also available in the complementary GitHub repository for this chapter: https://github.com/PacktPublishing/Mastering-Linux-Administration/blob/main/15/src/ansible/hosts. After saving the inventory file, we can validate it with the following command:

```
ansible-inventory -i ./hosts –list --yaml
```

Here's a brief explanation of the command's parameters:

- -i (--inventory): Specifies the inventory file; that is, ./hosts
- --list: Lists the current inventory, as read by Ansible
- --yaml: Specifies the output format as YAML

Upon successfully validating the inventory, the command will show the equivalent YAML output. (The default output format of the ansible-inventory utility is JSON.)

So far, we've expressed the Ansible inventory in INI format, but we may as well use a YAML file instead. The following figure shows a side-by-side comparison between the INI and YAML files describing the same inventory:

```
packt@neptune:~/ansible$ cat hosts        packt@neptune:~/ansible$ cat hosts.yml
[webservers]                              all:
web1                                        children:
web2                                          webservers:
                                                hosts:
[databases]                                       web1
db1                                               web2
db2                                           databases:
                                                hosts:
[ubuntu]                                          db1
web1                                              db2
db1                                           platforms:
                                                children:
[centos]                                          ubuntu:
web2                                                 hosts:
db2                                                    web1
                                                       db1
[platforms:children]                              centos:
ubuntu                                               hosts:
centos                                                 web2
                                                       db2
```

Figure 15.5 – Side-by-side comparison of the INI and YAML inventory formats

The YAML representation could be somewhat challenging, especially with large configurations, due to the strict indentation and formatting requirements. We'll continue to use the INI inventory format throughout the rest of this chapter.

Next, we'll point Ansible to our inventory. Edit the `./ansible.cfg` configuration file and add the following lines:

```
[defaults]
inventory = ~/ansible/hosts
```

After saving the file, we're ready to run Ansible commands or tasks that target our managed hosts. There are two ways we can perform Ansible configuration management tasks: using one-off **ad hoc commands** and via **Ansible playbooks**. We'll look at ad hoc commands next.

Using Ansible ad hoc commands

Ad hoc commands execute a single Ansible task and provide a quick way to interact with our managed hosts. These simple operations are helpful when we're making simple changes and performing testing.

The general syntax of an Ansible ad hoc command is as follows:

```
ansible [OPTIONS] -m MODULE -a ARGS PATTERN
```

The preceding command uses an Ansible MODULE to perform a particular task on select hosts based on a PATTERN. The task is described via arguments (ARGS). You may recall that modules encapsulate a specific functionality, such as managing users, packages, and services. To demonstrate the use of ad hoc commands, we'll use some of the most common Ansible modules for our configuration management tasks. Let's start with the Ansible ping module.

Working with the ping module

One of the simplest ad hoc commands is the Ansible ping test:

```
ansible -m ping all
```

The command performs a quick test on all managed hosts to check their SSH connectivity and ensure the required Python modules are present. Here's an excerpt from the output:

```
web1 | SUCCESS => {
    "ansible_facts": {
        "discovered_interpreter_python": "/usr/bin/python3"
    },
    "changed": false,
    "ping": "pong"
}
```

Figure 15.6 – A successful ping test with a managed host

The output suggests that the command was successful (web1 | SUCCESS) and that the remote server (web1) responded with a "pong" to our ping request ("ping" : "pong"). Please note that the Ansible ping module doesn't use ICMP to test the remote connection with managed hosts.

Next, we'll look at ad hoc commands while using the Ansible user module.

Working with the user module

Here's another example of an ad hoc command. This one is checking if a particular user (packt) exists on all the hosts:

```
ansible -m user -a "name=packt state=present" all
```

A successful check provides the following output (excerpt):

```
web1 | SUCCESS => {
    "ansible_facts": {
        "discovered_interpreter_python": "/usr/bin/python3"
    },
    "append": false,
    "changed": false,
    "comment": "Packt",
    "group": 1000,
    "home": "/home/packt",
    "move_home": false,
    "name": "packt",
    "shell": "/bin/bash",
    "state": "present",
    "uid": 1000
}
```

Figure 15.7 – Checking if a user account exists

The preceding output also suggests that we could be even more specific when checking for a user account by also making sure they have a particular user and group ID:

```
ansible -m user -a "name=packt state=present uid=1000
group=1000" all
```

We can target ad hoc commands against a limited subset of our inventory. The following command, for example, would only ping the web1 host for Ansible connectivity:

```
ansible -m ping web1
```

Host patterns can also include wildcards or group names. Here are a few examples:

```
ansible -m ping web*
ansible -m ping webservers
```

Let's look at the available Ansible modules next. Before we do that, you may want to add the following line to ./ansible.cfg, under the [defaults] section, to keep the noise down about deprecated modules:

```
deprecation_warnings = False
```

To list all the modules available in Ansible, run the following command:

```
ansible-doc --list
```

You may search or grep the output for a particular module. For detailed information about a specific module (for example, user), you can run the following command:

```
ansible-doc user
```

Make sure you check out the EXAMPLES section in the ansible-doc output for a specific module. You will find hands-on examples of using the module with ad hoc commands and playbook tasks.

If we want to create a new user (webuser) on all our web servers, we can perform the related operation with the following ad hoc command:

```
ansible -bK -m user -a "name=webuser state=present" webservers
```

Let's explain the command's parameters:

- -b (--become): Changes the execution context to sudo (root).
- -K (--ask-become-pass): Prompts for the sudo password on the remote hosts; the same password is used on all managed hosts.
- -m: Specifies the Ansible module (user).
- -a: Specifies the user module arguments as key-value pairs; name=webuser represents the username, while state=present checks if the user account exists before attempting to create it.
- Webservers: The group of managed hosts targeted by the operation.

Creating a user account requires administrative (sudo) privileges on the remote hosts. Using the -b (--become) option invokes the related **privilege escalation** for the Ansible command to act as a *sudoer* on the remote system.

> **Important note**
> By default, Ansible does not enable *unattended* privilege escalation. For tasks requiring sudo privileges, you must explicitly set the -b (--become) flag. You can override this behavior in the Ansible configuration file.

To enable unattended privilege escalation by default, add the following lines to the ansible.cfg file:

```
[privilege_escalation]
become = True
```

Now, you don't have to specify the --b (--become) flag anymore with your ad hoc commands.

If the sudoer account on the managed hosts has the sudo login password enabled, we'll have to provide it to our ad hoc command. Here is where the -K (--ask-become-pass) option comes in handy. Consequently, we're asked for a password with the following message:

```
BECOME password:
```

This password is used across all managed hosts targeted by the command.

As you may recall, we disabled the sudo login password on our managed hosts. (See the *Setting up the lab environment section* earlier in this chapter.) Therefore, we can rewrite the previous ad hoc command without explicitly asking for privilege escalation and the related password:

```
ansible -m user -a "name=webuser state=present" webservers
```

There are some security concerns regarding privilege escalation, and Ansible has the mechanisms to mitigate the related risks. For more information on this topic, you may refer to https://docs.ansible.com/ansible/latest/user_guide/become.html.

The preceding command produces the following output (excerpt):

```
web1 | CHANGED => {
    "ansible_facts": {
        "discovered_interpreter_python": "/usr/bin/python3"
    },
    "changed": true,
    "comment": "",
    "create_home": true,
    "group": 1001,
    "home": "/home/webuser",
    "name": "webuser",
    "shell": "/bin/sh",
    "state": "present",
    "system": false,
    "uid": 1001
}
```

Figure 15.8 – Creating a new user using an ad hoc command

You may have noticed that the output text here is yellow as opposed to green, as it was with our previous ad hoc commands. Ansible marks the output in yellow if it corresponds to a *change* in the *desired state* of the managed host. If you run the same command a second time, the output will be green, suggesting that there's been no change since the user account has been already created. Here, we can see Ansible's *idempotent operation* at work.

With the previous command, we created a user without a password for demo purposes only. What if we want to add or modify the password? Glossing through the user module's documentation with `ansible-doc user`, we can use the password field inside the module arguments, but Ansible will only accept *password hashes* as input. For hashing the passwords, we'll use a helper Python module called `passlib`. Let's install it on the Ansible control node with the following command:

```
pip install passlib
```

You'll need the Python package manager (`pip`) to run the previous command. If you installed Ansible using `pip`, you should be fine. Otherwise, follow the instructions in the *Installing Ansible using pip* section to download and install `pip`.

With `passlib` installed, we can use the following ad hoc command to create or modify the user password:

```
ansible webservers -m user \
    -e "password=changeit!" \
    -a "name=webuser \
        update_password=always \
        password={{ password | password_hash('sha512') }}"
```

Here are the additional parameters helping with the user password:

- `-e` (`--extra-vars`): Specifies custom variables as key-value pairs; we set the value of a custom variable to `password=changeit!`.

- `update_password=always`: Updates the password if it's different from the previous one.

- `password={{...}}`: Sets the password to the value of the expression enclosed within double braces.

- `password | password_hash('sha512')`: Pipes the value of the `password` variable (`changeit!`) to the `password_hash()` function, thus generating a SHA-512 hash; `password_hash()` is part of the `passlib` module we installed earlier.

The command sets the password of `webuser` to `changeit!` and it's an example of using variables (`password`) in ad hoc commands. Here's the related output (excerpt):

```
web1 | CHANGED => {
    "ansible_facts": {
        "discovered_interpreter_python": "/usr/bin/python3"
    },
    "changed": true,
    "comment": "",
    "create_home": true,
    "group": 1001,
    "home": "/home/webuser",
    "name": "webuser",
    "password": "NOT_LOGGING_PASSWORD",
    "shell": "/bin/sh",
    "state": "present",
    "system": false,
    "uid": 1001
}
```

Figure 15.9 – Changing the user's password using an ad hoc command

Ansible won't show the actual password for obvious security reasons.

Now, you can try to SSH into any of the web servers (`web1` or `web2`) using the `webuser` account, and you should be able to authenticate successfully with the `changeit!` password.

To delete the `webuser` account on all web servers, we can run the following ad hoc command:

```
ansible -m user -a "name=webuser state=absent remove=yes
force=yes" webservers
```

The `state=absent` module parameter invokes the deletion of the `webuser` account. The `remove` and `force` parameters are equivalent to the `userdel -rf` command, deleting the user's home directory and any files within, even if they're not owned by the user.

The related output is as follows:

```
web1 | CHANGED => {
    "ansible_facts": {
        "discovered_interpreter_python": "/usr/bin/python3"
    },
    "changed": true,
    "force": true,
    "name": "webuser",
    "remove": true,
    "state": "absent",
    "stderr": "userdel: webuser mail spool (/var/mail/webuser) not found\n",
    "stderr_lines": [
        "userdel: webuser mail spool (/var/mail/webuser) not found"
    ]
}
```

Figure 15.10 – Deleting a user account using an ad hoc command

You may safely ignore `stderr` and `stderr_lines`, which were captured in the output. The message is benign since the user didn't create a mail spool previously.

We'll look at the `package` module next and run a few related ad hoc commands.

Working with the package module

The following command installs the **Nginx** web server on all the hosts within the `webserver` group:

```
ansible -m package -a "name=nginx state=present" webservers
```

Here's an excerpt from the output:

```
web2 | CHANGED => {
    "ansible_facts": {
        "discovered_interpreter_python": "/usr/libexec/platform-python"
    },
    "changed": true,
    "msg": "",
    "rc": 0,
    "results": [
        "Installed: nginx-mod-http-xslt-filter-1:1.14.1-9.module_el8.0.0+184+e34fea82.x86_64",
        "Installed: nginx-mod-mail-1:1.14.1-9.module_el8.0.0+184+e34fea82.x86_64",
        "Installed: nginx-mod-stream-1:1.14.1-9.module_el8.0.0+184+e34fea82.x86_64",
        "Installed: nginx-1:1.14.1-9.module_el8.0.0+184+e34fea82.x86_64",
        "Installed: nginx-all-modules-1:1.14.1-9.module_el8.0.0+184+e34fea82.noarch",
        "Installed: nginx-mod-http-image-filter-1:1.14.1-9.module_el8.0.0+184+e34fea82.x86_64",
        "Installed: nginx-mod-http-perl-1:1.14.1-9.module_el8.0.0+184+e34fea82.x86_64"
    ]
}
```

Figure 15.11 – Installing the nginx package on the web servers

We use a similar ad hoc command to install the **MySQL** database server on all the hosts within the `databases` group:

```
ansible -m package -a "name=mysql-server state=present"
databases
```

Here's an excerpt from the command's output:

```
db2 | CHANGED => {
    "ansible_facts": {
        "discovered_interpreter_python": "/usr/libexec/platform-python"
    },
    "changed": true,
    "msg": "",
    "rc": 0,
    "results": [
        "Installed: mysql-server-8.0.21-1.module_el8.2.0+493+63b41e36.x86_64"
    ]
}
```

Figure 15.12 – Installing the mysql-server package on the database servers

If we wanted to *remove* a package, the ad hoc command would be similar, but would feature `state=absent` instead.

Although the `package` module provides a good OS-level abstraction across various platforms, certain package management tasks are best handled with platform-specific package managers. We'll show you how to use the `apt` and `yum` modules next.

Working with platform-specific package managers

The following ad hoc command installs the latest updates on all Ubuntu machines in our managed environment. This command targets the `ubuntu` group in our inventory:

```
ansible -m apt -a "upgrade=dist update_cache=yes" ubuntu
```

Similarly, we can install the latest updates on RHEL/CentOS machines by targeting the `centos` group with the following ad hoc command:

```
ansible -m yum -a "name=* state=latest update_cache=yes" centos
```

The platform-specific package management modules (`apt`, `yum`, and so on) match all the capabilities of the system-agnostic `package` module, featuring additional OS-exclusive functionality.

Let's look at the `service` module next and a couple of related ad hoc commands.

Working with the service module

The following command restarts the `nginx` service on all the hosts in the `webservers` group:

```
ansible -m service -a "name=nginx state=restarted" webservers
```

Here's a relevant excerpt from the output:

```
web1 | CHANGED => {
    "ansible_facts": {
        "discovered_interpreter_python": "/usr/bin/python3"
    },
    "changed": true,
    "name": "nginx",
    "state": "started",
    "status": {
        "ActiveEnterTimestamp": "Tue 2021-04-06 04:12:23 UTC",
        "ActiveEnterTimestampMonotonic": "10249321",
        "ActiveExitTimestampMonotonic": "0",
        "ActiveState": "active",
```

Figure 15.13 – Restarting the nginx service on the web servers

In the same way, we can restart the `mysql` service on all database servers, but there's a trick to it! On Ubuntu, the MySQL service is named `mysql`, while on RHEL and CentOS systems, it is called `mysqld`. We could, of course, target each host with the appropriate service name, but if you had many database servers, both on Ubuntu and RHEL/CentOS, it would be a laborious task. Alternatively, we can use the *exclusion pattern* (`!`) when targeting multiple hosts or groups.

The following command will restart the `mysql` service on all the hosts in the `databases` group, except the ones that are members of the `centos` group:

```
ansible -m service -a "name=mysql state=restarted"
'databases:!centos'
```

Similarly, we can restart the `mysqld` service on all the hosts in the `databases` group, except for the ones that are members of the `ubuntu` group, with the following ad hoc command:

```
ansible -m service -a "name=mysqld state=restarted"
'databases:!ubuntu'
```

Always use single quotes (`' '`) when you're targeting multiple hosts or groups with an exclusion pattern; otherwise, the `ansible` command will fail.

Let's look at one last Ansible module and the related ad hoc command, which is frequently used in upgrade scenarios.

Working with the reboot module

The following ad hoc command reboots all the hosts in the `webservers` group:

```
ansible -m reboot -a "reboot_timeout=3600" webservers
```

Slower hosts may take longer to reboot, especially during substantial upgrades, hence the increased reboot timeout of `3600` seconds. (The default timeout is `600` seconds.)

In our case, the reboot only took a few seconds. The output is as follows:

```
web1 | CHANGED => {
    "changed": true,
    "elapsed": 14,
    "rebooted": true
}
web2 | CHANGED => {
    "changed": true,
    "elapsed": 25,
    "rebooted": true
}
```

Figure 15.14 – Rebooting the webservers group

In this section, we showed a few examples of ad hoc commands using different modules. The next section will give you a brief overview of some of the most common Ansible modules and how to explore more.

Exploring Ansible modules

Ansible has a vast library of modules. As we noted previously, you may use the `ansible-doc --list` command to browse the available Ansible modules on the command-line Terminal. You can also access the same information online, on the Ansible modules index page, at `https://docs.ansible.com/ansible/2.9/modules/modules_by_category.html`.

The online catalog provides module-by-category indexing to help you quickly locate a particular module you're looking for. Here are some of the *most typical modules* used in everyday system administration and configuration management tasks with Ansible:

Packaging modules:

- `apt`: Performs APT package management
- `yum`: Performs YUM package management
- `dnf`: Performs DNF package management

System modules:

- `users`: Manages users
- `services`: Controls services
- `reboot`: Restarts machines
- `firewalld`: Performs firewall management

File modules:

- `copy`: Copies local files to the managed hosts
- `synchronize`: Synchronizes files and directories using `rsync`
- `file`: Controls file permissions and attributes
- `lineinfile`: Manipulates lines in text files

Net Tools modules:

- `nmcli`: Controls network settings
- `get_url`: Downloads files over HTTP, HTTPS, and FTP
- `uri`: Interacts with web services and API endpoints

Commands modules (not idempotent!):

- `raw`: Simply runs a remote command via SSH (unsafe!); doesn't need Python installed on the remote host
- `command`: Runs commands securely using Python's remote execution context
- `shell`: Executes shell commands on the managed hosts

We should note that ad hoc commands always execute a *single operation* using a *single module*. This feature is an advantage (for quick changes) but also a limitation. For more complex configuration management tasks, we use Ansible playbooks. The following section will take you through the process of authoring and running Ansible playbooks.

Using Ansible playbooks

An Ansible playbook is essentially a list of tasks that's executed automatically. Ansible configuration management workflows are primarily driven by playbooks. More precisely, a **playbook** is a YAML file containing one or more **plays**, each with a list of **tasks** executed in the order they are listed. Plays are execution units that run the associated tasks against a set of hosts, and they are selected via a group identifier or a pattern. Each task uses a single module that's executing a specific action targeted at the remote host. You may think of a task as a simple Ansible ad hoc command. As the majority of Ansible modules comply with idempotent execution contexts, playbooks are also idempotent. Running a playbook multiple times always yields the same result.

Well-written playbooks can replace laborious administrative tasks and complex scripts with relatively simple and maintainable manifests, running easily repeatable and predictable routines.

We'll create our first Ansible playbook next.

Creating a simple playbook

We'll build our playbook based on the ad hoc command we used for creating a user (webuser). As a quick refresher, the command was as follows:

```
ansible -m user -a "name=webuser state=present" webservers
```

As we write the equivalent playbook, you may notice some resemblance to the ad hoc command parameters.

While editing the playbook YAML file, please be aware of the YAML formatting rules:

- Use only space characters for indentation (no tabs).
- Keep the indentation length consistent (for example, two spaces).
- Items at the same level in the hierarchy (for example, list items) must have the same indentation.
- A child item's indentation is one indentation more than its parent.

Now, using a Linux editor of your choice, add the following lines to a `create-user.yml` file. Make sure you create the playbook in the `~/ansible` project directory, which is where we have our current inventory (`hosts`) and Ansible configuration file (`ansible.cfg`):

```
1    ---
2    - name: Create a specific user on all web servers
3      hosts: webservers
4      become: yes
5      tasks:
6        - name: Create the 'webuser' account
7          user:
8            name: webuser
9            state: present
```

Figure 15.15 – A simple playbook for creating a user

Let's look at each line in our `create-user.yml` playbook:

- `---`: Marks the beginning of the playbook file.
- `- name:`: Describes the name of the play; we can have one or more plays in a playbook.
- `hosts: webservers`: Targets the hosts in the `webservers` group.
- `become: yes`: Enables privilege escalation for the current task; you can leave this line out if you enabled unattended privilege escalation in your Ansible configuration file (with `become = True` in the `[privileged_escalation]` section).
- `tasks:`: The list of tasks in the current play.
- `- name:`: The name of the current task; we can have multiple tasks in a play.
- `user:`: The module being used by the current task.
- `name: webuser`: The name of the user account to create.
- `state: present`: The desired state upon creating the user – we want the user account to be *present*.

Let's run our `create-user.yml` playbook:

```
ansible-playbook create-user.yml
```

Here's the output we get after a successful playbook run:

```
packt@neptune:~/ansible$ ansible-playbook create-user.yml

PLAY [Create a specific user on all web servers] *************************************

TASK [Gathering Facts] *************************************************************
ok: [web2]
ok: [web1]

TASK [Create the 'webuser' account] ***********************************************
changed: [web1]
changed: [web2]

PLAY RECAP ***********************************************************************
web1                       : ok=2    changed=1    unreachable=0    failed=0    skipped=0
    rescued=0    ignored=0
web2                       : ok=2    changed=1    unreachable=0    failed=0    skipped=0
    rescued=0    ignored=0
```

Figure 15.16 – Running the create-user.yml playbook

Most of the `ansible-playbook` command-line options are similar to the ones for the `ansible` command. Let's look at some of these parameters:

- `-i` (`--inventory`): Specifies an inventory file path.

- `-b` (`--become`): Enables privilege escalation to `sudo` (`root`).

- `-C` (`--check`): Produces a dry run without making any changes and anticipating the end results – a useful option for validating playbooks.

- `-l` (`--limit`): Limits the action of the command or playbook to a subset of the managed hosts.

- `--syntax-check`: Validates the playbook's syntax without making any changes; this option is only available for the `ansible-playbook` command.

Let's experiment with a second playbook, this time for deleting a user. We'll name the playbook `delete-user.yml` and add the following content:

```
1   ---
2   - name: Delete a specific user on all web servers
3     hosts: webservers
4     become: yes
5     tasks:
6       - name: Delete the 'webuser' account
7         user:
8           name: webuser
9           remove: yes
10          force: yes
11          state: absent
```

Figure 15.17 – A simple playbook for deleting a user

Now, let's run this playbook in a selective way: we want to limit its action to only targeting Ubuntu `webservers` group. In other words, we will **limit** the playbook's targets to the `ubuntu` host group:

```
ansible-playbook delete-user.yml --limit ubuntu
```

The output of the preceding command is as follows:

```
packt@neptune:~/ansible$ ansible-playbook delete-user.yml --limit ubuntu

PLAY [Delete a specific user on all web servers] *********************************

TASK [Gathering Facts] **********************************************************
ok: [web1]

TASK [Delete the 'webuser' account] *********************************************
ok: [web1]

PLAY RECAP **********************************************************************
web1                       : ok=2    changed=0    unreachable=0    failed=0    skipped=0
    rescued=0    ignored=0
```

Figure 15.18 – Limiting the delete-user.yml playbook to the ubuntu host group

As the preceding output suggests, the playbook only targeted the hosts in the `webservers` group that are members of the `ubuntu` group (`web1`). Although the `delete-user.yml` playbook internally targets all the hosts in the `webserver` group, with the `--limit` flag, we restrict this action exclusively to the hosts of both the `webserver` and `ubuntu` groups.

Let's rerun the `delete-user.yml` playbook, this time without a limited scope:

```
ansible-playbook delete-user.yml
```

Notice the idempotent operation on `web1` (the Ubuntu web server), where the user (`webuser`) has already been removed with our previous playbook run. You can also see the color coding in the output: green for the unchanged state on `web1` and yellow for the change that was made on `web2`:

```
packt@neptune:~/ansible$ ansible-playbook delete-user.yml

PLAY [Delete a specific user on all web servers] ***********************************

TASK [Gathering Facts] ************************************************************
ok: [web1]
ok: [web2]

TASK [Delete the 'webuser' account] ***********************************************
ok: [web1]
changed: [web2]

PLAY RECAP ************************************************************************
web1                       : ok=2    changed=0    unreachable=0    failed=0    skipped=0
     rescued=0    ignored=0
web2                       : ok=2    changed=1    unreachable=0    failed=0    skipped=0
     rescued=0    ignored=0
```

Figure 15.19 – Rerunning the delete-user.yml playbook (without a limited scope)

Next, we'll look at ways to further streamline our configuration management workflows, starting with the use of variables in playbooks.

Using variables in playbooks

Ansible provides a flexible and versatile model for working with variables in both playbooks and ad hoc commands. Through variables, we are essentially *parameterizing* a playbook, making it reusable or dynamic. Take our previous playbook, for example, to create a user. We hard-coded the username (`webuser`) in the playbook. We can't really reuse the playbook to create another user (for example, `webadmin`), unless we add the related task to it. But then, if we had many users, our playbook would grow proportionally, making it harder to maintain. And what if we want to specify a password for each user as well? The complexity of the playbook would grow even more.

Here's where variables come into play. We can substitute the hard-coded values with variables, making the playbook dynamic. In terms of pseudocode, our example of using a `Playbook` to create a `User` with a specific `username` and `password` would look like this:

```
User = Playbook(username, password)
```

Variables in Ansible are enclosed in double braces; for example, `{{ username }}`. Let's see how we can leverage variables in our playbooks. Edit the `create-user.yml` playbook we worked on in the previous section and adjust it as follows:

```
1   ---
2   - name: Create a specific user on all web servers
3     hosts: webservers
4     become: yes
5     tasks:
6       - name: Create the '{{ username }}' account
7         user:
8           name: "{{ username }}"
9           state: present
```

Figure 15.20 – Using the username variable in a playbook

Lines 6 and 8 have the `{{ username }}` variable substituting our previously hard-coded value (`webuser`). In line 8, we surrounded the double braces with quotes to avoid syntax interference with the YAML dictionary notation. Variable names in Ansible must begin with a letter and only contain alphanumerical characters and underscores.

Next, we'll explain *how* and *where* to set the values for variables. Ansible implements a hierarchical model for assigning values to variables:

1. **Global variables**: The values are set for all the hosts, either via the `--extra-vars` `ansible-playbook` command-line parameter or the `./group_vars/all` file.

2. **Group variables**: The values are set for the hosts in a specific group, either in the inventory file or the local `./group_vars` directory in files named after each group.

3. **Host variables**: The values are set for a particular host, either in the inventory file or the local `./host_vars` directory in files named after each host. Host-specific variables are also available from Ansible **facts** via the `gather_facts` directive. You can learn more about Ansible facts at `https://docs.ansible.com/ansible/latest/user_guide/playbooks_vars_facts.html#ansible-facts`.

4. **Play variables**: The values are set in the context of the current play for the hosts targeted by the play; examples are the `vars` directive in a play or `include_vars` tasks.

In the previous numbered list, the order of precedence for a variable's value increases with each number. In other words, a variable value defined in a play will overwrite the same variable value specified at the host, group, or global level.

As an example, you may recall the peculiarity related to the MySQL service name on Ubuntu and RHEL/CentOS platforms. On Ubuntu, the service is `mysql`, while on CentOS, the service is `mysqld`. Suppose we want to restart the MySQL service on all the hosts in our `databases` group. We ran into this problem in the *Using Ansible ad hoc commands* section. Assuming most of our database servers run CentOS, we can define a group-level `service` variable as `service: mysqld`. We set this variable in the local project's `./group_vars/databases` file. Then, in the play where we control the service status, we can override the `service` variable value with `mysql` when the remote host's OS platform is Ubuntu.

Let's look at a few examples to illustrate what we've learned so far about placing variables and setting their values. Back in our `create-user.yml` playbook, we can define the `username` variable at the play level with the following directive:

```
vars:
    username: webuser
```

Here's what it looks like in the overall playbook (lines 5-6):

```
1    ---
2    - name: Create a specific user on all web servers
3      hosts: webservers
4      become: yes
5      vars:
6        username: webuser
7      tasks:
8        - name: Create the '{{ username }}' account
9          user:
10           name: "{{ username }}"
11           state: present
```

Figure 15.21 – Defining a variable at the play level

Let's run our playbook with the following command:

```
ansible-playbook create-user.yml
```

Here's the relevant excerpt from the output:

```
TASK [Create the 'webuser' account] ****************************************************
changed: [web1]
changed: [web2]
```

Figure 15.22 – Creating a user with a playbook using variables

To delete the user accounts, we can readjust our previous `delete-user.yml` file so that it looks as follows:

```
1   ---
2   - name: Delete a specific user on all web servers
3     hosts: webservers
4     become: yes
5     vars:
6       username: webuser
7     tasks:
8       - name: Delete the '{{ username }}' account
9         user:
10          name: "{{ username }}"
11          remove: yes
12          force: yes
13          state: absent
```

Figure 15.23 – Deleting a user with a playbook using variables

After saving the file, run the following command to delete the `webuser` account on all web servers:

```
ansible-playbook delete-user.yml
```

The relevant output from the preceding command run is as follows:

```
TASK [Delete the 'webuser' account] ************************************************
changed: [web1]
changed: [web2]
```

Figure 15.24 – Deleting a user with a playbook using variables

We can improve our `create-user` and `delete-user` playbooks even further. Since the play exclusively targets the `webservers` group, we can define the `username` variable in the `./group_vars/webservers` file instead. This way, we can keep the playbooks more compact. Let's remove the variable definition (lines 5-6) from both files. The `create-user.yml` file will look identical to *Figure 15.20*.

Next, create the `./group_wars` folder in the local directory (`~/ansible`) and add the following lines to a file named `webservers.yml`:

```
---

username: webuser
```

We could also just call the file `webservers` so that it matches the group we're targeting. However, we should prefer to use the `.yml` extension so that we're consistent with the file's YAML format. Ansible accepts both naming conventions. Here's the current tree structure of our project directory:

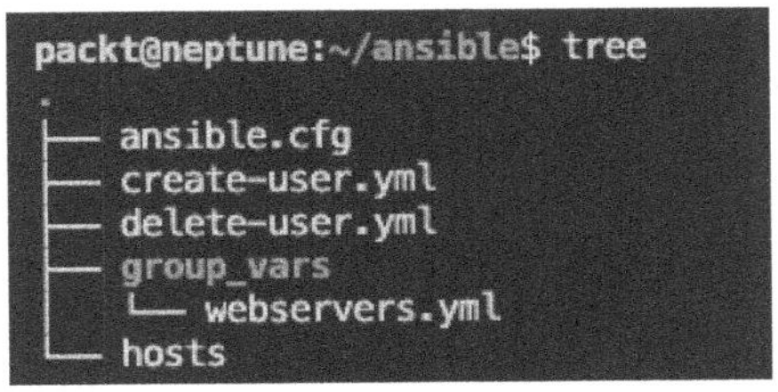

```
packt@neptune:~/ansible$ tree
.
├── ansible.cfg
├── create-user.yml
├── delete-user.yml
├── group_vars
│   └── webservers.yml
└── hosts
```

Figure 15.25 – The directory tree, including the group_vars folder

If we run our playbooks, the results should be identical to our previous runs:

```
ansible-playbook create-user.yml
ansible-playbook delete-user.yml
```

Now, let's add one more variable to our `create-user` playbook: the user's `password`. You may recall the ad hoc command we created for the same purpose. See the *Using Ansible ad hoc commands* section earlier in this chapter for more information.

Add the following lines to the `create-user.yml` file of the `user` task, at the same level as `name`:

```
password: "{{ password | password_hash('sha512') }}"
update_password: always
```

You may notice how these changes are similar to the related ad hoc command. The updated playbook contains the following content:

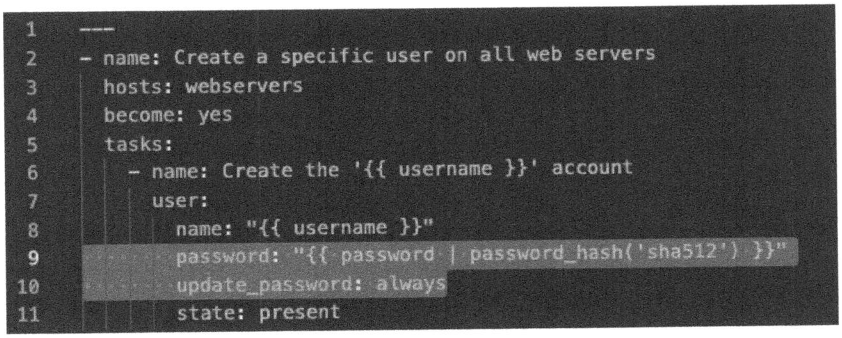

```
 1  ---
 2  - name: Create a specific user on all web servers
 3    hosts: webservers
 4    become: yes
 5    tasks:
 6      - name: Create the '{{ username }}' account
 7        user:
 8          name: "{{ username }}"
 9          password: "{{ password | password_hash('sha512') }}"
10          update_password: always
11          state: present
```

Figure 15.26 – The playbook with username and password variables

Next, edit the `./group_vars/webservers.yml` file and add the `password` variable with the `changeit!` value. Your updated file should have the following content:

```
---
username: webuser
password: changeit!
```

Let's run the playbook:

```
ansible-playbook create-user.yml
```

The command's output is identical to our similar previous commands. You may test the new username (`webuser`) and password (`changeit!`) by trying to SSH into one of the web servers (for example, `web1`):

```
ssh webuser@web1
```

The SSH authentication should succeed. Make sure to exit the remote Terminal before proceeding with the next steps. Let's remove the `webuser` account on the webservers with the following command, to get back to our initial state:

```
ansible-playbook delete-user.yml
```

Suppose we want to reuse the `create-user` playbook to create a different user with a different password. Let's name this user `webadmin`; we'll set the password to `changeme!`. One way to accomplish this task is to use the `-e` (`--extra-vars`) option parameter with `ansible-playbook`:

```
ansible-playbook -e '{"username": "webadmin", "password":
"changeme!"}' create-user.yml
```

The preceding command will create a new user (`webadmin`) with the related password. You can test the credentials with the following command:

```
ssh webadmin@web1
```

The SSH authentication should succeed. Make sure you exit the remote Terminal before continuing.

As you can see, the `-e` (`--extra-vars`) option parameter takes a JSON string featuring the `username` and `password` fields, along with the corresponding values. These values will *override* the values of the same variables defined at the group level in the `./group_vars/webservers.yml` file.

Let's remove the `webuser` and `webadmin` accounts before we proceed with the next steps. Let's run the `delete-user` playbook, first without any parameters:

```
ansible-playbook delete-user.yml
```

The preceding command removes the `webuser` account. Next, we'll use the `-e` (`--extra-vars`) option parameter to delete the `webadmin` user:

```
ansible-playbook -e '{"username": "webadmin"}' delete-user.yml
```

Using `--extra-vars` with our `create-user` and `delete-user` playbooks, we can act on multiple user accounts by running the playbooks manually or in a loop and feeding the JSON blob with the required variables. While this method could easily be scripted, Ansible provides even more ways to improve our playbooks by using task iteration with loops. We'll look at loops later in this chapter, but first, let's handle our passwords more securely with Ansible's encryption and decryption facilities for managing secrets.

Working with secrets

Ansible has a dedicated module for managing secrets called **Ansible Vault**. With Ansible Vault, we can encrypt and store sensitive data such as variables and files that are referenced in playbooks. Ansible Vault is essentially a password-protected secure key-value data store.

To manage our secrets, we can use the **ansible-vault** command-line utility. Regarding our playbook, where we're creating a user with a password, we want to avoid storing the password in clear text. It is currently in the `./group_vars/webservers.yml` file. As a reminder, our `webservers.yml` file has the following content:

Figure 15.27 – The sensitive data stored in the password variable

Line 3 contains sensitive data: the password is shown in plain text. We have a few of options here to protect our data:

- Encrypt the `webservers.yml` file.
- Encrypt the `password` variable only.
- Store the password in a separate protected file.

Let's briefly discuss each of these options. If we choose to encrypt the `webservers.yml` file, we could possibly incur the overhead of encrypting non-sensitive data, such as `username` or other general-purpose information. If we have many users, encrypting and decrypting non-sensitive data would be highly redundant.

The second option – encrypting only the password variable – would work fine for a single user. But with a growing number of users, we'll have multiple password variables to deal with, each with its own encryption and decryption. Performance would once again be an issue if we have a large number of users.

Ideally, we should have a separate file for storing all sensitive data. This file would be decrypted only once during the playbook run, even with multiple passwords stored. So, let's pursue this option and create a separate file to keep our user passwords in. We'll name the file `passwords.yml` and add the following content to it:

```
1    ---
2    webuser:
3      password: changeit!
```

Figure 15.28 – The passwords.yml file storing sensitive data

We added a YAML dictionary (or hash) item matching the `webuser` username related to the password. This item contains another dictionary as a key-value pair: `password: changeit!`. The equivalent YAML representation is as follows:

```
webuser: { password: changeit! }
```

This approach will allow us to add passwords that correspond to different users, like so:

```
webuser: { password: changeit! }
webadmin: { password: changeme! }
```

We'll explain the concept behind this data structure and its use when we consume the `password` variable in the playbook, later in this section.

Now, since we keep our password in a different file, we'll remove the corresponding entry from `webusers.yml`. Let's add some other user-related information using the `comment` variable. Here's what our `webusers.yml` file looks like:

```
1    ---
2    username: webuser
3    comment: Regular web user
```

Figure 15.29 – The webusers.yml file storing non-sensitive user data

Next, let's protect our secrets by encrypting the `passwords.yml` file using Ansible Vault:

```
ansible-vault encrypt passwords.yml
```

You'll be prompted to create a vault password for protecting the file. Remember the password, as we'll be using it throughout this section. Once you're done, check the `passwords.yml` file with the following command:

```
cat passwords.yml
```

The output shows our file is encrypted:

```
packt@neptune:~/ansible$ cat passwords.yml
$ANSIBLE_VAULT;1.1;AES256
6433356332643563313532336231373336337313839653033646132373633623264326466632636265
3065613334333738313866633363861316361636332346630a3965666130326235316531366633535
3736393238316632386663632376230666232343564346335326132373934653938316461393139
3838633263343535380a3862646464313139303833336313034616437376634643732306135396263
6230623838303763613235626432623166343861383239306133362346532613865616635353163463
61663232353933373432346232613333653764396136306633937
```

Figure 15.30 – The encrypted passwords.yml file

We can view the content of the `passwords.yml` file with the following command:

```
ansible-vault view passwords.yml
```

You'll be prompted for the vault password we created previously. The output shows the streamlined YAML content corresponding to our protected file:

```
packt@neptune:~/ansible$ ansible-vault view passwords.yml
Vault password:

webuser: { password: changeit! }
```

Figure 15.31 – Viewing the content of the protected file

If you need to make changes, you can edit the encrypted file with the following command:

```
ansible-vault edit passwords.yml
```

After authenticating with the vault password, the command will open a local editor (`vi`) to edit your changes. If you want to re-encrypt your protected file with a different password, you can run the following command:

```
ansible-vault rekey passwords.yml
```

You'll be prompted for the current vault password, followed by the new password.

Now, let's learn how to reference secrets in our playbook. First, let's make sure we can read our password from the vault. Make the following changes in the `create-user.yml` file:

```
 1  ---
 2  - name: Create a specific user on all web servers
 3    hosts: webservers
 4    become: yes
 5    tasks:
 6      - name: Get the password for {{ username }} from Vault
 7        include_vars:
 8          file: passwords.yml
 9
10      - name: Debug password for {{ username }}
11        debug:
12          msg: "{{ vars[username]['password'] }}"
```

Figure 15.32 – Debugging vault access

We've added a couple of tasks:

- `include_vars` (lines 6-8): Reading variables from the `passwords.yml` file
- `debug` (lines 10-12): Debugging the playbook and logging the password that was read from the vault

None of these tasks are *aware* that the `passwords.yml` file is protected. Line 12 is where the *magic* happens:

```
msg: "{{ vars[username]['password'] }}"
```

We use the `vars[]` dictionary to query a specific variable in the playbook. `vars[]` is a *reserved* data structure for storing all the variables that were created via `vars` and `include_vars` in an Ansible playbook. We can query the dictionary based on a key appointed by `username`:

```
{{ vars[username] }}
```

Our playbook gets `username` from the `./group_vars/webservers.yml` file, and its value is `webuser`. Consequently, the `vars[webuser]` dictionary item reads the corresponding entry from the `passwords.yml` file:

```
webuser: { password: changeit! }
```

To get the password value from the corresponding key-value pair, we specify the `'password'` key in the `vars[username]` dictionary:

```
{{ vars[username]['password'] }}
```

Let's run this playbook with the following command:

```
ansible-playbook --ask-vault-pass create-user.yml
```

We invoked the `--ask-vault-pass` option to let Ansible know that our playbook needs vault access. Without this option, we'll get an error when running the playbook. Here's the relevant output for our debug task:

```
TASK [Get the password for webuser from Vault] **********************************
ok: [web1]
ok: [web2]

TASK [Debug password for webuser] **********************************
ok: [web1] => {
    "msg": "changeit!"
}
ok: [web2] => {
    "msg": "changeit!"
}
```

Figure 15.33 – The playbook successfully reading secrets from the vault

Here, we can see that the playbook successfully retrieves the password from the vault. Let's wrap up our `create-user.yml` playbook by adding the following code:

```
1    ---
2    - name: Create a specific user on all web servers
3      hosts: webservers
4      become: yes
5      vars:
6        password: "{{ vars[username]['password'] }}"     1
7      tasks:
8        - name: Get the password for {{ username }} from Vault
9          include_vars:                                    2
10           file: passwords.yml
11
12       - name: Debug password for {{ username }}
13         debug:                                           3
14           msg: "{{ password }}"
15         no_log: true    ←
16
17       - name: Create the '{{ username }}' account
18         user:
19           name: "{{ username }}"
20           comment: "{{ comment }}"                       4
21           password: "{{ password | password_hash('sha512') }}"
22           update_password: always
23           state: present
```

Figure 15.34 – The playbook creating a user with a password retrieved from the vault

Here are a few highlights about the current implementation:

- We've added the `vars` block (lines 5-6) to define a local `password` variable (at the play scope) for reading the password from the vault; we are reusing the `password` variable in multiple tasks.

- The `include_vars` task (lines 8-10) adds an external reference to the variables defined in the protected `passwords.yml` file.

- The `debug` task (lines 12-15) helped with the initial debugging effort to make sure we can read the password from the vault. You may choose to remove this task or leave it in place for future use. If you keep the task, make sure you have `no_log: true` enabled (line 15) to avoid logging sensitive information in the output. When debugging, you can temporarily set `no_log: false`.

- The `user` task reads the `password` variable and hashes the corresponding value. This hashing is required by the Ansible `user` module for security reasons. We also added the `comment` field with additional user information. This field maps to the Linux GECOS record of the user. See the *Managing users* section of *Chapter 4, Managing Users and Groups*, for related information.

Let's run the playbook with the following command:

```
ansible-playbook --ask-vault-pass create-user.yml
```

After the command completes successfully, you can verify the new user account in a couple of ways:

- Use SSH to connect to any of the web servers using the related username and password, like so:

```
ssh webuser@web1
```

- Look for the `webuser` record in `/etc/passwd`:

```
tail -n 10 /etc/passwd
```

You should see the following line in the output (you'll notice the GECOS field also):

```
webuser:x:1001:1001:Regular web user:/home/webuser:/bin/
sh
```

You may want to run the `ansible-playbook` command without supplying a vault password, as required by `--ask-vault-pass`. Such functionality is essential in scripted or automated workflows when using Ansible Vault. To make your vault password automatically available when you're running a playbook using sensitive data, start by creating a regular text file, preferably in your home directory; for example, `~/vault.pass`. Add the vault password to this file in a single line. Then, you can choose *either of* the following options to use the vault password file:

- Create the following environment variable:

```
export ANSIBLE_VAULT_PASSWORD_FILE=~/vault.pass
```

- Add the following line to the `ansible.cfg` file's `[defaults]` section:

```
vault_password_file = ~/vault.pass
```

Now, you can run the `create-user` playbook without the `--ask-vault-pass` option:

```
ansible-playbook create-user.yml
```

Sometimes, protecting multiple secrets with a single vault password raises security concerns. Ansible supports multiple vault passwords through vault IDs.

Using vault IDs

A **vault ID** is an identifier, or a label, associated with one or more vault secrets. Each vault ID has a unique password to unlock the encryption and decryption of the corresponding secrets. To illustrate the use of vault IDs, let's look at our `passwords.yml` file. Suppose we want to secure this file using a vault ID. The following command creates a vault ID labeled `passwords` and prompts us to create a password:

```
ansible-vault create --vault-id passwords@prompt passwords.yml
```

The `passwords` vault ID protects the `passwords.yml` file. Now, let's assume we also want to secure some API keys associated with users. If we stored these secrets in the `apikeys.yml` file, the following command would create a corresponding vault ID called `apikeys`:

```
ansible-vault create --vault-id apikeys@prompt apikeys.yml
```

Here, we created two vault IDs, each with its own password and protecting different resources. Vault IDs provide an improved security context when managing secrets. If one of the vault ID passwords becomes compromised, the resources that have been secured by the other vault IDs are still protected. With vault IDs, we can also leverage different access levels to vault secrets. For example, we can define `admin`, `dev`, and `test` vault IDs for the related groups of users. Alternatively, we can have multiple configuration management projects, each with their own dedicated vault IDs and secrets; for example, `user-config`, `web-config`, and `db-config`.

You can associate a vault ID with multiple secrets. For example, the following command creates a `user-config` vault ID that secures the `passwords.yml` and `api-keys.yml` files:

```
ansible-vault create --vault-id user-config@prompt passwords.yml apikeys.yml
```

When using vault IDs, we can also specify a password file to supply the related vault password. The following command encrypts the `apikeys.yml` file, which is reading the corresponding vault ID password from the `apikeys.pass` file:

```
ansible-vault encrypt --vault-id apikeys@apikeys.pass apikeys.yml
```

You can name your vault password files anything you want, but keeping a consistent naming convention, possibly one that matches the related vault ID, will make your life easier when you're managing multiple vault secrets.

Similarly, you can pass a vault ID (`passwords`) to a playbook (`create-users.yml`) with the following command:

```
ansible-playbook --vault-id passwords@passwords.pass create-users.yml
```

For more information about Ansible Vault, you may refer to the related online documentation at `https://docs.ansible.com/ansible/latest/user_guide/vault.html`.

So far, we have created a single user account with a password. What if we want to onboard multiple users, each with their own password? As we noted previously, we could call the `create-user` playbook and override the `username` and `password` variables using the `--extra-vars` option parameter. But this method is not a very efficient one, not to mention the difficulty of maintaining it. In the next section, we'll show you how to use task iteration in Ansible playbooks.

Working with loops

Loops provide an efficient way of running a task repeatedly in Ansible playbooks. There are several loop implementations in Ansible, and we can classify them into the following categories based on their keyword or syntax:

- `loop`: The recommended way of iterating through a collection.
- `with_<lookup>`: Collection-specific implementations of loops; examples include `with_list`, `with_items`, and `with_dict`, to name a few.

In this section, we'll keep our focus on the `loop` iteration (equivalent to `with_list`), which is best suited for simple loops. Let's expand our previous use case and adapt it to creating multiple users. First, we'll start by making a quick comparison between running repeated tasks *with* and *without* using loops.

As a preparatory step, make sure `~/ansible` is your current working directory. Also, you can delete the `./group_vars` folder, as we're not using it anymore. Now, let's create a couple of playbooks, `create-users1.yml` and `create-users2.yml`, as suggested in the following figure:

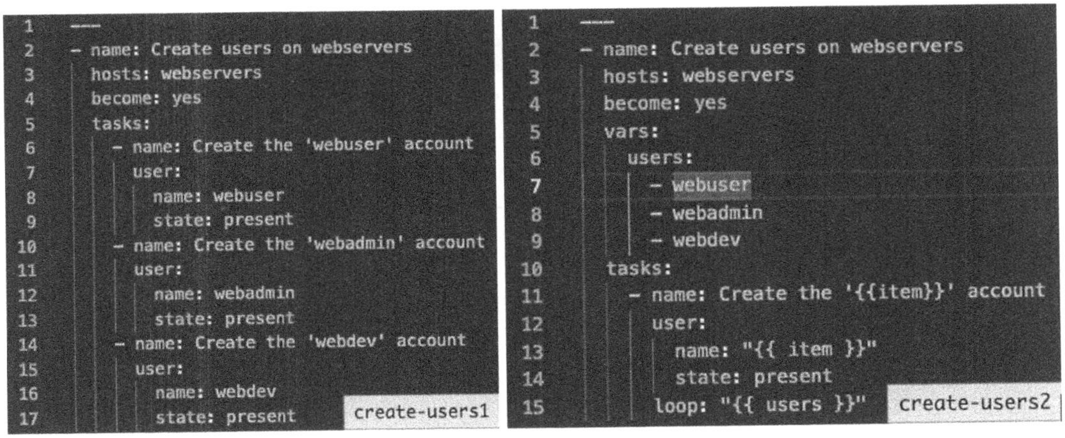

Figure 15.35 – Playbooks with multiple versus iterative tasks

Both playbooks create three users: `webuser`, `webadmin`, and `webdev`. The `create-users1` playbook has three distinct tasks, one for creating each user. On the other hand, `create-users2` implements a single task iteration using the `loop` directive (in line 15):

```
loop: "{{ users }}"
```

The loop iterates through the items of the users list, defined as a play variable in lines 6-9. The user task uses the {{ item }} variable, referencing each user while iterating through the list.

Before running any of these playbooks, let's also create one for deleting the users. We'll name this playbook delete-users2.yml, and it will have a similar implementation to create-users2.yml:

```
1   ---
2   - name: Delete users on webservers
3     hosts: webservers
4     become: yes
5     vars:
6       users:
7         - webuser
8         - webadmin
9         - webdev
10    tasks:
11      - name: Delete user
12        user:
13          name: "{{ item }}"
14          state: absent
15          remove: yes
16          force: yes
17        loop: "{{ users }}"
```

Figure 15.36 – A playbook using a loop for deleting users

Now, let's run the create-user1 playbook while targeting only the web1 web server:

```
ansible-playbook create-users1.yml --limit web1
```

In the output, we can see that three tasks have been executed, one for each user:

```
PLAY [Create users on webservers] ***************************************************

TASK [Gathering Facts] **************************************************************
ok: [web1]

TASK [Create the 'webuser' account] *************************************************
changed: [web1]

TASK [Create the 'webadmin' account] ************************************************
changed: [web1]

TASK [Create the 'webdev' account] **************************************************
changed: [web1]

PLAY RECAP **************************************************************************
web1                       : ok=4    changed=3    unreachable=0    failed=0    skipped=0
    rescued=0    ignored=0
```

Figure 15.37 – The output of the create-user1 playbook, with multiple tasks

Let's delete the users by running the `delete-users2.yml` playbook:

```
ansible-playbook delete-users2.yml --limit web1
```

Now, let's run the `create-user2` playbook, again targeting only the `web1` web server:

```
ansible-playbook create-users2.yml --limit web1
```

This time, the output shows a single task iterating through all the users:

```
PLAY [Create users on webservers] ***************************************************

TASK [Gathering Facts] **************************************************************
ok: [web1]

TASK [Create the '{{item}}' account] ************************************************
changed: [web1] => (item=webuser)
changed: [web1] => (item=webadmin)
changed: [web1] => (item=webdev)

PLAY RECAP **************************************************************************
web1                       : ok=2    changed=1    unreachable=0    failed=0    skipped=0
    rescued=0    ignored=0
```

Figure 15.38 – The output of the create-user2 playbook, with a single task iteration

The difference between the two playbook runs is significant. The first playbook executes a task for each user. While forking a task is not an expensive operation, you can imagine that creating hundreds of users would incur a significant load on the Ansible runtime. On the other hand, the second playbook runs a single task, loading the user module three times, to create each user. Loading a module takes significantly fewer resources than running a task.

Now that we know how to implement a simple loop, we'll make our playbook more compact and maintainable. We'll also try to come closer to a real-world scenario by storing the users and their related passwords in a reusable and secure fashion. We will keep the web user's information in the users.yml file. The related passwords are in the passwords.yml file. Here are the two files, along with some example user data:

```
# Passwords are stored in passwords.yml
webusers:
  - username: webuser
    comment: Regular web user

  - username: webadmin
    comment: Web administrator

  - username: webdev
    comment: Web developer          users.yml
```
```
# Usernames match the records in webusers.yml.
# The order doesn't matter!
webuser:
    password: bb37e5d1

webadmin:
    password: 7705b8a4

webdev:
    password: 8365b176          passwords.yml
```

Figure 15.39 – The users.yml and passwords.yml files

The users.yml file contains a dictionary with a single key-value pair:

- *Key*: webusers
- *Value*: A list of username and comment tuples

The passwords.yml file contains a nested dictionary with multiple key-value pairs, as follows:

- *Key*: <username> (for example, webuser, webadmin, and so on)
- *Value*: A nested dictionary with the password: <value> key-value pair

You may use the ansible-vault edit command to update the passwords.yml file. Alternatively, you can create the file from scratch and then encrypt it, following the steps previously described in the *Working with secrets* section.

The `create-users.yml` playbook file has the following implementation:

```yaml
1  ---
2  - name: Create users on webservers
3    hosts: webservers
4    become: yes
5    tasks:
6      - name: Load users
7        include_vars:
8          file: users.yml
9          name: users
10     - name: Load passwords
11       include_vars:
12         file: passwords.yml
13         name: passwords
14     - name: Create user accounts
15       user:
16         name: "{{ item.username }}"
17         comment: "{{ item.comment }}"
18         password: "{{ passwords[item.username]['password'] | password_hash('sha512') }}"
19         update_password: always
20         state: present
21       loop: "{{ users.webusers }}"
```

Figure 15.40 – The create-users.yml playbook

These files are also available in the GitHub repository for this book, in the related chapter folder. Let's quickly go over the playbook's implementation. We have three tasks:

- **Load web users** (lines 6-9): Reads the web user information from the `users.yml` file and stores the related values in the `users` dictionary.

- **Load passwords** (lines 10-13): Reads the passwords from the encrypted `passwords.yml` file and stores the corresponding values in the `passwords` dictionary.

- **Create the user account** (lines 14-21): Iterates through the `users.webusers` list and for each `item`, creates a user account with the related parameters; the task performs a password lookup in the `passwords` dictionary based on `item.username`.

Run the playbook with the following command:

```
ansible-playbook create-users.yml
```

Here's the output:

```
TASK [Gathering Facts] ******************************************************************
ok: [web2]
ok: [web1]

TASK [Load users] ***********************************************************************
ok: [web1]
ok: [web2]

TASK [Load passwords] *******************************************************************
ok: [web1]
ok: [web2]

TASK [Create user accounts] *************************************************************
changed: [web1] => (item={'username': 'webuser', 'comment': 'Regular web user'})
changed: [web1] => (item={'username': 'webadmin', 'comment': 'Web administrator'})
changed: [web2] => (item={'username': 'webuser', 'comment': 'Regular web user'})
changed: [web1] => (item={'username': 'webdev', 'comment': 'Web developer'})
changed: [web2] => (item={'username': 'webadmin', 'comment': 'Web administrator'})
changed: [web2] => (item={'username': 'webdev', 'comment': 'Web developer'})
```

Figure 15.41 – Running the create-users.yml playbook

We can see the following playbook tasks at work:

- **Gathering Facts**: Discovering the managed hosts and related system variables (facts); we'll introduce Ansible facts later in this chapter.

- **Load users**: Reading the users from the users.yml file.

- **Load passwords**: Reading the passwords from the encrypted passwords.yml file.

- **Create user accounts**: The task iteration loop creating the users.

You may verify the new user accounts using the methods presented earlier in the *Working with secrets* section. As an exercise, create the delete-users.yml playbook using a similar implementation with the create-users playbook.

For more information about loops, you may refer to the related online documentation at https://docs.ansible.com/ansible/latest/user_guide/playbooks_loops.html.

Now, let's look at how we can improve our playbook and reuse it to seamlessly create users across all the hosts in the inventory, web servers, and databases alike. We'll use conditional tasks to accomplish this functionality.

Running conditional tasks

Conditionals in Ansible playbooks decide when to run a task, depending on a condition (or a state). This condition can be the value of a **variable, fact,** or the **result** of a previous task. Ansible uses the when task-level directive to define a condition.

We learned about variables and how to use them in playbooks. Facts and results are essentially variables of a specific type and use. We'll look at each of these variables in the context of conditional tasks. Let's start with facts first.

Using Ansible facts

Facts are variables that provide specific information about the *remote* managed hosts. Fact variable names start with the `ansible_` prefix.

Here are a few examples of Ansible facts:

- `ansible_distribution`: The OS distribution (for example, `CentOS` or `Ubuntu`)
- `ansible_all_ipv4_addresses`: The IPv4 addresses
- `ansible_architecture`: The platform architecture (for example, `x86_64` or `i386`)
- `ansible_processor_cores`: The number of CPU cores
- `ansible_memfree_mb`: The available memory (in MB)

You may recall the ad hoc commands we used in the *Working with platform-specific package managers* section for installing the latest updates on the Ubuntu and RHEL/CentOS machines in our inventory. Let's look at the ad hoc commands.

You can update the Ubuntu hosts with the following command:

```
ansible -m apt -a "upgrade=dist update_cache=yes" ubuntu
```

You can update the RHEL/CentOS hosts with the following command:

```
ansible -m yum -a "name=* state=latest update_cache=yes" centos
```

In both commands, we targeted the related group of hosts, `ubuntu` and `centos`. Now, what if we didn't have groups explicitly created for classifying our hosts in Ubuntu and CentOS systems? In this case, we could *gather* the facts about our managed hosts, detect their OS type, and perform the conditional update task, depending on the underlying platform. Let's implement this functionality in a playbook using Ansible facts.

We'll name our playbook `install-updates.yml` and add the following content to it:

```
1    ---
2    - name: Install system updates
3      hosts: all
4      become: yes
5
6      tasks:
7      - name: Install CentOS system updates
8        yum: name=* state=latest update_cache=yes
9        when: ansible_distribution == "CentOS"
10
11     - name: Install Ubuntu system updates
12       apt: upgrade=dist update_cache=yes
13       when: ansible_distribution == "Ubuntu"
```

Figure 15.42 – The install-updates.yml playbook

The playbook targets all hosts and has two conditional tasks, based on the `ansible_distribution` fact:

- **Install CentOS system updates** (lines 7-9): Runs exclusively on CentOS hosts based on the `ansible_distribution == "CentOS"` condition (line 9)
- **Install Ubuntu system updates** (lines 11-13): Runs exclusively on Ubuntu hosts based on the `ansible_distribution == "Ubuntu"` condition (line 13)

Let's run our playbook:

```
ansible-playbook install-updates.yml
```

Here's the corresponding output:

```
TASK [Gathering Facts] ************************************************************
ok: [web2]
ok: [web1]
ok: [db1]
ok: [db2]

TASK [Install CentOS system updates] *********************************************
skipping: [web1]
skipping: [db1]
ok: [db2]
ok: [web2]

TASK [Install Ubuntu system updates] *********************************************
skipping: [web2]
skipping: [db2]
ok: [web1]
ok: [db1]
```

Figure 15.43 – Running conditional tasks

There are three tasks illustrated in the preceding output:

- **Gathering Facts**: The default discovery task that's executed by the playbook to gather facts about the remote hosts
- **Install CentOS system updates**: The conditional task for skipping all Ubuntu hosts (web1, and db1) and running on CentOS hosts (web2, and db2)
- **Install Ubuntu system updates**: The conditional task for skipping all CentOS hosts (web2, and db2) and running on Ubuntu hosts (web1, and db1)

Next, we'll look at how to use Ansible's environment-specific variables in conditional tasks.

Using magic variables

Magic variables describe the local Ansible environment and its related configuration data. Here are a few examples of magic variables:

- `ansible_playhosts`: A list with the active hosts in the current play
- `group_names`: A list of all the groups the current host is a member of
- `vars`: A dictionary with all the variables in the current play
- `ansible_version`: The version of Ansible

To see the magic variables in action while using conditional tasks, we'll improve our `create-users` playbook even further and create specific groups of users on different host groups. So far, the playbook only creates users on the hosts that belong to the `webservers` group (web1, web2). The playbook creates the `webuser`, `webadmin`, and `webdev` user accounts on all web servers. What if we want to create a similar group of users – dbuser, dbadmin, and dbdev – on all our database servers?

Start by adding the new user accounts and passwords to the `users.yml` and `passwords.yml` files, respectively. Here's what we have after adding the database's user accounts and passwords:

```
# Passwords are stored in passwords.yml
webusers:
  - username: webuser
    comment: Regular web user

  - username: webadmin
    comment: Web administrator

  - username: webdev
    comment: Web developer

dbusers:
  - username: dbuser
    comment: Regular database user

  - username: dbadmin
    comment: Database administrator

  - username: dbdev                    users.yml
    comment: Database developer
```

```
# Usernames should match the records in
# the webusers.yml file. The order
# doesn't matter.

# Web user passwords.
webuser:
  password: bb37e5d1
webadmin:
  password: 7705b8a4
webdev:
  password: 8365b176

# Database user passwords.
dbuser:
  password: 4695b3db
dbadmin:
  password: 99057ee9
dbdev:                                 passwords.yml
  password: a966ada8
```

Figure 15.44 – The users.yml and passwords.yml files

Note that you can edit the `passwords.yml` file using the `ansible-vault edit` command. Alternatively, you can decrypt, edit the file, and re-encrypt it. Now, let's update the `create-user` playbook with the required conditional tasks to handle both groups – webusers and databases – selectively. Update the `create-users.yml` file with the following content:

```
 1   ---
 2   - name: Create users
 3     hosts: all
 4     become: yes
 5     tasks:
 6       - name: Load users
 7         include_vars:
 8           file: users.yml
 9           name: users
10       - name: Load passwords
11         include_vars:
12           file: passwords.yml
13           name: passwords
14       - name: Create web user accounts
15         user:
16           name: "{{ item.username }}"
17           comment: "{{ item.comment }}"
18           password: "{{ passwords[item.username]['password'] | password_hash('sha512') }}"
19           update_password: always
20           state: present
21         loop: "{{ users.webusers }}"
22         when: "'webservers' in group_names"
23       - name: Create database user accounts
24         user:
25           name: "{{ item.username }}"
26           comment: "{{ item.comment }}"
27           password: "{{ passwords[item.username]['password'] | password_hash('sha512') }}"
28           update_password: always
29           state: present
30         loop: "{{ users.dbusers }}"
31         when: "'databases' in group_names"
```

Figure 15.45 – The create-users.yml playbook with conditional tasks

Here are the essential changes we made compared to the previous version of the playbook:

- Modifying line 3, hosts: all: To target all hosts

- Adding line 22, when: "'webservers' in group_names": To run the web user task conditionally, but only for hosts belonging to the webservers group

- Copy/pasting the web user task (lines 14-22) into the corresponding database user task (lines 23-31)

- Adjusting the loop and when clauses in lines 30-31 to use the database-specific variables

Let's run the playbook:

```
ansible-playbook create-users.yml
```

The output shows the web user task skipping the database servers and the database user task skipping the web servers, suggesting that the web and database users have been created successfully:

```
TASK [Create web user accounts] *******************************************************
skipping: [db1] => (item={'username': 'webuser', 'comment': 'Regular web user'})
skipping: [db1] => (item={'username': 'webadmin', 'comment': 'Web administrator'})
skipping: [db1] => (item={'username': 'webdev', 'comment': 'Web developer'})
skipping: [db2] => (item={'username': 'webuser', 'comment': 'Regular web user'})
skipping: [db2] => (item={'username': 'webadmin', 'comment': 'Web administrator'})
skipping: [db2] => (item={'username': 'webdev', 'comment': 'Web developer'})
changed: [web1] => (item={'username': 'webuser', 'comment': 'Regular web user'})
changed: [web1] => (item={'username': 'webadmin', 'comment': 'Web administrator'})
changed: [web2] => (item={'username': 'webuser', 'comment': 'Regular web user'})
changed: [web1] => (item={'username': 'webdev', 'comment': 'Web developer'})
changed: [web2] => (item={'username': 'webadmin', 'comment': 'Web administrator'})
changed: [web2] => (item={'username': 'webdev', 'comment': 'Web developer'})

TASK [Create database user accounts] ***************************************************
skipping: [web1] => (item={'username': 'dbuser', 'comment': 'Regular database user'})
skipping: [web1] => (item={'username': 'dbadmin', 'comment': 'Database administrator'})
skipping: [web1] => (item={'username': 'dbdev', 'comment': 'Database developer'})
skipping: [web2] => (item={'username': 'dbuser', 'comment': 'Regular database user'})
skipping: [web2] => (item={'username': 'dbadmin', 'comment': 'Database administrator'})
skipping: [web2] => (item={'username': 'dbdev', 'comment': 'Database developer'})
changed: [db1] => (item={'username': 'dbuser', 'comment': 'Regular database user'})
changed: [db1] => (item={'username': 'dbadmin', 'comment': 'Database administrator'})
changed: [db2] => (item={'username': 'dbuser', 'comment': 'Regular database user'})
changed: [db1] => (item={'username': 'dbdev', 'comment': 'Database developer'})
changed: [db2] => (item={'username': 'dbadmin', 'comment': 'Database administrator'})
changed: [db2] => (item={'username': 'dbdev', 'comment': 'Database developer'})
```

Figure 15.46 – The web and database user tasks running selectively

For a complete list of Ansible special variables, including magic variables, please visit `https://docs.ansible.com/ansible/latest/reference_appendices/special_variables.html`. For more information about facts and magic variables, check out the online documentation at `https://docs.ansible.com/ansible/latest/user_guide/playbooks_vars_facts.html`.

Next, we'll look at variables for tracking task results, also known as register variables.

Using register variables

Ansible registers capture the output of a task in a variable called a **register variable**. Ansible uses the `register` directive to capture the task's output in a variable. A typical example of using register variables is collecting the result of a task for debugging purposes. In more complex workflows, a particular task may or may not run, depending on a previous task's result.

Let's consider a hypothetical use case. As we onboard new users and create different accounts on all our servers, we want to make sure the number of users doesn't exceed the maximum number allowed. If the limit is reached, we may choose to launch a new server, redistribute the users, and so on. Let's start by creating a playbook named `count-users.yml` with the following content:

```
1   ---
2   - name: Detect if the number of users exceeds the limit
3     hosts: all
4     become: yes
5     vars:
6       max_allowed: 30
7     tasks:
8     - name: Count all users
9       shell: "getent passwd | wc -l"
10      register: count
11    - name: Debug number of users
12      debug:
13        msg: "Number of users: {{ count.stdout }}. Limit: {{ max_allowed }}"
14    - name: Detect limit
15      debug:
16        msg: "Maximum number of users reached!"
17      when: count.stdout | int > max_allowed
```

Figure 15.47 – The count-users.yml playbook

We created the following tasks in the playbook:

- **Count all users** (lines 8-10): A task that uses the `shell` module to count all users; we register the `count` variable by capturing the task output (line 10).

- **Debug number of users** (line 11-13): A simple task for debugging purposes that logs the number of users and the maximum limit allowed (line 13).

- **Detect limit** (lines 14-17): A conditional task that's run when the limit has been reached; the task checks the value of the `count` register variable and compares it with the `max_allowed` variable (line 17).

Line 17 in our playbook needs some further explanation. Here, we take the actual standard output of the register variable; that is, `count.stdout`. As-is, the value would be a string, and we need to cast it to an integer; that is, `count.stdout | int`. Then, we compare the resulting number with `max_allowed`. Let's run the playbook while targeting only the `web1` host:

```
ansible-playbook count-users.yml --limit web1
```

The output is as follows:

```
TASK [Gathering Facts] *******************************************************
ok: [web1]

TASK [Count all users] *******************************************************
changed: [web1]

TASK [Debug number of users] *************************************************
ok: [web1] => {
    "msg": "Number of users: 36. Limit: 30"
}

TASK [Detect limit] **********************************************************
ok: [web1] => {
    "msg": "Maximum number of users reached!"
}
```

Figure 15.48 – The conditional task (Detect limit) is executed

Here, we can see that the number of users is 36, thus exceeding the maximum limit of 30. In other words, the *Detect limit task* ran as it's supposed to.

Now, let's edit the `count-users.yml` playbook and change line 6 to the following:

```
max_allowed: 50
```

Save and rerun the playbook. This time, the output shows that the **Detect limit** task was skipped:

```
TASK [Gathering Facts] *******************************************************
ok: [web1]

TASK [Count all users] *******************************************************
changed: [web1]

TASK [Debug number of users] *************************************************
ok: [web1] => {
    "msg": "Number of users: 36. Limit: 50"
}

TASK [Detect limit] **********************************************************
skipping: [web1]
```

Figure 15.49 – The conditional task (Detect limit) is skipped

To learn more about conditional tasks in Ansible playbooks, please visit https://docs. ansible.com/ansible/latest/user_guide/playbooks_conditionals. html. By combining conditional tasks with Ansible's all-encompassing facts and special variables, we can write extremely powerful playbooks and automate a wide range of system administration operations.

In the following sections, we'll explore additional ways to make our playbooks more reusable and versatile. We'll look at dynamic configuration templates next.

Using templates with Jinja2

One of the most common configuration management tasks is copying files to managed hosts. Ansible provides the `copy` module for serving such tasks. A typical file copy operation has the following syntax in Ansible playbooks:

```
- copy:
    src: motd
    dest: /etc/motd
```

The copy task takes a source file (`motd`) and copies it to a destination (`/etc/motd`) on the remote host. While this model would work for copying static files to multiple hosts, it won't handle host-specific customizations in these files *on-the-fly*. Take, for example, a network configuration file featuring the IP address of a host. Attempting to copy this file on all hosts to configure the related network settings could render all but one host unreachable. Ideally, the network configuration file should have a *placeholder* for the dynamic content (for example, IP address) and adapt the file accordingly, depending on the target host.

To address this functionality, Ansible provides the **template** module with the **Jinja2** templating engine. Jinja2 uses Python-like language constructs for variables and expressions in a template. The `template` syntax is very similar to `copy`:

```
- template:
    src: motd.j2
    dest: /etc/motd
```

The source, in this case, is a Jinja2 template file (`motd.j2`) with host-specific customizations. Before copying the file to the remote host, Ansible reads the Jinja2 template and replaces the dynamic content with the host-specific data. This processing happens on the Ansible control node.

To illustrate some of the benefits and internal workings of Ansible templates, we'll work with a couple of use cases and create a Jinja2 template for each. Then, we'll create and run the related playbooks to show the templates in action.

Here are the two templates we'll be creating in this section:

- A *message-of-the-day* template: For displaying a customized message to users about a scheduled system maintenance
- A *hosts file* template: For generating a custom /etc/hosts file on each system with the hostname records of the other managed hosts

Let's start with the message-of-the-day template.

Creating a message-of-the-day template

In our introductory notes, we used the /etc/motd (message of the day) file as an example. On a Linux system, the content of this file is displayed when a user logs into the Terminal. Suppose you plan to upgrade your web servers on Thursday night and would like to give your users a friendly reminder about the upcoming outage. Your motd message could be something like this:

```
This server will be down for maintenance on Thursday night.
```

There's nothing special about this message, and the motd file could be easily deployed with a simple copy task. In most cases, such a message would probably do just fine, apart from the rare occasion when users may get confused about which exactly is *this server*. You may also consider that Thursday night in the US could be Friday afternoon on the other side of the world, and it would be nice if the announcement were more specific.

Perhaps a better message would state, on the web1 web server, the following:

```
web1 (172.16.191.12) will be down for maintenance on Thursday,
 April 8, 2021, between 2 - 3 AM (UTC-08:00).
```

On the web2 web server, the message would reflect the corresponding hostname and IP address. Ideally, the template should be reusable across multiple time zones, with playbooks running on globally distributed Ansible control nodes. Let's see how we can implement such a template. We'll assume your current working directory is ~/ansible.

First, create a templates folder in your local Ansible project directory:

```
mkdir -p ~/ansible/templates
```

Ansible will look for template files in the local directory, which is where we'll create our playbook, or in the `./templates` folder. Using a Linux editor of your choice, create a `motd.j2` file in `./templates` with the following content:

```
1    {# We could pass these variables from the playbook #}
2    {% set date = '2021-04-08' %}
3    {% set start_time = date ~ 'T02:00:00-0800' %}
4    {% set end_time = date ~ 'T03:00:00-0800' %}
5
6    {% set fmt = '%Y-%m-%dT%H:%M:%S%z' %}
7    {% set start = (start_time | to_datetime(format=fmt)) %}
8    {% set end = (end_time | to_datetime(format=fmt)) %}
9
10   {# We could also use the {{ inventory_hostname }} special variable #}
11   {{ ansible_facts.fqdn }} ({{ ansible_facts.default_ipv4.address }}) will
     be down for maintenance on {{ start.strftime('%A, %B %-d, %Y') }}, between
     {{ start.strftime('%-I') }} - {{ end.strftime('%-I %p (%Z)') }}.
```

Figure 15.50 – The motd.j2 template file

Note some of the particularities of the Jinja2 syntax:

- Comments are enclosed in {# ... #} (lines 1 and 10).
- Expressions are surrounded by {% ... %} (for example, lines 2, 3, 4, and so on).
- External variables are referenced with {{ ... }} (for example, line 11).

Here's what the script does:

- Lines 1-3 define the initial set of local variables for storing the time boundaries for the outage: the day of the outage (`date`), the starting time (`start_time`), and the ending time (`end_time`).
- Line 6 defines the input date-time format (`fmt`) for our starting and ending time variables.
- Lines 7-8 build the `datetime` objects that correspond to `start_time` and `end_time`. These Python `datetime` objects are formatted according to our needs in the custom message.
- Line 11 prints the custom message, featuring the user-friendly time outputs and a couple of Ansible facts, namely the FQDN (`ansible_facts.fqdn`) and IPv4 address (`ansible_facts.default_ipv4.address`) of the host where the message is displayed.

Now, let's create the playbook running the template. We will name the playbook update-motd.yml and add the following content:

```
 1    ---
 2    - name: Update the message of the day
 3      hosts: all
 4      become: yes
 5      tasks:
 6        - name: Deploy the 'motd' template
 7          template:
 8            src: motd.j2
 9            dest: /etc/motd
10            owner: root
11            group: root
12            mode: 0644
```

Figure 15.51 – The update-motd.yml playbook

The template module reads and processes the motd.j2 file (line 8), generating the related dynamic content, then copies the file to the remote host in /etc/motd (line 9) with the required permissions (lines 10-12).

Now, we're ready to run our playbook:

```
ansible-playbook update-motd.yml
```

The command should complete successfully. You can immediately verify the motd message on any of the hosts (for example, web1) with the following command:

```
ansible web1 -a "cat /etc/motd"
```

The preceding command runs remotely on the web1 host and displays the content of the /etc/motd file:

```
packt@neptune:~/ansible$ ansible web1 -a "cat /etc/motd"
web1 | CHANGED | rc=0 >>

web1 (172.16.191.12) will be down for maintenance on Thursday, April 8, 2021,
 between 2 - 3 AM (UTC-08:00).
```

Figure 15.52 – The content of the remote /etc/motd file

We can also SSH into any of the hosts to verify the `motd` prompt:

```
ssh packt@web1
```

The Terminal shows the following output:

```
web1 (172.16.191.12) will be down for maintenance on Thursday, April 8, 2021,
  between 2 - 3 AM (UTC-08:00).
Last login: Thu Apr  8 08:33:37 2021 from 172.16.191.11
packt@web1:~$ █
```

Figure 15.53 – The motd prompt on the remote host

Now that we know how to write and handle Ansible templates, let's improve `motd.j2` to make it a bit more reusable. We'll *parameterize* the template by replacing the hard-coded local variables for date and time with input variables that are passed from the playbook. This way, we'll make our template reusable across multiple playbooks and different input times for maintenance. Here's the updated template file (`motd.j2`):

```
1    {% set start_time_ = date ~ 'T' ~ start_time ~ utc %}
2    {% set end_time_ = date ~ 'T' ~ end_time ~ utc %}
3
4    {% set fmt = '%Y-%m-%dT%H:%M:%S%z' %}
5    {% set start = (start_time_ | to_datetime(format=fmt)) %}
6    {% set end = (end_time_ | to_datetime(format=fmt)) %}
7
8    {{ ansible_facts.fqdn }} ({{ ansible_facts.default_ipv4.address }}) will
     be down for maintenance on {{ start.strftime('%A, %B %-d, %Y') }}, between
     {{ start.strftime('%-I') }} - {{ end.strftime('%-I %p (%Z)') }}.
```

Figure 15.54 – The modified motd.j2 template with input variables

The relevant changes are in lines 1-2, where we build the `datetime` objects using the `date`, `start_time`, `end_time`, and `utc` input variables. Notice the difference between the *local* variables – `start_time_`, `end_time_` (suffixed with `_`) – and the corresponding *input* variables; that is, `start_time`, `end_time`. You may choose any naming convention for the variables, assuming they are Ansible-compliant.

Now, let's look at our modified playbook (`update-motd.yml`):

```
1    ---
2    - name: Update the message of the day
3      hosts: all
4      become: yes
5      vars:
6        date: "2021-04-08"
7        start_time: "02:00:00"
8        end_time: "03:00:00"
9        utc: "-0800"
10     tasks:
11     - name: Deploy the 'motd' template
12       template:
13         src: motd.j2
14         dest: /etc/motd
15         owner: root
16         group: root
17         mode: 0644
```

Figure 15.55 – The modified update-motd.yml playbook with variables

The relevant changes are in lines 5-9, where we added the variables serving the input for the `motd.j2` template. Running the modified playbook should yield the same result as the previous implementation. We'll leave the related exercise to you.

Next, we'll look at another use case featuring template-based deployments: updating the `/etc/hosts` files on the managed hosts with the host records of all the other servers in the group.

Creating a hosts file template

Let's start by creating a new template file, named `hosts.j2`, in the `./templates` folder. Add the following content:

```
1    # This file is autogenerated!
2
3    127.0.0.1 {{ inventory_hostname }} localhost
4
5    {% for host in groups['all'] %}
6    {% if host != inventory_hostname %}
7    {{ hostvars[host].ansible_facts.default_ipv4.address }} {{ host }}
8    {% endif %}
9    {% endfor %}
```

Figure 15.56 – The hosts.j2 template file

Here's how the template script works:

- Line 3 adds a `localhost` record corresponding to the current host referenced by the `inventory_hostname` Ansible special variable.

- Lines 5-9 execute a loop through all the hosts in the inventory referenced by the `groups['all']` list (special variable).

- Line 6 checks if the current host in the loop matches the target host, and it will only execute line 7 if the hosts are *different*.

- Line 7 adds a new host record by reading the default IPv4 address (`default_ipv4.address`) of the current host from the related Ansible facts (`hostvars[host].ansible_facts`).

Now, let's create the `update-hosts.yml` playbook file referencing the `hosts.j2` template. Add the following content:

```
 1   ---
 2   - name: Update the hosts file
 3     hosts: all
 4     become: yes
 5     tasks:
 6       - name: Deploy the 'hosts' template
 7         template:
 8           src: hosts.j2
 9           dest: /etc/hosts
10           owner: root
11           group: root
12           mode: 0644
```

Figure 15.57 – The update-hosts.yml playbook file

This playbook is very similar to `update-motd.yml`. Line 9 targets the `/etc/hosts` file. With the playbook and template files ready, let's run the following command:

```
ansible-playbook update-hosts.yml
```

After the command completes, we can check the `/etc/hosts` file on any of the hosts (for example, `web1`) by using the following command:

```
ansible web1 -a "cat /etc/hosts"
```

The output shows the expected host records:

```
# This file is autogenerated!

127.0.0.1 web1 localhost

172.16.191.14 web2
172.16.191.13 db1
172.16.191.15 db2
```

Figure 15.58 – The auto-generated /etc/hosts file on web1

You can also SSH into one of the hosts (for example, `web1`) and ping any of the other hosts by name (for example, `db2`):

```
ssh packt@web1
ping db2
```

You should get a successful ping response:

```
packt@web1:~$ ping db2
PING db2 (172.16.191.15) 56(84) bytes of data.
64 bytes from db2 (172.16.191.15): icmp_seq=1 ttl=64 time=0.322 ms
64 bytes from db2 (172.16.191.15): icmp_seq=2 ttl=64 time=0.641 ms
64 bytes from db2 (172.16.191.15): icmp_seq=3 ttl=64 time=0.588 ms
```

Figure 15.59 – Successful ping by hostname from one host to another

This concludes our study of Ansible templates. The topics we covered in this section barely scratch the surface of the powerful features and versatility of Jinja2 templates. We strongly encourage you to explore the related online help resources at https://docs.ansible.com/ansible/latest/user_guide/playbooks_templating.html, as well as the titles mentioned in the *Future reading* section at the end of this chapter.

Now, we will turn our attention to another essential feature of modern configuration management platforms: sharing reusable and flexible modules for a variety of system administration tasks. Ansible provides a highly accessible and extensible framework to accommodate this functionality – Ansible Galaxy and its roles. In the next section, we'll look at roles for automation reuse.

Using roles with Ansible Galaxy

With Ansible roles, you can bundle your automated workflows into reusable units. A role is essentially a package containing playbooks and other resources that have been adapted to a specific configuration using variables. An arbitrary playbook would invoke a role by providing the required parameters and run it just like any other task. Functionally speaking, roles encapsulate a generic configuration management behavior, making it reusable across multiple projects and even shareable with others.

Here are the key benefits of using roles:

- Encapsulating functionality provides standalone packaging that can easily be shared with others.

- Encapsulation also enables separation of concerns: multiple DevOps and system administrators can develop roles in parallel.

- Roles can make larger automation projects more manageable.

Let's describe the process of creating a role and how to use it in a sample playbook.

Creating roles

When authoring roles, we usually follow these steps and practices:

- Create or initialize the role directory's structure. The directory contains all the resources required by the role in a well-organized fashion.

- Implement the role's content. Create the related playbooks, files, templates, and so on.

- Always start from simple to more advanced functionality. Test your playbooks as you add more content.

- Make your implementation as generic as possible. Use variables to expose the related customizations.

- Don't store secrets in your playbooks or related files. Provide input parameters for them.

- Create a dummy playbook with a simple play running your role. Use this dummy playbook to test your role.

- Design your role with user experience in mind. Make it easy to use and share it with others if you think it would bring value to the community.

At a high level, creating a role involves the following steps:

1. Initializing the role directory structure

2. Authoring the role's content

3. Testing the role

We'll use the `create-users.yml` playbook we created earlier in the *Using Ansible playbooks section* as our example for creating a role. Before proceeding with the next steps, let's add the following line to our `ansible.cfg` file in the `[defaults]` section:

```
roles_path = ~/ansible
```

This configuration parameter sets the default location for our roles.

Now, let's start by initializing the role directory.

Initializing the role directory's structure

Ansible has a strict requirement regarding the folder structure of the role directory. The directory must have the same name as the role; for example, `create-users`. We can create this directory manually or by using a specialized command-line utility for managing roles, called `ansible-galaxy`.

To create the skeleton of our role directory, run the following command:

```
ansible-galaxy init create-users
```

The command completes with the following message:

```
- Role create-users was created successfully
```

You can display the directory structure using the `tree` command:

```
tree
```

You'll have to manually install the `tree` command-line utility using your local package manager (`apt`, `yum`, and so on). The output shows the `create-users` directory structure of our role:

```
packt@neptune:~/ansible$ tree
├── ansible.cfg
├── create-users    ◄──
│   ├── defaults
│   │   └── main.yml
│   ├── files
│   ├── handlers
│   │   └── main.yml
│   ├── meta
│   │   └── main.yml
│   ├── README.md
│   ├── tasks
│   │   └── main.yml
│   ├── templates
│   ├── tests
│   │   ├── inventory
│   │   └── test.yml
│   └── vars
│       └── main.yml
└── hosts
```

Figure 15.60 – The create-users role directory

Here's a brief explanation of each folder and the corresponding YAML file in the role directory:

- `defaults/main.yml`: The default variables for the role. They have the lowest priority among all the available variables and can be overwritten by any other variable.

- `files`: The static files referenced in the role tasks.

- `handlers/main.yml`: The handlers used by the role. Handlers are tasks that are triggered by other tasks. You can read more about handlers at `https://docs.ansible.com/ansible/latest/user_guide/playbooks_handlers.html`.

- `README.md`: Explains the intended purpose of the role and how to use it.

- `meta/main.yml`: Additional information about the role, such as the author, licensing model, platforms, and dependencies on other roles.

- `tasks/main.yml`: The tasks played by the role.

- `Templates`: The template files referenced by the role.

- `tests/test.yml`: The playbook for testing the role. The `tests` folder may also contain a sample `inventory` file.

- `vars/main.yml`: The variables used internally by the role. These variables have high precedence and are not meant to be changed or overwritten.

Now that we are familiar with the role directory and the related resource files, let's create our first role.

Authoring the role's content

It is a common practice to start from a previously created playbook and evolve it into a role. We'll take the `create-users.yml` playbook we authored in the *Working with loops* section as the boilerplate code for our future role. We can see the related implementation in *Figure 15.40*. The playbook also references the `users.yml` and `passwords.yml` files shown in *Figure 15.39*. Let's refactor these files to make them more generic.

Here are the modified `users.yml` and `passwords.yml` files:

```
---                                          1  ---
# Example user accounts.                     2  # Example user passwords.
# Passwords are stored in passwords.yml      3  # Usernames match the records in users.yml.
list:                                        4  # The order doesn't matter.
  - username: testuser                       5  testuser:
    comment: Test user                       6    password: bb37e5d1
                                             7
  - username: testadmin                      8  testadmin:
    comment: Test admin                      9    password: 7705b8a4
                                            10
  - username: testdev                       11  testdev:
    comment: Test dev          users.yml    12    password: 8365b176     passwords.yml
```

Figure 15.61 – The modified users.yml and passwords.yml files

As you may have noticed, we renamed the example user accounts and gave them more generic names. We also changed the user dictionary key name from `webusers` to `list` in the `users.yml` file. Remember that Ansible requires root-level dictionary entries (key-value pairs) in the YAML files that provide variables.

Let's look at the updated `create-users.yml` playbook:

```
1    ---
2    - name: Create users with passwords
3      hosts: all
4      become: yes
5      vars:
6        users_file: users.yml
7        passwords_file: passwords.yml
8      tasks:
9        - name: Load users
10         include_vars:
11           file: "{{ users_file }}"
12           name: users
13       - name: Load passwords
14         include_vars:
15           file: "{{ passwords_file }}"
16           name: passwords
17       - name: Create user accounts
18         user:
19           name: "{{ item.username }}"
20           comment: "{{ item.comment }}"
21           password: "{{ passwords[item.username]['password'] | password_hash('sha512') }}"
22           update_password: always
23           state: present
24         loop: "{{ users.list }}"
```

Figure 15.62 – The modified create-users.yml file

We made the following modifications:

- We readjusted the `loop` directive (line 24) to read `users.list` instead of `users.webusers` due to the name change of the related dictionary key in the `users.yml` file.

- We refactored the `include_vars` file references to use variables instead of hard-coded file names in lines 11 and 15.

- We added the `vars` section, with the `users_file` and `passwords_file` variables pointing to the corresponding YAML files.

With these changes in the playbook, we're now ready to implement our role. Looking at the `create-users` role directory, as illustrated in *Figure 15.60*, we'll do the following:

- Copy/paste the variables in the `vars` section of `create-users.yml` (lines 6-7) into `defaults/main.yml`.

- Copy/paste the tasks from `create-users.yml` (lines 9-24) into `tasks/main.yml`. Make sure you keep the relative indentations.

- Create a simple playbook using the role. Use the `tests/test.yml` file for your test playbook. Copy/move `users.yml` and `passwords.yml` to the `tests/` folder.

The following screenshot captures all these changes:

```
1    ---
2    users_file: users.yml
3    passwords_file: passwords.yml                                          defaults/main.yml
```

```
1    ---
2    - name: Load users
3      include_vars:
4        file: "{{ users_file }}"
5        name: users
6    - name: Load passwords
7      include_vars:
8        file: "{{ passwords_file }}"
9        name: passwords
10   - name: Create user accounts
11     user:
12       name: "{{ item.username }}"
13       comment: "{{ item.comment }}"
14       password: "{{ passwords[item.username]['password'] | password_hash('sha512') }}"
15       update_password: always
16       state: present
17     loop: "{{ users.list }}"                                             tasks/main.yml
```

```
1    ---
2    - hosts: all
3      become: yes
4      roles:
5      - role: create-users
6        vars:
7          users_file: users.yml
8          passwords_file: passwords.yml                                    tests/test.yml
```

Figure 15.63 – The files that we changed in the create-users role directory

We also recommend updating the README.md file in the `create-users` directory with notes about the purpose and usage of the role. You should also mention the requirement of having the `users.yml` and `passwords.yml` files with the related data structures. The names of these files can be changed via the `users_file` and `passwords_file` variables in `defaults/main.yml`. You can also provide some examples of how to use the role. We also created an additional `test2.yml` playbook using a task to run the role:

```
1    ----
2    - hosts: all
3      become: yes
4      tasks:
5      - name: Create users
6        include_role:
7          name: create-users
8        vars:
9          users_file: users.yml
10         passwords_file: passwords.yml
```

Figure 15.64 – Running a role using a task

At this point, we've finished making the required changes for implementing the role. You may choose to remove all the empty or unused folders in the create-users role directory.

Now, let's test our role.

Testing the role

To test our role, we will use the playbooks in the tests/ folder and run them with the following commands:

```
ansible-playbook create-users/tests/test.yml
ansible-playbook create-users/tests/test2.yml
```

Both commands should complete successfully. You can also test with a third playbook (test3.yml), where you don't specify users_file and passwords_file in the task running the role:

```
1    ----
2    - hosts: all
3      become: yes
4      tasks:
5      - name: Create users
6        include_role:
7          name: create-users
```

Figure 15.65 – Running the role with default variables

The following command should also complete successfully:

```
ansible-playbook create-users/tests/test3.yml
```

Now, let's test with different filenames. Copy the `users.yml` and `passwords.yml` files in the `tests/` folder to `myusers.yml` and `mypasswords.yml`, respectively:

```
cp create-users/tests/users.yml create-users/tests/myusers.yml
cp create-users/tests/passwords.yml create-users/tests/
mypasswords.yml
```

Change the corresponding variables in `defaults/main.yml` to reflect the new filenames:

```
users_file: myusers.yml
passwords_file: mypasswords.yml
```

Running the `test3.yml` playbook should complete successfully:

```
ansible-playbook create-users/tests/test3.yml
```

Make sure you revert the previous changes if you still plan on testing with `test.yml` and `test2.yml`. These playbooks override the role variables in the play.

Now that we know how to create and use a role, it's time for us to look at Ansible Galaxy, an online community for managing and sharing roles. Think of any configuration management operation, and there's a good chance you'll find a role for it on Ansible Galaxy. So, let's look at how to select and retrieve roles from Ansible Galaxy and use them in our playbooks.

Introducing Ansible Galaxy

Ansible Galaxy is essentially a public library of Ansible roles written by a community of professionals. You can access the Ansible Galaxy web portal at `https://galaxy.ansible.com/`. The home page provides a few useful links, such as general documentation, popular topics, a community page, and a search button:

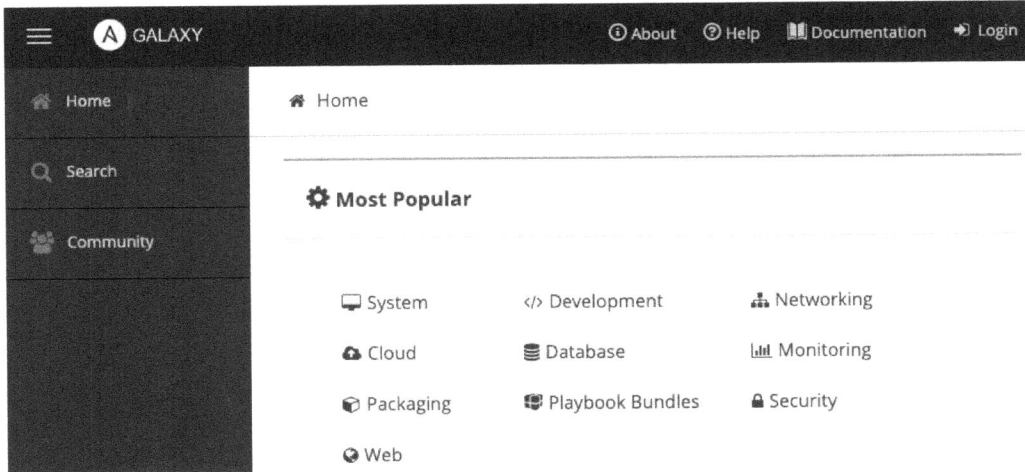

Figure 15.66 – The Ansible Galaxy web portal

Let's click on the **Search** button and look for a particular role. We'll try to find a role for installing and configuring NGINX. The search is very flexible and supports various filtering and sort options.

Working with the NGINX role

Entering `nginx` in our search criteria for roles yields the following top results:

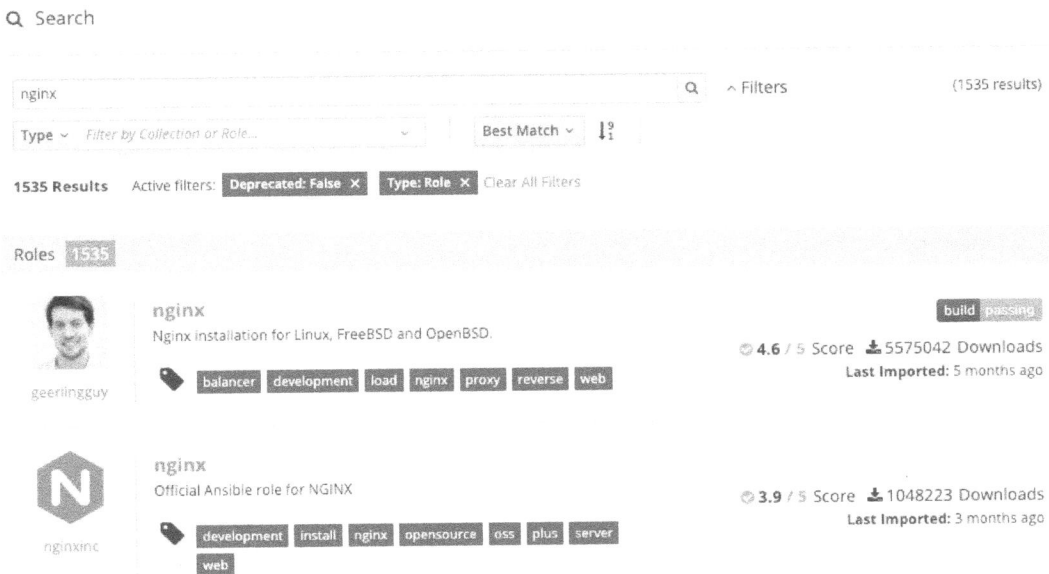

Figure 15.67 – Searching for NGINX

Notice the score and the number of downloads for each role. We'll choose **Official Ansible role for NGINX**. The following screenshot shows the related **Details** page:

Figure 15.68 – The official Ansible role for NGINX

Perhaps the most relevant pieces of information on this page are the **Installation** command and the **GitHub Repo** button. Let's copy the installation command and run it in our Terminal. We'll use ~/ansible as our working directory:

```
ansible-galaxy install nginxinc.nginx
```

After successfully downloading the role, you can immediately verify it with the following command:

```
ansible-galaxy list
```

Here, we can see the default directory for the roles (/home/packt/ansible) and the current roles that have been installed in our Ansible environment:

```
# /home/packt/ansible
- create-users, (unknown version)
- nginxinc.nginx, 0.19.1
```

Notice our *home-made* role, create-users, and the NGINX role we just downloaded.

We mentioned earlier that the GitHub repo contains relevant information about the role. The related GitHub link for the NGINX role is `https://github.com/nginxinc/ansible-role-nginx`. In general, you will find valuable information by visiting a role's GitHub page. You should look for the role's variables and usage examples. We can also infer this information by simply browsing the related role directory in our Ansible environment – for example, `~/ansible/nginxinc.nginx` – and look at the implementation details of the role.

By doing minimal research into the role's GitHub page and implementation, we crafted the following `nginx.yml` playbook for installing the NGINX server on our web server hosts:

```
 1  ---
 2  - name: Install NGINX on all webservers
 3    hosts: webservers
 4    become: yes
 5    tasks:
 6      - name: Install NGINX
 7        include_role:
 8          name: nginxinc.nginx
 9        vars:
10          nginx_debug_output: true
11          nginx_selinux: true
12          nginx_selinux_tcp_ports:
13            - "80"
14            - "443"
```

Figure 15.69 – The nginx.yml playbook using the official NGINX role

Notice the task running our `nginxinc.nginx` role with the related configuration variables. The following command will install and configure the NGINX server on all the web servers in our inventory:

```
ansible-playbook nginx.yml
```

We have the NGINX web servers installed and running, but we may not have HTTP access enabled yet on the servers due to restrictive firewall rules. Let's turn to Ansible Galaxy again in search of some fitting firewall roles.

Working with firewall roles

We have Ubuntu and RHEL/CentOS web servers in our inventory, so we need firewall managers of both flavors; that is, `ufw` and `firewalld`. Let's use the `ansible-galaxy` command-line utility this time to look for a `ufw` role. The following command searches the Ansible Galaxy repository for roles that contain `ufw` in their name and `configures ufw` in their description:

```
ansible-galaxy search ufw | grep 'configures ufw'
```

The output shows a few results. Let's pick one of them (`weareinteractive.ufw`) and get some related information:

```
ansible-galaxy info weareinteractive.ufw
```

Among other information, we should also look for the `download_count` attribute:

```
download_count: 51546
```

It turns out that the `weareactive.ufw` role has the most downloads among our finds. A quick lookup of the role in the Galaxy web portal also shows good reviews, so we'll install it with the following command:

```
ansible-galaxy install weareinteractive.ufw
```

The GitHub repository for the role is `https://github.com/weareinteractive/ansible-ufw` and contains information about how to use it.

Using the same approach, we will look for a `firewalld` role. You may feel more comfortable using the Ansible Galaxy web portal with your search. Our choice for the `firewalld` role is `flatkey.firewalld`:

```
ansible-galaxy install flatkey.firewalld
```

The corresponding GitHub repository is `https://github.com/FlatKey/ansible-firewalld-role`.

With both these firewall roles and the related resources, we came up with the following implementation of our `firewall.yml` playbook:

```
 1    ---
 2    - hosts: webservers
 3      become: true
 4
 5      tasks:
 6        - name: Configure Ubuntu firewall
 7          include_role:
 8            name: weareinteractive.ufw
 9          vars:
10            ufw_rules:
11              - logging: "full"
12              - rule: allow
13                to_port: "80"
14                proto: tcp
15          when: ansible_distribution == "Ubuntu"
16
17        - name: Configure CentOS firewall
18          include_role:
19            name: flatkey.firewalld
20          vars:
21            default_zone: public
22            firewalld_service_rules:
23              http:
24                state: enabled
25                zone: public
26                permanent: true
27                immediate: true
28          when: ansible_distribution == "CentOS"
```

Figure 15.70 – The firewall.yml playbook with the ufw and firewalld roles

The playbook contains two tasks – one using the `weareactive.ufw` role (lines 6-15) and the other using the `flatkey.firewalld` role (lines 17-28). Both tasks have conditional statements in lines 15 and 28, respectively, to run exclusively on the platform they support. The firewalls enable HTTP access on port 80.

Let's run the playbook with the following command:

```
ansible-playbook firewall.yml
```

Now, we can access all our web servers over HTTP. Here's a quick test for the web1 web server using the curl command:

```
curl http://web1
```

With that, we've provided an exploratory view of Ansible roles and Galaxy. Roles are a powerful feature of Ansible, and Galaxy brings community support with them. Together, they enable modern system administrators and DevOps to move quickly from concept to implementation, accelerating the deployment of everyday configuration management workflows.

Summary

In this chapter, we covered significant ground in terms of Ansible. Due to this chapter's limited scope, we couldn't capture all of Ansible's vast number of features. However, we tried to provide an overarching view of the platform, from Ansible's architectural principles to configuring and working with ad hoc commands and playbooks. You learned how to set up an Ansible environment, with several managed hosts and a control node, thereby emulating a real-world deployment at a high level. You also became familiar with writing Ansible commands and scripts for typical configuration management tasks. Most of the commands and playbooks presented throughout this chapter closely resemble everyday administrative operations.

Whether you are a systems administrator or a DevOps, a seasoned professional, or on the way to becoming one, we hope this chapter brought new insights to your everyday Linux administration tasks and automation workflows. The tools and techniques you've learned here will give you a good start for scripting and automating larger portions of your daily administrative routines.

The same closing thoughts also apply to this book in general. You have come a long way in terms of learning and mastering some of the most typical Linux administration tasks in on-premises and cloud environments alike.

We hope that you enjoyed our journey together.

Questions

Let's try to wrap up some of the essential concepts we learned about in this chapter by completing the following quiz:

1. What are idempotent operations or commands in Ansible?

2. You want to set up passwordless authentication with your managed hosts. What steps should you follow?

3. What is the ad hoc command for checking the communication with all your managed hosts?

4.   Enumerate a few Ansible modules. Try to think of a configuration management scenario where you could use each module.

5.   What is the ad hoc command for running an arbitrary shell operation or process on a remote host, such as `cat /etc/passwd`?

6.   Write a simple playbook with a single task using the `ping` module.

7.   Write a playbook for copying your current directory to a remote host.

8.   You must deploy secret API keys to every host into a well-defined location, each host with its own API key. How would you design and implement this functionality?

9.   Think of a simple playbook that monitors the memory that's available on your hosts and will notify you if that memory is above a given threshold.

10.   Find an Ansible Galaxy role for creating users on a Linux system. Use this role in a simple playbook to create a few arbitrary users.

Further reading

Here are a few resources we found helpful for learning more about Ansible internals:

- Ansible Documentation: `https://docs.ansible.com/`

- *Ansible Use Cases*, by Red Hat: `https://www.ansible.com/use-cases`

- *Dive into Ansible – From Beginner to Expert in Ansible [Video]*, by *James Spurin, Packt Publishing* (`https://www.packtpub.com/product/dive-into-ansible-from-beginner-to-expert-in-ansible-video/9781801076937`)

- *Practical Ansible 2*, by *Daniel Oh, James Freeman, Fabio Alessandro Locati, Packt Publishing* (`https://www.packtpub.com/product/practical-ansible-2/9781789807462`)

Packt>

Packt.com

Subscribe to our online digital library for full access to over 7,000 books and videos, as well as industry leading tools to help you plan your personal development and advance your career. For more information, please visit our website.

Why subscribe?

- Spend less time learning and more time coding with practical eBooks and Videos from over 4,000 industry professionals

- Improve your learning with Skill Plans built especially for you

- Get a free eBook or video every month

- Fully searchable for easy access to vital information

- Copy and paste, print, and bookmark content

Did you know that Packt offers eBook versions of every book published, with PDF and ePub files available? You can upgrade to the eBook version at packt.com and as a print book customer, you are entitled to a discount on the eBook copy. Get in touch with us at customercare@packtpub.com for more details.

At www.packt.com, you can also read a collection of free technical articles, sign up for a range of free newsletters, and receive exclusive discounts and offers on Packt books and eBooks.

Other Books You May Enjoy

If you enjoyed this book, you may be interested in these other books by Packt:

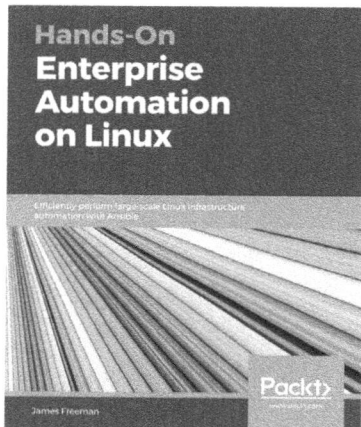

Hands-On Enterprise Automation on Linux

James Freeman

ISBN: 978-1-78913-161-1

- Perform large-scale automation of Linux environments in an enterprise
- Overcome the common challenges and pitfalls of extensive automation
- Define the business processes needed to support a large-scale Linux environment
- Get well-versed with the most effective and reliable patch management strategies
- Automate a range of tasks from simple user account changes to complex security policy enforcement
- Learn best practices and procedures to make your Linux environment automatable

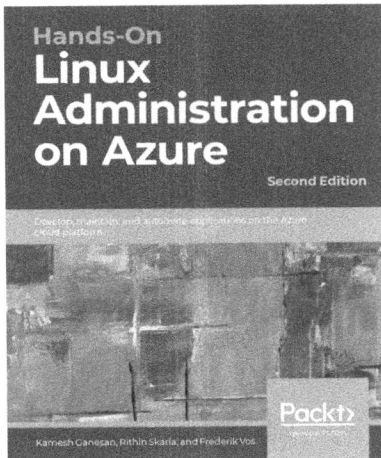

Hands-On Linux Administration on Azure - Second Edition

Kamesh Ganesan, Rithin Skaria, Frederik Vos

ISBN: 978-1-83921-552-0

- Grasp the fundamentals of virtualization and cloud computing
- Understand file hierarchy and mount new filesystems
- Maintain the life cycle of your application in Azure Kubernetes Service
- Manage resources with the Azure CLI and PowerShell
- Manage users, groups, and filesystem permissions
- Use Azure Resource Manager to redeploy virtual machines
- Implement configuration management to configure a VM correctly
- Build a container using Docker

Packt is searching for authors like you

If you're interested in becoming an author for Packt, please visit authors.packtpub.com and apply today. We have worked with thousands of developers and tech professionals, just like you, to help them share their insight with the global tech community. You can make a general application, apply for a specific hot topic that we are recruiting an author for, or submit your own idea.

Leave a review - let other readers know what you think

Please share your thoughts on this book with others by leaving a review on the site that you bought it from. If you purchased the book from Amazon, please leave us an honest review on this book's Amazon page. This is vital so that other potential readers can see and use your unbiased opinion to make purchasing decisions, we can understand what our customers think about our products, and our authors can see your feedback on the title that they have worked with Packt to create. It will only take a few minutes of your time, but is valuable to other potential customers, our authors, and Packt. Thank you!

Index

www.ingramcontent.com/pod-product-compliance
Lightning Source LLC
Chambersburg PA
CBHW080338220326
41598CB00030B/4535